WORLD ENERG

2005

Middle East and North Africa Insights

INTERNATIONAL ENERGY AGENCY

INTERNATIONAL ENERGY AGENCY

The International Energy Agency (IEA) is an autonomous body which was established in November 1974 within the framework of the Organisation for Economic Co-operation and Development (OECD) to implement an international energy programme.

It carries out a comprehensive programme of energy co-operation among twenty-six of the OECD's thirty member countries. The basic aims of the IEA are:

- to maintain and improve systems for coping with oil supply disruptions;
- to promote rational energy policies in a global context through co-operative relations with non-member countries, industry and international organisations;
- to operate a permanent information system on the international oil market;
- to improve the world's energy supply and demand structure by developing alternative energy sources and increasing the efficiency of energy use;
- to assist in the integration of environmental and energy policies.

The IEA member countries are: Australia, Austria, Belgium, Canada, the Czech Republic, Denmark, Finland, France, Germany, Greece, Hungary, Ireland, Italy, Japan, the Republic of Korea, Luxembourg, the Netherlands, New Zealand, Norway, Portugal, Spain, Sweden, Switzerland, Turkey, the United Kingdom, the United States. The European Commission takes part in the work of the IEA.

ORGANISATION FOR ECONOMIC CO-OPERATION AND DEVELOPMENT

The OECD is a unique forum where the governments of thirty democracies work together to address the economic, social and environmental challenges of globalisation. The OECD is also at the forefront of efforts to understand and to help governments respond to new developments and concerns, such as corporate governance, the information economy and the challenges of an ageing population. The Organisation provides a setting where governments can compare policy experiences, seek answers to common problems, identify good practice and work to co-ordinate domestic and international policies.

The OECD member countries are: Australia, Austria, Belgium, Canada, the Czech Republic, Denmark, Finland, France, Germany, Greece, Hungary, Iceland, Ireland, Italy, Japan, Korea, Luxembourg, Mexico, the Netherlands, New Zealand, Norway, Poland, Portugal, the Slovak Republic, Spain, Sweden, Switzerland, Turkey, the United Kingdom and the United States. The European Commission takes part in the work of the OECD.

This book is dedicated to Pierre Lefèvre, who died on 28 August 2005.
His outstanding commitment, wisdom and humanity shaped the
IEA Public Information Office from 2002 to 2005.

The *World Energy Outlook (WEO)* series is designed to give a framework of coherent future visions to assist government and industry in their decision making.

This year's *WEO*, while maintaining the global coverage, examines in detail the prospects and issues in the major oil- and gas-producing countries of the Middle East and North Africa (MENA). This publication has involved a huge data gathering exercise and in depth analysis. We have been aided by experts throughout the MENA region and from elsewhere in the world. I am indebted to them. I also wish to pay tribute to the IEA team, led by Dr. Fatih Birol, for its remarkable endeavor and results.

The picture which emerges is one of sustained growth in demand for the region's oil and gas, coupled with sustained economic and energy-demand growth in the region. The conclusion appears straightforward. In a well-functioning global economy, a suitably responsive investment in oil and gas, resulting in supply growth from the MENA region will benefit both MENA's and the global economy.

But the global economy does not function perfectly – mistaken perceptions of risk and opportunity may erroneously impact or delay decisions, and these effects can also be long lasting. Moreover, rational commercial decisions in a particular market may cause profound global damage if insufficient weight is given to environmental values.

Deferred investment in oil and gas supply is one risk. Our analysis shows that a failure by MENA governments to commit or facilitate the necessary timely investment would diminish economic welfare not only among oil-consuming countries but also in the region. Oil and gas prices would rise, but demand would fall, resulting in a net loss to producers and consumers alike.

As we go to press, unduly high oil prices, sustained in part by inadequate upstream and downstream oil-supply infrastructure, are giving rise to new cries of alarm in consuming countries. Energy-intensive developing countries are harder hit than the diverse economies of the OECD.

Governments have other reasons to act in the energy sector. They are increasingly aware of the need to take new measures to increase efficiency and to cut the growth in demand from those forms of energy which contribute disproportionately to the burden of greenhouse-gas emissions.

These considerations led us to develop the World Alternative Policy Scenario. The policy and market responses to the oil shocks of the 1970s show that radical changes to the energy economy can be made. We have not assumed drastic measures, merely that governments in consuming countries put into effect those measures which they are already contemplating. How far might the energy future be altered by such policies? The answer is that demand still grows – albeit more slowly, but still very substantially. Whichever scenario, the central message of our analysis is unchanged: the world needs sustained investment in oil and gas in any plausible future, a major part of it in the countries of the Middle East and North Africa.

We have set out the policy issues faced by both consumer countries and producers. It is certain that both will make better decisions if they freely exchange their information, intentions and anxieties.

Claude Mandil
Executive Director

ACKNOWLEDGEMENTS

This study was prepared by the Economic Analysis Division of the International Energy Agency in co-operation with other divisions of the IEA. The Deputy Executive Director and Director of the Non-Member Countries Office, William C. Ramsay, and the Director of the Long-Term Office, Noé van Hulst, provided guidance and encouragement during the project. The study was designed and managed by Fatih Birol, Head of the Economic Analysis Division.

Other members of EAD who were responsible for bringing the study to completion include: Maria Argiri, Marco Baroni, Amos Bromhead, François Cattier, Laura Cozzi, Hideshi Emoto, Lisa Guarrera, Teresa Malyshev, Trevor Morgan, Nicola Pochettino and Maria T. Storeng. Claudia Jones provided essential support. Robert Priddle carried editorial responsibility.

The study also benefited from input provided by other IEA colleagues, particularly: Chris Besson, Jeff Brown, Dunia Chalabi, Viviane Consoli, Sylvie Cornot-Gandolphe, Gordon Duffus, Lawrence Eagles, Jason Elliott, David Fyfe, Rebecca Gaghen, Dagmar Graczyk, Klaus Jacoby, Tom O'Gallagher, Pawel Olejarnik, Riccardo Quercioli, James Ryder, Atsushi Suda, Harry Tchilinguirian, Nancy Turck, and Fritdjof Unander. Pierre Lefèvre, Angela Costrini, Muriel Custodio, Loretta Ravera, Bertrand Sadin and Tyna Wynaendts of the Public Information Office provided substantial help in producing this book.

The work could not have been achieved without the substantial support provided by many government bodies, international organisations and energy companies worldwide, notably the US Department of Energy, the UK Foreign & Commonwealth Office, the Japanese Ministry of Economy, Trade and Industry, the Organization of the Petroleum Exporting Countries (OPEC), Statoil ASA, Schlumberger Ltd, IHS Energy and the International Monetary Fund (IMF). Special thanks go to the Kuwaiti Ministry of Energy for hosting the 3rd Joint IEA-OPEC Workshop in Kuwait City in May 2005, which provided key input to this study.

An Advisory Panel provided overall guidance for the work, shaped the analysis and drew out key messages. Its members are:

Mr. Mohamed A. Awad	Chairman, Middle East and Asia, Schlumberger, UAE
Prof. Sadek Boussena	Université Pierre Mendès France de Grenoble, France, and former Minister of Energy, Algeria
Mr. Guy Caruso	Administrator, Energy Information Administration, US
Prof. Musa Essayyad	King Fahd University of Petroleum & Minerals, Saudi Arabia
Dr. Fereidun Fesharaki	President, FACTS, Inc., Hawaii, US
Mr. Walid Khadduri	Chief Economics Editor, Dar Al Hayat, Lebanon
Mr. Mohsin Khan	Director, Middle East and Central Asia, International Monetary Fund, US
Dr. Hisham Khatib	WEC Honorary Vice Chairman, and former Minister of Energy, Jordan
Dr. Edward L. Morse	Executive Adviser, Hess Energy Trading Company LLC, and former Deputy Assistant Secretary of State, US
Dr. Masahisa Naitoh	President, The Institute of Energy Economics, Japan
Dr. Ramzi Salman	Advisor to the Minister of Energy and Industry, Qatar and Chairman of the Executive Board of the International Energy Forum Secretariat, Saudi Arabia
Ms. Nemat Shafik	Director General, Department for International Development, UK and former Vice-President, Infrastructure, World Bank, US
Dr. Adnan Shihab-Eldin	Acting Secretary General, OPEC, Austria

Many international experts commented on the underlying analytical work and reviewed early drafts of each chapter. Their comments and suggestions were of great value. Prominent contributors by chapter include:

Oil Outlook in the Middle East and North Africa

Thomas Ahlbrandt	United States Geological Survey, US
Ken Chew	IHS Energy, Switzerland
Will Davie	Simmons & Company International Ltd., UK
Mark J. Finley	BP Plc., UK
Guy Gantley	Foreign & Commonwealth Office, UK
Dermot Gately	New York University, US
Nadir Gürer	OPEC, Austria
Troy Hansen	Halliburton, Norway
Sigurd Heiberg	Statoil ASA, Norway
Jostein Dahl Karlsen	Ministry of Petroleum and Energy, Norway
David Knapp	Energy Intelligence Group, US
Fikri Kuchuk	Schlumberger, UAE
Valerie Marcel	Royal Institute of International Affairs, UK
Yves Mathieu	Institut Français du Pétrole, France
Peter Nicol	Tristone Capital, UK
S. Hossein Samiei	International Monetary Fund, US
Robert G. Skinner	Oxford Institute for Energy Studies, UK
Mahendra K. Verma	United States Geological Survey, US

Natural Gas Outlook in the Middle East and North Africa

Thomas Ahlbrandt	United States Geological Survey, US
Albert Bressand	Shell, UK
Ken Chew	IHS Energy, Switzerland
Jostein Dahl Karlsen	Ministry of Petroleum and Energy, Norway
Robert G. Skinner	Oxford Institute for Energy Studies, UK
Jonathan Stern	Oxford Institute for Energy Studies, UK
Wim Thomas	Shell, UK
Mahendra K. Verma	United States Geological Survey, US

Electricity and Water Outlook in the Middle East and North Africa

Pierre Audinet	World Bank, US
Alex Bakalian	World Bank, US
Dirk Beeuwsaert	Tractebel S.A, Belgium
Franz Gerner	World Bank, US
Ahmed Irej Jalal	International Atomic Energy Agency, Austria
Keith Miller	Abu Dhabi Water and Electricity Company, UAE
Agata Pawlik	Institute for International Research - Middle East, UAE
Koussai Quteishat	Middle East Desalination Research Center, Oman
Jonathan Walters	World Bank, US

Deferred Investment Scenario

Brad Bourland	Samba Financial Group, Saudi Arabia
Ådne Cappelen	Statistics Norway, Norway
Will Davie	Simmons & Company International Ltd., UK
Mark J. Finley	BP Plc., UK
Herman Franssen	International Energy Associates, US
Dermot Gately	New York University, US
Nadir Gürer	OPEC, Austria
Alan S. Hegburg	Center for Strategic and International Studies, US
Ken Koyama	The Institute of Energy Economics, Japan
Alessandro Lanza	Eni SpA, Italy
Giacomo Luciani	European University Institute, Italy
Knut Einar Rosendahl	Statistics Norway, Norway
S. Hossein Samiei	International Monetary Fund, US
Robert G. Skinner	Oxford Institute for Energy Studies, UK
Frank Verrastro	Center for Strategic and International Studies, US
David Victor	Stanford University, US

Global Refinery Outlook

Jan Ban	OPEC, Austria
Benoît Chagué	Total, France
Kevin Goodwin	BP Plc., UK
Jean-François Gruson	Institut Français du Pétrole, France
Frédéric Lantz	Institut Français du Pétrole, France
Thomas O'Connor	ICF Consulting Group, US
Stuart Simpson	UOP Limited, US
Peter Snowdon	Shell, UK
Thomas Stenvoll	Hess Energy Trading Company LLC, US
Toshiaki Tanaka	The Institute of Energy Economics, Japan
Peter Theunissen	Total, France

Algeria

Ali Aissaoui	Arab Petroleum Investment Corporation, Saudi Arabia
Mabrouka Bouziane	Institut Français du Pétrole, France
Hamid Dahmani	Ministry of Energy and Mining, Algeria
Will Davie	Simmons & Company International Ltd., UK
Valerie Marcel	Royal Institute of International Affairs, UK

Egypt

Hussein Abdallah	Former Secretary-General of the Ministry of Petroleum, Egypt
Mark J. Finley	BP Plc., UK
Geoffrey Frewer	European Investment Bank, Luxembourg
Sharif Ghalib	Energy Intelligence Group, US
Peter Hughes	BG Group, UK

Iran

Mohammad Sadegh Ahadi	National Climate Change Office, Iran
Thomas Ahlbrandt	United States Geological Survey, US
Katrin Elborgh-Woytek	International Monetary Fund, US
Guy Gantley	Foreign & Commonwealth Office, UK
Mohammad Mazraati	OPEC, Austria
Rolf Ødegård	Statoil ASA, Norway

S. Hossein Samiei	International Monetary Fund, US
Alimorad Sharifi	Isfahan University, Iran
Jawad Yarjani	Ministry of Petroleum, Iran

Iraq

Thomas Ahlbrandt	United States Geological Survey, US
Saadalla Al-Fathi	Dome International LLC, UAE, and former Advisor, Ministry of Oil, Iraq
Guy Gantley	Foreign & Commonwealth Office, UK
David Knapp	Energy Intelligence Group, US
Fikri Kuchuk	Schlumberger, UAE
Hasim Sultan	Ministry of Oil, Iraq
Frank Verrastro	Center for Strategic and International Studies, US
Muhammad-Ali Zainy	Center for Global Energy Studies, UK

Kuwait

Mohammed Al Shatti	Kuwait Petroleum Corporation, Kuwait
Valerie Marcel	Royal Institute of International Affairs, UK
Abdulaziz Al-Attar	OPEC, Austria

Libya

Will Davie	Simmons & Company International Ltd., UK
Mark J. Finley	BP Plc, UK
Alan S. Hegburg	Center for Strategic and International Studies, US
Alessandro Lanza	Eni SpA, Italy
David Knapp	Energy Intelligence Group, US
Frank Verrastro	Center for Strategic and International Studies, US

Qatar

Ali Al-Hammadi	Qatar Liquefied Gas Company Ltd., Qatar
Albert Bressand	Shell, UK
Rusty Martin	Qatargas Operating Company Ltd., Qatar
Salman Saif Ghouri	Qatar Petroleum, Qatar

Wim Thomas	Shell, UK
Nick Wallace	Qatargas Operating Company Ltd., Qatar

Saudi Arabia

Thomas Ahlbrandt	United States Geological Survey, US
Brad Bourland	Samba Financial Group, Saudi Arabia
Ådne Cappelen	Statistics Norway, Norway
Fernando Delgado	International Monetary Fund, US
David Knapp	Energy Intelligence Group, US
Fikri Kuchuk	Schlumberger Oilfield Services, UAE
Giacomo Luciani	European University Institute, Italy
Sakura Sakakibara	Mitsui Global Strategic Studies Institute, Japan
Robert G. Skinner	Oxford Institute for Energy Studies, UK
Frank Verrastro	Center for Strategic and International Studies, US

United Arab Emirates

Ali Obaid Al Yabhouni	Abu Dhabi National Oil Company, UAE
Saadalla Al-Fathi	Dome International LLC, UAE, and former Advisor, Ministry of Oil, Iraq
Kazuhiko Chou	The Institute of Energy Economics, Japan
Mangal Goswami	International Monetary Fund, US
Mark J. Finley	BP Plc, UK
Fikri Kuchuk	Schlumberger Oilfield Services, UAE
Keith Miller	Abu Dhabi Water and Electricity Company, UAE

All errors and omissions are solely the responsibility of the IEA.

Comments and questions are welcome and should be addressed to:

Dr. Fatih Birol
Chief Economist
Head, Economic Analysis Division
International Energy Agency
9, rue de la Fédération
75739 Paris Cedex 15
France

Telephone: 33 (0) 1 4057 6670
Fax: 33 (0) 1 4057 6659
Email: Fatih.Birol@iea.org

TABLE OF CONTENTS

Annexes

List of Figures

Chapter 1: The Context

Chapter 2: Global Energy Trends

Chapter 3: Energy Trends in the Middle East and North Africa

Chapter 5: Natural Gas Market Outlook in the Middle East and North Africa

Chapter 6: Electricity and Water Outlook in the Middle East and North Africa

Chapter 14: Libya

Chapter 15: Qatar

Chapter 16: Saudi Arabia

List of Tables

Chapter 14: Libya

Chapter 15: Qatar

Chapter 16: Saudi Arabia

List of Boxes

World Energy Outlook Series

World Energy Outlook 1993
World Energy Outlook 1994
World Energy Outlook 1995
World Energy Outlook 1996
World Energy Outlook 1998
World Energy Outlook: 1999 Insights
    Looking at Energy Subsidies: Getting the Prices Right
World Energy Outlook 2000
World Energy Outlook: 2001 Insights
    Assessing Today's Supplies to Fuel Tomorrow's Growth
World Energy Outlook 2002
World Energy Investment Outlook: 2003 Insights
World Energy Outlook 2004
World Energy Outlook: 2005
    Middle East and North Africa Insights
World Energy Outlook 2006 (forthcoming)

The oil and gas resources of the Middle East and North Africa (MENA) will be critical to meeting the world's growing appetite for energy. The greater part of the world's remaining reserves lie in that region. They are relatively under-exploited and are sufficient to meet rising global demand for the next quarter century and beyond. The export revenues they would generate would help sustain the region's economic development. But there is considerable uncertainty about the pace at which investment in the region's upstream industry will occur, how quickly production capacity will expand and, given rising domestic energy needs, how much of the expected increase in supply will be available for export. The implications for both MENA producers and consuming countries are profound. This *Outlook* seeks to shed light on these very complex issues.

Global energy needs are likely to continue to grow steadily for at least the next two-and-a-half decades. If governments stick with current policies – the underlying premise of our Reference Scenario – the world's energy needs would be more than 50% higher in 2030 than today. Over 60% of that increase would be in the form of oil and natural gas. MENA's share of global oil and gas output would grow substantially, as long as MENA countries invest enough in energy production and transportation infrastructure. But the global trends in the Reference Scenario would raise several serious concerns. Climate-destabilising carbon-dioxide emissions would continue to rise, calling into question the long-term sustainability of the global energy system. And the sharply increased dependence of consuming regions on imports from a small number of MENA countries would exacerbate worries about the security of energy supply.

More vigorous government policies in consuming countries could, and no doubt *will*, steer the world onto a different energy path. The leaders of the G8 and several large developing countries, meeting at Gleneagles in July 2005, acknowledged as much when they called for stronger action to combat rising consumption of fossil fuels and related greenhouse-gas emissions. Most OECD governments have declared their intention to do more and other countries around the world can be expected to follow suit. Such policies are all the more likely to be implemented if energy prices remain high.

Consuming-country policies could curb demand growth and reduce the world's reliance on MENA oil and gas. A World Alternative Policy Scenario demonstrates that if governments around the world were to implement new policies they are considering today, aimed at addressing environmental and energy-security concerns, fossil-fuel demand and carbon-dioxide emissions would be significantly lower. But even in this scenario, global energy demand in 2030 would still be 37% higher than today and the volume of MENA hydrocarbon exports would still grow significantly. Far more radical policy action and technology breakthroughs would be needed to reverse these trends.

A critical uncertainty is whether the substantial investments needed in the upstream hydrocarbons sector in MENA countries will, in fact, be forthcoming. In a Deferred Investment Scenario, much lower MENA oil production drives up the international price of oil and, with it, the price of gas. Higher energy prices, together with slower economic growth, would choke off energy demand in all regions and would, therefore, reduce demand for oil and gas compared with the Reference Scenario. MENA exports, nonetheless, continue to grow. Current market instability and the recent surge in oil prices demonstrate the vital importance of adequate investment in upstream and downstream capacity and the threat posed by surging global demand.

The prospects for MENA's role in global energy supply developments have far-reaching implications for the global economy. The governments of producing and consuming countries alike have a mutual interest in addressing the concerns highlighted in this *Outlook*. The information and analysis presented here can provide a solid quantitative framework for understanding the challenges, deepening the dialogue between producers and consumers and devising appropriate policy responses.

World Energy Demand will Grow Inexorably, Absent New Policies

In the absence of new government policies, the world's energy needs will rise inexorably. In the Reference Scenario, world primary energy demand is projected to expand by more than half between now and 2030, an average annual growth rate of 1.6%. By 2030, the world will be consuming 16.3 billion tonnes of oil equivalent – 5.5 billion toe more than today. More than two-thirds of the growth in world energy use will come from the developing countries, where economic and population growth are highest. The international energy prices that underpin these projections have been revised upwards from last year's *Outlook*. The average IEA crude oil import price is now assumed to ease to around $35 per barrel in 2010 (in year-2004 dollars) as new crude oil production and refining capacity come on stream. It is then assumed to rise slowly to $37 in 2020 and $39 in 2030. In nominal terms, the price will reach $65 in 2030.

Fossil fuels will continue to dominate energy supplies, meeting more than 80% of the projected *increase* in primary energy demand. Oil remains the single most important fuel, with two-thirds of the increase in oil use coming from the transport sector. Demand reaches 92 mb/d in 2010 and 115 mb/d in 2030. The lack of cost-effective substitutes for oil-based automotive fuels will make oil demand more rigid. Natural gas demand grows faster, driven mainly by power generation. It overtakes coal as the world's second-largest primary energy source around 2015. The share of coal in world primary demand falls a little, with demand growth concentrated in China and India. The share of nuclear power declines marginally, while that of hydropower remains broadly constant. The share of biomass declines slightly, as it is replaced with modern commercial fuels in developing countries. Other renewables, including geothermal, solar and wind energy, grow faster than any other energy source, but still account for only 2% of primary energy demand in 2030.

The world's energy resources are adequate to meet the projected growth in energy demand in the Reference Scenario. Global oil reserves today exceed the cumulative projected production between now and 2030, but reserves will need to be "proved up" in order to avoid a peak in production before the end of the projection period. Exploration will undoubtedly be stepped up to ensure this happens. The exact cost of finding and exploiting those resources over the coming decades is uncertain, but will certainly be substantial. Cumulative energy-sector investment needs are estimated at about $17 trillion (in year-2004 dollars) over 2004-2030, about half in developing countries. Financing the required investments in non-OECD countries is one of the biggest challenges posed by our energy-supply projections.

The global oil-refining industry has an urgent need for more distillation and upgrading capacity. As a result of strong growth in demand for refined products in recent years, spare capacity has been rapidly diminishing and flexibility has fallen even faster. Effective capacity today is almost fully utilised, so growing demand for refined products can only be met with additional capacity. Upgrading capacity will be needed even more than distillation capacity, since demand will continue to shift to lighter products, while crude oil production is becoming heavier, with a higher sulphur content.

MENA Domestic Energy Demand is Set to Surge...

Rapidly expanding populations, steady economic growth and heavy subsidies will continue to drive up MENA energy demand. In the Reference Scenario, demand is projected to grow on average by 2.9% per year between now and 2030. As a result, demand more than doubles. By 2030 the MENA region will account for 7.5% of global primary energy demand, two percentage points more than today. The biggest contributors to demand growth will be Saudi Arabia and Iran. These two countries will account for some 45% of

MENA energy demand in 2030, about the same as today. The fastest *rate* of energy-demand growth will occur in Qatar.

Most MENA countries will continue to rely almost exclusively on oil and natural gas to meet their energy needs. Gas will overtake oil after 2020 as the region's main energy source for domestic use, thanks to policies aimed at freeing up oil for export. The use of other fuels increases, but together they account for less than 4% of primary energy demand in 2030 – hardly more than at present.

Despite rapid growth in MENA energy use, per capita consumption projected for 2030 will still be barely half the current level in OECD countries. Large discrepancies in per capita energy use among MENA countries will remain. In most of the Gulf countries, per capita *electricity* consumption will remain among the highest in the world – mainly the consequence of heavy price subsidies which lead to inefficient energy use and of the hot climate which necessitates considerable air-conditioning.

The power and water sectors will absorb a growing share of the region's total primary energy use as electricity and desalinated water needs expand rapidly. Heavy subsidies to both services are accentuating this trend. Gas-fired power plants, mostly using combined-cycle gas-turbine technology, will meet 71% of new generating-capacity needs. Water desalination, an energy-intensive process usually integrated with power production, will account for more than one-quarter of the increase in total fuel use in the power and water sector in Saudi Arabia, the United Arab Emirates, Kuwait, Qatar, Algeria and Libya combined.

...but Even Faster Growth in MENA Output will Boost Exports

Output of oil and natural gas in the MENA region is poised for rapid expansion. In the Reference Scenario, oil production (including natural gas liquids) is projected to rise from 29 mb/d in 2004 to 33 mb/d in 2010 and to 50 mb/d by 2030. In some countries, this may require opening up the upstream sector to foreign investment. The contribution of giant oilfields to total production will drop sharply, from 75% today to 40% in 2030, as mature giant fields decline and new developments focus more on smaller fields. Production in MENA countries, especially in the Middle East, increases more rapidly than elsewhere because their resources are greater and their production costs lower. Growth in aggregate production outside MENA is expected to slow over the *Outlook* period. Saudi Arabia, which has the largest proven reserves of oil in the world, will remain by far the largest supplier. Its output will rise from 10.4 mb/d in 2004 to 11.9 mb/d in 2010 and just over 18 mb/d in 2030. Iraq is expected to see the fastest rate of production growth, and the biggest increase in volume terms after Saudi Arabia. In some countries, including Iraq, increased production will hinge on large-scale foreign investment.

On this basis, MENA's share of world oil production would jump from 35% in 2004 to 44% in 2030. Almost all the increase comes from the Middle East. Saudi Arabia's share of total MENA oil output in 2030 will be much the same as today, at about 36%. Four countries will see their share in MENA output increase: Iraq, Kuwait, the UAE and Libya.

MENA production outpaces growth in domestic demand, allowing the region's net oil exports to rise by three-quarters over the *Outlook* period, from 22 mb/d in 2004 to 25 mb/d in 2010 and 39 mb/d by 2030. Most exports will still be as crude oil in 2030, but refined products will account for a growing share. Exports to developing Asian countries will increase most, but will grow to all the major consuming regions.

MENA gas production is projected to grow even more rapidly than oil, trebling over the projection period to 1 210 billion cubic metres in 2030. This is faster than almost any other major world region. The biggest volume increases in the region occur in Qatar, Iran, Algeria and Saudi Arabia. A third of MENA gas output comes from North Field/South Pars, a field shared by Qatar and Iran, and Hassi R'Mel in Algeria. This share will increase as they are further developed. Demand for MENA gas will be driven by strong global demand and dwindling output in many other gas-producing regions. The bulk of the increase in output will be exported, mostly as liquefied natural gas. Net exports from MENA countries to other regions are projected to more than quadruple to 440 bcm in 2030, with a marked shift in sales to Western markets. Europe will remain the primary destination for North African gas exports. Major gas importers, including most OECD countries and developing Asia, will become ever more dependent on imports from MENA countries.

MENA oil- and gas-export revenues, which have surged in the last few years, will remain high. Aggregate MENA oil and gas revenues are projected to rise from about $310 billion in 2004 to $360 billion in 2010 and $635 billion in 2030. Natural gas will make a growing contribution. Cumulative revenues will far exceed the investment needed to make them possible. Total oil and gas investment is projected to amount to about $1 trillion over the period 2004-2030 (in year-2004 dollars), or $39 billion per year.

The need for more comprehensive and transparent data on oil and gas reserves in all regions is a pressing concern. The preparation of this *Outlook* involved an extensive effort to collect the best available data on reserves from official and informal sources. But there are inconsistencies in the way reserves are defined and measured, and a lack of verifiable data on reserves and of a universally recognised reporting system makes it difficult to assess the quality of data on reported proven reserves in many regions, including MENA. Uncertainties about just how big reserves are and the true costs of developing them are casting shadows over the oil market outlook and heightening fears of higher costs and prices in future.

Lower MENA Oil Investment would Radically Alter the Global Energy Balance

A major shortfall in MENA investment in upstream oil would radically alter the global energy balance. In recent years, global investment, crude oil production capacity and refining capacity have lagged the rise in demand, driving up oil prices. Our projections in the Reference Scenario involve a doubling of the level of annual upstream investment in MENA countries. It is far from certain that all that investment will be forthcoming: MENA governments could choose *deliberately* to develop production capacity more slowly than we project in our Reference Scenario. Or external factors such as capital shortages could *prevent* producers from investing as much in expanding capacity as they would like. The Deferred Investment Scenario analyses how energy markets might evolve if upstream investment in each MENA country were to remain constant as a share of GDP at the average level of the past decade. This would result in a $110 billion, or 23%, drop in cumulative upstream MENA oil investment over 2004-2030.

Lower investment on this scale causes MENA oil production to drop by almost a third by 2030 compared with the Reference Scenario. Production falls further than investment by the end of the projection period because of the *cumulative* effect over the projection period. In 2030, total MENA output reaches 35 mb/d, compared with 50 mb/d in the Reference Scenario. Saudi Arabia's production, at 14 mb/d in 2030, is more than 4 mb/d lower than in the Reference Scenario. MENA's share of world oil production drops from 35% in 2004 to 33% in 2030 (against a rise to 44% in the Reference Scenario). As a result, MENA oil exports are almost 40% lower in 2030. By contrast, higher prices stimulate an 8% increase in non-MENA oil production compared to the Reference Scenario. Natural gas production in MENA countries also falls significantly, due to lower global demand and lower output of associated gas. Gas exports fall by 46% in 2030, with Qatar's falling furthest in absolute terms.

In the Deferred Investment Scenario, the international crude oil price is significantly higher than in the Reference Scenario over the projection period. In the Reference Scenario, the average IEA import price is assumed to fall back from recent highs to around $35 (in year-2004 dollars) in 2010, and then to rise slowly to $39 in 2030. In the Deferred Investment Scenario, the price increases gradually over time, relative to the Reference Scenario. It is about $13 higher in 2030, or $21 in nominal terms – an increase of almost one-third. Natural gas prices rise broadly in line with oil prices. The coal price also increases slightly. Energy prices would become more volatile.

As a result of higher prices and lower world GDP, global energy demand is reduced by about 6% in 2030, compared with the Reference Scenario. World GDP growth, the main driver of energy demand, is on average 0.23 percentage points per year lower. Lower oil and gas revenues and higher prices cause primary energy-demand growth in MENA countries to slow, but less markedly than in non-MENA regions. Among the primary fuels, global demand for oil falls most. Global oil demand, at 105 mb/d in 2030, is 10 mb/d lower than in the Reference Scenario. Demand for both gas and coal also falls, mainly as a result of lower demand for fuel inputs to power generation.

Our analysis suggests that MENA producers would lose out financially were investment to be deferred in the way assumed in the Deferred Investment Scenario. The increase in prices fails to compensate for lower export volumes. Over 2004-2030, the cumulative value of aggregate MENA oil and gas export revenues would be more than a trillion dollars lower (in year-2004 prices) than in the Reference Scenario. The loss of revenues is almost five times more than the reduction in oil and gas investment. Revenues also fall in terms of net present value. Oil accounts for about 70% of the fall in revenues.

Consuming-Country Policies could Reduce MENA Export Demand

The World Alternative Policy Scenario depicts the energy future that might emerge if consuming-country governments press ahead with the vigorous new policy measures already being contemplated. They involve promoting more efficient energy use and switching away from fossil fuels, for environmental or energy-security reasons. The basic assumptions about macroeconomic conditions and population are the same as in the Reference Scenario. But energy prices change, because of the new level at which an equilibrium between supply and demand is established.

In the World Alternative Policy Scenario, global primary energy demand is about 10% lower in 2030 than in the Reference Scenario. Primary energy demand grows by 1.2% per year, 0.4 percentage points less than in the Reference Scenario. Nonetheless, demand in 2030 is still 37%

above the current level. Oil remains the leading energy source. Its share of global primary energy demand – just over one-third – is only slightly lower than in the Reference Scenario in 2030. By contrast, the share of coal in primary energy demand falls sharply in all regions. On the other hand, the use of non-hydro renewables, excluding biomass, is almost 30% higher in 2030 than in the Reference Scenario. Biomass and nuclear energy also grow. The effect of energy-efficiency and fuel-diversification policies on energy demand grows over the projection period, as the stock of energy capital goods is gradually replaced and new measures are introduced.

The fall in oil and gas demand in the main consuming regions leads to a reduction in MENA production and exports, and drives down prices. By 2030, MENA oil production reaches 45 mb/d – almost 6 mb/d less than in the Reference Scenario. But it is still more than 50% higher than in 2004. The oil price is on average about 15% lower compared with the Reference Scenario. Lower demand and prices cut cumulative MENA oil and gas export revenues by 21% over the projection period compared with the Reference Scenario. Revenues also fall in terms of net present value. Nonetheless, revenues in 2030 are $160 billion, or just over 50%, higher than in 2004.

Lower overall energy consumption and a larger share of less carbon-intensive fuels in the primary energy mix yield a 5.8 gigatonne, or 16%, reduction in global carbon-dioxide emissions in 2030 compared to the Reference Scenario. This is comparable to the current combined emissions of the United States and Canada. The bulk of the reduction comes from lower coal use, especially in power generation in non-OECD countries. This results mainly from the reduction in electricity demand brought about by new end-use efficiency policies. Emissions, nonetheless, still rise 28% over current levels.

Deepening the Consumer-Producer Dialogue would Bring Mutual Benefits

The policies of producing and consuming countries will change over time in response to each other, to market developments and to shifts in market power. If MENA upstream investment falters and prices rise, the more likely it becomes that consuming countries will adopt additional policies to curb demand growth and reliance on MENA. This would have the effect of tempering the long-term impact on prices of lower MENA investment. It would also amplify the depressive effect of higher prices on oil and gas demand. The more successful the importing countries' policies are, the more likely it is that the producing countries will adopt policies to sustain their production and their global market share. Lower prices would result.

These interactions illustrate the case for improving market transparency, for more effective mechanisms for exchanging information between oil producers and consumers, and for a more profound dialogue between them. Concerns among consuming countries about security of supply are matched by those among producing countries about security of demand. Consuming countries will continue to seek to diversify their energy mix, while producing countries will continue to seek to diversify their economies. Together, consumer and producer governments can improve the mechanisms by which they seek to reconcile their interests and achieve mutually beneficial outcomes.

THE CONTEXT

HIGHLIGHTS

- The Middle East and North Africa (MENA) region is exceptionally well endowed with energy resources, holding 61% of the world's proven oil reserves and about 45% of gas reserves. These resources are relatively under-exploited. Further development of them will be critical to meeting global energy needs in the coming decades.
- The region's macroeconomic progress will determine the pace of domestic energy demand growth and, therefore, export availability. In the Reference Scenario, the rate of growth of GDP in MENA is expected to average 3.4% in 2003-2030. The pace of economic growth will fluctuate, due partly to movements in international oil prices.
- In most MENA countries, market reforms aimed at increasing the role of private and foreign investors in the development of energy infrastructure and the rest of the economy are expected to continue to move ahead, but slowly. Difficult challenges include the removal of subsidies on domestic energy use, which are large in most countries, and economic diversification away from hydrocarbons.
- Demographic factors will continue to play a major role in economic and energy development in the region. Total population is projected to grow by 1.7% per year over 2003-2030, to over 500 million. The labour force will grow even more rapidly, as half the current population is under the age of 20. Unemployment – and underemployment – is a serious and growing problem in many MENA countries.
- The near-term outlook for international oil prices remains highly uncertain. In our Reference Scenario, the average price for IEA crude oil imports is assumed to fall back from recent highs of over $60 a barrel (in year-2004 dollars) to around $35 in 2010 and then climb to $39 in 2030.
- A Deferred Investment Scenario analyses how global energy markets might evolve if investment in the upstream oil industry of MENA countries were to be substantially lower than projected in the Reference Scenario. That might occur because of domestic production policies or difficulties in securing capital in some countries, leading to higher prices and lower energy demand than in the Reference Scenario.
- Higher prices make it more likely that consuming countries would adopt policies to curb energy use and to promote switching away from fossil fuels. The effects of introducing such policies are analysed in a World Alternative Policy Scenario.

Purpose and Scope of the Study

Recent editions of the *World Energy Outlook* have highlighted the growing role that the Middle East is expected to play in world energy markets over the coming decades. Development of the region's vast oil and gas resources will be essential to meet future energy needs. The Middle East is already the world's largest oil-exporting region and is expected to become even more dominant in the future. But, as the *World Energy Investment Outlook 2003* and the *World Energy Outlook 2004* point out, there is considerable uncertainty about the pace at which investment in the region's upstream industry will occur, how quickly production capacity will expand and, given rising domestic energy needs, how much of the expected increase in supply will be available for export. The implications for the rest of the world are profound.

This *Outlook* seeks to shed light on these very complex issues. Against the background of an updated assessment of global supply and demand, it sets out in detail the prospects for energy demand and supply in the Middle East and North Africa (MENA) region, focusing on the factors that will determine the availability of the region's hydrocarbon resources for export and the implications of projected trends in Middle East energy trade for the rest of the world. It seeks to answer the question, how much oil and gas can we reasonably expect MENA countries to export to the rest of the world in the next two-and-a-half decades?

To answer that question, we have prepared detailed sets of projections of the region's energy market, country by country and fuel by fuel. These projections cover how much energy will be produced, consumed locally and made available for export. The aim is to provide policy-makers and other interested parties with a rigorous quantitative framework for analysing possible future developments in MENA energy supply and their implications for global energy markets. Our hope is that this will add depth and momentum to the consumer-producer dialogue and improve understanding on both sides of the issues surrounding the expansion of the region's energy infrastructure. To our knowledge, this is the first time medium- to long-term prospects for the region and their global consequences have been analysed in such a detailed way.

In 2004, MENA countries exported 22 million barrels per day of oil (including natural gas liquids), equal to 27% of world production and 62% of net inter-regional trade (Figure 1.1). MENA plays a less important role in gas supply, both globally and in the OECD. MENA holds 61% of the world's proven oil reserves and about 45% of gas reserves (Figure 1.2), yet its share of global production in 2004 was only 35% for oil and 15% for gas. The region's vast reserves and its low production-to-reserves ratios mean that it has the potential to provide much of the additional oil and gas that the energy-importing countries – including the majority of OECD countries – will need in the coming decades. Whether and how these resources are developed will affect the evolution of energy prices and the global energy balance, with significant implications for the world economy. While the main imponderable is the level of MENA production, the prospects for domestic demand will have a major bearing on export availability. Primary energy demand is growing faster in MENA than in any other major region outside East Asia.

Figure 1.1: **MENA Share in World Oil and Gas Supply, 2004**

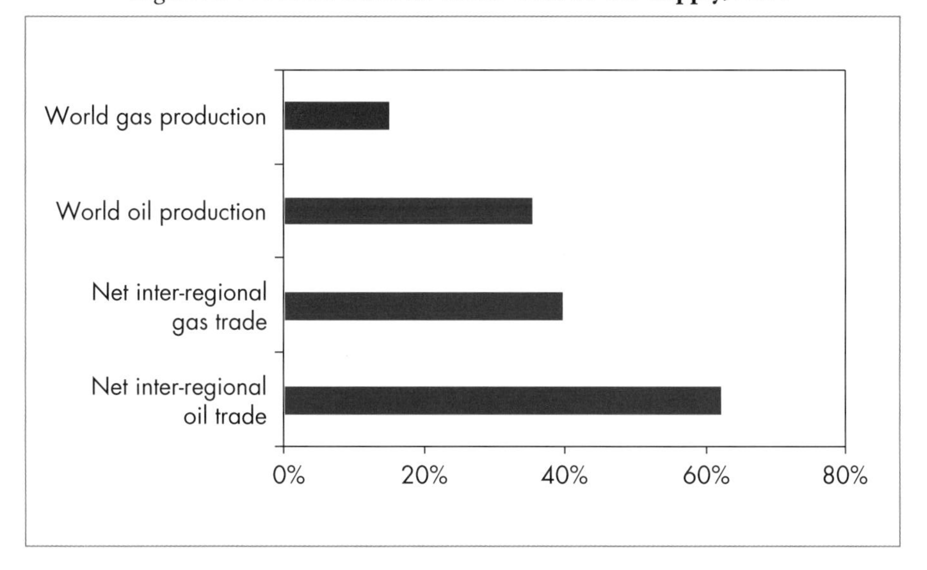

MENA has been the scene of periodic internal and external conflicts, resulting from a complex mix of political, economic, demographic and cultural factors beyond the scope of this study. In some cases, these conflicts have disrupted the flow of hydrocarbon exports, most notably during the 1973-1974 oil crisis, at the start of the Iranian revolution in 1979, throughout the 1980-1988 Iran/Iraq war and, most recently, at the time of the 2003 Iraqi invasion. They have also held back the development of reserves and export capacity. Parts of the region remain unstable, casting doubt on future energy-infrastructure developments and the reliability of energy supplies. Yet the majority of hydrocarbon-exporting MENA countries have been reliable suppliers of oil and gas to OECD and other markets for many years and have established good long-term relationships.

Inevitably, the expectation of increased reliance on MENA hydrocarbons raises questions for importing countries about their energy security. A central

Figure 1.2: **Share of MENA in World Proven Oil and Gas Reserves, End-2004**

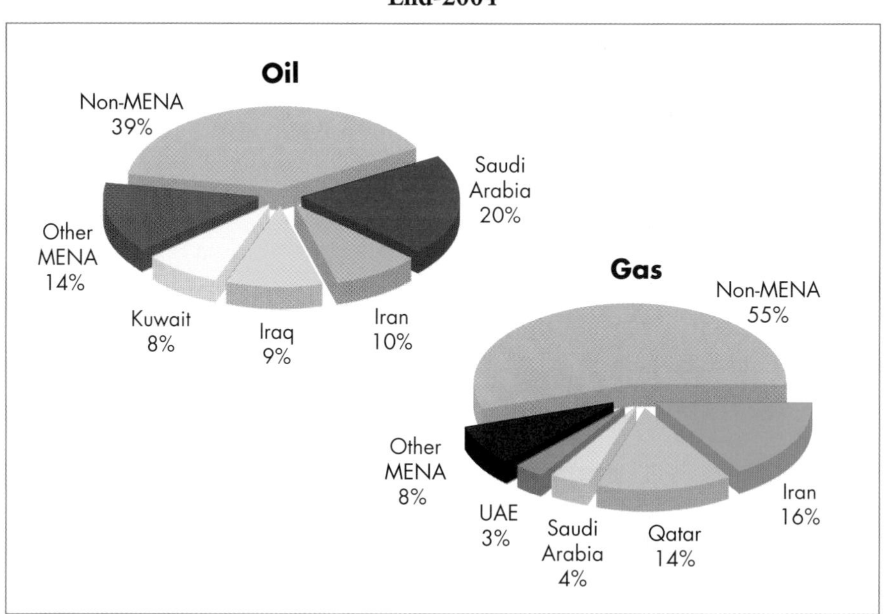

Sources: *Oil and Gas Journal* (20 December 2004); Cedigaz (2005).

message of *WEO-2005* is that short-term risks to energy security will grow, due largely to growing imports from this region. Expanding trade strengthens mutual dependence between exporting and importing countries, but it also increases the risk of a supply disruption resulting from piracy, terrorist attacks, accidents or political action. And long-term energy security hinges on sufficient investment in energy infrastructure. A shortfall or postponement in investment in the key MENA countries would have serious implications for energy prices and the economic health of importing countries – especially in the developing world. It would also damage the long-term interest of producing countries (Chapter 7).

This study focuses on the six largest energy-producing countries in the Middle East – Iran, Iraq, Kuwait, Qatar, Saudi Arabia and the United Arab Emirates – and the three most important North African countries – Algeria, Egypt and Libya. The other countries that make up the MENA region (Bahrain, Israel, Jordan, Lebanon, Oman, Syria and Yemen in the Middle East, as well as Morocco and Tunisia in North Africa) are assessed in a more aggregated fashion. The MENA region in total has a population of 323 million, equal to about 5% of the world total. Its gross domestic product accounts for 3.7% of the world total. The share of the oil and gas sectors in the region's GDP, at 17.9% in 2003, is the highest of any world region.

The Methodological Approach

As in past editions of the *Outlook*, a scenario approach has been adopted to analyse the possible evolution of MENA energy markets. The goal is to identify and quantify the primary drivers of energy supply and demand. The first year of the projections is 2004, as 2003 is the last year for which historical energy data are available for all countries in the region.[1] Projections run to 2030, but results are also presented for 2010 and 2020.

The projections have been underpinned by an extensive data-collection and modelling effort. Separate energy-demand and power generation models were developed for the nine main countries listed above, while the other Middle Eastern and North African countries are treated as two single groups. These models are used to generate detailed projections of energy demand by sector and fuel. One distinctive feature is the explicit treatment of energy use for water desalination, either as an end-use sector or integrated into the power generation modelling, in those countries where it accounts for an important share of overall energy use. The MENA demand models were integrated into the IEA's World Energy Model[2] in order to derive global demand projections. Oil and gas supply is modelled differently according to each scenario (see below). Coal supply in the region is trivial and so is not modelled explicitly. A major concern throughout the modelling exercise was to obtain the most accurate data available (Box 1.1).

The demand and supply modelling, notably the analysis of future production profiles for existing oil and gas fields and of new upstream developments, involved compiling and processing large quantities of data. A large number of energy companies, financial institutions, international organisations and consultants – acknowledged at the beginning of this report – made a major contribution to this work. The contribution of OPEC and the governments and state companies in providing information, verifying our analysis and commenting on the results has been particularly vital. The preparation of this study benefited considerably from a joint OPEC/IEA workshop on the energy outlook for the MENA region, held in Kuwait on 15 May 2005.

The results of three scenarios are presented in detail in this *Outlook.* The Reference Scenario represents a baseline or business-as-usual vision of how energy markets could evolve in the absence of new government policy initiatives, assuming the huge oil and gas resources of the MENA region are

1. Data on economic indicators, population and prices, however, are available for most countries for 2004. Preliminary energy data are also available in some cases for that year.

2. A description of the World Energy Model, a large-scale mathematical model that the IEA has developed over many years, can be found at www.worldenergyoutlook.org.

Box 1.1: **The Accuracy and Quality of MENA Energy Data**

The compilation of country by country statistics on energy supply, infrastructure, demand and prices was a key pillar of our analysis. The energy data used in this *Outlook* are gathered from the most reliable statistical sources available worldwide. The IEA Statistics Division provided much of the energy demand and price data for all countries. For the purpose of this *Outlook,* the sectoral disaggregation and the data accuracy of energy consumption were enriched and enhanced. Up-to-date technical and economic information on existing and planned energy infrastructure in the region were collected. Additional information on energy consumption, domestic energy prices and capital stock was obtained from energy companies operating in MENA countries, ministries and international organisations, including the Organization of the Petroleum Exporting Countries (OPEC), the United Nations, the International Monetary Fund and the World Bank.

Creating and maintaining a complete, accurate and consistent database on oil and gas production by field was a major component of our supply analysis. The database includes field-by-field oil and gas production time series, reserves, oil quality, and geological characteristics of the reservoirs. The data compilation drew heavily on production and reserve data in the IHS Energy database, as well as information obtained on actual and planned projects from national and international oil companies, oil field service companies and consultants. All the field-by-field production numbers have been reviewed by international experts.[3]

However, all statistical information is subject to error. In particular, the accuracy, reliability and completeness of the field-specific information used for analysing oil and gas production prospects are highly variable. In many cases, production time series are not complete. In some MENA countries, statistics on domestic energy consumption have been collected only in recent years. Their completeness and accuracy is not yet up to international standards. Although the quality of the data will undoubtedly be improved in the future, we believe that our analysis is based on the best information currently available.

3. The IEA is working with several other international organisations to enhance the quality and transparency of international oil data. The Joint Oil Data Initiative, which brings together the Asia Pacific Energy Research Centre, the statistics office of the European Union (Eurostat), the Latin-American Energy Organization (OLADE), OPEC, the UN Statistical Division and the IEA, is aimed at improving the availability, accuracy and timeliness of monthly oil market data. The IEA is also working with several organisations on improving the definition and classification of oil reserves (see Chapter 4).

mobilised to meet rising global demand for oil and gas. The Reference Scenario projections are derived from exogenous assumptions about macroeconomic conditions, population growth, energy prices and technology in the MENA countries and other world regions. The Deferred Investment Scenario analyses the implications for the global energy balance of lower investment in MENA countries in upstream oil and gas projects. The World Alternative Policy Scenario examines the consequences of possible new policy initiatives by the governments of consuming countries around the world.

The Reference Scenario

The Reference Scenario is defined in the same way as in *WEO-2004*. It takes account of those government policies and measures in both MENA and non-MENA countries that have already been enacted or adopted, though some have not yet been fully implemented and so their effects are not apparent in historical energy data. These initiatives cover a wide array of sectors and a variety of policy instruments. They include policies on curbing energy demand and related emissions of noxious and greenhouse gases, on reforming electricity and gas markets and on developing and producing oil and gas. The Reference Scenario does not include possible, potential or even likely future policy initiatives. Major new energy-policy initiatives will inevitably be implemented during the projection period, but it is impossible to know today which measures will eventually be adopted and how they will be implemented, especially towards the end of the projection period. Thus, the Reference Scenario projections should be considered a baseline vision of how energy markets would evolve if governments took no further action to affect their evolution beyond that which they have already committed themselves to as of mid-2005. By contrast, the World Alternative Policy Scenario (Box 1.2), presented in Chapter 8, analyses the impact of new demand-side policies that could be adopted during the projection period.

The energy-market implications of policies and measures in all regions have been updated from *WEO-2004* to take account of major new initiatives and market developments. In addition, this *Outlook* incorporates a much more detailed analysis of policies that had been adopted in MENA countries as of mid-2005. Details of these policies can be found in the individual country chapters. The assumptions for non-MENA regions for the rate of growth in gross domestic product – by far the most important driver of energy demand – are broadly unchanged from last year's *Outlook,* though new data for 2003 and 2004 have been incorporated. The GDP assumptions for the Middle East as a whole differ markedly, as assumptions have been developed for each of the six largest countries and for the remaining countries in the region on the basis of a more detailed assessment of economic prospects. Similarly, the overall growth assumption for

Box 1.2: **The World Alternative Policy Scenario**

The Alternative Policy Scenario analyses how the global energy market could evolve were countries around the world to adopt a set of new policies and measures (already under discussion) to reduce air pollution and greenhouse-gas emissions and to enhance energy security. Such policies, called for at the G8 summit in Gleneagles in the United Kingdom in July 2005, are more likely to be implemented if energy prices remain high. This analysis builds on and updates that of *WEO-2004*. New policies taken into consideration include, among others, the extension and strengthening of motor-vehicle fuel-efficiency standards and the promotion of renewables in China, increased nuclear capacity in China and India, the prolongation of the new emissions-trading scheme in the European Union, some of the measures detailed in a new EU green paper on energy efficiency and new financial incentives for renewables in Korea.

Only those policies that governments are either currently discussing or that they might reasonably be expected to implement over the projection period have been taken into account. The most important are measures to improve energy efficiency and increase the use of renewables. The measures analysed have not been selected strictly according to their cost-effectiveness, but rather as a reflection of the current energy-policy debate. The basic assumptions about macroeconomic conditions and population are the same as in the Reference Scenario. But energy prices change, because of the new level at which an equilibrium between supply and demand is established.

The rates of efficiency gain achieved vary with local conditions. They take into account past efforts to encourage more efficient energy use and to reduce environmental damage. On average, the improvements in energy efficiency are assumed to be higher in countries or regions in the developing world than in OECD countries, reflecting their greater potential for efficiency improvements and adoption of new technologies. Depending on the region, the measures taken into account include the strengthening of existing policies and broadening of their coverage, as well as the introduction of new policies. Many of the policies considered here promote faster deployment of more efficient and less polluting technologies. As these technologies are deployed in OECD countries, their unit costs fall, eventually making them affordable for *all* countries.

Many of the policies considered have effects at a very detailed level in the economy. The effects of mandatory efficiency standards, for example, cannot be estimated from past patterns of energy use, since they impose new technical constraints on the energy system. To analyse such measures, we have incorporated detailed "bottom-up" sub-models of the energy system into the World Energy Model. A key aspect of these models is the explicit representation of energy efficiency, of the different types of activity that drive energy demand and of the physical stock of capital. Capital-stock turnover is a key issue. The very long life of power plants, buildings and even cars limits the rate at which more efficient technology can be deployed.

total Africa is different, because detailed assumptions have been developed for the three main North African countries and other North Africa. The assumptions for economic growth for MENA countries have been derived using a methodology, based on our production projections, to ensure their consistency with projected oil and gas revenues (see below). Assumptions about technology are generally the same for all regions as for *WEO-2004.* The assumptions for global population growth and international energy prices have been revised from last year's *Outlook,* and are summarised below.

Geopolitical Developments

Political developments in MENA are of great importance to the energy security of the rest of the world, because of the region's large and growing share of world energy supply. MENA has often been the scene of civil unrest and the cause of instability of the international oil market. The world has an obvious common interest in supporting the pursuit and maintenance of political and social stability and peace in the region.

Predicting geopolitical developments in MENA is beyond the scope of this book. And yet they will inevitably influence energy market trends, directly and indirectly. Implicitly, our analysis assumes no major change in the geopolitical situation and that MENA countries will achieve some degree of success in addressing the political, economic and security challenges that lie ahead. Although prospects for peace and stability vary markedly among countries, we assume no general worsening of the situation. Critically, we assume that MENA producers are able and willing to undertake the large investments in new oil and gas production capacity that are called for in the Reference Scenario.

Energy Policy Developments in the Middle East and North Africa

The main energy-policy goal of the resource-rich MENA countries is to develop their oil and gas reserves in such a way so as to maximise their value to both current and future generations. This entails maintaining reliable supplies to the global market and judicious management of the balance between short-term revenue needs and long-term objectives. For most of these countries, short-term production decisions are taken within the framework of OPEC agreements on production ceilings. How rapidly they plan to develop their oil and gas resources varies according to the size of their resources, depletion policy and expectations about the future call on their supply. A detailed discussion of policies on short-term oil-production management and on the long-term development of oil and gas resources can be found in Chapters 4 and 5.

Most MENA oil-producing countries seek to influence oil prices collectively through participation in OPEC policies to manage supply. One requirement of any supply-management system is the maintenance of a certain amount of spare capacity. Saudi Arabia's official goal is to maintain 1.5 mb/d to 2 mb/d of spare crude-oil production capacity. As Saudi capacity is officially expected to reach 12.5 mb/d (excluding natural gas liquids) by 2009, this implies that the Saudi government expects the call on their oil by that time to reach 10.5 to 11 mb/d – little higher than at present. We believe the call will be higher (see Chapter 16). Other MENA countries collectively are expected to expand their capacity at a pace that ensures some "operational" spare capacity, though no country other than Saudi Arabia has a formal target. MENA oil producers are also expected to continue to engage in dialogue with consuming states in an effort to build confidence through an exchange of views on oil-market developments and prospects.

Several MENA countries will give more emphasis to the development of natural gas reserves as a way of meeting rising domestic needs, diversifying away from dependence on oil in order to free up more oil for export. Attracting investment in the gas sector will accordingly take on increasing importance. Iran has the second-largest proven gas reserves in the world and Qatar the third-largest. Although Qatar has a significant head-start over Iran in developing its reserves, Iran is expected to put more emphasis on gas projects in the medium term. Political support for building an integrated gas grid in the Persian Gulf states is expected to grow.

In most MENA countries, resource-rich and poor alike, there is a commitment to market reforms, involving an increase in the role of private and foreign investors in the development of energy infrastructure and establishing a more rational energy pricing. However, there are difficulties in reducing subsidies, ideological sensitivities over foreign participation and resistance to change from entrenched interests. Saudi Arabia has recently opened up its gas and electricity sectors to foreign investment, but upstream oil remains off-limits. Iran has attracted foreign investment in upstream oil and gas projects through buy-back contracts, under which the resources remain in the ownership of the National Iranian Oil Company, but there are doubts about the attractiveness of the terms on offer and the prospects for further deals. Iraq is seeking large amounts of foreign investment in the upstream and downstream oil and gas sectors as well as in the power sector, but achieving this will hinge on a major improvement in security. In the UAE, foreign ownership of upstream assets is allowed under concession agreements. Kuwait is aiming to bring in international oil companies to assist in upstream oil development. Egypt, Algeria and, more recently, Libya are pursuing policies to attract private and foreign investment in a range of energy projects.

Reducing and eliminating subsidies to domestic energy supplies is a major challenge throughout the region. In almost every MENA country, oil product prices are held to well below international market levels. Fuel oil and gas inputs to power generation are typically subsidised, as are electricity prices to final consumers. In several Gulf states, electricity is priced at a fraction of its true supply cost or simply given away. Such policies distort the economy: they encourage inefficient use of energy and put a heavy burden on public budgets. In poorer MENA countries, subsidies hinder attempts to attract private funding for domestic projects, especially in the power sector. There is widespread recognition among policy-makers of the need to lower subsidies, but consumers are very resistant to price increases. As a result, progress in most cases is slow or non-existent. Subsidies remain particularly large in Iran and Egypt. Further progress in reducing subsidies is likely to be very slow throughout the region and large subsidies will probably remain at the end of the projection period.

International Energy Prices

The projections of energy demand and supply in MENA countries and other world regions are derived from assumptions about average end-user prices for oil, natural gas and (where relevant) coal, based on assumed price trends on international markets. Formal policies to modify or remove energy subsidies are taken into account. Tax rates are assumed to remain unchanged over the projection period. Final electricity prices move in line with marginal power-generation costs.

The assumed price trends in the Reference Scenario, summarised in Table 1.1, reflect our judgment about how high prices will need to be to ensure sufficient supply to meet projected demand for each fuel over the *Outlook* period. The concomitant investment is assumed to be forthcoming. Prices in the Deferred Investment Scenario are higher. Although the assumed oil-price paths follow smooth trends, this should not be interpreted as a prediction of market stability, but rather as long-term trends around which prices will fluctuate. Indeed, oil prices may become more volatile in the future.

Our assumptions about international energy prices have been revised upwards from last year's *Outlook*, as a result of the prevailing change in market

Table 1.1: **Fossil-Fuel Price Assumptions in the Reference Scenario**
(in year-2004 dollars unless otherwise stated)

	2004	2010	2020	2030
IEA crude oil imports ($/barrel)	36	35	37	39
In nominal terms	*36*	*40*	*50*	*65*
Natural gas ($/MBtu):				
US imports	5.70	5.80	5.90	6.20
European imports	4.20	5.00	5.20	5.60
Japanese LNG imports	5.20	6.00	6.10	6.20
OECD steam coal imports ($/tonne)	55	49	50	51

Note: Prices in the first column are historical data. Gas prices are expressed on the basis of gross calorific value.

expectations after more than three years of high prices. The average IEA crude oil import price, a proxy for international prices, averaged $36.33 per barrel in 2004 and peaked at over $60 (in year-2004 dollars) in September 2005. In the Reference Scenario, the price is assumed to ease to around $35 in 2010 as new crude oil production and refining capacity comes on stream. It is then assumed to rise slowly, in a more or less linear way, to close to $37 in 2020 and $39 in 2030 (Figure 1.3). In nominal terms, the price will reach $65 in 2030 assuming inflation of 2% per year. Prices of the major benchmark crude oils, West Texas Intermediate and Brent, will be correspondingly higher. In 2004, the average IEA crude oil import price was $5.11 per barrel lower than first-month WTI and $1.89 lower than dated Brent.

This higher oil-price path assumed here than in *WEO-2004* reflects, in part, a recent shift in producing countries' price objectives. It also takes into account the results of our analysis of upstream and downstream oil projects that are scheduled to be completed in the next few years, as well prospects for energy demand. They suggest that crude oil and product markets will be tighter, in the first two or three years of the projection period, than had been expected last year.

The near-term outlook for oil prices remains unusually uncertain, complicating the analysis of overall energy-market trends. Prices continued to increase in 2005. The price in nominal terms of WTI reached at an all-time intra-day high of just over $70 in late August in the wake of Hurricane Katrina, which disrupted oil and gas industry operations in southern US states. OPEC agreed to increase its output targets at meetings in March and June, and Saudi Arabia has indicated a desire to meet market needs. A spate of new upstream projects in OPEC and non-OPEC countries is expected to raise spare production capacity in the next few years. But strong demand and limited spare

Figure 1.3: **Average IEA Crude Oil Import Price in the Reference Scenario**

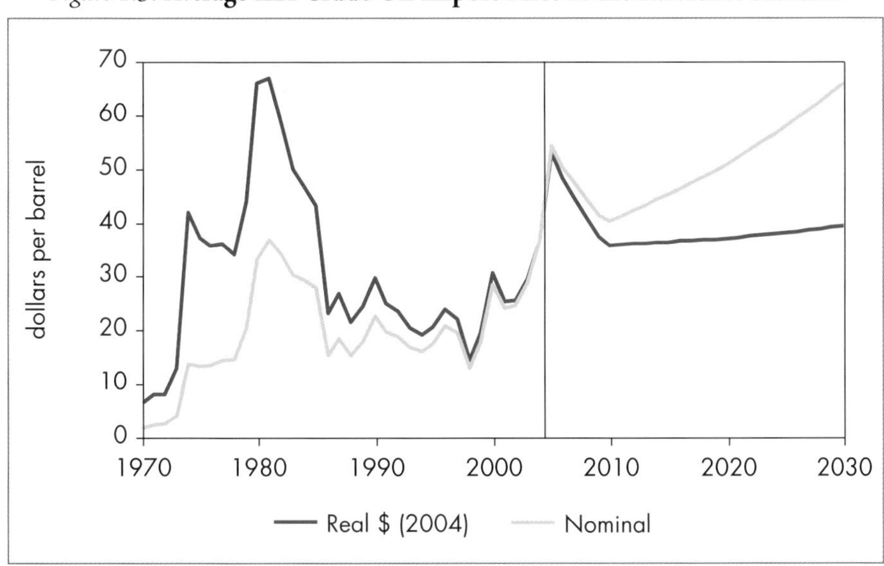

crude-oil production and refining capacity could drive prices higher again and make them more volatile, especially if there were a large and sustained supply disruption.

The assumed slowly rising trend in real prices after 2010 reflects an expected increase in marginal production costs outside OPEC, an increase in the market share of a small number of major producing countries and lower spare capacity. Reflecting the global distribution of oil and gas resources, most of the additional production capacity that will be needed over the projection period is expected to come from OPEC countries, mainly in the Middle East. The resulting growing concentration of production in these countries will increase their market dominance and, therefore, their ability to impose higher prices through their collective production and investment policies. But the Reference Scenario assumes that they will seek to avoid prices rising too much and too quickly, for fear of depressing global demand and of accelerating the development of alternatives to hydrocarbons.

Natural gas prices are also assumed to decline in all regions over the second half of the current decade and then to begin to rise from around the middle of the 2010s. These trends reflect the assumptions about oil prices, which will remain a key determinant of gas prices because of competition between gas and oil products. They also reflect a judgment of the prices that will be needed to stimulate investment in replacing and expanding supply infrastructure, as well as the impact of increasing gas-to-gas competition. Rising supply costs will contribute to higher gas prices from the end of the current decade in North America and Europe. This is expected to offset the downward pressure on gas prices relative to oil prices in Europe that might otherwise arise as a result of increasing gas-to-gas competition. Increased short-term trading in liquefied natural gas (LNG), which permits arbitrage among regional markets, is expected to cause regional prices to converge to some degree over the projection period (Figure 1.4). Lower unit pipeline-transport costs might also contribute to this convergence.

Figure 1.4: **Natural Gas Price Assumptions**

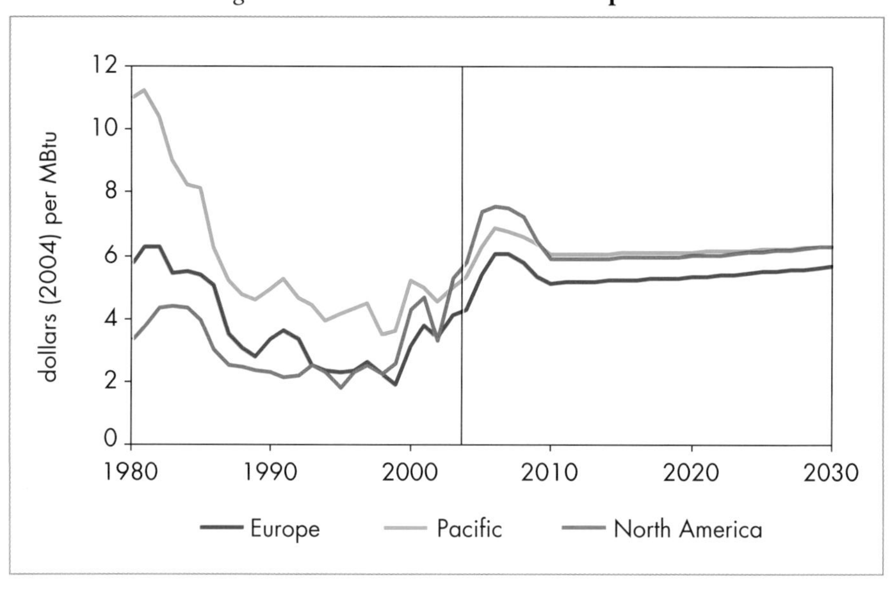

Gas prices in North America are assumed to fall from an average of over $7.50/MBtu in 2005 to about $5.80/MBtu on average (in year-2004 dollars) by the end of the decade, recovering to about $6.20/MBtu by 2030. Prices in Europe are assumed to peak at about $6.20/MBtu in 2006 in lagged response to high oil prices in 2005, falling to $5/MBtu by 2010 and then rising back to $5.60/MBtu by the end of the projection period. LNG prices in Japan, a proxy for gas prices in the Middle East, drop from $7.50/MBtu in 2006 to $6/MBtu in 2010, recovering slightly to $6.20/MBtu in 2030.

International steam coal prices have risen steadily in recent years on the back of rising oil prices and strong demand. The price of OECD steam coal imports jumped from an average of $36 per tonne in 2000 to $55 in 2004 (in year-2004 dollars). It is assumed to fall back to around $49/tonne by 2010, and rise slightly through to 2030 to $51/tonne.

Demographic Trends

Population growth is an important determinant of the size and pattern of energy demand, both directly and through its effect on economic development. Our population growth-rate assumptions have been revised in this *Outlook*, based on the latest UN projections (UNDESA, 2005). The differences in aggregate are extremely small: world population is still expected to reach almost 8.1 billion in 2030, an increase of 1.8 billion over the 2003 level. The average annual projected rate of increase is still about 1% (Table 1.2).

Demographic factors will continue to play a particularly important role in the economic and energy development of MENA countries. The region as a whole has one of the highest rates of population growth in the world, averaging 2.7% per annum between 1971 and 2003. Population has more than doubled since 1971, reaching 323 million in 2003 – 177 million in the Middle East and 145 million in North Africa (Figure 1.5). Rapid population expansion has led to a drop in the average age: about half of the total MENA population is now under the age of 20. Population-growth rates in all MENA countries are

Table 1.2: **Population Growth Assumptions** (average annual percentage change)

	1971-2003	1990-2003	2003-2010	2010-2020	2020-2030	2003-2030
OECD	**0.8**	**0.8**	**0.5**	**0.4**	**0.3**	**0.4**
Transition economies	**0.4**	**–0.1**	**–0.2**	**–0.2**	**–0.4**	**–0.3**
Developing countries*	**2.0**	**1.7**	**1.4**	**1.2**	**0.9**	**1.1**
Middle East	**3.1**	**2.4**	**2.2**	**2.0**	**1.6**	**1.9**
Iran	2.6	1.5	1.3	1.4	0.9	1.2
Iraq	3.0	2.5	2.7	2.3	1.8	2.2
Kuwait	3.5	0.9	2.7	1.8	1.4	1.9
Qatar	5.8	3.2	2.9	1.5	1.1	1.7
Sauti Arabia	4.2	2.8	2.7	2.2	1.8	2.2
UAE	9.0	6.5	3.4	2.0	1.6	2.2
Other Middle East	3.1	3.2	2.7	2.4	2.2	2.4
North Africa	**2.3**	**1.9**	**1.8**	**1.4**	**1.0**	**1.4**
Algeria	2.6	1.9	1.6	1.3	0.9	1.2
Egypt	2.2	2.0	2.0	1.6	1.2	1.6
Libya	3.1	2.0	1.9	1.5	1.0	1.4
Other North Africa	2.0	1.7	1.4	1.2	0.9	1.1
MENA	**2.7**	**2.2**	**2.0**	**1.8**	**1.4**	**1.7**
World	**1.6**	**1.4**	**1.2**	**1.0**	**0.8**	**0.9**

* Including MENA.

expected to fall over the projection period, in line with historical trends. Our population assumptions are drawn from the most recent United Nations projections contained in *World Population Prospects: the 2004 Revision*, originally released in 2005 and subsequently revised. According to that report, total MENA population is projected to grow by 1.7% per year over 2003-2030 – well above the average of 1.1% for developing regions and 0.9% for the world as a whole. MENA population will reach just over 500 million in 2030, close to the current population of OECD Europe.

Population growth rates differ widely across the region, reflecting economic and cultural factors. Population has grown fastest in the past three decades in the oil-producing Gulf countries, due partly to a massive influx of foreign workers and increased spending on the provision of health services, which boosted fertility rates and cut mortality rates sharply. Inflows of foreign workers have slowed markedly since 1990. Nonetheless, today nationals make up less than half the population in Kuwait, Qatar and the UAE (Figure 1.6). A significant share of the foreign population comes from outside the Middle East, mainly South Asia.

Rising populations with a growing number of young people are putting considerable strain on social infrastructure, notably educational services. This problem is likely to become more acute in the future. The quality of education has been declining in many countries, creating a widening gap between rich and poor and discontent among the latter. Rising unemployment among

Figure 1.5: **Population of MENA Countries**

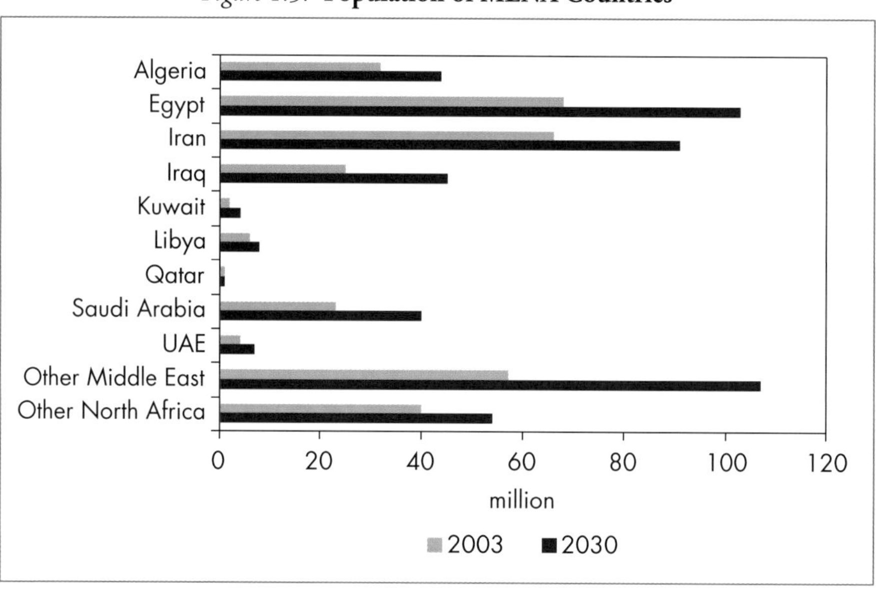

Source: UNDESA (2005).

Figure 1.6: **Population of Persian Gulf Countries by Category, 2002**

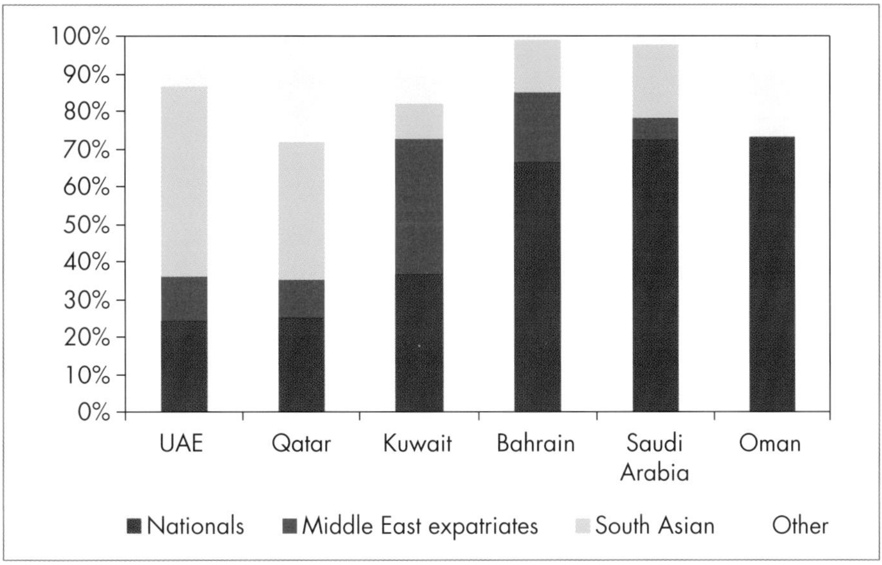

Source: Based on Cordesman (2004).

young people is adding to these tensions. Growing demand for social services could constrain the availability of capital that might otherwise be used for productive economic activities.

Macroeconomic Trends in the Middle East and North Africa

Economic and Social Development

Economically, MENA is an extremely diverse region. Countries are at differing stages of development, with different resource endowments and income levels. Average per capita income in the MENA region is high, but this disguises enormous differences between countries. In general, the countries with the largest per capita hydrocarbon-resource endowments have the highest per capita gross domestic product (Figure 1.7). Exploitation of their resources has allowed most of the Persian Gulf countries to achieve relatively high levels of per capita income, but their economies remain heavily reliant on energy exports and are, therefore, vulnerable to fluctuations in oil prices. Incomes are much lower in North Africa, Iran and Iraq. In Egypt, for example, 44% of the population lives on less than $2 a day (UNDP, 2005). Several non-oil economies rely heavily on aid, capital inflows and remittances from workers in the oil-producing countries. There are also major differences in household incomes within countries. Average incomes in Qatar and the UAE are among the highest in the world, but many foreign workers employed in menial jobs in those countries are poorly paid. Poverty is widespread in several countries.

The MENA economy has grown strongly in recent years, largely due to higher oil prices and hydrocarbon production. Oil-export revenues rose by more than two-thirds between 2001 and 2004, reaching about $300 billion in 2004. They are expected to exceed $400 billion in 2005. This surge in revenue has financed a sharp acceleration in domestic spending, largely via government consumption and investment. This factor more than offsets the negative impact of the 2003 war in Iraq. Non-oil export growth has also been strong, aided by the fall in the dollar, to which most of the oil-exporting countries' currencies are pegged. The non-oil exporting countries have also benefited from investment and remittance flows from the oil exporters. MENA GDP growth averaged 5.1% in 2003 and an estimated 5.7% in 2004. Per capita incomes have risen much less, however, because of rapid population growth.

The oil boom in exporting countries has led to a surge in the prices of equities and property, as was the case after the oil-price hikes of the 1970s. The stock market capitalisation of the six Gulf Cooperation Council countries – Bahrain, Kuwait, Oman, Qatar, Saudi Arabia and the UAE – has tripled since 2001, reaching $875 billion in 2005. But governments are not spending their additional revenues in the same way as in the past. Public spending has risen much less quickly than revenues since the start of the decade, reflecting a more prudent stance by governments. The World Bank estimates that oil-exporting Middle Eastern countries ran a budget surplus of almost 7.9% of GDP in 2004,

Figure 1.7: **Per Capita GDP and Proven Oil Reserves in MENA, 2004**

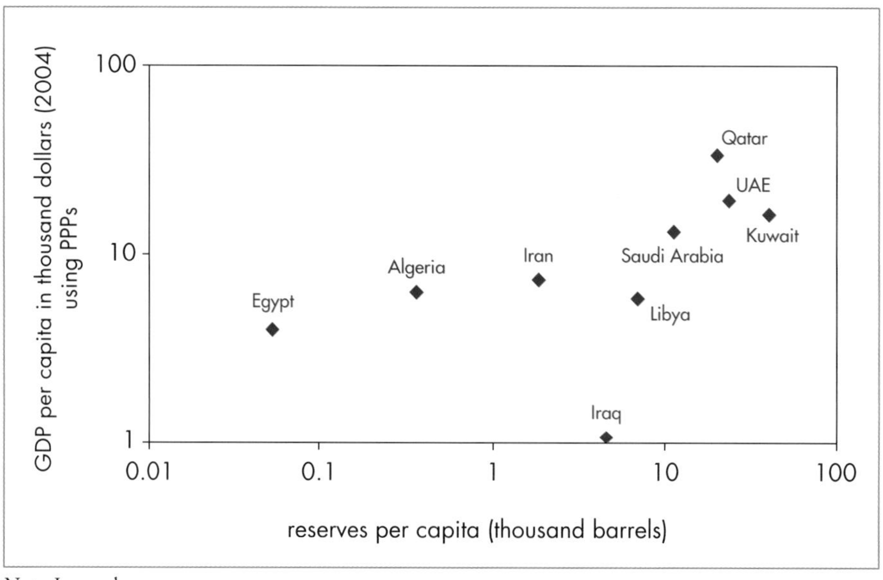

Note: Log scale.
Sources: Reserves - *Oil and Gas Journal* (20 December 2004); GDP - IEA databases.

compared with a deficit of 3.5% in the 1990s. This has allowed governments to pay off debt: Saudi Arabia alone paid off $12 billion in debt in 2004, reducing the total by 7% to $164 billion. And much of the increase in spending has gone to infrastructure projects, boosting local construction and contracting industries.

The oil boom is providing new resources for regional development, but will not by itself solve the structural economic problems of the region. More rigorous economic reforms are needed. Despite the recent economic upturn, unemployment – especially among the young – remains a pressing concern for most MENA countries. The average rate of unemployment is estimated at around 13% in MENA as a whole (World Bank, 2005),[4] but in some countries the rate is as high as 30% (Figure 1.8). Unemployment has fallen in the last few years, from 14.9% in 2000, thanks to the oil-led economic boom, but strong demographic pressures mean that huge numbers of new jobs will need to be created in the coming years. In general, governments are attempting to lower the expectation that the state will provide jobs for university graduates and to put in place policies that will create more private-sector jobs. They are also encouraging nationals to seek employment in activities that have been shunned up to now. This is a particular problem in some of the richer oil-exporting countries, which have grown heavily dependent on foreign workers. If

Figure 1.8: **Unemployment Rates in Selected MENA Countries, End-2004**

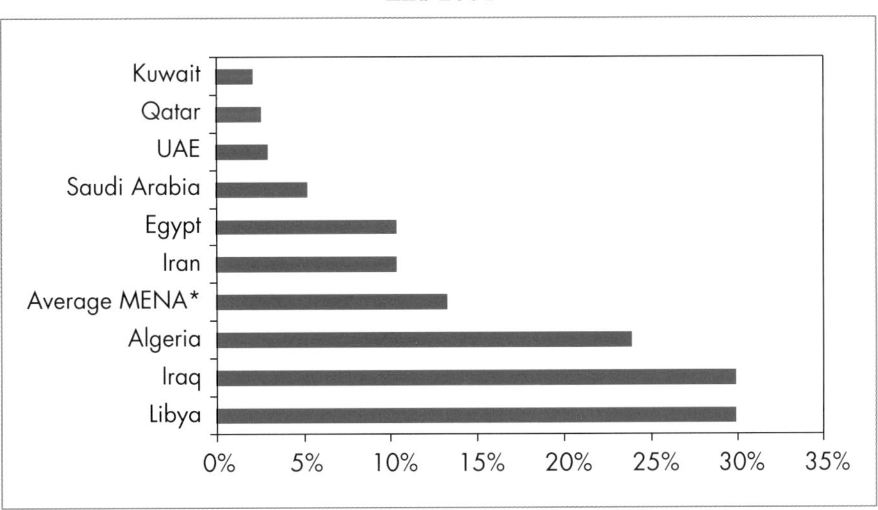

* Excluding Iraq, Israel, Libya and Qatar.
Sources: MENA average - World Bank (2005); other countries – see country chapters.

4. Excluding Iraq, Israel, Libya and Qatar.

successful, these policies will help to reduce social tensions, as well as raise productivity and long-term sustainable economic growth rates.

The oil-rich Persian Gulf states generally have the highest rankings of the UN human development index (Table 1.3). Nonetheless, their scores are generally lower than their per capita GDP would suggest. In large part this is because they have enjoyed high levels of income only since the 1970s and other indicators of human development have not yet caught up. In all MENA countries, the index has, in fact, advanced significantly since 1990. Yemen has the lowest value of all MENA countries covered in this *Outlook*, followed by Morocco and Egypt. Out of a total of 177 countries covered by the index, nine of the 17 MENA countries (excluding Iraq, for which no score is available) are in the top half of the rankings.

Table 1.3: **Human Development Indicators in MENA Countries, 2003**

	Life expectancy at birth (years)	**Adult literacy** (% at 15 and above)	**Gross school enrolment** (%)	**GDP per capita** ($ in PPPs)	**Human development index value**	**Ranking among 177 countries worldwide**
Israel	79.7	96.9	91	20 033	0.915	23
Qatar	72.8	89.2	82	19 844	0.849	40
UAE	78.0	77.3	74	22 420	0.849	41
Bahrain	74.3	87.7	81	17 479	0.846	43
Kuwait	76.9	82.9	74	18 047	0.844	44
Libya	73.6	81.7	96	n.a.	0.799	58
Oman	74.1	74.4	63	13 584	0.781	71
Saudi Arabia	71.8	79.4	57	13 226	0.772	77
Lebanon	72.0	86.5	79	5 074	0.759	81
Tunisia	73.3	74.3	74	7 161	0.753	89
Jordan	71.3	89.9	78	4 320	0.753	90
Iran	70.4	77.0	69	6 995	0.736	99
Algeria	71.1	69.8	74	6 107	0.722	103
Syria	73.3	82.9	62	3 576	0.721	106
Egypt	69.8	55.6	74	3 950	0.659	119
Morocco	69.7	50.7	58	4 004	0.631	124
Yemen	60.6	49.0	55	889	0.489	151
Iraq	58.8	n.a.	63	n.a.	n.a.	n.a.

Note: The GDP per capita numbers in this table are from UNDP (2005). They differ from the GDP per capita numbers in the country chapters which are from the IEA database.
Source: UNDP (2005).

Economic Policy

Although the recent economic upturn is reminiscent of the oil booms of the 1970s and early 1980s, there are signs that oil exporters are adopting more prudent macroeconomic policies. According to the World Bank, only a quarter of the additional export revenues during the current period of high prices has been spent, compared with 60% during the 1973 boom (World Bank, 2005). Much of the spending – notably in Saudi Arabia – has been used to reduce external debt, with the rest accumulating as foreign reserves. Several countries divert oil-export revenues to stabilisation funds. Kuwait was the first country in the region to establish a Fund for Future Generations, to which 10% of the country's total revenues are channelled in order to prepare for the day when oil revenues diminish. A plan to increase this share to 15% and to give every Kuwaiti citizen a $680 grant is under discussion. Iran, Algeria, Qatar and Oman have also created funds.

The vulnerability of the oil-exporting countries' economies to oil-price fluctuations became particularly evident in 1998 and 1999, when oil prices dropped to less than $10 a barrel. The slump in revenues put enormous strain on government finances, because of continuing obligations to large state sectors and expensive welfare systems, including free health care, education and housing, as well as heavily subsidised energy and water. That experience led many oil-exporting countries to initiate structural economic reforms aimed at accelerating economic diversification away from over-reliance on oil, though progress in implementing them has been patchy.

In the late 1990s, Saudi Arabia announced a series of measures aimed at reducing public spending to sustainable levels and encouraging private-sector investment. Private firms have been allowed to invest in some sectors and opportunities for foreign investors have been increased with the adoption of a new law in 2000 permitting full foreign ownership of Saudi property and licensed projects in certain sectors (excluding upstream oil). The General Investment Authority has been set up to streamline applications by foreign investors, and taxes on company profits have been lowered. Provincial governments have also devised measures to attract foreign investment. Saudi Arabia's expected accession to the World Trade Organization in late 2005 or early 2006, which will require the opening-up of domestic markets and the removal of price controls, is giving additional impetus to economic reform. But privatisation has yet to get underway: not one state asset has yet been sold off to private buyers. Economic reform in Saudi Arabia, as in other MENA countries, requires time-consuming consensus-building. Reforms are expected to continue to move ahead, albeit slowly, though there is a danger that the recent surge in oil revenues will weaken commitment to them.

Moves to diversify away from oil and gas have had some success in other countries. Bahrain has been successful in promoting the expansion of its banking sector, while the UAE has emphasised the development of banking and tourism. Other countries are seeking to promote tourism, notably Egypt, which already depends on this sector for a large share of its national income. Libya, which until recently had been completely closed, has begun to implement economic reforms aimed at attracting foreign investment. Several countries, notably Saudi Arabia, have promoted the development of the petrochemical industry.

GDP Growth Assumptions

Our assumptions about GDP growth in MENA countries take oil and gas revenues explicitly into account. In countries with a large oil and gas sector, nominal GDP is closely correlated with oil prices. Higher oil prices in recent years have boosted the sector's share of GDP and, therefore, made GDP even more sensitive to fluctuations in international oil prices. For example, the share in 2003 is estimated at 32% in Saudi Arabia, 59% in Qatar and 76% in Iraq.

Since the domestic energy-demand projections for MENA countries depend primarily on assumed rates of future GDP growth, it is important that these assumptions are fully consistent with the assumptions about international oil and gas prices and production projections. In other words, for a given set of oil and gas prices, the assumptions about GDP growth need to reflect our expectations about the future share of oil and gas in GDP, as well as trends in production. These, in turn, will reflect how successful these countries are expected to be in diversifying their economies away from oil – a major policy objective in many of them.

We developed a model to calibrate our GDP assumptions to our oil- and gas-price assumptions.[5] The approach treats GDP as the sum of two components:

- *Hydrocarbon-related GDP*, based on our projections of oil and gas production. Since the GDP deflator for all countries assumes constant prices, real hydrocarbon GDP is not affected by changes in prices.
- *Non-hydrocarbon-related GDP*, derived from three factors:
  - Productivity, measured as non-oil GDP per person of the active labour force (including foreign workers). Growth in productivity is based on historic trends, adjusted according to factors specific to each country – notably the pace of economic reform – and independent projections of economic growth.
  - The size of the labour force. This is assumed to change with the growth in population aged between 15 and 64 years, as projected by the United Nations Population Division (UNDESA, 2005), as well as with assumptions about female participation in the labour force.

5. We are grateful to the Research Department of the International Monetary Fund for its help in refining the methodology.

- The impact of changes in oil and gas revenues, via government spending and oil industry investment. Increases in revenues stimulate economic activity in the rest of the economy.

In the Reference Scenario, the rate of growth of GDP in MENA is assumed to average 3.4% over 2003-2030 (Table 1.4). The pace of economic growth varies over the projection period, due mainly to fluctuations in international oil prices, oil and gas production and the size of the labour force. GDP growth averages 4.3% per year over 2003-2010 (boosted in part by very high rates in 2004 and 2005), dropping to 3.3% in 2010-2020 and 2.9% in 2020-2030.

GDP is assumed to grow slightly faster in the Middle East than in North Africa. The Iraqi economy is expected to grow most quickly by far, at an average of almost 7%, as it recovers from the 2003 war and two decades of economic stagnation. Combining these assumptions with those for population yields an average increase in per capita income of 1.7% per annum in MENA as a whole, from $6 100 in 2003 to $9 650 in 2030 in year-2004 dollars (Figure 1.10). Yet, by 2030, average per capita income in MENA will still be under one-quarter that of OECD countries, about the same ratio as today. Per capita incomes grow quickest in Iraq, at an average rate of 4.6%.

The projected rates of growth in economic activity are not expected to be sufficient to reduce unemployment significantly in most MENA countries. The World Bank estimates that 100 million new jobs would be needed to keep

Table 1.4: **MENA GDP Growth Assumptions in the Reference Scenario**
(average annual growth rates, in %)

	1971-2003	2003-2010	2010-2020	2020-2030	2003-2030
Middle East	**3.2**	**4.3**	**3.4**	**3.0**	**3.5**
Iran	3.3	4.5	3.4	3.0	3.6
Iraq	–3.5	10.8	5.7	5.5	6.9
Kuwait	0.6	3.6	3.1	2.6	3.1
Qatar	2.5	6.6	3.7	2.4	3.9
Saudi Arabia	4.4	4.2	3.5	3.0	3.5
UAE	7.2	4.6	3.0	2.6	3.3
Other Middle East	4.4	3.3	3.1	2.9	3.1
North Africa	**4.0**	**4.1**	**3.2**	**2.6**	**3.2**
Algeria	4.1	4.4	2.7	2.3	3.0
Egypt	5.3	4.2	3.6	3.1	3.6
Libya	–0.7	3.6	3.3	2.7	3.1
Other North Africa	4.2	3.9	3.1	2.2	3.0
MENA	**3.4**	**4.3**	**3.3**	**2.9**	**3.4**

Figure 1.9: **Per Capita Income in MENA Countries in the Reference Scenario**

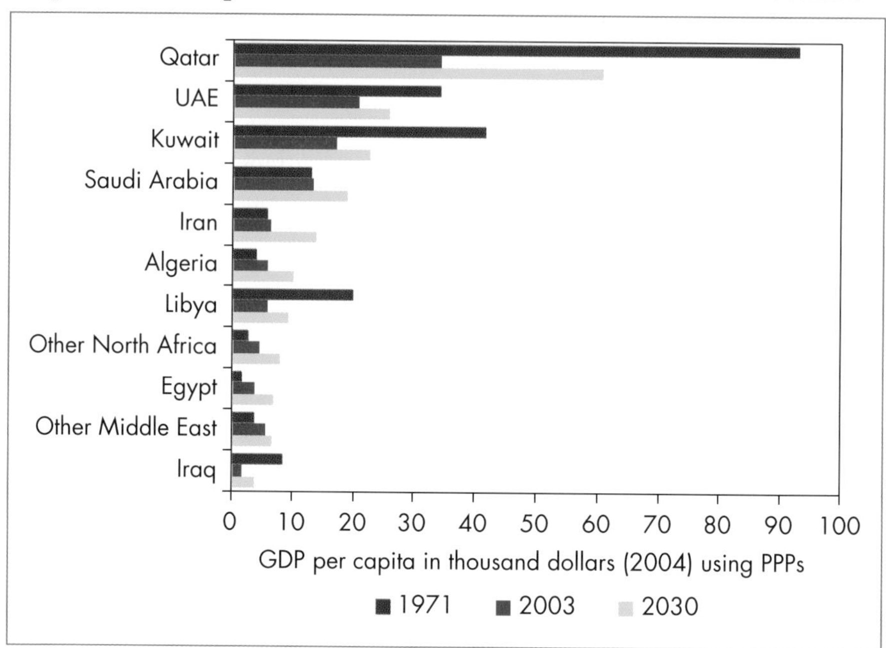

pace with new entrants to the labour force and to absorb the currently unemployed in the twenty years to 2025 (World Bank, 2005). This would require a doubling of the total number of jobs. Meeting the goal of full employment would require average GDP growth of 6% to 7% per year. This could be achieved only with radical economic reforms.

The Deferred Investment Scenario

The Deferred Investment Scenario analyses how global energy markets might evolve if investment in the upstream oil and gas industry of MENA countries were to be substantially lower than implicitly assumed in the Reference Scenario. Investment, as a proportion of GDP, is assumed to be constant over the projection period at the level of the last ten years in each MENA country (investment is assumed to rise sharply relative to GDP in the Reference Scenario). This equates to a 23% reduction in total MENA upstream oil investment compared with the Reference Scenario, leading to slower development of hydrocarbon resources and, therefore, slower growth in actual supply.

Lower investment may occur for various reasons, such as restrictions on foreign access to resources or unattractive fiscal terms. For social reasons, governments may give priority to domestic spending over capital allocations to national oil companies for their exploration and production programmes. Or producer governments might deliberately limit investment in new production capacity, either in order to drive up short-term international prices and boost revenues or in response to a conservative forecast of the future call on their oil and gas. Whatever the possible causes of lower investment, the Deferred Investment Scenario assesses the consequences for global energy markets. The assumptions on other aspects of government policy, on population and on technology are the same in the Deferred Investment Scenario as in the Reference Scenario. However, energy prices are driven higher. GDP growth also differs, due to the change in prices and production levels.

The methodology for preparing the Deferred Investment Scenario projections involved several steps:

- The impact of the reduction in investment in oil and gas production in each country was modelled on the basis of the estimated costs of new field developments and of estimated decline rates both for existing and new fields.
- The impact of reduced MENA production on the oil price was quantified, using a world oil-equilibrium model. Gas prices are assumed to increase broadly in the same proportion as oil prices. Coal prices also increase.
- GDP growth rates for each MENA country were estimated by taking into account higher prices and lower investment and production. GDP growth rates for non-MENA countries were changed in line with the new oil-price trajectory using the world oil-equilibrium model.
- The World Energy Model was run to generate new energy-demand projections for the MENA countries and all other regions.
- The new energy-demand projections were combined with the new MENA oil and gas production trajectories to yield new profiles for non-MENA oil and gas production. This assumes that non-MENA production would rise in response to higher prices to meet the shortfall in global supply resulting from lower investment and production in MENA countries. This is different from the approach adopted in the Reference Scenario, in which MENA countries are assumed to be the residual supplier to the world market.

Inter-regional oil and gas trade flows, together with investment by region, were recalculated on the basis of the new projections for demand and supply in MENA countries and non-MENA regions.

A more detailed explanation of the methodology and assumptions of the Deferred Investment Scenario, together with the results for each country, can be found in Chapter 7.

The World Alternative Policy Scenario

The assumptions and results of this scenario, which reflect active intervention by consuming governments, are discussed in Chapter 8. A fuller discussion of the methodology used can be found in the 2004 edition of the *World Energy Outlook*.

CHAPTER 2

GLOBAL ENERGY TRENDS

HIGHLIGHTS

- World primary energy demand in the Reference Scenario is projected to expand by more than half between 2003 and 2030, reaching 16.3 billion tonnes of oil equivalent. More than two-thirds of the increase will come from the developing countries.
- Fossil fuels will continue to dominate energy supplies. Oil, natural gas and coal will meet 81% of primary energy demand by 2030, one percentage point higher than in 2003. Other renewables – a group that includes geothermal, solar, wind, tidal and wave energy – will expand at the fastest rate. The share of nuclear power in total primary demand falls. The transport and power-generation sectors will continue to absorb a growing share of global energy.
- The world's energy resources are adequate to meet the projected growth in energy demand in the Reference Scenario but they are far from evenly distributed geographically. Almost all the increase in energy production will occur in non-OECD countries. In response, energy exports from non-OECD to OECD countries will increase substantially.
- The Reference Scenario trends imply that global energy-related carbon-dioxide emissions will increase by 52% between 2003 and 2030. Developing countries will account for 73% of this increase.
- Cumulative energy-sector investment of $17 trillion (in year-2004 dollars) will be required through to 2030 – about half in developing countries. Financing the required investments in non-OECD countries is one of the biggest sources of uncertainty surrounding our energy-supply projections.
- In the Reference Scenario, global oil refining capacity rises from 83 mb/d in 2004 to 93 mb/d in 2010 and 118 mb/d by 2030 at an investment cost of $487 billion. Over the projection period, product demand shifts to lighter products and crude oil becomes progressively heavier and more sour. Any shortfall in capacity would put upward pressure on petroleum product and crude oil prices, as it has done recently.
- In a World Alternative Policy Scenario, global primary energy demand is about 10% lower in 2030 than in the Reference Scenario, thanks to policies that promote more efficient use of energy and switching away from fossil fuels.

This chapter summarises the projections of global energy demand and supply in the Reference Scenario, which assumes that no government policies beyond those adopted as of mid-2005 are introduced. The chapter also highlights refinery issues. Chapters 3 to 6, which focus on MENA energy markets, also present the results of the Reference Scenario. The results of a Deferred Investment Scenario, which considers the impact on global energy markets of lower upstream oil investment in MENA countries, are presented in Chapter 7, as well as in the country chapters. The results of a World Alternative Policy Scenario, which analyses the impact of a set of new policies that are assumed to be adopted during the projection period, are summarised in Box 2.1 and presented in more detail in Chapter 8.

Energy Demand

Primary Fuel Mix

Global primary energy demand in the Reference Scenario is projected to increase by 52% from 2003 to 2030, reaching 16.3 billion tonnes of oil equivalent (toe). The increase in demand will average 1.6% per year, compared to 2.1% per year over the period 1971-2003. The projected rate is 0.1 percentage point lower than that for 2002-2030 in *WEO-2004*. The difference is largely explained by the effect of the higher prices assumed in this *Outlook.*

Fossil fuels will continue to meet the overwhelming bulk of the world's energy needs (Figure 2.1). Oil, natural gas and coal will between them account for 83% of the *increase* in world primary demand over 2003-2030. By the end of the projection period, they will make up 81% of energy demand, up from 80% in 2003. The share of nuclear power falls from 6.4% to 4.7%, while the share of renewable energy sources – including traditional biomass – is projected to increase from 13% to 14%. The share of non-biomass, non-hydro renewables increases from 0.5% in 2003 to 1.7% in 2030.

Box 2.1: **Summary of World Alternative Policy Scenario Projections**

The World Alternative Policy Scenario analyses the impact of a range of potential government policies and measures which are already being contemplated, but are not yet adopted. They aim to address energy-security and environmental concerns in all regions. Global primary energy demand in 2030 reaches 14.7 billion toe – 1.6 billion toe, or about 10%, less than in the Reference Scenario. At 1.2% per year, the average annual rate of demand growth is 0.4 percentage points less than in the Reference Scenario. The effect of energy-saving and fuel diversification policies on energy demand grows throughout the projection period, as the stock of energy capital is gradually replaced and new measures are introduced.

Oil and gas demand in the Alternative Policy Scenario are both about 10% lower in 2030 than in the Reference Scenario. Coal use falls much more, by 23%, due to lower demand in power generation: the use of more efficient technology reduces the demand for electricity, and generators choose to use more carbon-free fuels. On the other hand, the use of non-hydro renewables, excluding biomass, is 27% higher in 2030 than in the Reference Scenario. Biomass and nuclear energy also grow. Most of the net increase in renewable use results from OECD government policies aimed at promoting their use in the power sector and in transport. Lower overall energy consumption and a larger share of carbon-free fuels in the primary energy mix yield a 16%, or 5.8 gigatonnes, reduction in global carbon-dioxide emissions compared with the Reference Scenario.

Figure 2.1: **World Primary Energy Demand by Fuel in the Reference Scenario**

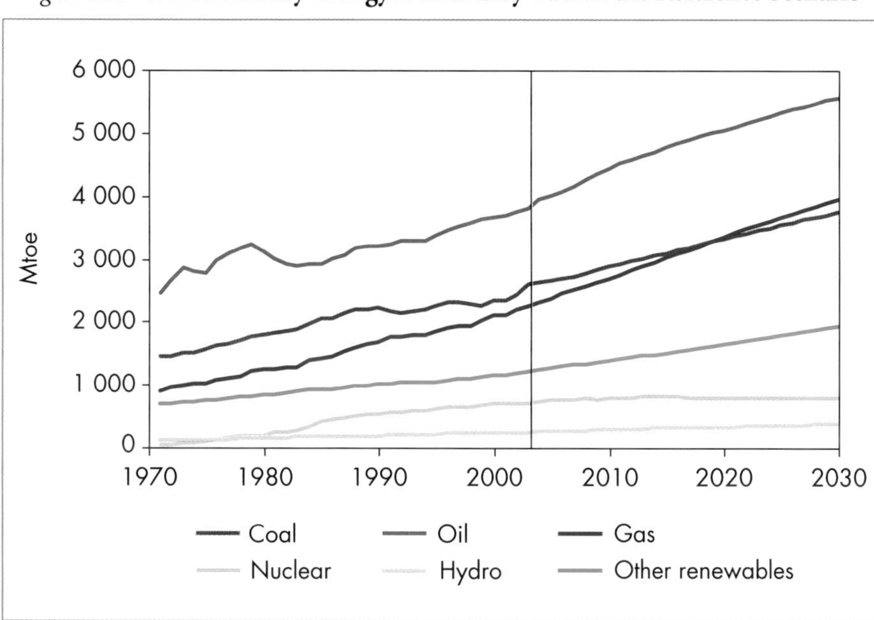

Oil will remain the single largest fuel in the global primary energy mix in the Reference Scenario (Table 2.1). Its share will nonetheless fall marginally, from 35% in 2003 to 34% in 2030. Oil demand is projected to grow by 1.4% per year, from 79 mb/d in 2003 to 92 mb/d in 2010 and 115 mb/d in 2030. Two-thirds of the total increase in oil use will come from the transport sector, where oil will remain the main fuel. Oil will remain a marginal fuel in power generation, with its share declining in every region. Industrial, commercial and residential demand for oil is projected to increase moderately, with all of the growth coming from non-OECD countries. Oil products will remain the main source of modern commercial energy for cooking and heating in developing countries, especially in rural areas. The use of oil in non-transport sectors in OECD countries will decline markedly.

Table 2.1: **World Primary Energy Demand in the Reference Scenario** (Mtoe)

	1971	**2003**	**2010**	**2020**	**2030**	**2003 - 2030***
Coal	1 439	2 582	2 860	3 301	3 724	1.4%
Oil	2 446	3 785	4 431	5 036	5 546	1.4%
Gas	895	2 244	2 660	3 338	3 942	2.1%
Nuclear	29	687	779	778	767	0.4%
Hydro	104	227	278	323	368	1.8%
Biomass and waste	683	1 143	1 273	1 454	1 653	1.4%
Other renewables	4	54	107	172	272	6.2%
Total	**5 600**	**10 723**	**12 389**	**14 402**	**16 271**	**1.6%**

* Average annual growth rate.

After registering strong growth of 2% in 2003, world oil consumption in 2004 increased even more quickly, by 3.6%. This is the fastest rate of growth since 1978. China, which saw a jump of 16% or nearly 0.9 mb/d in its oil use in 2004, accounted for 30% of the global demand increase (Table 2.2). This surge came despite record oil prices. The average IEA crude oil import price averaged over $36 a barrel in 2004, a jump of almost 30% from 2003. The price of first-month West Texas Intermediate (WTI) averaged $41.49 and that of dated Brent $38.27 in 2004. In 2005, prices have continued to rise, reaching more than $70 for WTI in late summer – a record in nominal terms. Adjusted for inflation, prices are still below the levels of the 1970s. The surge in prices has so far not cooled demand much, though there are signs that high oil prices are starting to dampen economic growth and energy demand, especially in Asia. Oil demand will continue to grow most quickly in developing countries over the projection period, particularly in China and Africa.

Some 95% of the increase in demand over the projection period will be for middle distillates and light fuels (Figure 2.2). The slight increase in heavy fuel demand in developing countries, mainly in industry and for bunkers, will be almost entirely offset by a fall in demand for these fuels in OECD countries. Global demand for middle distillates – diesel for road transport and jet kerosene for aviation – reaching almost 49 mb/d, an increase over 2003 of

Table 2.2: **World Oil Demand in the Reference Scenario** (million barrels per day)

	2003	2004	2010	2020	2030	2004 - 2030*
OECD	**47.0**	**47.6**	**50.5**	**53.2**	**55.1**	**0.6%**
OECD North America	24.1	24.9	26.9	29.1	30.6	0.8%
OECD Europe	14.5	14.5	15.0	15.4	15.7	0.3%
OECD Pacific	8.4	8.3	8.6	8.7	8.8	0.3%
Transition economies	**4.2**	**4.4**	**4.9**	**5.6**	**6.2**	**1.3%**
Russia	2.5	2.6	2.9	3.3	3.5	1.2%
Developing countries	**25.0**	**27.0**	**33.9**	**42.9**	**50.9**	**2.5%**
China	5.4	6.2	8.7	11.2	13.1	2.9%
India	2.5	2.6	3.3	4.3	5.2	2.8%
Other Asia	5.1	5.4	6.6	8.3	9.9	2.3%
Latin America	4.5	4.7	5.4	6.5	7.5	1.9%
Brazil	*2.0*	*2.1*	*2.4*	*3.0*	*3.5*	*2.0%*
Africa	2.6	2.6	3.3	4.5	5.7	3.0%
North Africa	*1.2*	*1.3*	*1.5*	*2.0*	*2.4*	*2.4%*
Middle East	5.1	5.4	6.5	8.1	9.4	2.2%
International marine bunkers	3.0	3.1	3.1	3.2	3.3	0.3%
World	**79.2**	**82.1**	**92.5**	**104.9**	**115.4**	**1.3%**

*Average annual growth rate.

18 mb/d. The increase in demand for light and middle distillates will be about ten times bigger than the increase in heavy fuels in developing countries. Transport fuels will account for the bulk of the increase in oil demand over the *Outlook* period. In the five years to 2004, most of the increase in oil demand in the OECD came from the transport sector.

Primary demand for **natural gas** will grow by 2.1%, meaning that gas will overtake coal by around 2020 as the world's second-largest primary energy source. Gas consumption will increase by three-quarters between 2003 and 2030, reaching 4 789 billion cubic metres (Table 2.3). The share of gas in world energy demand will rise from 21% in 2003 to 24% in 2030 – mostly at the expense of coal and nuclear energy (Figure 2.3). Power generation will account for most of the increase in gas demand over the projection period because, in many parts of the world, gas will be the preferred fuel in new power stations for economic and environmental reasons. A small but increasing share of gas demand will come from gas-to-liquids plants and from the production of hydrogen for fuel cells.

Figure 2.2: **Incremental Global Oil Demand, 2004-2030**

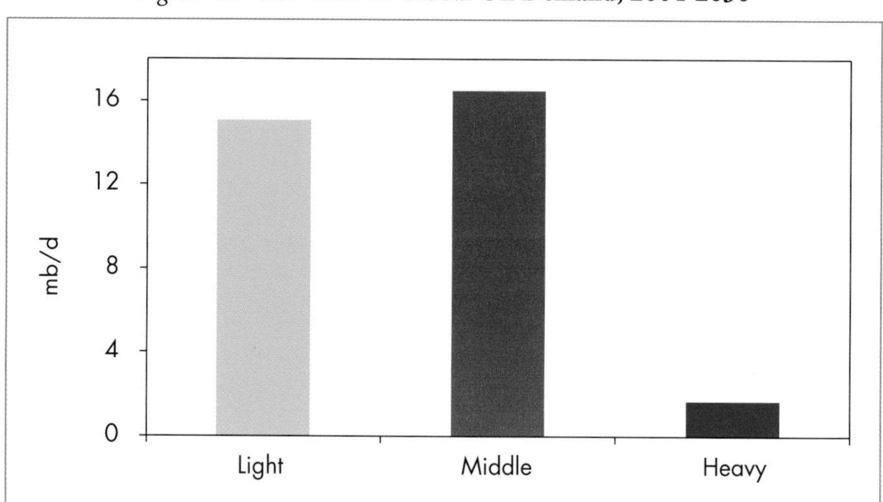

Table 2.3: **World Natural Gas Demand in the Reference Scenario** (bcm)

	2003	2010	2020	2030	2003-2030*
OECD	**1 436**	**1 617**	**1 872**	**2 061**	**1.3%**
OECD North America	775	848	964	1039	1.1%
OECD Pacific	141	176	217	244	2.1%
OECD Europe	520	593	691	778	1.5%
Transition economies	**637**	**705**	**815**	**925**	**1.4%**
Russia	417	460	525	591	1.3%
Developing countries	**636**	**893**	**1 374**	**1 803**	**3.9%**
China	39	60	106	152	5.1%
India	28	42	71	98	4.7%
Other Asia	162	215	305	387	3.3%
Latin America	107	145	220	318	4.1%
Africa	74	107	165	232	4.3%
North Africa	*62*	*85*	*121*	*152*	*3.4%*
Middle East	226	324	507	615	3.8%
World	**2 709**	**3 215**	**4 061**	**4 789**	**2.1%**

*Average annual growth rate.

Demand for **coal** is projected to rise from almost 5 200 million tonnes (Mt) in 2003 to almost 7 300 Mt in 2030, an average annual rate of increase of 1.4%. Its share in world primary demand will still fall a little, from 24% in 2003 to

23% in 2030. China and India, which both have large coal resources, will together account for about two-thirds of the increase in world coal demand. Power generation will remain the main driver of world coal demand, though coal will continue to lose market share in power generation in all OECD regions and in some developing regions. The industrial, residential and commercial sectors in non-OECD regions will burn more coal, more than offsetting a drop in final use in the OECD.

Figure 2.3: **Fuel Shares in World Primary Energy Demand**

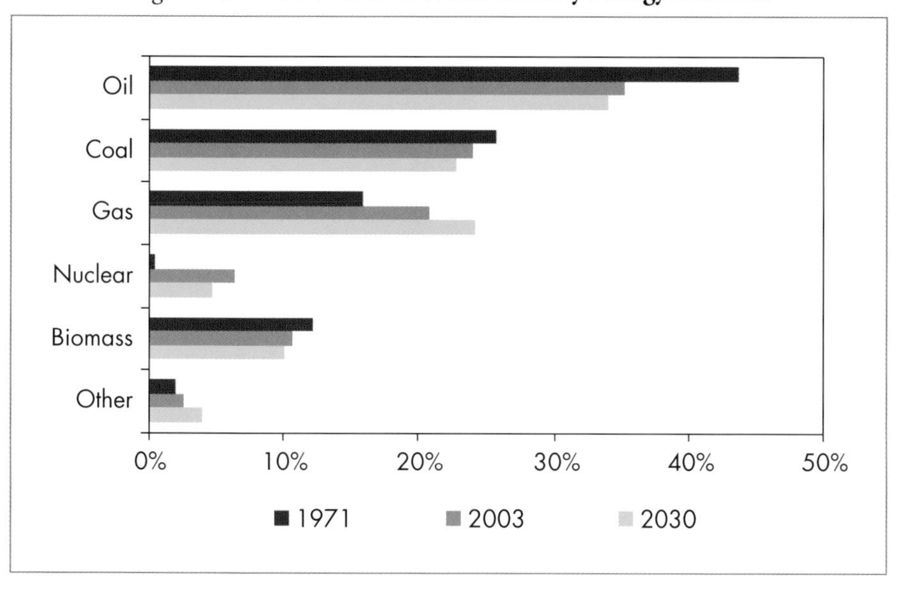

The share of **nuclear power** in global primary energy demand will decline over the projection period. Few new reactors are expected to be built and several will be retired. Nuclear power will struggle to compete with other technologies and many countries have restrictions on new construction or policies to phase out nuclear power. As a result, nuclear production is projected to peak around 2015 and then decline gradually. Its share of world primary demand will remain flat, at about 6%, through 2010 and then fall to less than 5% by 2030. Nuclear output will increase in only a few countries, mostly in Asia. The largest declines in nuclear production are expected to occur in Europe. However, these projections remain very uncertain. Shifts in government policies and public attitudes towards nuclear power could mean that this energy source plays a much more important role than projected here.

Hydropower will remain an important source of electricity production. Much of the OECD's low-cost hydroelectric resources have already been exploited, but there are still opportunities for adding capacity in developing countries. World hydropower production is projected to grow by an average 1.8% a year through 2030 and its share of primary demand will remain broadly constant at 2% over the *Outlook* period. The developing countries will account for over three-quarters of the increase in hydropower production.

The role of **biomass and waste**, much of which is used in traditional ways in developing countries, will decline slightly over the projection period. Their share of world primary energy demand will fall from 11% in 2003 to 10% in 2030, as they are replaced by modern commercial fuels. In absolute terms, the consumption of traditional biomass in developing countries will continue to grow, but will slow over the projection period. The use of biomass and waste will increase in power generation, particularly in OECD countries.

Other **renewables** – a group that includes geothermal, solar, wind, tidal and wave energy – will grow faster than any other energy source, at an average rate of 6.2% per year over the projection period. But they will still make only a small contribution to meeting global energy demand in 2030, because they start from a very low base. Their share in primary demand will grow from 0.5% in 2003 to 1.7% in 2030. Most of the increase in the use of renewables will be in the power sector. The increase in the use of renewables will be much bigger in OECD countries, many of which have adopted strong measures aimed at encouraging the take-up of new renewable-energy technologies.

Global energy intensity, measured as total primary energy use per unit of gross domestic product, is projected to fall by 1.6% per year over 2003-2030. Intensity will fall most quickly in the non-OECD regions, largely because of improved energy efficiency in power generation and end-uses, but also because of structural economic changes away from heavy industry towards lighter industry and services. The transition economies, in particular, will become much less energy-intensive as more energy-efficient technologies are introduced, wasteful energy practices are tackled and energy markets are reformed.

Regional Demand Trends

More than two-thirds of the increase in world primary energy demand between 2003 and 2030 in the Reference Scenario will come from the developing countries. Their demand growth will be more rapid than in the industrialised and transition economies, because their economies and populations will grow more quickly. Industrialisation, urbanisation and a shift in energy use from traditional non-commercial biomass to commercial fuels will also boost demand. OECD countries will account for almost a quarter of the global increase and the transition economies for the remaining 7%. As a result, the OECD's share of world demand will decline, from 51% in 2003 to 42% in 2030, while that of the developing countries will increase, from 39% to 49% (Figure 2.4). The transition economies' share will fall slightly from 10% to 9%.

Figure 2.4: **Regional Shares in World Primary Energy Demand in the Reference Scenario**

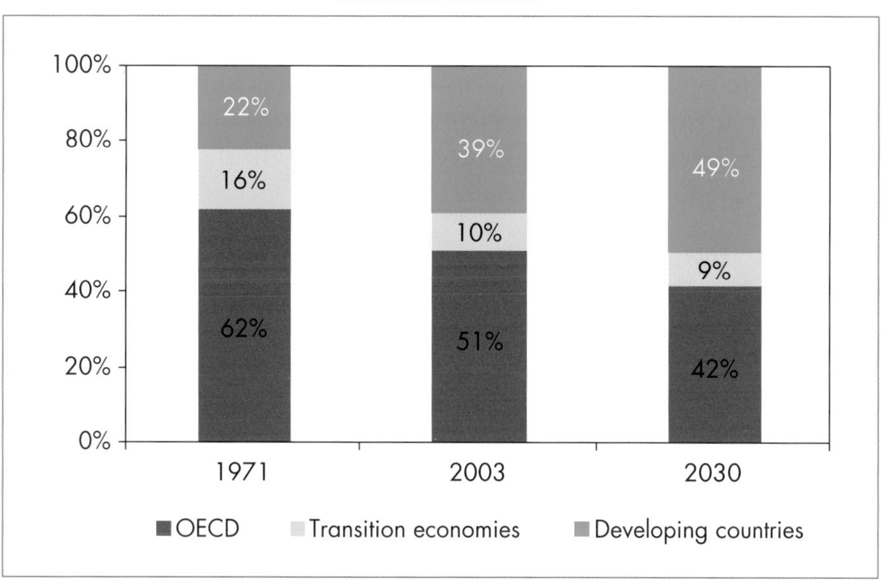

The developing countries' share of global demand will increase for all primary energy sources, except biomass. The increase in share will be pronounced for nuclear power, production of which will fall in the OECD while expanding in China and other parts of developing Asia. Nuclear output will increase marginally in the rest of the world. The developing regions' share of world coal consumption is also projected to increase sharply, mainly because of booming demand in China and India. Coal demand will increase more rapidly in those two countries, which have large, low-cost resources, than anywhere else in the world. By 2030, China and India will account for 48% of total world coal demand, up from 40% in 2003. Coal demand will grow slowly in OECD countries and the transition economies.

Nearly three-quarters, or 26 mb/d, of the 36-mb/d increase in global oil demand between 2003 and 2030 will come from developing regions, especially in Asia. Oil demand in China is projected to increase almost 2.5 times over the projection period, to 13.1 mb/d in 2030.

Natural gas demand will grow strongly globally and the share of gas in the primary fuel mix will increase in every region. The fastest rates of growth will occur in China and India, where gas consumption is currently relatively low.

Energy Production and Trade

Overview

The world's economically exploitable energy resources are adequate to meet the projected growth in energy demand in the Reference Scenario. Global proven oil reserves today exceed the cumulative projected production between 2003 and 2030, but additional reserves will need to be moved up into the proven category in order to avoid a peak in production before the end of the projection period. Exploration will undoubtedly be stepped up to ensure this happens.

Nonetheless, there will be a pronounced shift in the geographical breakdown of sources of energy production over the projection period, in response to a combination of cost, geopolitical and technical factors. Almost all the net increase in energy production will occur in non-OECD countries, compared to just 70% from 1971 to 2003. Consequently, the developing countries will emerge as the leading energy-producing region.

In the Reference Scenario, the Middle East and the former Soviet Union are projected to provide much of the growth in world oil and gas supply. Latin America, especially Venezuela and Brazil, and Africa will also see their output of both oil and gas rise substantially. Oil production will decline almost everywhere else. Production of natural gas, resources of which are more widely dispersed than oil, will increase in every region other than Europe.

The growing regional mismatch between demand and production will result in a major expansion of international trade in oil and gas, both in absolute terms and as a share of supply (Figure 2.5). Trade between countries within each grouping will also expand.

Oil will remain the most traded fuel. The volume of oil traded will increase by two-thirds. As a result, 52% of all the oil consumed worldwide will be traded between the main *WEO* regions in 2030, compared with 44% in 2003. The share of total gas supply that is traded between regions will also grow strongly, from 14% at present to 19% in 2030.

Figure 2.5: **Share of Inter-Regional Trade in World Primary Oil and Gas Demand in the Reference Scenario**

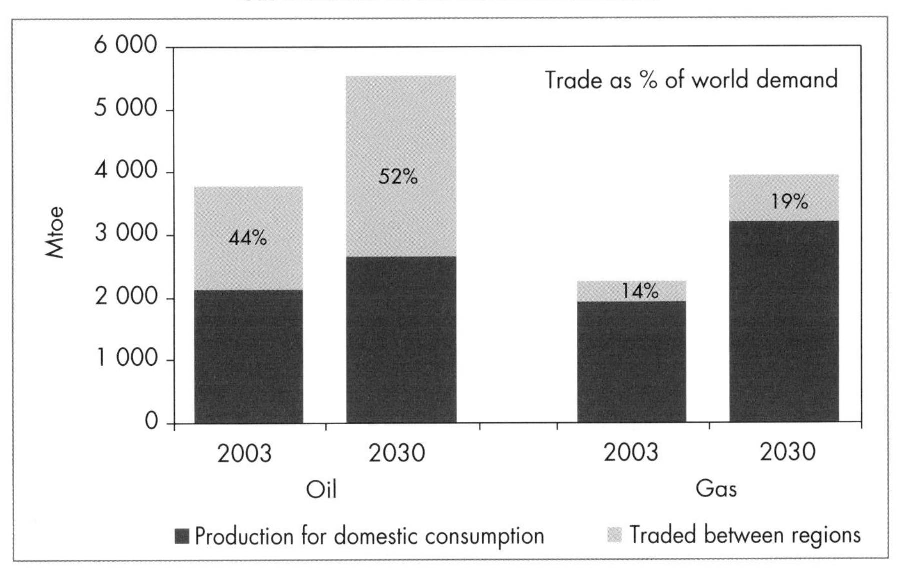

Oil Supply

World oil supply in the Reference Scenario is projected to grow from 82.1 mb/d in 2004 to 115.4 mb/d in 2030 (Table 2.4). Non-MENA countries will contribute most of the increase in global production over the rest of the current decade. High oil prices have started to stimulate increased development of reserves in those countries in recent years. Production is expected to continue to grow particularly strongly in transition economies, West Africa and Latin America. The output of the transition economies, which has soared in recent years, thanks to rapid growth in Russia, will continue to rise, with Caspian countries making a larger contribution. It will reach 14.5 mb/d in 2010, compared to 11.4 mb/d in 2004.

In the longer term, production in OPEC countries, especially in the Middle East, is expected to increase more rapidly than in other regions because their resources are much larger and their production costs are generally lower. OPEC's market share is projected to rise from 39% in 2004 to 50% in 2030, close to its historical peak in 1973. Global oil production is not expected to peak before 2030, although output in most regions will already be in decline by then. OPEC's market share would be lower if its members' policies have the effect of limiting production and driving up prices, thereby stimulating non-OPEC production of conventional and non-conventional oil, and encouraging alternative energy technologies. This is analysed in the Deferred Investment Scenario (Chapter 7).

Table 2.4: **World Oil Production in the Reference Scenario**
(million barrels per day)

	2004	2010	2020	2030	2004-2030*
Non-OPEC	**46.7**	**51.4**	**49.4**	**46.1**	**0.0%**
OECD	**20.2**	**19.2**	**16.1**	**13.5**	**–1.5%**
OECD North America	13.6	14.4	12.6	10.8	–0.9%
US and Canada	*9.7*	*10.5*	*8.8*	*7.4*	*–1.1%*
Mexico	*3.8*	*3.9*	*3.7*	*3.4*	*–0.5%*
OECD Europe	6.0	4.4	3.1	2.3	–3.7%
OECD Pacific	0.6	0.5	0.4	0.4	–1.4%
Transition economies	**11.4**	**14.5**	**15.6**	**16.4**	**1.4%**
Russia	9.2	10.7	10.9	11.1	0.7%
Developing countries	**15.2**	**17.7**	**17.6**	**16.3**	**0.3%**
China	3.5	3.5	3.0	2.4	–1.5%
India	0.8	0.9	0.8	0.6	–1.2%
Other Asia	1.9	2.1	1.7	1.3	–1.7%
Latin America	3.8	4.7	5.5	6.1	1.8%
Brazil	*1.5*	*2.5*	*3.3*	*4.1*	*3.8%*
Africa	3.3	4.9	5.2	4.7	1.4%
Middle East	1.9	1.7	1.5	1.4	–1.3%
OPEC	**32.3**	**36.9**	**47.4**	**57.2**	**2.2%**
OPEC Middle East	22.8	26.6	35.3	44.0	2.6%
Other OPEC	9.6	10.3	12.1	13.2	1.3%
Non-conventional oil	**2.2**	**3.1**	**6.5**	**10.2**	**6.1%**
of which GTLs	*0.1*	*0.3*	*1.3*	*2.3*	*13.9%*
Miscellaneous**	0.9	1.1	1.6	1.9	2.9%
World	**82.1**	**92.5**	**104.9**	**115.4**	**1.3%**
MENA	**29.0**	**33.0**	**41.8**	**50.5**	**2.2%**
Middle East	24.6	28.3	36.8	45.3	2.4%
North Africa	4.3	4.7	5.0	5.1	0.7%

*Average annual growth rate.
**Includes processing gains and stock changes.
Note: Includes NGLs and condensates.

Natural Gas Supply

Gas resources can easily meet the projected increase in global demand through the projection period, as proven reserves are now equal to 66 years of production at current rates. The regional outlook for production stems largely from the proximity of reserves to markets and production costs. Despite substantial unit cost reductions in recent years, gas transportation remains very expensive and usually represents most of the overall cost of gas delivered to consumers.

Production is projected to grow most strongly in volume terms in the Middle East and the transition economies (Figure 2.6), which between them have most of the world's proven reserves. Most of the incremental output in these regions will be exported to North America, Europe and Asia, where indigenous output will fall behind demand. Latin America and Africa will experience the fastest rates of increased gas production.

Figure 2.6: **Natural Gas Production by Region in the Reference Scenario**

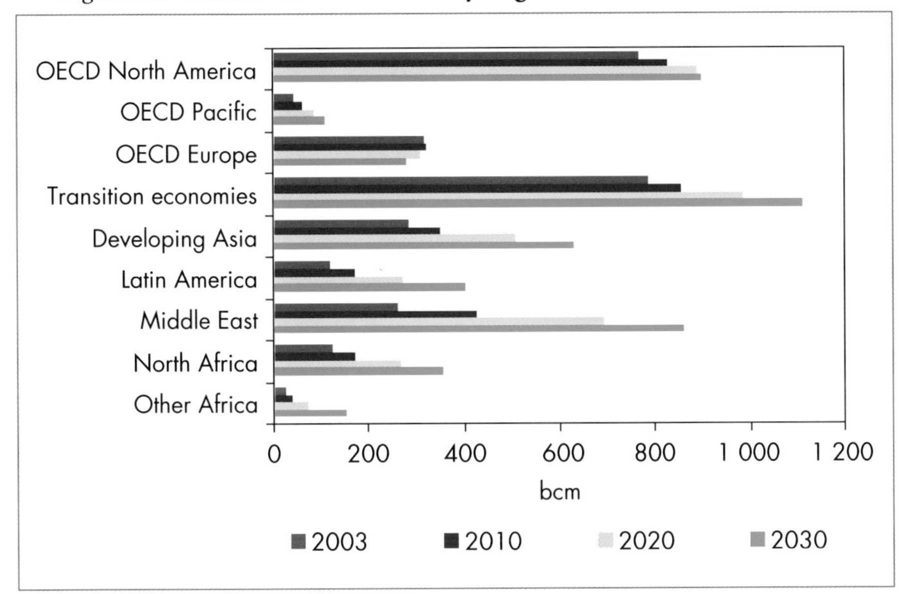

The largest volume increases in net imports are expected to arise in Europe and North America, where Canadian exports to the United States will not be able to keep pace with rising US import needs. OECD Europe will see its net imports grow from 203 bcm in 2003 to 499 bcm in 2030, meeting 64% of the region's total gas demand. North America, which is largely self-sufficient in gas at present, will see imports surge to 142 bcm in 2030, or 14% of total gas demand. Net imports will also rise further in OECD Pacific, where rising gas demand in Japan and Korea (which import virtually all their gas) will outstrip exports from Australia. China and India will also emerge as important gas importers over the *Outlook* period (Figure 2.7).

Figure 2.7: **Net Natural Gas Imports by Major Region in the Reference Scenario**

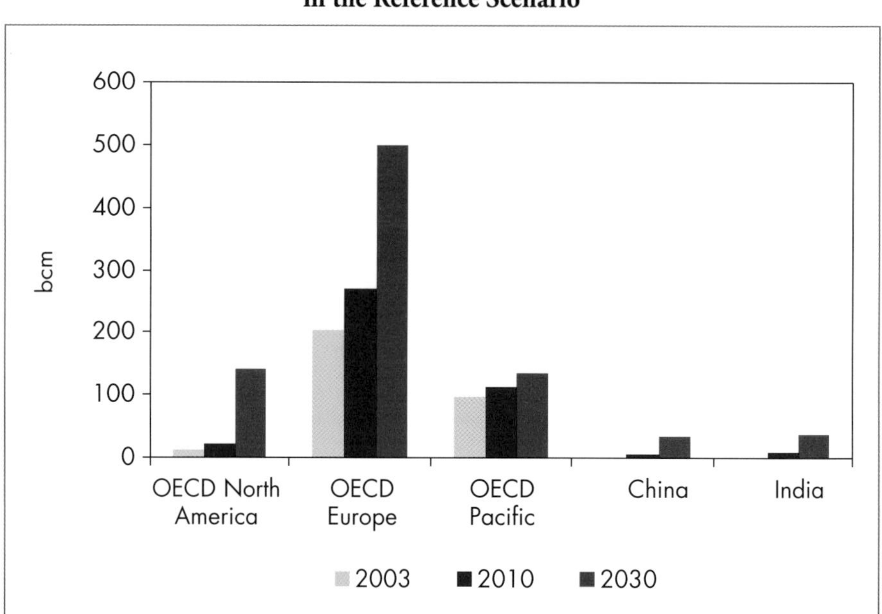

Energy-Related CO_2 Emissions

The projected trends in energy use in the Reference Scenario imply that global energy-related carbon-dioxide emissions will increase by 1.6 % per year over 2003-2030. Emissions will exceed 37 gigatonnes in 2030, an increase of almost 13 billion tonnes, or 52%, over the 2003 level (Table 2.5). By 2010, energy-related CO_2 emissions will be 38% higher than in 1990. Power generation is expected to contribute around half the increase in global emissions from 2003 to 2030. Transport will contribute a quarter.

Developing countries will be responsible for almost three-quarters of the increase in global CO_2 emissions from 2003 to 2030. They will overtake the OECD as the leading contributor to global emissions early in the 2020s. The increase in emissions from China alone will exceed the increase in all OECD countries and Russia combined. OECD countries accounted for 53% of total emissions in 2003, developing countries for 37% and transition economies for 11%. By 2030, the developing countries will account for 49%, the OECD countries for 42% and the transition economies for 9% (Figure 2.8).

Table 2.5: **Energy-Related CO_2 Emissions** (million tonnes)

	1990	2003	2010	2020	2030	2003 - 2030*
OECD	**11 026**	**12 776**	**13 794**	**14 824**	**15 341**	**0.7%**
OECD North America	5 556	6 620	7 338	7 949	8 387	0.9%
OECD Pacific	1 519	2 025	2 163	2 314	2 319	0.5%
OECD Europe	3 951	4 131	4 293	4 561	4 636	0.4%
Transition economies	**3 731**	**2 537**	**2 841**	**3 169**	**3 414**	**1.1%**
Russia	2 326	1 515	1 694	1 885	2 003	1.0%
Developing countries	**5 319**	**8 815**	**11 063**	**14 525**	**18 113**	**2.7%**
China	2 289	3 760	4 646	5 895	7 173	2.4%
India	598	1 050	1 296	1 736	2 283	2.9%
Other Asia	682	1 291	1 696	2 371	3 052	3.2%
Latin America	602	850	1 065	1 395	1 779	2.8%
Brazil	*192*	*303*	*369*	*492*	*626*	*2.7%*
Africa	547	763	949	1 276	1 635	2.9%
North Africa	*193*	*295*	*372*	*493*	*604*	*2.7%*
Middle East	601	1 102	1 411	1 852	2 191	2.6%
World incl. bunkers	**20 439**	**24 587**	**28 176**	**33 006**	**37 372**	**1.6%**
European Union	*3 733*	*3 789*	*3 941*	*4 157*	*4 219*	*0.4%*

*Average annual growth rate.

Figure 2.8: **Energy-Related CO_2 Emissions by Region in the Reference Scenario**

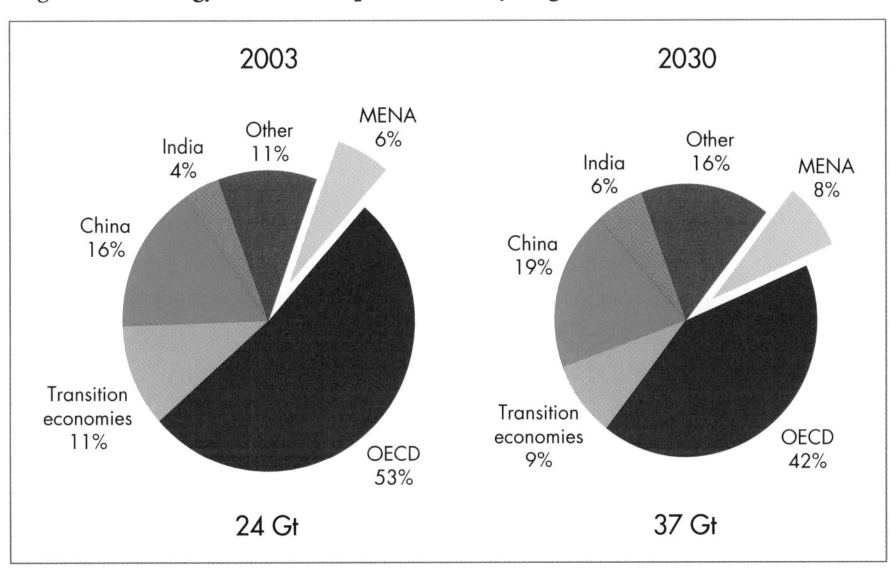

Energy Investment

The global energy supply projections in this *Outlook* will call for cumulative infrastructure investment of $17 trillion (in year-2004 dollars) over 2004-2030.[1] This investment will be needed to expand supply capacity and to replace existing and future supply facilities that will be closed during the projection period. More than half of the investment in production will go simply to maintain the present level of supply. Most of the world's current production capacity for oil, gas and coal will need to be replaced by 2030. Indeed, much of the new production capacity brought on stream in the early years of the projection period will itself need to be replaced before 2030. Some power plants and transmission and distribution infrastructure will also need to be replaced or refurbished, particularly in OECD countries.

More than 60% of total energy supply investments will go into the power sector. Power generation, transmission and distribution will require more than $10 trillion. If investment in the fuel chain to meet the fuel needs of power stations is included, electricity's share rises to more than 70%. More than half of the investment in the electricity industry will go to transmission and distribution networks. Total investment in the oil and gas sectors will each amount to about $3 trillion. More than three-quarters of the total investment in the oil industry will be in upstream projects. Upstream gas investment will account for around 60% of total investment in the gas industry. Coal investment will amount to almost $400 billion, or 2%. To produce and transport a given amount of energy, coal is about a sixth as capital-intensive as gas.

Developing countries will require about half of global energy investments because their demand and supply will increase rapidly (Figure 2.9). China alone will need to invest $2.5 trillion – 15% of the world total and more than in all the other developing Asian countries put together. Africa and the Middle East will require about $1.2 trillion each. North Africa alone will account for $300 billion. The greater part of investment in the Middle East and Africa will be in upstream oil and gas development. OECD countries will account for almost 40% of global investment and Russia and other transition economies for the remaining 10%. Over 40% of total non-OECD investment in the oil, gas- and coal-supply chains will go to provide fuels for export to OECD countries.

Globally, there is enough money to finance the projected energy investment. Domestic savings alone are much larger than the capital required for energy projects. But in some regions, those capital needs represent a very large share of

1. This figure compares with the estimate in *WEO-2004* of $16 trillion (in year-2000 dollars) for the period 2003-2030. The higher estimate in this year's *Outlook* is because of the change in the base year (from 2000 to 2004) and increases in some capital costs. These factors more than outweigh the one-year shorter projection period.

Figure 2.9: **Cumulative Energy Investment in the Reference Scenario, 2004-2030**

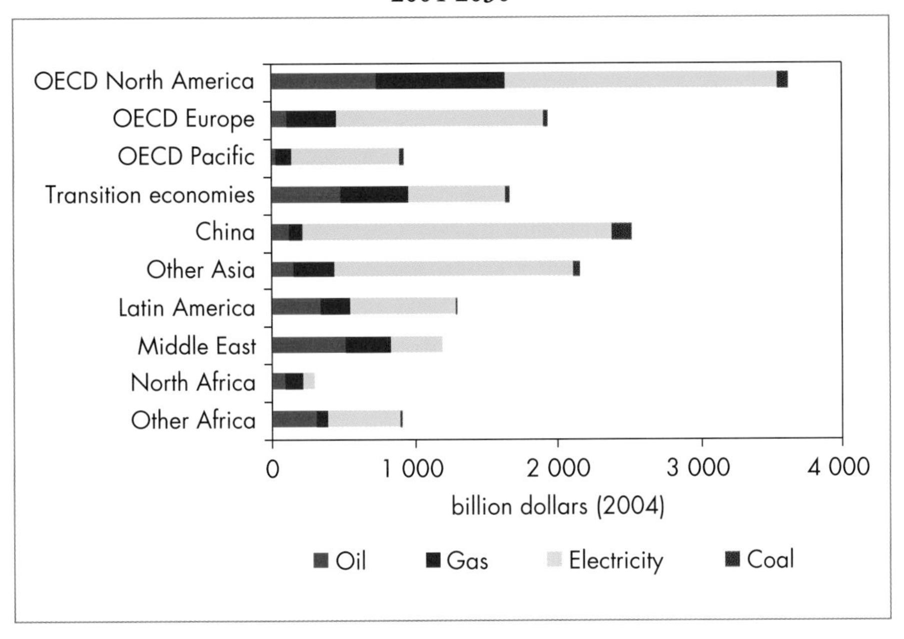

2

total savings. In Africa, for example, the share is about half. Although sufficient capital will be available overall from domestic and international sources, it is far from certain that all the infrastructure needed in the future will be fully financed in all cases. Mobilising the investment required will depend on whether returns are high enough to compensate for the risks involved. More than in the past, capital needed for energy projects will have to come from private sources, as governments continue to withdraw from the provision of energy services. Foreign direct investment is expected to become an increasingly important source of capital in non-OECD regions.

Financing the required investments in non-OECD countries is the biggest challenge and the main source of uncertainty surrounding our energy-supply projections. The financial needs in the transition economies and developing regions are much bigger, relative to the size of their economies, than in OECD countries. In general, investment risks are also greater in these regions, particularly for domestic electricity and downstream gas projects. Few governments could fully fund the necessary investment, even if they wanted to. Raising private finance will depend critically on the establishment by governments of an attractive investment framework and climate.

Focus on Oil Refining

Challenges Facing the Global Refining Industry

The oil-refining industry is at a crossroad in its history. World refining capacity was in surplus from the mid-1970s through to the mid-1990s, due to over-investment in the early 1970s, a downturn in demand after the oil shocks of 1973-74 and 1979-80, and the Asian financial crisis. As a result of strong growth in demand for refined products in recent years, spare capacity is rapidly diminishing and production flexibility even faster. Throughput capacity of 83 mb/d (Figure 2.10) was 85% utilised on average in 2004. Allowing for maintenance, there is now little scope for further increasing utilisation, so that capacity additions will be essential to meet growing demand for refined

Figure 2.10: **World Crude Oil Distillation Capacity and Oil Demand by Region, 2004**

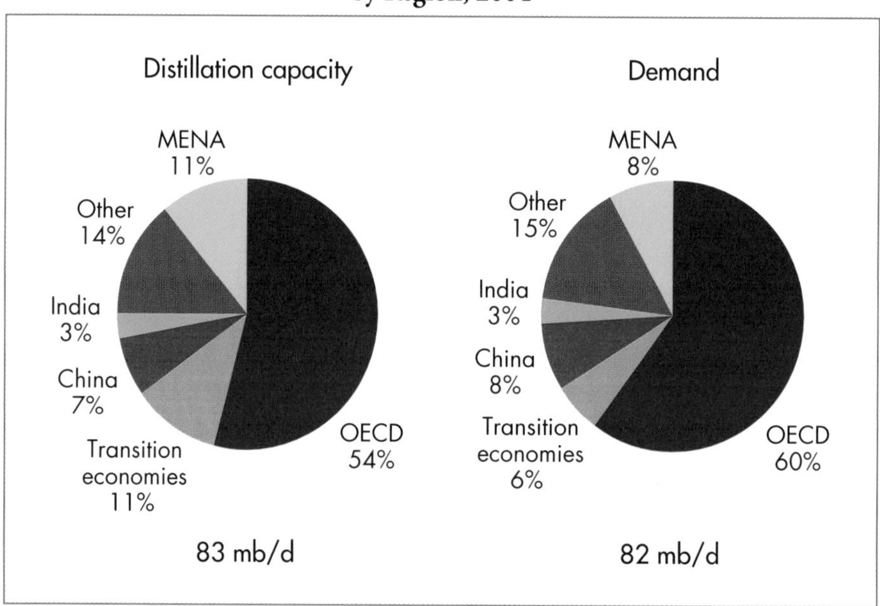

products. In the Reference Scenario, global refining distillation capacity will need to rise to 93 mb/d in 2010 and 118 mb/d in 2030. The near-term outlook for refining capacity is based on an assessment of projects that are either planned or under construction. This assessment, together with our demand projections, suggests that capacity utilisation rates are likely to remain high through to 2010 (Figure 2.11).

Figure 2.11: **World Crude Oil Distillation Capacity and Refined Product Demand in the Reference Scenario**

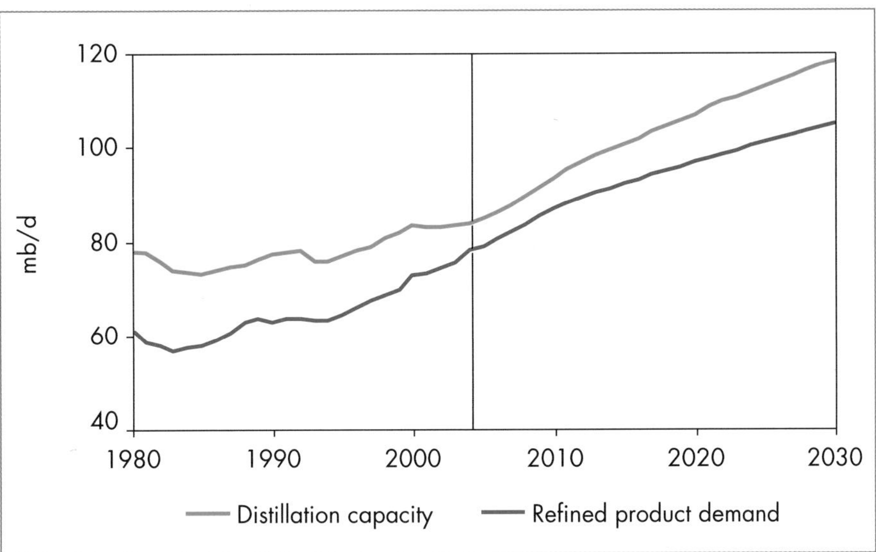

A number of factors complicate the outlook for refining and raise concerns that the industry will struggle to provide the amount and types of capacity needed to meet demand:

- It is normally more economic to build refineries close to market, mainly because this minimises transport and logistical costs. However, environmental restrictions and local resistance are hampering new additions and upgrades in many parts of the world. For example, in OECD North America, where much of the increase in demand is occurring, it is now virtually impossible to build a grassroots refinery. The other OECD regions face similar problems.
- A shift in demand towards lighter products, mainly for road and air transport, will require major investment in upgrading units. The share of light products and middle distillates in global oil consumption has risen from 65% in the early 1980s to 80% today. The conversion units that are needed are costly and take several years to plan and build.
- Adding to these challenges is the shift in the quality of crude oil that refiners have to process. On average, crude oil production is becoming heavier and more sour (containing more sulphur). Heavy, sour crudes are more difficult to process and yield more heavy products, which need additional processing.
- Tighter product specifications require refiners either to use better-quality crude oil or to invest in new units to upgrade product streams. New rules on

sulphur content in diesel and gasoline are particularly onerous. Europe will have a maximum sulphur content of 10 parts per million by 2009. India has enforced a 500 ppm sulphur-content limit on diesel since April 2005 in selected cities and China is expected to follow suit by 2008. Many other regions of the world will also have moved to low and ultra-low sulphur fuels by the end of this decade.

- Uncertainty about future investment returns discourages much-needed investments in refining. Financial returns over the past three decades have usually been very low, though they have improved significantly in recent years. The average margin worldwide in 2004 was around $8 per barrel, easily high enough to cover the capital cost. But a refinery usually takes around five years to build, increasing the risk that future margins may not cover investment costs. This factor makes integrated companies reluctant to embark on major projects. Even if refinery investments are judged to be profitable, upstream margins are often higher.

Refining Capacity Outlook

World crude oil distillation capacity was estimated at about 83.1 mb/d at the end of 2004. To meet demand in 2010, capacity will need to reach almost 93 mb/d. This represents an average growth rate of 1.8% per year. New refinery additions, however, will struggle to keep pace with demand growth over this period on current construction plans. As a result, utilisation rates will increase to over 86% in 2010. Refining capacity will need to expand further to 118 mb/d in 2030 (Table 2.6). Additions to capacity, as well as higher crude runs, will be primarily

Table 2.6: **Global Crude Oil Distillation Capacity in the Reference Scenario** (mb/d)

	2004	2010	2020	2030
OECD	**44.3**	**45.9**	**49.6**	**51.8**
OECD North America	20.5	22.1	24.6	25.6
OECD Pacific	8.1	8.1	8.9	9.7
OECD Europe	15.7	15.7	16.1	16.6
Transition economies	**9.5**	**9.5**	**9.7**	**10.2**
Developing countries	**29.3**	**37.2**	**47.1**	**55.7**
China	6.1	8.8	12.5	14.6
India	2.4	3.2	4.2	5.2
Middle East	7.0	8.8	10.7	12.9
North Africa	1.8	2.1	2.6	3.1
Other developing countries	12.0	14.4	17.1	19.9
World	**83.1**	**92.7**	**106.4**	**117.8**

directed at meeting middle distillate demand. Any shortfall in investment, leading to shortages in various types of refining capacity, would put upward pressure on petroleum product and crude oil prices (Box 2.2).

Box 2.2: **Impact of Refining Capacity Constraints on Crude Oil Prices**

In the coming years, capacity for producing middle distillates and gasoline is set to become more and more constrained. Limited expansion in conversion capacity over the next couple of years will support strong refinery margins and put upward pressure on the prices of lighter sweet crudes. Constraints in the refining sector are essentially centred on upgrading capacity – units that convert heavy products, such as fuel oil, produced in crude distillation, to lighter products, such as gasoline, diesel and jet kerosene.

Upgrading capacity comes in different types of units, from the common catalytic cracking units to the more sophisticated and expensive coking units. Depending on the number and type of units, a refinery is said to be more or less complex. A refinery that has little upgrading capacity is referred to as a simple or hydroskimming refinery. Hydroskimming requires more crude than would be required with a complex refinery to produce a given amount of light products.

Constraints in upgrading capacity will lead to a greater increase in crude oil demand, regardless of quality, than would be required if more upgrading capacity was available. So, in a situation in which upgrading capacity is constrained, strong demand for light products will push up crude oil prices. This increase will be broadly proportional to the rise in the value of the product slate. Refiners typically use hydro-treating units to reduce sulphur levels to meet environmental requirements. If a refinery does not have enough of this type of capacity, it will prefer sweet crude (with low-sulphur content) over sour crude (with high-sulphur content).

Despite a projected 25 mb/d increase in distillation capacity from 2010 to 2030, the average refinery utilisation rate will remain well above 80%, dropping only slightly, to just under 84%, in 2030. In the second half of the projection period, the increasing production of light and middle distillates from sources other than refineries (mainly NGLs and GTL plants) will bolster supply. By 2030, such products will amount to 11.1 mb/d, meeting 9% of global oil demand (Table 2.7).

Capacity additions over the projection period will be concentrated in developing countries, because of the difficulties in building new refineries or in

Table 2.7: **Oil Product Supply Shares by Source in the Reference Scenario**

	1971	1990	2004	2010	2020	2030
Refinery output*	93.9%	93.6%	94.5%	93.7%	92.0%	90.9%
NGL**	6.1%	6.4%	5.4%	6.0%	6.6%	7.0%
GTL	0.0%	0.0%	0.1%	0.3%	1.4%	2.2%

* Includes all oil products produced in refineries, including NGL processed through condensate splitters.
** Includes all NGL processed in plants other than refineries, such as petrochemical plants.
Note: Excludes biofuels.

expanding capacity at existing ones in the OECD (Figure 2.12).[2] The producing countries of the Middle East, in particular, will play an increasingly important role in meeting global demand for refined products. *China's* installed capacity is projected to more than double over the projection period, to reach 14.6 mb/d in 2030, as it seeks to keep up with its growing demand. But it will still be a net importer of refined products in 2030. Unlike China,

Figure 2.12: **Global Refining Capacity Additions in the Reference Scenario, 2004-2030**

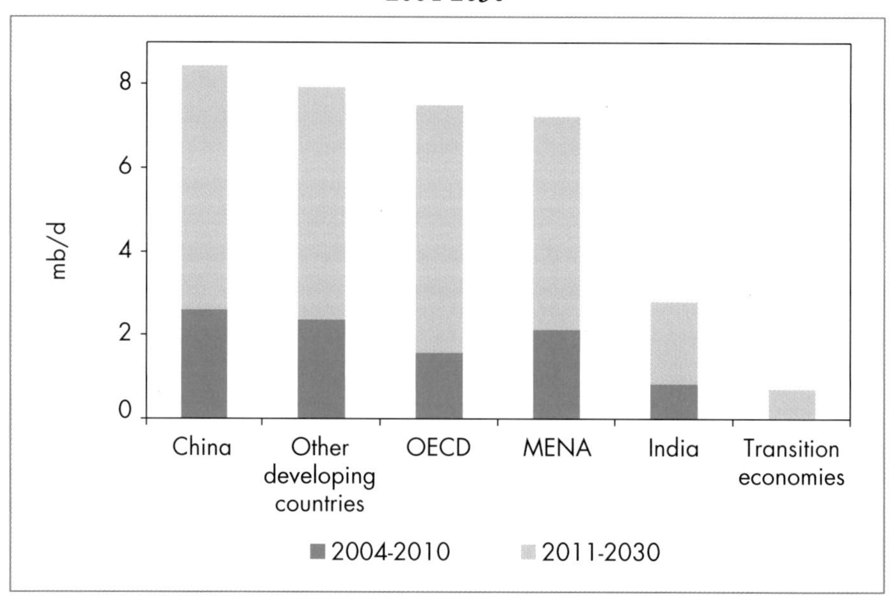

2. In the United States and in most OECD countries, capacity creep will be the only way to increase distillation capacity.

India is keeping up with the rise in its demand but availability for export is becoming slimmer. India has several projects involving debottlenecking and further additions, and is expected to bring installed capacity up to 5.2 mb/d in 2030. Despite this large increase in capacity, India will be only a marginal net exporter of refined products in 2030.

In *North America*, capacity creep will result in a small increase in capacity, to 22.1 mb/d in 2010 and 25.6 mb/d by 2030. Despite these additions and massive conversion of existing capacity to meet the new fuel specifications, OECD North America will still be a net importer of refined products in 2030. Installed distillation capacity in Europe will reach 16.6 mb/d in 2030. Environmental constraints are hampering refineries from converting capacity to meet projected demand for middle distillates. By 2010, Europe is likely to become a net importer of middle distillates.

With current utilisation rates of less than 60%, Russia and other transition economies will not add much distillation capacity over the projection period, but they are expected to increase their refinery-utilisation rates. Investing in refineries has not been a priority in these countries.[3] The scale of investment needed to produce lower-sulphur products to meet European specifications will discourage large-scale refinery upgrading. Russia will continue to export gasoil to Europe, for blending with jet fuel in order to meet product specifications.

Refined Product Trade

In most countries, local refineries supply the domestic market. However, with growing regional imbalances, the volume of trade in petroleum products is projected to increase by 1.8% per year over 2003-2030 in the Reference Scenario. Inter-regional trade is projected to grow from around 10 mb/d in 2004 to just under 16 mb/d in 2030.

The current pattern of gasoline flows from Europe to the United States is expected to continue until the mid-2020s. By the end of the projection period, Europe's gasoline exports are expected to diminish in volume, to be partially replaced by products from the Middle East. Europe will have a diesel deficit of up to 1 mb/d by 2010.

In Asia, India will remain a marginal net exporter of oil products, but it will be outpaced by Saudi Arabia. Despite a substantial decrease in its oil imports and expected refinery additions, China will still be a net oil-product importer by 2030.

3. Due to lack of maintenance and deterioration of infrastructure, a significant portion of the underutilised capacity in the transition economies is probably not usable. The utilisation rate of the refineries of this region is therefore underestimated.

Refinery Investment

The total requirement for refinery investment over the *Outlook* period is projected to amount to $487 billion (in year-2004 dollars), or $18.7 billion per year. Two-thirds of this investment will need to be made in developing countries, over half of this in the Middle East and China alone (Figure 2.13). MENA will account for nearly a quarter of total refinery investments – more than $110 billion – over the *Outlook* period. Saudi Arabia alone will account for 30% (see Chapter 16).

Figure 2.13: **Cumulative Refinery Investment in the Reference Scenario, 2004-2030**

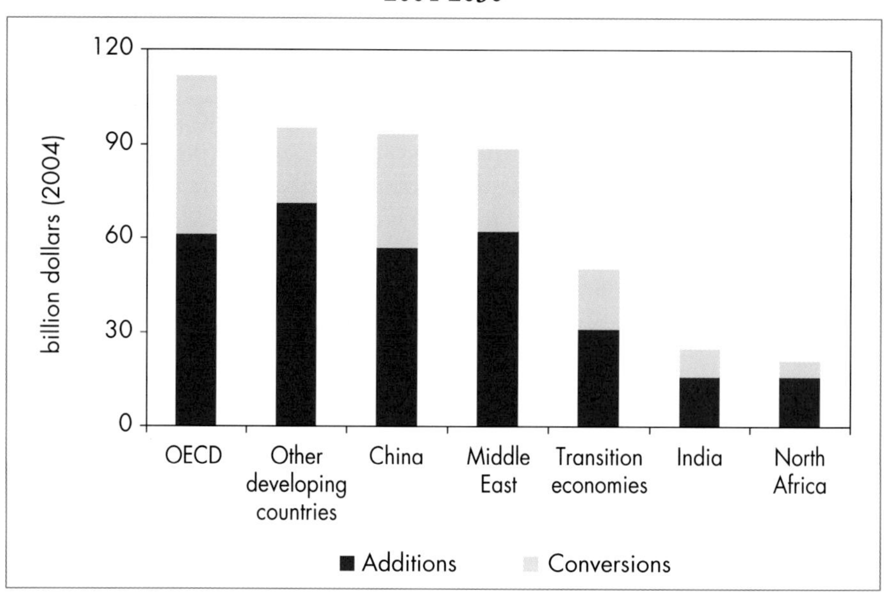

Nearly two-thirds of refinery investment will go to crude oil distillation capacity additions and maintenance of existing capacity, while the remaining 35% will cover conversion capacity (Table 2.8). OECD countries will account for 30% of total conversion investment. Sizeable investment in conversion will also be needed in China and India, where specifications are being tightened, and also in the MENA region, which aims to supply OECD markets. Overall conversion investments will represent around 45% of the total investment needs in the OECD, compared to 31% of total investment in developing countries.

Investment requirements in refining capacity will vary over time. Over the rest of the current decade, nearly half of the global investment needs will be required for capacity conversion (Figure 2.14). Conversions will represent 69% of total investment from 2004 to 2010 in OECD and over 38% of the investment in developing countries. Global investment for capacity extension

Table 2.8: **Cumulative Refinery Investment in the Reference Scenario, 2004-2030** ($ billion in year-2004 dollars)

	Additions/ upgrades	Conversions	Total investment
OECD	**61**	**51**	**112**
OECD North America	44	25	69
OECD Europe	10	8	18
OECD Pacific	7	18	26
Transition economies	**31**	**19**	**51**
Developing countries	**222**	**102**	**324**
China	57	37	94
India	16	9	25
Middle East	62	26	89
North Africa	16	6	21
Other developing countries	71	24	95
World	**315**	**172**	**487**

Figure 2.14: **Global Refinery Investment by Decade in the Reference Scenario**

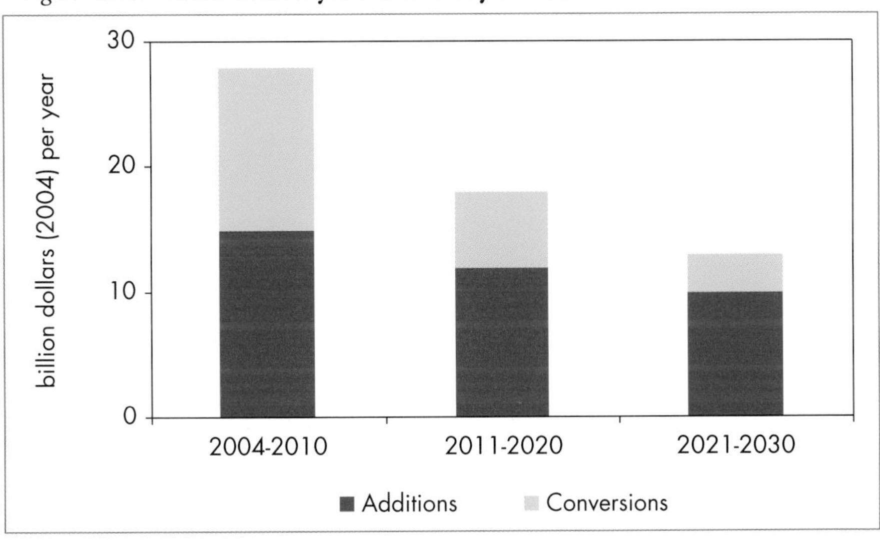

will slightly decrease in the last decade of the projection period. Most additions will be needed in developing countries, where costs are lower, and a larger share of capacity additions will arise from creep and additions to existing refineries, rather than through building new refineries. As a result, global refinery investment needs will fall by about a quarter, from $181 billion between 2011 and 2020 to $137 billion dollars from 2021 to 2030.

CHAPTER 3

ENERGY TRENDS IN THE MIDDLE EAST AND NORTH AFRICA

HIGHLIGHTS

- MENA primary energy demand in the Reference Scenario more than doubles between 2003 and 2030, to 1.2 billion tonnes of oil equivalent. Despite brisk growth in energy use in the MENA region, the per capita consumption projected for 2030 is about half the current level in OECD countries. MENA energy intensity will be broadly flat through to 2020, and will then start to decline gradually.
- MENA countries will continue to rely almost exclusively on oil and natural gas to meet their domestic energy needs. Gas will overtake oil after 2020 to become the region's key energy source, partly in response to policies aimed at freeing up oil for export. Coal, hydropower, renewables and nuclear will all increase in volume but will total less than 4% of primary energy demand by 2030.
- Oil production in the MENA region is projected to surge from 29 mb/d today to almost 50 mb/d by 2030. It will outpace growth in domestic demand, allowing the region's oil exports to expand by almost 75% to around 39 mb/d by 2030. MENA natural gas production will triple by 2030, reaching 1 210 bcm. Domestic demand will not grow as fast as output, enabling exports, mostly as LNG, to increase fourfold.
- These projections call for investment of some $1.5 trillion (in year-2004 dollars) over the period 2004-2030, or $56 billion per year. The oil and gas sectors will require around 70% of the total and the electricity sector the rest.
- In real terms, MENA oil-export revenues in 2005 are expected to approach the record levels seen during 1980, in the wake of the second oil price shock. But on a per capita basis, they are well below previous peaks. We expect MENA oil export revenues to increase over the projection period. Natural gas will make a growing contribution to overall export revenues, increasing from 6% in 2004 to 13% in 2030.
- MENA energy-related carbon-dioxide emissions will double by 2030. By that time, the region's share of global emissions, if the Reference Scenario's projections are realised, will be 7%, up from around 6% today.
- In the Deferred Investment Scenario, MENA domestic energy demand grows slightly less rapidly. Oil and gas production falls substantially compared with the Reference Scenario, but still grows in volume terms.

MENA Domestic Energy Demand

The countries of the MENA region – Iran, Iraq, Kuwait, Qatar, the UAE, Saudi Arabia, Bahrain, Israel, Jordan, Lebanon, Oman, Syria, Yemen, Algeria, Egypt, Libya, Morocco and Tunisia – currently represent around 5% of global energy demand. This share has increased more than fourfold since 1971, in line with robust growth in gross domestic product and a more than doubling of the population. Despite this rapid growth, domestic energy consumption still absorbs only a moderate portion of the region's overall energy production. Oil and gas have supplied almost all the MENA region's needs (Table 3.1).

Table 3.1: **MENA Key Energy Indicators**

	1971	2003	1971-2003*
Total primary energy demand (Mtoe)	69.2	569.6	6.8%
Total primary energy demand per capita (toe)	0.5	1.8	4.0%
Total primary energy demand/GDP**	0.1	0.3	3.3%
Share of oil in total primary energy demand (%)	74.9	54.2	–
Net oil exports (mb/d)	19.7	21.0	0.2%
Share of oil exports in production (%)	95	77	–
Share of gas in total primary energy demand (%)	19	42	–
Net gas exports (bcm)	7.1	96.9	8.5%
Share of gas exports in production (%)	31	25	–
CO_2 emissions (Mt)	170	1 397	6.8%

* Average annual growth rate.
** Toe/thousand dollars of GDP in year-2004 dollars and PPPs.

MENA energy demand is set to grow faster than in any other major world region. In our Reference Scenario, MENA primary energy demand increases by an average 2.9% per year from 2003 to 2030, reaching 1.2 billion tonnes of oil equivalent. The rate of growth in demand is significantly below the 6.8% seen over the last three decades, but by 2030 the MENA region will nonetheless account for 7.5% of global total primary energy demand, more than a two percentage point increase on current levels.

In absolute terms, the biggest contributors to demand growth will be Saudi Arabia and Iran. Between them, they will account for about 45% of MENA energy demand in 2030, slightly less than today. The fastest *rates* of energy demand growth will occur in Qatar (Table 3.2).

Despite brisk growth in energy use in the MENA region, the per capita consumption projected for 2030 of 2.4 toe is only about half the current level in OECD countries. Within the region, there will remain large discrepancies in per capita energy use; in general, it will remain much higher in the Middle Eastern countries than in those of North Africa. In most of the Gulf countries, per capita *electricity* consumption will remain close to the highest levels seen anywhere in the world. This is due to factors such as heavy price subsidies, which lead to inefficient use, and the hot climate, which necessitates considerable air-conditioning.

Table 3.2: **MENA Primary Energy Demand by Country** (Mtoe)

	2003	2010	2020	2030	2003-2030*
Middle East	**446**	**597**	**807**	**963**	**2.9%**
Iran	136.4	172.9	224.8	271.5	2.6%
Iraq	25.8	35.3	47.4	62.0	3.3%
Kuwait	22.9	30.2	41.1	48.7	2.8%
Qatar	15.2	31.9	59.4	67.3	5.7%
Saudi Arabia	130.8	181.0	246.8	289.1	3.0%
United Arab Emirates	39.2	54.3	71.8	84.4	2.9%
Other Middle East	75.5	91.7	115.5	139.5	2.3%
North Africa	**124**	**160**	**213**	**262**	**2.8%**
Algeria	33.0	42.5	57.9	69.7	2.8%
Egypt	53.9	68.2	87.8	108.6	2.6%
Libya	18.0	25.3	35.9	46.2	3.6%
Other North Africa	19.0	24.2	31.5	37.5	2.6%
MENA	**570**	**757**	**1 020**	**1 225**	**2.9%**
MENA share in world demand	*5.3%*	*6.1%*	*7.1%*	*7.5%*	–

* Average annual growth rate.

Primary Fuel Mix

The share of oil in primary energy supply within MENA will fall from 54% in 2003 to 47% in 2030. Nevertheless, oil demand is projected to grow at a healthy 2.3% per year, from 309 Mtoe (6.3 mb/d) in 2003 to 576 Mtoe (11.8 mb/d) in 2030 (Figure 3.1). By 2030, the MENA region will consume 23% of its oil production domestically, about the same as it does today. Half of the increase in oil use will be in the transport sector, which will account for 42% of the region's total oil demand in 2030, up from 36% at present. The share of oil in energy use will decline in all other sectors, as it is increasingly substituted by natural gas. The largest loss of market share will be in the power sector, although reliance on oil-fired generation will remain high in comparison to other parts of the world.

MENA primary demand for natural gas will grow by 3.5% per year from 240 Mtoe (288 bcm) in 2003 to around 600 Mtoe (767 bcm) in 2030. Natural gas will overtake oil just after 2020 to become the region's major primary energy source. In many countries in the region, gas is widely available at a lower economic cost than oil and the authorities are encouraging its use domestically to free up oil for export. At present, the domestic MENA market consumes about 75% of its overall gas production. By 2030, this will have decreased to 63%, as production increases more rapidly than domestic demand. Almost half the increase in gas demand will come from the power generation and water desalination sector. Industrial use of gas will also increase rapidly.

Figure 3.1: **MENA Primary Energy Demand by Fuel**

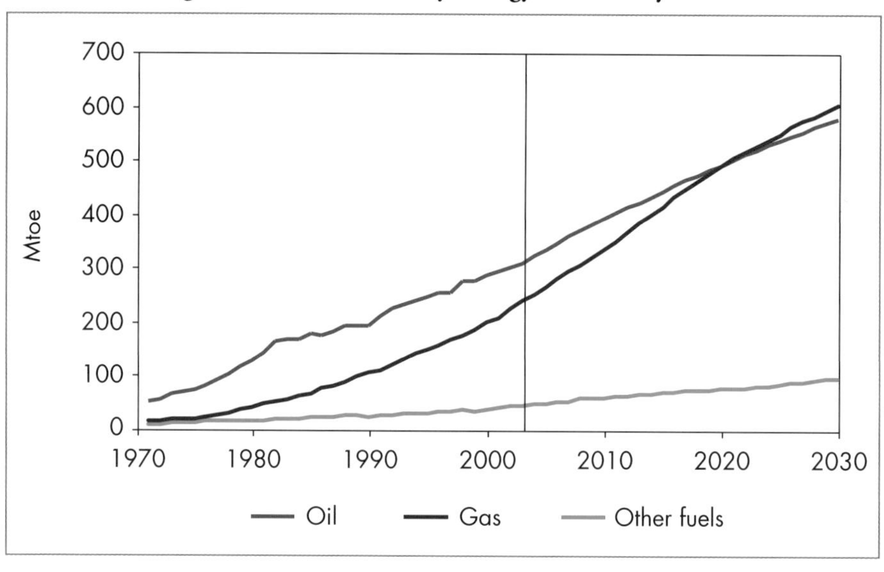

Coal, hydropower, other renewables and nuclear will each contribute growing volumes of energy but, as a group, they will contribute just 3.9% of MENA primary energy demand in 2030, compared to 3.7% at present. This is a reflection of the abundance of low-cost oil and gas throughout the region.

MENA energy intensity, or total primary energy use per unit of GDP, is projected to fall by 0.5% per year over the projection period, though in 2030 it will still be one-and-a-half times the current level of OECD Europe (Figure 3.2). The expected reduction in intensity in the region contrasts with growth in energy intensity in the region over the last three decades. Most of the improvement will occur in the second half of the *Outlook* period, largely as a result of energy efficiency improvements. As with per capita consumption, significant differences persist in energy intensity among the countries that make up the region, reflecting differences in their stage of economic development, the energy efficiency of end-use technologies, economic structure, energy prices, climate, geography, culture and lifestyles. For instance, energy intensity in Qatar will actually increase by around 60% through the *Outlook* period, as a consequence of the rapid expansion of the country's energy-intensive LNG and GTL businesses.

Figure 3.2: **MENA Primary Energy Intensity**

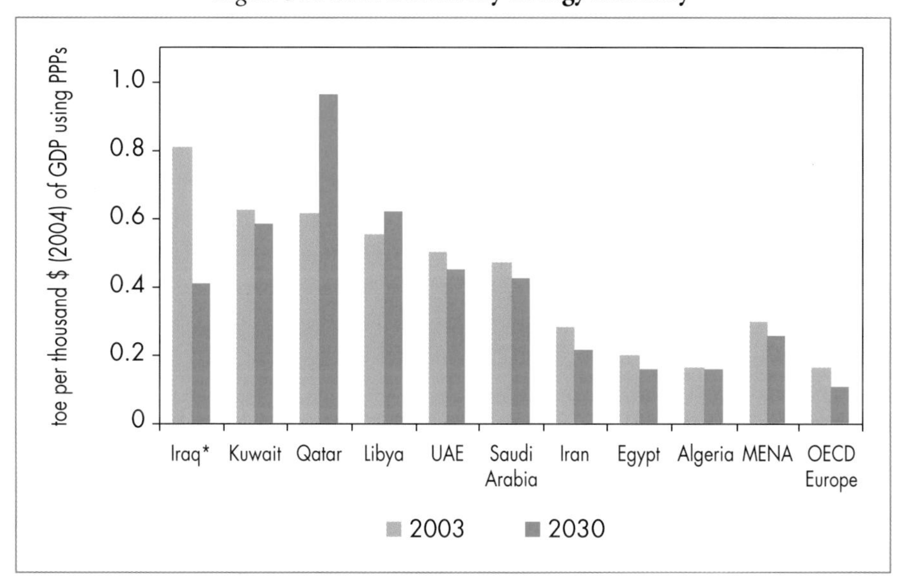

* Figure shown for Iraq is for 2002.

Demand Trends by Sector

The share of the power generation and water desalination sector in MENA total primary energy demand will increase slightly, from 32% at present to 33% by 2030. Energy inputs to these facilities will increase by 3.0% per year, below the projected demand growth for electricity of 3.5%. This reflects expected improvements in the thermal efficiency of power plants. Most new power plants are expected to be gas-fired, increasing the share of gas in electricity generation from 56% in 2003 to 58% in 2010 and 69% in 2030. Oil's share will fall from 33% to 20% between 2003 and 2030.

Water desalination – an energy-intensive process – is already relied upon extensively in Saudi Arabia, the UAE, Kuwait, Qatar, Algeria and Libya. Energy for water desalination will account for more than a quarter of the total increase of the power and water desalination sector's fuel requirements in these six countries through to 2030. MENA primary energy requirements for water desalination will increase from 26 Mtoe in 2003 to 61 Mtoe, growth of 3.2% per year.[1] Gas will be the preferred fuel for water desalination in MENA countries over the projection period, as it is more efficient, less costly and less environmentally damaging than oil.

Energy use in final sectors – industry, transport, households, services, agriculture and non-energy uses – will grow by 2.8% per year over 2003-2030. This is marginally slower than primary demand. As a result, the share of final consumption in primary demand will drop by a couple of percentage points, to 64%.

Industry will remain the largest end-use sector, with a share of total final energy consumption of 36% in 2030, marginally less than today. Petrochemicals manufacturing, which many countries in the region view as an important means to diversify their economies away from basic oil and gas production, will continue to constitute the major part of industrial energy demand (Box 3.1). Among the Middle Eastern countries, Saudi Arabia is expected to remain the largest producer of petrochemicals, followed by the UAE and Iran. Natural gas and oil will continue to dominate industrial energy use in the region, their share remaining close to 90%. But use of electricity, which currently accounts for just 9% of the total energy consumed by industry in MENA (compared with 25% in the EU), will grow most quickly.

The transport sector will account for the second-largest component of the increase in total final energy consumption. Growth will be underpinned by subsidised fuel prices, increased transportation of goods and a young and rapidly growing population. Transport's share of total final demand will rise slightly to 32%, fuelled almost exclusively by oil. There is considerable scope both for growth in the vehicle fleet and for improving fuel efficiency standards. At present, there are only 64 cars for every thousand people in the MENA region, compared with 770 in North America and 500 in Europe.

1. The primary energy requirements include both the fuel used to generate the electricity consumed by the desalination plants and the fuel requirements for the steam production.

The residential sector's share of total final consumption will remain just over 20% throughout the projection period. Electricity will rapidly take a larger share of the energy mix in this sector, as households continue to switch away from LPG and kerosene because of the price advantage, convenience and an increasing uptake of electrical appliances. Natural gas will also be increasingly used in households for heating and cooking, supported by an extension of the distribution network in several key countries, such as Iran. Despite strong growth, per capita residential energy use in the MENA region will remain significantly lower than in OECD countries.

Box 3.1: **Petrochemicals in the Middle East**

The Middle East has experienced a huge expansion of its petrochemical sector as several countries in the region have made efforts to diversify away from sole reliance on oil and gas production. Saudi Arabia led efforts in the late 1970s to add value to gas produced during oil production, which had previously been flared. After witnessing their success, other countries in the region followed suit, mainly with export markets in mind. Rapid growth followed, underpinned by the region's vast, low-cost hydrocarbon resources, which provide the necessary feedstock. Access to state-of–the-art technologies through joint ventures with well-established international companies contributed.

By 2004, petrochemical exports from the Gulf region had grown to 30 Mt and they are expected to top 40 Mt in 2005. More than 80% of current output is from Saudi Arabia, with much of the rest from Iran, the UAE and Qatar. The bulk of production is based on ethane recovered from associated gas during oil production and the main products include polypropylene, ethylene and ethylene dichloride.

In the global petrochemical sector, demand is booming in the fast-growing markets in the Asia-Pacific region and the demand slate is slowly shifting away from ethylene to products such as polyethylene, mercapto-ethylguanidine and polyvinyl chloride. The Middle East is well placed to respond to these changes. Its geographical proximity to the Asia-Pacific region – where demand is expected to match the combined size of the mature markets of the United States and Europe by 2010 – is a distinct advantage. It is also expected to increase trade to OECD regions, where local output will struggle to keep up with demand growth. Among the Middle Eastern countries, Saudi Arabia will remain the largest producer of petrochemicals: energy demand in its petrochemical sector is projected to grow by 2.8% per year from 2003 to 2030. Demand in the UAE, Qatar and Kuwait are expected to grow at an even faster rate, but from lower bases (Figure 3.3).

Figure 3.3: **Energy Demand in the Petrochemical Sector in the Middle East**

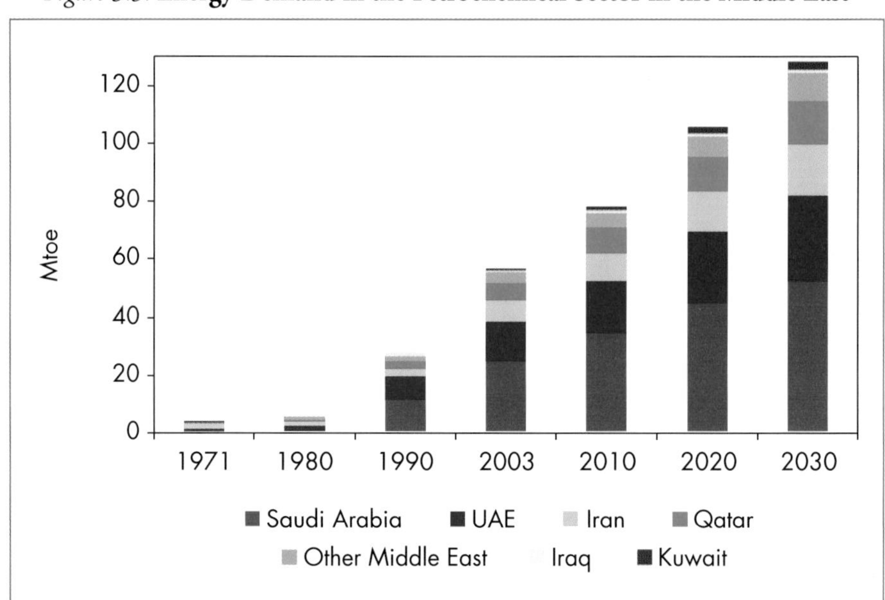

Crude Oil Production and Trade[2]

In the Reference Scenario, MENA production of crude oil is projected to increase from 29.0 mb/d in 2004 to 50.5 mb/d in 2030. These projections assume that the countries of the region are willing and able to develop new production capacities to meet the projected call on MENA oil, at the prices assumed: in some cases, this may mean opening up to foreign investment in upstream oil. Production in MENA countries, especially in the Middle East, is projected to increase more rapidly than in other regions, both because resources are larger and production costs generally lower. In recognition of the many uncertainties that influence these projections, a Deferred Investment Scenario analyses how global energy markets might evolve if investment in the upstream oil industry of MENA countries were not to grow so rapidly (see Chapter 7).

In the Reference Scenario, MENA's share of world oil production is projected to rise from 35% in 2004 to 44% in 2030. Within MENA, the Middle East will account for almost all of the oil production increase. Saudi Arabia will

2. See Chapter 4 for detailed oil sector analysis and results. All the oil-production figures cited in this section include crude oil, natural gas liquids (NGLs) and condensates.

remain by far the largest supplier, its output rising from 10.4 mb/d in 2004 to 18.2 mb/d in 2030. Its share of total MENA oil output then will be much the same as today, at about 36%. Three countries will see their shares rise: Iraq, Kuwait and Libya.

MENA countries as a group will play an increasing role in international oil trade to 2030. Net exports from MENA countries to other regions will rise from 22 mb/d in 2004 to 25 mb/d in 2010 and 39 mb/d in 2030. Most exports will still be as crude oil in 2030, but refined products will account for a growing share of exports. The region will increase its exports to all the major consuming regions. Exports to developing Asian countries will increase most.

Natural Gas Production and Trade[3]

MENA gas production is projected to treble over the projection period from 385 bcm in 2003 to 1 211 bcm in 2030. The rate of growth will be faster than that of any other major world region. The biggest volume increases in the region will occur in Qatar, Iran, Algeria and Saudi Arabia.

The call on MENA gas supply will increase rapidly over the projection period, a result of strong global demand and declining output in many other gas-producing regions. Net exports from MENA countries to other regions are projected to climb from 97 bcm in 2003 to 444 bcm in 2030. Most of the increase will be in the form of LNG. There will be a marked shift in the balance of Middle East exports from eastern to western markets. Europe will remain the primary destination for North African gas exports.

Investment Needs

The Reference Scenario's projected increase in MENA energy supply calls for cumulative infrastructure investment of about $1.5 trillion (in year-2004 dollars) over the period 2004-2030, or $56 billion per year (Figure 3.4). This investment will be needed both to expand supply capacity and to replace existing and future supply facilities that will be retired during the projection period.

The oil sector represents 41% of the total energy-investment requirement. Its needs amount to $614 billion, or $23 billion a year. This is more than twice the average annual amount of investment seen over the last decade. The upstream industry will absorb four-fifths of the total. National oil companies are expected to invest the most, but investment by international oil companies can be expected to rise as investment needs grow.

3. See Chapter 5 for detailed gas sector analysis and results.

Figure 3.4: **Cumulative MENA Energy Investment, 2004-2030**

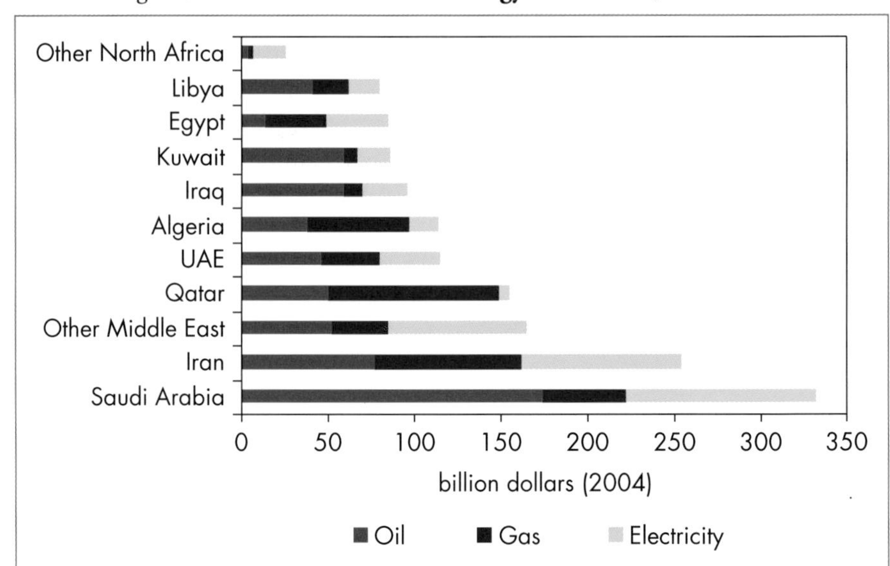

The gas-supply projections for MENA countries will entail a cumulative investment of $436 billion over 2004-2030, or $16 billion per year. Around 60% of this investment – $269 billion – will be needed upstream. The rest will go to liquefaction plants, transmission and distribution networks and storage facilities.

MENA investment in power generation, transmission and distribution is expected to total $458 billion. This is close to one-third of the region's total energy investment, or just slightly less than investment in upstream oil. If the sector is not reformed, electricity-sector investments will continue to be a burden on state budgets. Governments may find it difficult to finance these investments if the generous price subsidies provided in many parts of the region continue. MENA countries are opening their power sector to private investment, though progress has been slow in most countries and the impact of recent reforms, aiming at attracting private investment, remains unclear.

Hydrocarbon Revenues[4]

In the Reference Scenario, MENA oil and natural gas revenues – the value of oil and gas exports[5] – is projected to rise from $313 billion in 2004 to $635 billion by 2030 (in year-2004 dollars). Saudi Arabia will remain the

4. The impact on hydrocarbon export revenues of a deferral of investment in the MENA region's upstream oil and gas sector is analysed in Chapter 7.

5. Revenues are calculated using the oil and gas price assumptions outlined in Chapter 1 as a proxy for the actual value of exports.

biggest export earner in absolute terms, though its share of the region's total hydrocarbon revenues will fall while that of Iraq will increase significantly (Figure 3.5). The contribution natural gas makes to the region's overall hydrocarbon-export revenues will increase rapidly in line with the fast growing gas-export market. By 2030 natural gas will represent 13% of total hydrocarbon export revenues, compared to 5% in 2004.

3

Figure 3.5: **MENA Oil and Natural Gas Export Revenues, Share by Country**

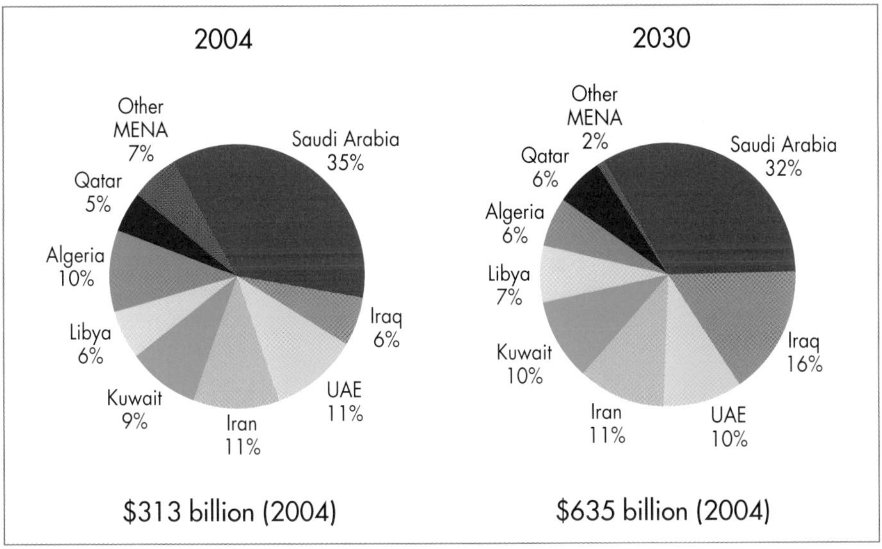

2004

Saudi Arabia 35%
Iraq 6%
UAE 11%
Iran 11%
Kuwait 9%
Libya 6%
Algeria 10%
Qatar 5%
Other MENA 7%

$313 billion (2004)

2030

Saudi Arabia 32%
Iraq 16%
UAE 10%
Iran 11%
Kuwait 10%
Libya 7%
Algeria 6%
Qatar 6%
Other MENA 2%

$635 billion (2004)

MENA oil-export revenues have been highly volatile and, in many countries in the region, this has made budgetary planning extremely complex. In real terms, revenues are currently at historically high levels and, over 2005 as a whole, are expected to exceed the annual revenues seen in the wake of the first oil shock in 1974. Nonetheless, they remain below the record levels of 1980. The Reference Scenario projections of oil production, together with the assumed trend of international oil prices, point to a rising trend in MENA oil revenues through to 2030 but, in the short term, revenues decline in line with the assumption that oil prices will fall back temporarily. We expect MENA oil revenues to rise from $296 billion in 2004 to $323 billion in 2010 and then increase steadily to $551 billion in 2030 (Figure 3.6). During the same period, gas revenues are projected to increase almost fivefold, reaching $84 billion by 2030. Unlike oil, the expected fall in the natural gas price through to 2010 will be more than offset by the increase in the volume exported. The Middle Eastern countries will take a growing share of the region's gas-export revenues, as their exports increase more rapidly than those from North Africa.

As a result of rapid population growth, on a per capita basis, MENA oil revenues are now well below levels seen between 1973 and the early 1980s. For

Figure 3.6: **MENA Oil Export Revenues**

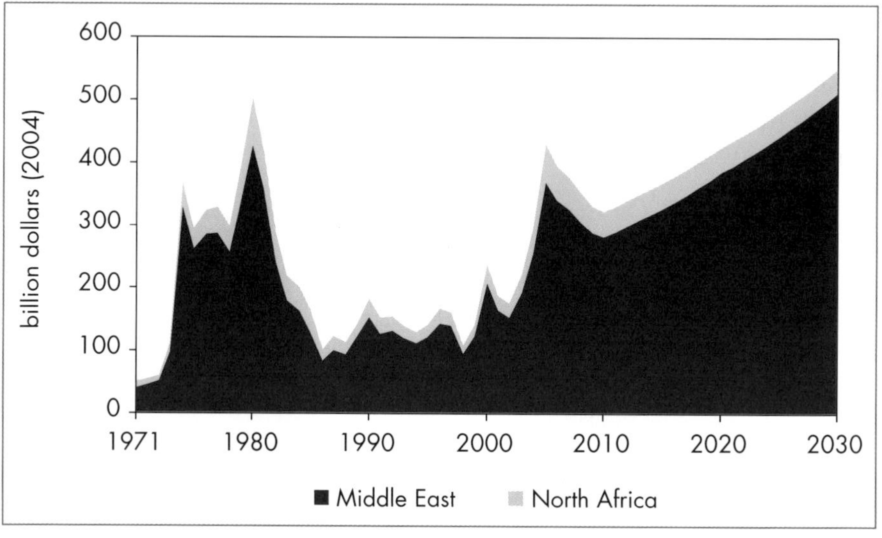

the region as a whole, export earnings in 2004 were about $900 per capita, about 32% of the peak level in 1980 (in year-2004 dollars). Obviously, there are significant differences in export revenues from country to country. Qatar has long headed the list and in 2004 its oil-export revenues approached $16 000 per capita. In the Reference Scenario, average per capita oil-export revenues for MENA as a whole reach about $1 100 by 2030 (Figure 3.7).

Figure 3.7: **MENA Oil Export Revenues per Capita**

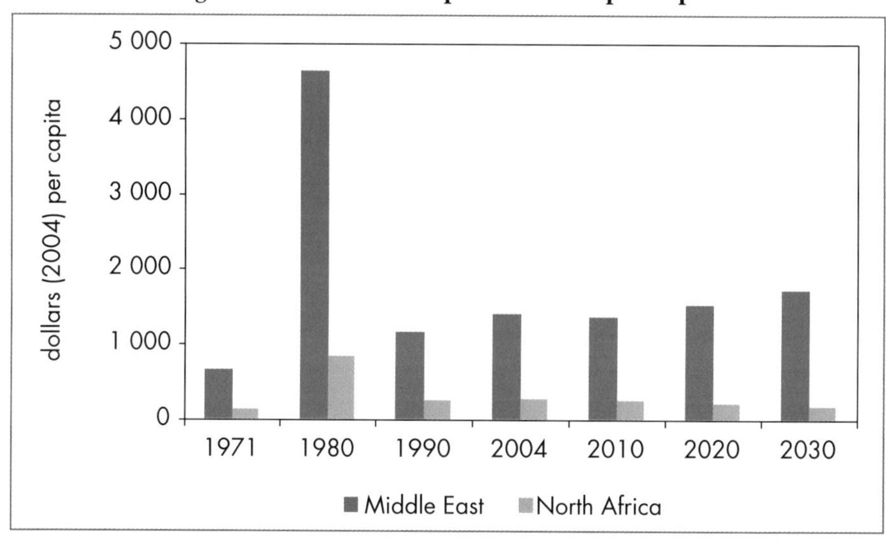

Energy-Related CO_2 Emissions

The projected trends in energy use in the Reference Scenario mean that energy-related carbon-dioxide emissions from the MENA region will increase by 2.6% per year over 2003-2030, reaching 2 795 million tonnes in 2030, just over twice current levels. By that time, the region's share of global carbon-dioxide emissions, if the Reference Scenario's projections are realised, will be 7%, up from around 6% today. Emissions per capita will remain well below those of OECD regions.

3

Table 3.3: **MENA Energy-Related CO_2 Emissions** (million tonnes)

	Middle East		North Africa		MENA	
	2003	**2030**	**2003**	**2030**	**2003**	**2030**
Power sector	383	800	98	200	481	1000
Industry	232	459	54	90	286	549
Transport	239	507	66	163	305	670
Other*	248	425	77	151	325	576
Total	**1 102**	**2 191**	**295**	**604**	**1 397**	**2 795**

*Includes other transformation, residential and services and non-energy use.

Power generation is expected to make the largest contribution to the growth in carbon-dioxide emissions, followed by the transport sector (Table 3.3). Carbon-dioxide emissions from power generation will more than double over the period 2003-2030, growing from 481 million tonnes of CO_2 to 1 000 million tonnes. The share of the power sector in the region's total CO_2 emissions will increase from 34% to 36%.

CHAPTER 4

OIL OUTLOOK IN THE MIDDLE EAST AND NORTH AFRICA

HIGHLIGHTS

- MENA primary oil demand will increase from 6.7 mb/d in 2004 to 11.8 mb/d in 2030, an average annual rate of growth of 2.2% per year. Oil intensity – the amount of oil used per dollar of gross domestic product – is now high in MENA compared to the rest of the world, but will decline through to 2030. Transport will be the fastest growing oil-consuming sector.
- MENA has the largest proven oil reserves of any world region. Saudi Arabia, with 262 billion barrels, has the most, equivalent to 69 years of production at current rates. A large share of the region's oil reserves is in giant and super-giant fields. Their contribution to total MENA production will drop sharply, from 75% today to 40% in 2030, as mature giant fields decline and new developments focus more on smaller fields.
- In the Reference Scenario, MENA oil production is projected to rise from 29 mb/d in 2004 to 33 mb/d in 2010 and to 50 mb/d in 2030, on the assumption that the countries of the region are willing to develop new production capacities despite some fall-back from current high oil prices. In the longer term, MENA oil production increases more rapidly than in other regions because resources are larger and production costs lower.
- MENA's share of global oil production is projected to rise from 35% in 2004 to 44% in 2030. In the Deferred Investment Scenario, MENA production grows much less quickly, reaching only 35 mb/d in 2030.
- MENA refining capacity is projected to expand rapidly, from 8.8 mb/d today to 10.9 mb/d in 2010 and 16.0 mb/d in 2030. Middle Eastern countries will account for most of the additions to MENA capacity. Large investments in new distillation and upgrading capacity are planned in several countries, notably Saudi Arabia and Kuwait.
- MENA is set to play an increasing role in international oil trade over the *Outlook* period. Net exports from MENA countries to other regions will jump from 22 mb/d in 2004 to 25 mb/d in 2010 and 39 mb/d in 2030. Most exports will still be as crude oil in 2030, but refined products will account for a growing share.
- About $614 billion of MENA oil investment, or $23 billion per year, will be needed between 2004 and 2030. The upstream industry will absorb more than three-quarters of this spending. National oil companies will continue to invest the most, but investment opportunities for international companies are expected to grow.

Oil Demand

Primary oil consumption in the MENA region will increase from 6.7 mb/d in 2004 to 11.8 mb/d in 2030, an average annual rate of growth of 2.2% per year (2.2% in the Middle East and 2.4% in North Africa). Demand will nonetheless rise much more slowly than in the past three decades, when it grew by 5.7% per year. Less rapid population growth and a continuing shift in energy use towards natural gas will contribute to this slower demand growth.

Domestic oil demand will also grow more slowly than GDP over the *Outlook* period – the opposite of what happened in 1971-2003 (Figure 4.1). Oil intensity, expressed as the amount of primary oil consumed per unit of GDP, is now high in MENA compared to the rest of the world. But it will decline over the projection period, reducing the gap. As the region develops, it will become generally less energy-intensive as the services, which consume relatively little energy, increase their share in the economy. Increasing use of natural gas – especially in power generation, water desalination and manufacturing – and electricity will also drive down oil intensity. Nonetheless, oil intensity in MENA countries will still be two-and-a-half times higher than in other developing countries in 2030 and more than twice higher than in OECD countries (Figure 4.2).

Figure 4.1: **MENA GDP and Oil Demand Average Annual Growth Rate**

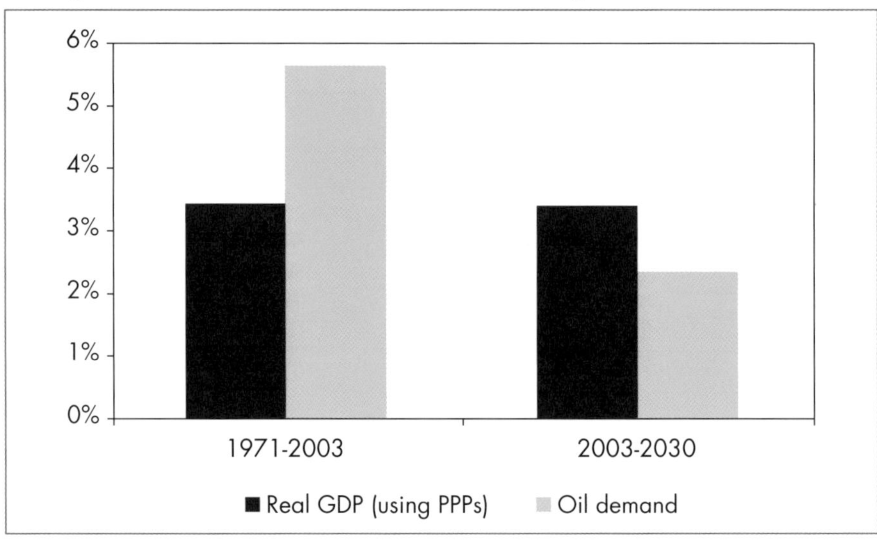

The biggest increases in domestic oil demand, in volume terms, will come from Saudi Arabia and Iran, which are already the largest consumers. Between them, they will account for about half of total MENA demand in 2030, roughly the same share as today. The fastest rate of demand growth is projected in Qatar (Table 4.1).

Figure 4.2: **Primary Oil Intensity**

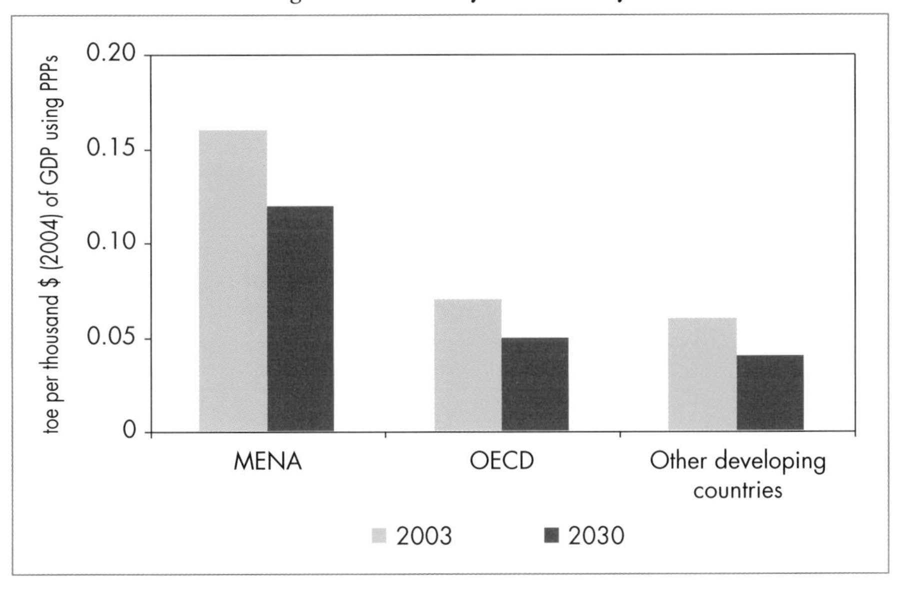

4

Table 4.1: **MENA Oil Demand** (mb/d)

	2004	2010	2020	2030	2004-2030*
Middle East	**5.4**	**6.5**	**8.1**	**9.4**	**2.2%**
Iran	1.4	1.6	2.0	2.3	1.8%
Iraq	0.6	0.6	0.8	0.9	1.9%
Kuwait	0.3	0.4	0.5	0.5	2.3%
Qatar	0.1	0.2	0.2	0.3	4.3%
Saudi Arabia	2.0	2.6	3.3	3.7	2.4%
United Arab Emirates	0.2	0.2	0.3	0.3	1.6%
Other Middle East	0.7	0.9	1.0	1.2	2.0%
North Africa	**1.3**	**1.5**	**2.0**	**2.4**	**2.4%**
Algeria	0.2	0.3	0.4	0.5	2.6%
Egypt	0.5	0.6	0.8	1.0	2.3%
Libya	0.3	0.3	0.4	0.5	2.6%
Other North Africa	0.3	0.3	0.4	0.5	2.3%
Total MENA	**6.7**	**8.0**	**10.0**	**11.8**	**2.2%**
World	**82.1**	**92.5**	**104.9**	**115.4**	**1.3%**

* Average annual growth rate.

The rate of growth in MENA oil consumption will be lower than in most other developing regions, but higher than in the OECD. In all sectors, the share of oil in MENA countries is at present higher than in most other regions. Oil dominates because it is an abundant resource and is often heavily subsidised.

The share of oil in MENA energy use will decline in all sectors except transport for two main reasons. First, the use of natural gas, which is in many countries widely available at a lower economic cost than oil, will grow rapidly, substituting for oil. Most of the increase in primary energy demand will come from the power sector and most of that increase will be met by gas. The share of gas in final energy use will also grow, especially in industry. Second, there will be a strong incentive in some countries to switch away from oil in order to increase oil exports – especially where there is limited potential to expand oil production and where domestic prices are heavily subsidised.

As in other regions, MENA demand for automotive fuels will continue to grow briskly. The passenger-car fleet in the MENA region grew by almost 5% per year in the past 20 years, while GDP grew by only 2.6% per year and GDP per capita grew by 0.2% per year. Car ownership is more strongly correlated with population than with GDP, but is still low by developed country standards. There are less than 100 cars for every thousand people in the MENA region (Figure 4.3) compared with 780 in North America and 500 in Europe. With population set to rise rapidly and large numbers of young people reaching driving age, the region's car fleet and, therefore, transport demand will expand steadily. The share of transportation in total primary oil consumption in MENA is projected to grow from an average of 36% in 2003 to 42% in 2030.

Figure 4.3: **Car Ownership Rates, 2002**

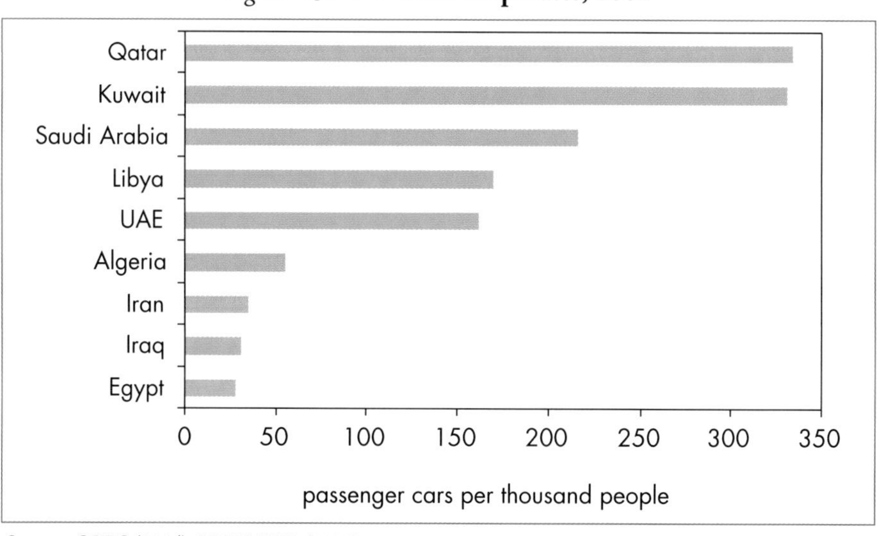

Sources: OPEC (2004); UNESCWA (2005).

Oil use in transportation in MENA is boosted by the dominant role roads play in transport generally throughout the region. Freight is transported almost entirely by road, which uses solely gasoline and diesel. The few railway systems that exist in the region use diesel. Air transport, which also relies exclusively on oil products, is also growing very quickly. Both freight and passenger traffic have doubled in the last 10 years. Steady economic growth and rising household incomes will boost demand for air travel over the *Outlook* period.

Oil Resources

Estimates of Proven Reserves[1]

Estimates of proven oil reserves vary somewhat depending on the source. According to the *Oil and Gas Journal*, remaining proven reserves in MENA countries totalled 784 billion barrels. According to all the leading sources of oil reserves data, the bulk of the world's oil is to be found in the MENA region, though the exact share quoted varies, according to definition, from 46% based on IHS Energy estimates[2] to 65% based on *World Oil* data. Where a common definition is used, most of the variations arise in figures for Middle Eastern countries. The share of North African countries is relatively constant across sources at around 4-5%.

All sources also agree that Saudi Arabia has the world's largest remaining proven oil reserves. The *Oil and Gas Journal* puts them at 262 billion barrels, while the BP figure is 1 billion barrels higher. They are twice the size of Iran's reserves, which are the third-largest in the world. Iraq and Kuwait both have reserves in excess of 100 billion barrels. In the world ranking of reserves (using *Oil and Gas Journal* data, which include all proven non-conventional reserves in Canada and Venezuela), Middle Eastern countries hold four of the first five places. Of the 19 countries with reserves of more than 5 billion barrels, six are in the Middle East and two are in North Africa (Figure 4.4). There are doubts about the reliability of official MENA reserves estimates, which have not been audited by independent auditors. For this reason, the country-by-country analysis of MENA oil-production trends in this *Outlook* is based on information drawn from several sources, including oil service companies, national and international oil companies, government bodies, consulting firms and other experts.

1. See *World Energy Outlook 2004* for a detailed discussion of oil and gas reserve issues.
2. The IHS Energy estimates related to "proven and probable technically recoverable resources". They include non-conventional oil very selectively: developed oil sands in Canada and certain developed extra-heavy reserves in the Orinoco Belt in Venezuela.

Box 4.1: **Definition of Oil in the *World Energy Outlook***

Oil is defined in the *World Energy Outlook* to include all liquid hydrocarbon fuels. Sources include conventional and non-conventional oil reserves, natural gas liquids (NGLs) and condensates, and refinery-processing gains. Oil is considered conventional if it is produced from underground reservoirs by means of wells. Oil is classified as non-conventional if it is produced in other ways, or requires additional processing to produce synthetic crude. Non-conventional oil includes shale oil, synthetic crude and products derived from oil or tar sands and extra-heavy oil, coal- and biomass-based liquids and the output of natural gas-to-liquids (GTL) plants.

The figures shown in this *Outlook* for resources and reserves include conventional oil and, depending on the source, may include some categories of non-conventional oil. Most non-conventional oil reserves are found in Canada and Venezuela; they are of minor importance in MENA countries. Oil-production figures include crude oil, NGLs and condensates, unless mentioned otherwise. Data for oil products supply and demand include output from refineries, natural gas processing plants, GTL plants and liquids derived from coal and biomass.

Figure 4.4: **Leading Countries for Proven Oil Reserves**

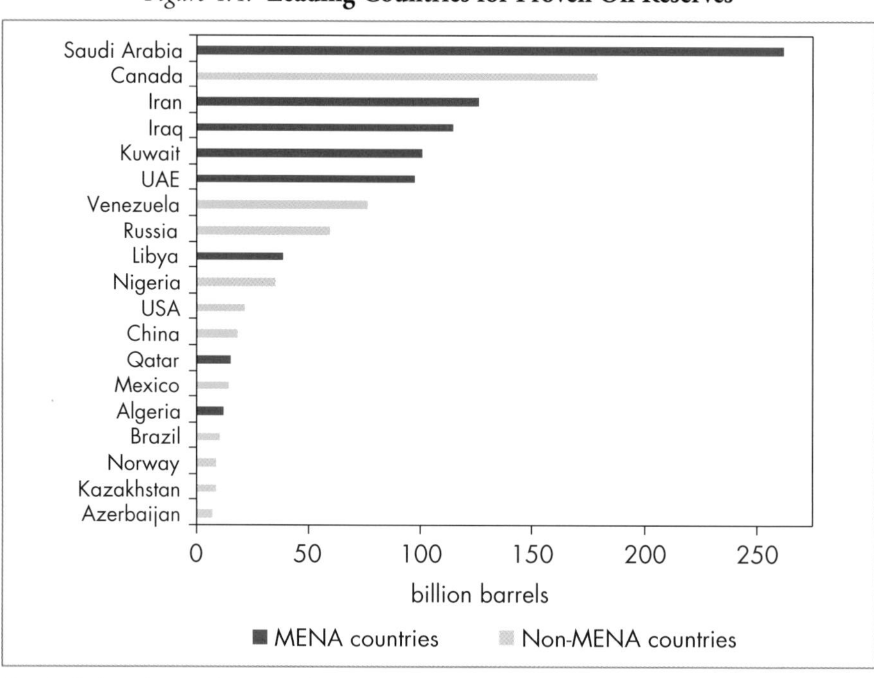

Note: Includes NGLs, condensates and non-conventional oil.
Source: *Oil and Gas Journal* (20 December 2004).

MENA proven oil reserves increased sharply in the 1980s and, after a period during which they hardly increased, rose further around the turn of the century. From around 400 billion barrels at the start of the 1980s, reserves ballooned to almost 700 billion barrels by 1989 and reached nearly 800 billion barrels at the end of 2004 (Figure 4.5). Most of these increases occurred in the Middle East. In the second half of the 1980s, Saudi Arabia and Kuwait revised their reserves upwards by about one-half. The United Arab Emirates and Iraq also recorded large upward revisions at that time. Total Middle East reserves jumped from 398 billion barrels in 1985 to 663 billion barrels in 1990. As a result, world oil reserves increased by more than 40%.

Figure 4.5: **MENA Proven Oil Reserves, 1984-2004**

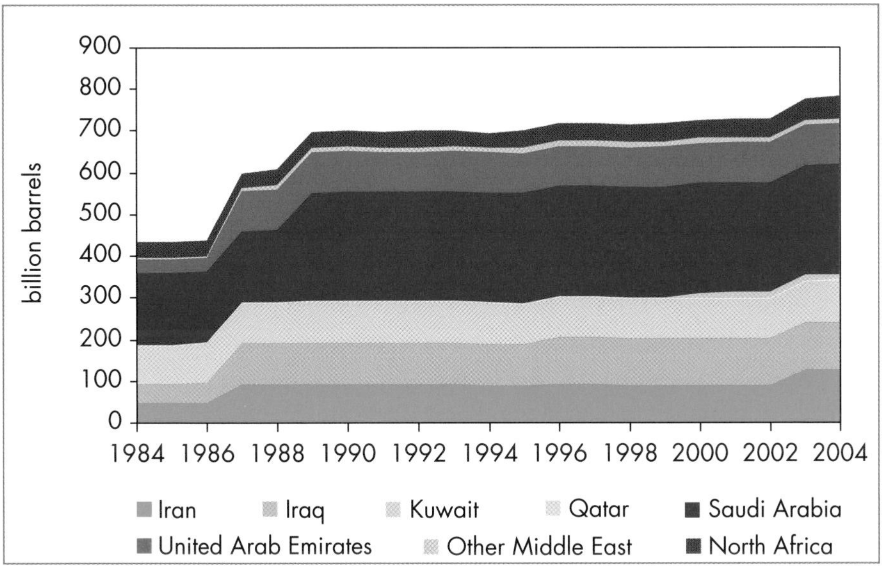

Source: *Oil and Gas Journal* (20 December 2004).

This dramatic and sudden revision in MENA reserves has been much debated. It reflected partly the shift in ownership of reserves away from international oil companies, some of which were obliged to report reserves under strict US Securities and Exchange Commission rules. The revision was also prompted by discussions among OPEC countries over setting production quotas based, at least partly, on reserves. What is clear is that the revisions in official data had little to do with the actual discovery of new reserves. Total reserves in many MENA countries hardly changed in the 1990s. Official reserves in Kuwait, for example, were unchanged at 96.5 billion barrels (including its share of the Neutral Zone) from 1991 to 2002, even though the country produced more than 8 billion barrels and did not make any important new discoveries during that period. The case of Saudi Arabia is even more striking, with proven reserves estimated at between 258 and 262 billion barrels in the past 15 years, a variation of less than 2%. A substantial rise in oil prices would lead to higher reserves estimates, as more oil reserves become economically recoverable.

Resources and Production Potential

As is the case for proven reserves, MENA also holds the bulk of the world's recoverable resources of conventional oil. The US Geological Survey estimates that worldwide "ultimately recoverable resources" of conventional oil and NGL total 3 345 billion barrels. Ultimately recoverable resources include cumulative production to date, identified remaining reserves, undiscovered recoverable resources and estimates of "reserves growth" in existing fields. Undiscovered resources in MENA are estimated to be 313 billion barrels (mean value), indicating with a high degree of confidence that the region has a particularly large production potential (Table 4.2). Reserves growth – increases in the estimated size of reserves in oilfields as they are developed and produced – accounts for around 9% of the region's estimated remaining ultimately recoverable resources, whereas remaining reserves and undiscovered resources account for about 65% and 26% respectively. As is the case with proven reserves, most of the other categories of ultimately recoverable resources of conventional oil are concentrated in the Persian Gulf countries, especially Saudi Arabia, Iran and Iraq.

It is necessary to emphasise, again, the enormous uncertainties surrounding estimates of resources in all regions. There are inconsistencies in the way they are defined and measured. A lack of independent auditing makes it impossible to

Table 4.2: **Estimates of Ultimately Recoverable Oil and NGL Resources**
(billion barrels)

	MENA	Rest of world	Total
Undiscovered	313	570	883
Reserves growth	109	199	308
Remaining reserves	784	322	1 106
Cumulative production	334	714	1 048
Total ultimately recoverable resources	1 541	1 804	3 345
Remaining ultimately recoverable resources	**1 206**	**1 090**	**2 297**

Sources: Undiscovered resources and total ultimately recoverable reserves – USGS (2000); remaining reserves – *Oil and Gas Journal* (20 December 2004); cumulative production – IHS Energy database; reserves growth – IEA analysis based on USGS.

Figure 4.6: **Oil Reserves and Production in MENA Countries, 2004**

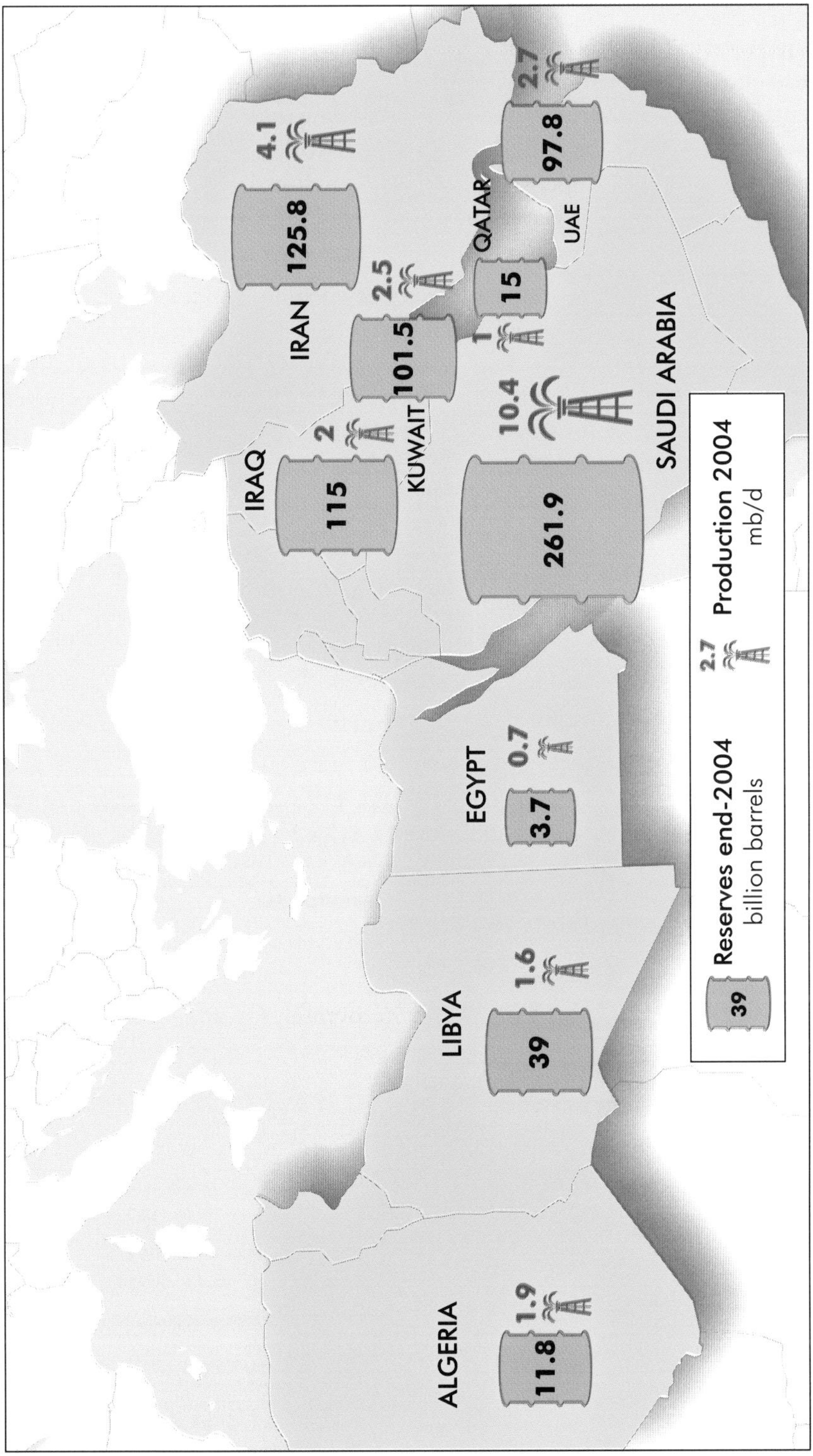

verify the data, even on reported proven reserves in many countries. The uncertainties about other categories of resources are even larger. Few estimates exist and, as indicated by USGS assessments, the estimated amount of oil which remains to be discovered varies widely according to the degree of probability that all of that oil will be found. The IEA is working with the UN Economic Commission for Europe, OPEC, the International Energy Forum Secretariat (IEFS), the Society of Petroleum Engineers (SPE) and other organisations to improve the definition and classification of energy reserves and resources.

A widely used indicator of the adequacy of remaining oil reserves is the reserves-to-production (R/P) ratio, or proven reserves divided by annual production. The R/P ratios of six countries of the MENA region – Iraq, Kuwait, the UAE, Iran, Saudi Arabia and Libya – are above 50 years (Figure 4.7). These countries clearly have little incentive for now to explore for oil in order to expand their reserve base. In three countries – Saudi Arabia, Iran and Iraq – undiscovered resources and reserves growth combined are equal to more than 30 years of current production: 36 years for Saudi Arabia, 44 for Iran and 70 for Iraq. This highlights the tremendous potential for adding to proven reserves in these countries, which would reinforce their dominant position as major oil producers. Other countries, including Kuwait and the UAE, have relatively small amounts of undiscovered resources compared with current proven reserves.

Figure 4.7: **Ratio of Ultimately Recoverable Oil and NGL Resources and Proven Reserves to Production Ratios in MENA Countries, End-2004**

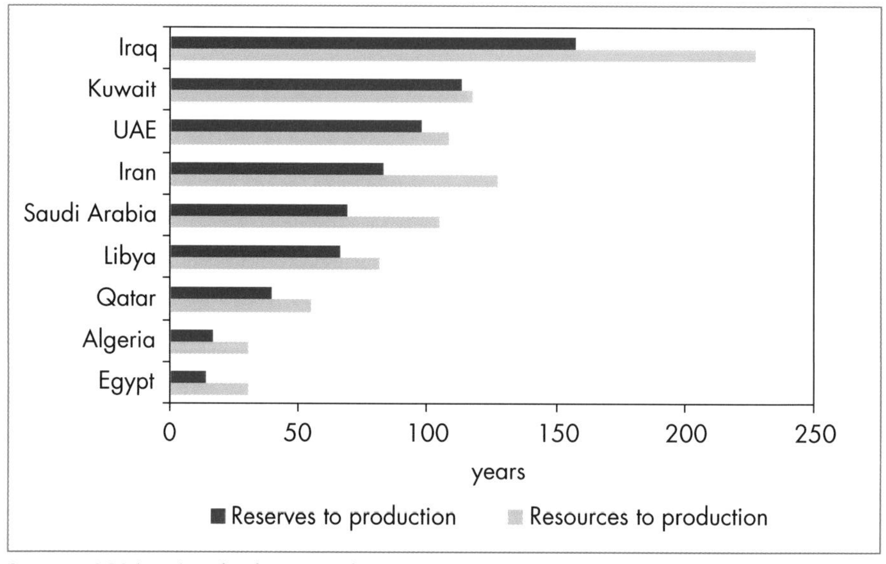

Sources: USGS (2000), *Oil and Gas Journal* (20 December 2004), IEA database and analysis.

Technology can have a big impact on the share of initial oil in place that can be recovered economically (the recovery factor) and, therefore, added to proven reserves. Each percentage point increase in the recovery factor typically leads to an increase of two to three percentage points in the recoverable reserve base. Put another way, a five percentage point increase in the global recovery rate would add more oil than all of Saudi Arabia's proven reserves. In the MENA region, average recovery factors, at between 30% and 35%, are generally lower than in the rest of the world. There is a large potential for improving these factors by introducing more advanced technology and modern production practices, especially in Iraq. An increase of one percentage point in the average MENA recovery rate would add 34 billion barrels – equal to more than the amount of oil consumed worldwide in 2004. However, geological factors make it hard to predict how much oil can eventually be recovered (see below).

Petroleum Geology and Reservoir Characteristics

The MENA region contains a relatively large proportion of the world's oil resources because of the region's favourable petroleum geology. The Arabian tectonic plate – comprising the Arabian Peninsula, Iran and Iraq – was a shallow sea for long geological periods, as the Tethys Ocean was slowly being closed by the tectonic movement of the African continent towards Eurasia.[3] As a result, the region has an abundance of source rocks dating back to various geological periods, notably the Silurian, Jurassic, Cretaceous and Tertiary periods (Box 4.2 briefly outlines the basic principles of petroleum geology). With the migration of the sea onto the Arabian plate, large amounts of reservoir rock, including a variety of carbonate sediments, were laid down over large areas, mainly during the Permian, Jurassic and Cretaceous periods. The subsequent formation of extensive cap rocks, which prevented hydrocarbons from leaking to the surface, enabled large volumes of oil and gas to accumulate in these reservoirs, giving rise to the hydrocarbon wealth of the Middle East today.

The geological history of the North African countries is a little different. They are situated at the northern edge of a "craton" – a continental plate that has been preserved over long periods of geological time. Several cycles of transgression and regression – movements in and out of the sea, as sea levels fluctuated or the land tilted over very long periods – created favourable conditions for sedimentary deposits to accumulate and hydrocarbon reservoirs to develop.

3. Today, the Arabian plate is still moving towards Asia, closing the Persian Gulf (and opening the Red Sea). The earlier collision of the two plates is what created the Zagros fold in Iran, a major petroleum region.

Box 4.2: A Guide to Basic Petroleum Geology

Hydrocarbons deposits result from the burial and transformation of biomass over periods of hundreds of millions of years. Fine sediments rich in organic matter are deposited, either at the bottom of the sea or on land, usually in lakes. As they are buried under additional sediments, they may form organic-rich shales. When these are buried deep enough, the organic matter they contain is slowly transformed into oil or gas, depending on the conditions of temperature and pressure.

Under the prevailing pressure of tectonic forces, some of these hydrocarbons get expelled from the "source rock" (typically shale) in which they were formed. Because they are less dense than the water that is also trapped within sedimentary rocks, they begin to migrate out of the source rock, typically vertically upwards, but also often laterally over distances of up to hundreds of kilometres. How they migrate and how far depends on the pathways available to them, such as permeable rocks, fissures, fractures and faults. The hydrocarbons continue to migrate upwards until they are trapped by an impermeable layer, "a cap rock". They then accumulate in the pores of a porous "reservoir rock" under the cap. If they do not meet a suitable cap rock, they eventually flow, or seep, all the way to the surface, where they are dissipated through evaporation or bacterial degradation. Sometimes, degradation by bacterial activity near the surface turns the oil into solid bitumen that stops flowing.

Extensive source rocks are typically found in areas that have at some point been under the sea. Reservoir rocks are generated in a variety of sedimentary settings such as shallow seas, river deltas and sand dunes.

As a result of these different geological developments, the MENA region has many different types of reservoirs. The region has no unique geological characteristic. Most of the region can be characterised as relatively immature from the point of view of the exploitation of its hydrocarbon resources:

- Although the extraction of hydrocarbon on an industrial scale started close to a century ago, a smaller proportion of the region's resources has been produced than in other major producing regions, such as North America, Indonesia or the North Sea.
- Today, production comes predominately from a small number of giant fields. The entire MENA region currently produces 29 mb/d from around 500 fields. By comparison, the United States produces about 7.6 mb/d from more than 60 000 fields.

- There are 14 652 producing wells in MENA, compared with 524 556 in the United States. In Saudi Arabia fewer than 9 000 wells have ever been drilled. The figure for the United States is about 4 million.
- MENA production has concentrated on the largest fields which are technically much easier and cheaper to develop. For example, the Ghawar field in Saudi Arabia – the world's largest – produces over 5 mb/d from less than 1 000 wells. By comparison, the Samotlor field in western Siberia, which produced as much as 3.5 mb/d in the 1980s, has 15 000 wells. Over time, development of the region's resources will shift to smaller fields as the giant and super-giant fields are depleted and their production costs rise.
- Several large fields, particularly in Kuwait, are still at the primary production stage. Secondary recovery, through the injection of water or gas to maintain reservoir pressure is, in some MENA countries, relatively recent or not used at all.
- The region is far from fully explored. Large parts of Iraq, Iran, Libya, Algeria and Saudi Arabia – particularly the Rub Al-Khali region (Empty Quarter) – are still underexplored.

Another important feature of the region is the relative abundance of carbonate reservoir rocks. This is in part due to geological history, with the formation of extensive carbonate platforms in shallow seas in the Arabian Peninsula and the Zagros fold region. Several of the giant fields now in production in Iraq, Iran and Saudi Arabia are in carbonates. This is important because production performance is much more difficult to predict for carbonate reservoirs than for the more common, siliclastic reservoirs. Carbonate rocks are more heterogeneous, with small features, such as fractures, fissures, and pronounced variations in permeability, that are difficult to detect using seismic or other measurement techniques. These features can radically affect the movement of liquids in the reservoir during production. In addition, carbonates tend to be "oil wet", whereby the oil tends to stick to the rock better than water – the opposite of "water wet" silicate rocks. This characteristic can reduce the recovery of oil from water injection. Consequently, the share of the oil that can be recovered from carbonate reservoirs can be smaller than from silicate rocks.

Reservoir heterogeneities are a particular problem in giant fields exploited with a relatively small number of wells. Indeed, wells are not just conduits for producing hydrocarbons or injecting water or gas; they also facilitate obtaining well-calibrated measurement of the reservoir properties. The small number of wells, or rather large spacing between wells, limits the amount of information gathered and increases the chance that small features, that could affect recoverability, are not detected. That is why some analysts have questioned the ability to maintain production at the Ghawar field.

Reserves Growth and Oil Discoveries

Additions to reserves come from discoveries of new oilfields and from "reserves growth" – increases in the total technically and economically recoverable petroleum reserves of a field in production or undergoing appraisal.[4] In the last few decades, an increasing share of reserve additions in MENA countries has come from reserves growth. The appraisal of known fields is increasingly carried out through seismic surveys and reservoir simulation, in combination with drilling wells.

Worldwide, the rate of reserve additions from discoveries has fallen sharply since the 1960s. In the last decade, discoveries have replaced only half the oil produced. Nowhere has the fall in oil discoveries been more dramatic than in the Middle East, where they plunged from 187 billion barrels in 1963-1972 to 16 billion barrels during the decade ending in 2002. Both the number of new discoveries and the size of fields discovered in MENA countries (measured by oil in place) have fallen sharply since the 1960s (Figure 4.8). This is particularly marked after 1980, when very few discoveries were made. This is partly because little exploration was carried out compared to other prospective regions, as reserves in most countries were already large and spare production capacity was growing. Exploration has been stepped up in recent years, stimulated by high oil prices, leading to a number of new discoveries, including Azadegan and Kushk in Iran. The number of wells drilled in the region increased by an estimated 7.5% in 2004. The number of drilling rigs in operation in the Middle East reportedly reached 248 in August 2005 – the most since 1988 and 100 more than the average for the 1990s.

The success rate for exploratory wells – the proportion of wells drilled that yield oil or gas – has increased sharply in recent years in MENA, in line with global trends. New techniques and technology have boosted the average success rate from less than 40% (23% worldwide) at the start of the 1960s to about 65% (over 40% worldwide) today (Alazard and Mathieu, 2005). The pace of improvement in success rates, however, varies considerably among MENA countries: that of Kuwait has been stable in recent years at about 40%. Iran has seen their success rates rise to around 60-65% on average. Technological advances have also reduced the cost of drilling wells. The more widespread application of advanced oilfield technology is expected to lead to a substantial increase in MENA proven reserves over the projection period.

4. Such revisions can occur for a variety of reasons. These include improved knowledge about the geological and operational characteristics of the field, the discovery of new reservoirs or pools and a higher recovery rate through the use of new, improved or enhanced recovery methods.

Figure 4.8: **MENA Oilfield Discoveries, 1960-2002**

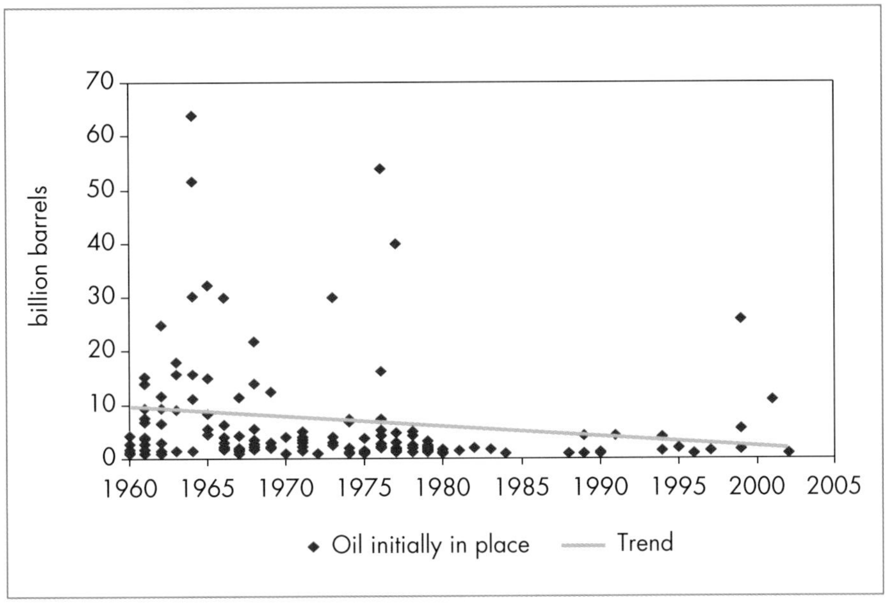

Note: All fields containing at least 1 billion barrels of oil in place.
Sources: IHS Energy database; IEA analysis.

Production Costs

Oil reserves in MENA countries are among the cheapest to find, develop and produce in the world. Total production costs – including exploration, development and lifting or running costs (including fuel to run pumps, chemicals to treat wells or effluents, equipment maintenance and staffing of production and processing facilities) – typically average between \$3 and \$5 per barrel of oil produced. This compares with over \$15 in the North Sea and \$12 in the Gulf of Mexico (Figure 4.9). In MENA countries, exploration and development accounts for about half the total cost of production, and lifting for the other half. Saudi Arabia, Kuwait and Iraq have the lowest costs in MENA and in the world.

The capital costs per barrel per day of new onshore production capacity in MENA countries (covering exploration and development capital expenditure) range from around \$4 000 in Iraq to \$12 000 in Qatar. This compares with more than \$10 000 worldwide and nearly \$16 000 in the OECD. Costs in MENA are generally lowest in the countries with the highest proven reserves (Figure 4.10). Saudi Arabia has among the lowest development costs and by far

Figure 4.9: **Indicative Crude Oil Production Costs in Selected MENA Countries and non-MENA Regions**

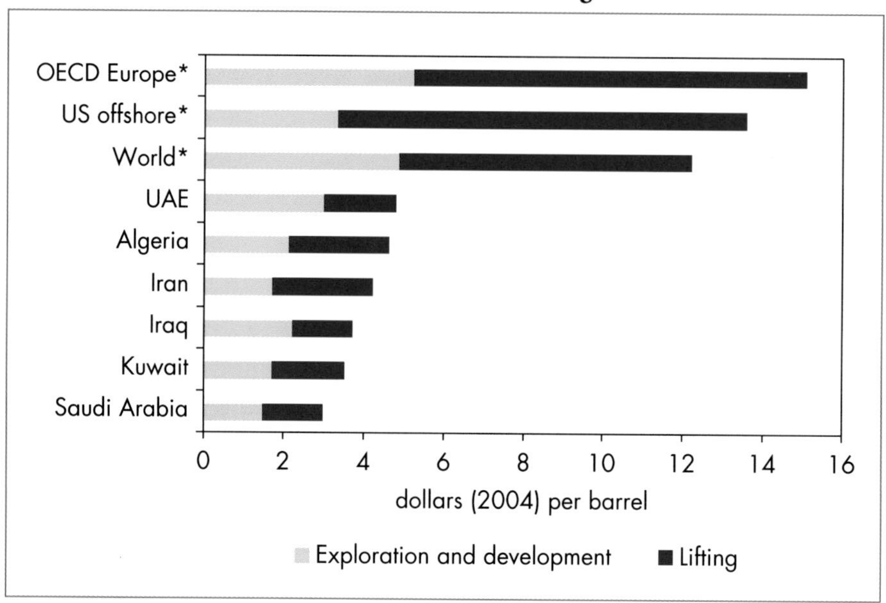

* Financial Reporting Survey companies only.
Source: EIA (2003); IEA analysis.

the largest reserves. A group of four countries – Iran, Iraq,[5] Kuwait and the UAE – have large reserves and relatively low costs. Another group of countries, comprising Libya, Algeria, Qatar and Egypt, have much lower reserves and slightly higher costs.

There are several reasons why production costs are generally low in MENA:

- *The current dominance of giant fields:* Finding and development costs are clearly much less for one giant field than for several smaller fields holding the same volume of hydrocarbons. For a given investment in wells and facilities, production can be maintained longer in a giant field than in a set of smaller ones, so the costs per barrel are reduced.
- *Geographical characteristics:* Most of the fields in MENA are onshore or shallow offshore and are therefore much cheaper to develop than deep offshore fields. The Persian Gulf has a maximum water depth of about

5 . The very low cost of exploration and development in Iraq reflects the potential for rehabilitating existing fields. This does not take account of the potentially large additional costs associated with ensuring security and the risk that poor reservoir management in recent years may have severely undermined production potential and raised the cost of expanding capacity. The cost of developing new fields is likely to be closer to that in Iran.

Figure 4.10: **Average Exploration and Development Costs versus Proven Oil Reserves in MENA Countries**

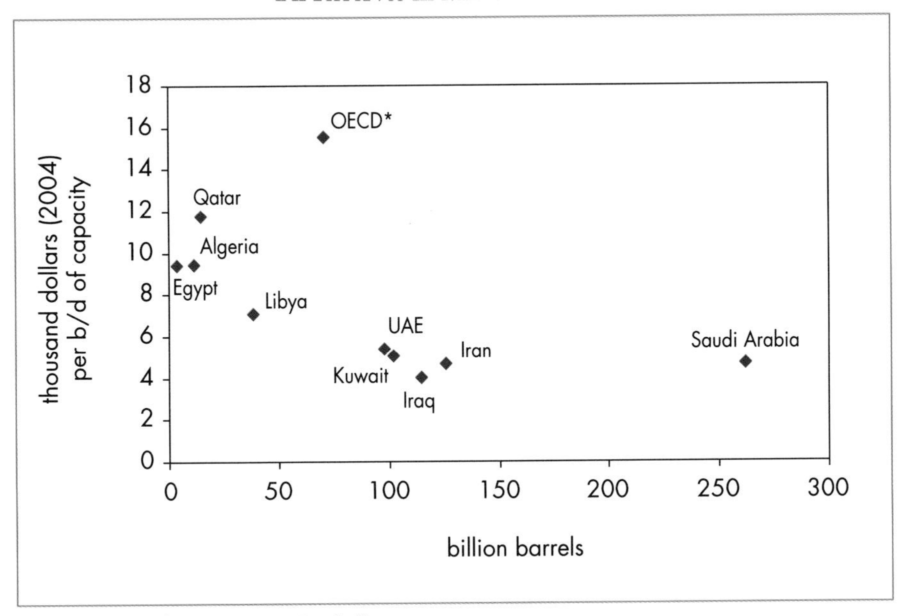

* Excluding Canadian non conventional oil reserves.
Sources: Oil and Gas Journal (20 December 2004); IEA analysis.

300 metres, enabling the use of jack-up offshore platforms, which are much less expensive than floating facilities. Even in the few deep-water projects, such as in Egyptian Mediterranean waters, the sea conditions are much more favourable than, for example, in the North Sea. The clustering of fields – such as those along the Persian Gulf coast – makes access for heavy equipment and transportation to export terminals easier and cheaper. In addition, low population density in most producing areas makes the installation of wells and surface facilities easier.

- *Favourable geology:* Reservoir depths are usually between 1 500 and 4 000 metres throughout most of the region. Most of the currently exploited reservoirs are in the shallower part of that range. There has been little exploration drilling beyond 4 000 metres. Depth has a very large impact on drilling costs, which increases at least fourfold when depth is doubled. Oilfields in the MENA region generally have favourable porosity and permeability, which allows relatively high oil-flow rates with few wells.

MENA's cost advantage over other producing regions is nonetheless expected to become less marked in the future. As the giant fields that dominate MENA production decline, they will need to be replaced with smaller, deeper accumulations, farther from major logistic hubs. Recent exploration activity in

Saudi Arabia, for example, is focusing on deeper targets or more frontier areas, such as the Rub Al-Khali region. MENA countries will seek to use technology to offset these trends and keep costs down.

Technology Deployment

MENA countries can be categorised in three ways with respect to their sources of technology to find, develop and produce oil:

- Some countries – including the UAE, Qatar, Egypt, Oman and, to some extent, Algeria – have traditionally relied on partnerships with international oil companies. This has ensured their access to the world's most sophisticated petroleum technology.
- A few countries, such as Iran, Iraq and Libya, have seen their access to international technology restricted, either because of trade sanctions or armed conflicts. As a result, they currently lag behind in the use of the latest techniques.
- A third group of countries – including Saudi Arabia and Kuwait – have chosen full national control over exploitation of their resources. Yet they have still been able to make use of the latest technologies, by buying assistance from international oil and oil-service companies. Saudi Aramco has also invested heavily in research and development.

The technological sophistication of each MENA country is not easy to assess quantitatively. There is no published data on the total amount spent on different types of technology in each country. The technological needs of the upstream oil and gas industry in MENA countries are generally similar to those in other parts of the world, but technological challenges of particular relevance to MENA include the following:

- *Optimising long-term recovery in giant fields:* Private companies generally strive to recover oil as quickly as possible to maximise net present value (NPV) of the assets. This can reduce the recovery factor. By contrast, national oil companies, the owners of all the giant reservoirs in MENA, are often more interested in maximising and prolonging recovery, even if this diminishes net present value. This approach calls for sophisticated reservoir management technologies, such as those employed by Saudi Aramco. Water management will become increasingly important, as the need for water flooding to maintain reservoir pressure grows in many fields in the region.[6]
- *Imaging between wells:* Production from giant fields is commonly characterised by large well spacing, which reduces information about heterogeneities in the reservoirs. This can lead to uncertainty about ultimate

6. See Box 16.2 in Chapter 16 for a discussion of water-management issues.

recovery. Improved imaging between wells is, accordingly, a key technological challenge for the MENA oil industry.

- *Carbonates:* Improving understanding of the characteristics of carbonate reservoirs, notably their heterogeneities and "wettability", is vitally important in the Middle East, given the prevalence of such reservoirs.
- *Improved seismic surveys in deserts:* Seismic surveys are one of the most important tools in exploring for new reservoirs and in optimising production from existing reservoirs. They are most economical offshore, where the acoustic sources and sensors can be dragged behind boats. On land they are generally less efficient and more costly. Surveying can be carried out almost as efficiently in desert environments as in offshore locations, but near-surface effects – for example, those caused by shifting sand dunes – tend to undermine data quality. This is a particular problem for deeper horizons, such as the Khuff formation under the Ghawar field in Saudi Arabia.
- *Developing small fields:* In order to contain costs, it will be important to adapt technology for use in deeper, less accessible, smaller fields.
- *Environmental protection:* The rapid development of MENA's hydrocarbon resources, as projected in our Reference Scenario, will raise serious concerns about its impact on the environment. Key issues include the treatment and disposal of effluents, the "footprint" left by the construction of large-scale industrial facilities and gas flaring (Mahroos, 2005).

4

Oil Production

Crude Oil[7]

In the Reference Scenario, MENA oil production is projected to increase from 29 mb/d in 2004 to 33 mb/d in 2010 and to 50 mb/d in 2030 (Figure 4.11).[8] These projections assume that there is no major supply crunch, that current high oil prices will decline during the current decade, then rise steadily to 2030, and that countries of the region are willing to develop new production capacities to meet the projected call on MENA oil at the prices assumed. The projected increases in production are roughly commensurate with their reserves.

The projections to 2010 take account of current and planned upstream development projects. Production is projected to rise by 4 mb/d between 2004 and 2010. As domestic demand is projected to grow by only 1.4 mb/d, net

7. Unless otherwise stated, all the oil-production figures cited in this section include crude oil, natural gas liquids and condensates.

8. MENA oil production is significantly lower in the Deferred Investment Scenario (Chapter 7) and the World Alternative Policy Scenario (Chapter 8).

Figure 4.11: **MENA Crude Oil Production by Country in the Reference Scenario**

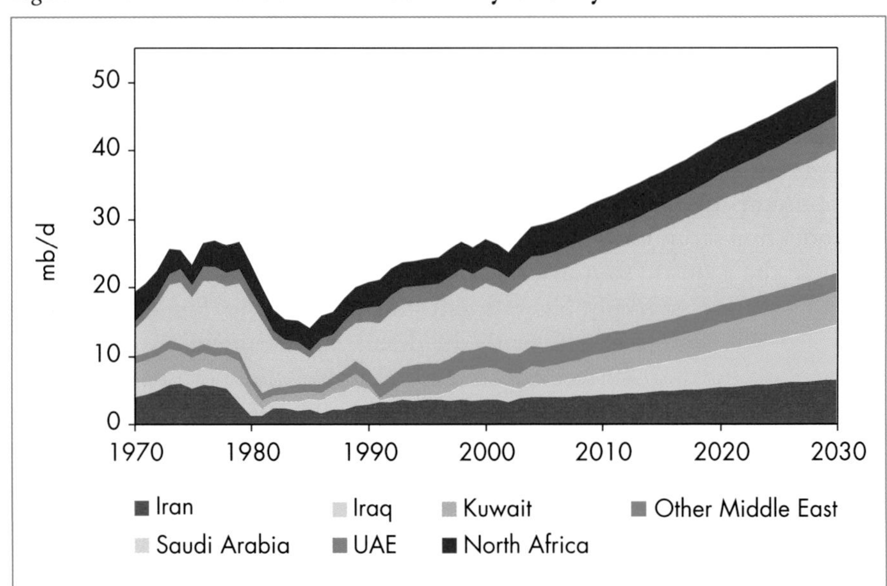

Notes: Includes NGLs and condensates.
Other Middle East includes Qatar in addition to Bahrain, Israel, Jordan, Lebanon, Oman, Syria and Yemen.

exports grow by 2.7 mb/d, to 25 mb/d (Figure 4.12). In the longer term, production in MENA countries, especially in the Middle East, is projected to increase more rapidly than in other regions, because their resources are much larger and their production costs are generally lower. MENA's share of world production is projected to rise from 35% in 2004 to 44% in 2030. The region will compensate for the decline of production in most other regions. Oil production is expected to peak in all non-MENA regions except the transition economies, Latin America and Africa before 2030.

Many MENA countries are members of OPEC, so the organisation's production and pricing policies will affect the level of MENA oil output as a whole. In the past three decades, MENA countries have held a significant amount of spare capacity – the largely unintended consequence of OPEC and national oil policies at a time of rising non-OPEC production (Figure 4.13).[9] The existence of this spare capacity has played an important role in mitigating supply disruptions and sudden, unexpected surges in demand. It has also allowed producers with spare capacity to influence oil prices. Today, only Saudi Arabia has any significant amount of spare capacity and has a policy of maintaining 1.5 to 2 mb/d of spare capacity. In the Reference Scenario, it is assumed that MENA production fills the gap between non-MENA production

9. See Chapter 8 for a discussion of the cost of strategic storage capacity in consuming countries.

Figure 4.12: **Incremental MENA Crude Oil Production and Net Exports in the Reference Scenario, 2004-2010**

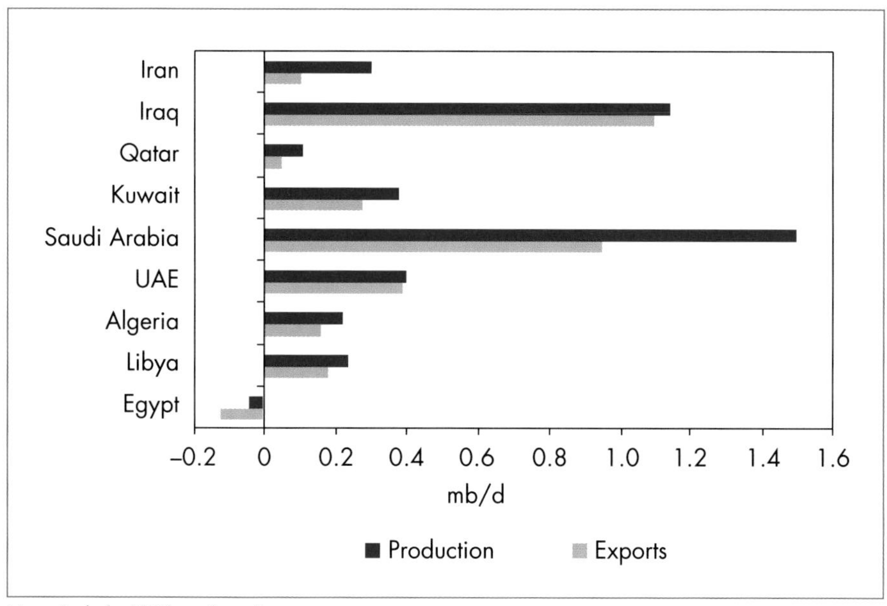

Note: Includes NGLs and condensates.

Figure 4.13: **MENA Crude Oil Production and Spare Maximum Sustainable Capacity**

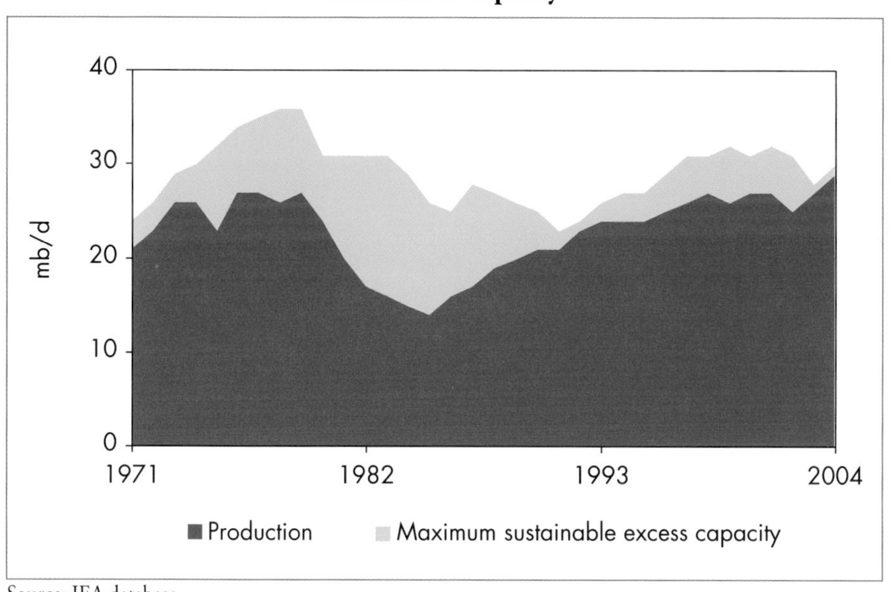

Source: IEA database.

Box 4.3: **When Will World and MENA Oil Production Peak?**

According to our Reference Scenario, global oil production will not reach its peak until some time after 2030. As a result, demand will not be constrained by supply availability before that date if required investments are forthcoming. Of course, oil production must peak one day. Precisely when that happens will depend on demand and production trends. The amount of oil that can ultimately be recovered technically and economically will be a key determinant. *WEO-2004* demonstrates that, under the extremely conservative assumption that ultimately recoverable resources are only 1 700 billion barrels, conventional oil production would peak at around 2015. Using a more optimistic assumption of 3 200 billion barrels pushes the production peak out to around 2035. In both cases, non-conventional sources, including tar sands in Canada, extra-heavy oil in Venezuela and gas-to-liquids output, fill the growing gap between conventional oil production and global oil demand.

The situation differs widely across regions. Some regions, including Europe, have already reached their production peak. But most have not. It is expected in North America around 2010. Output in sub-Saharan Africa will peak around 2020. Globally, non-OPEC conventional oil production is expected to hit a ceiling between 2010 and 2015. Production will continue to grow through to 2030 in the Middle East, North Africa, the transition economies and Latin America. The biggest increase is expected to occur in the Middle East. Consequently, the rate of expansion of installed production capacity in this region and the MENA region as a whole will determine when global production peaks. In both the Reference and Deferred Investment Scenarios, MENA production continues to grow, though at different rates throughout the *Outlook* period. In any event, MENA production will most likely peak some time after global production. How soon after will depend on investment.

and total world oil demand. Thus, it is assumed that OPEC policies do not hold back the expansion of MENA production capacity. In practice, however, MENA countries individually or collectively may not be willing to allow such rapid development of their resources. The Deferred Investment Scenario examines the consequences of a much slower increase in MENA oil output.

Within MENA, the Middle East will account for almost all of the production increase. Saudi Arabia will remain by far the largest supplier, its output rising from 10.4 mb/d in 2004 to 18.2 mb/d in 2030 in the Reference Scenario. Its share of total MENA oil output will remain stable at about 36%. Four countries will see their shares rise: Iraq, Kuwait, Libya and the UAE. Their combined share of MENA production will climb by 10 percentage points to 41% in 2030.

Table 4.3: **MENA Giant and Super-Giant Oilfields*, 2004**

	Number of fields	**Proven reserves at end-year** (billion barrels)	**Production** (mb/d)	**Production** (as % of total production)
Middle East	**80**	**540**	**20.1**	**81**
Iran	25	106	3.4	82
Iraq	13	67	1.9	92
Kuwait	6	50	1.2	47
Qatar	4	4	0.5	49
Saudi Arabia	19	261	10.0	96
United Arab Emirates	9	44	2.2	79
Other Middle East	4	8	1.0	53
North Africa	**21**	**27**	**1.7**	**38**
Algeria	5	10	0.7	34
Egypt	3	1	0.3	36
Libya	13	16	0.8	47
Total MENA	**101**	**567**	**21.7**	**75**

* A super-giant field contains at least 5 billion barrels of proven reserves; a giant field contains at least 1 billion barrels.
Note: Reserves and production include NGLs and condensates.
Sources: Verma et al (2004); IHS Energy (2005); IEA analysis.

Giant and super-giant fields account for 75% of reserves in MENA (Table 4.3) – a larger share than for most other regions. Those fields also account for a large share of current production, though this share has fluctuated somewhat since the 1970s as they bore the brunt of short-term production constraints imposed as a result of OPEC policies (Figure 4.14). Their contribution to regional supply is projected to continue to increase over the first half of the projection period as, despite their already long production life, the growing use of advanced secondary recovery techniques boosts output in the short to medium term. Later, as the natural decline rates of giant fields increase, a growing share of oil will come from smaller fields currently awaiting development. As a result, the share of giant and super-giant fields in total MENA production will drop to 40% in 2030. These trends are expected to raise the average cost of production. More exploration will also be needed, further adding to total production costs.

Output from fields already in production is projected to continue to grow over the rest of the current decade, with the drilling of new wells and the

Figure 4.14: **MENA Crude Oil Production by Size of Field**

Sources: IHS Energy database; IEA analysis.

Figure 4.15: **MENA Crude Oil Production by Category in the Reference Scenario**

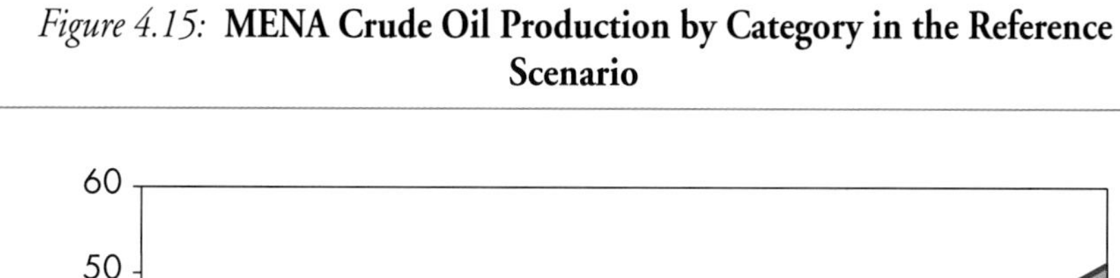

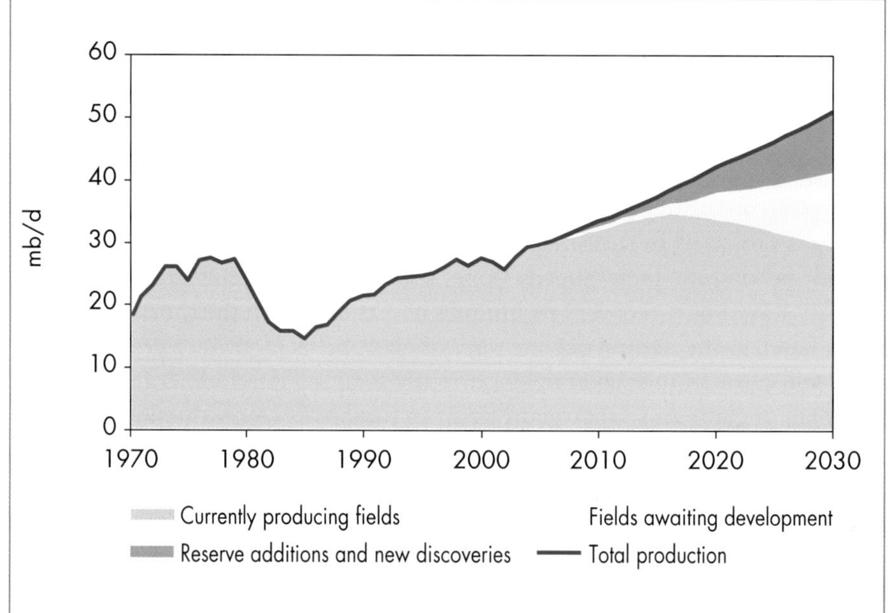

deployment of enhanced and improved recovery techniques. Their production will peak after 2010, unless more reserves are found in those fields or more of the oil in place becomes economically recoverable, but by 2030, currently producing fields will still account for nearly 60% of total MENA production in the Reference Scenario (Figure 4.15). Ghawar in Saudi Arabia is projected to produce more than 6 mb/d by 2010, but output will decline gradually thereafter, to 3.7 mb/d in 2030. Fields that have already been discovered but not yet developed will account for one-quarter, with fields yet to be discovered making up the rest. Box 4.4 describes the methodology used to derive field-by-field production projections. Table 4.4 details production trends for the largest MENA fields.

Table 4.4: **MENA Crude Oil Production Outlook in the Reference Scenario**
(mb/d)

	2004	2010	2020	2030
Currently producing fields	**29.0**	**32.0**	**33.6**	**29.2**
Ghawar (Saudi Arabia)	5.8	6.0	5.5	3.7
Greater Burgan (Kuwait)	1.4	1.5	1.6	1.5
Rumaila (Iraq)	1.2	1.1	1.1	0.8
Other	20.7	23.3	25.3	23.2
New developments	**0.0**	**1.1**	**8.2**	**21.3**
Fields awaiting development	0.0	0.6	4.3	12.0
Reserve additions and new discoveries	0.0	0.4	4.0	9.3
Total	**29.0**	**33.0**	**41.8**	**50.5**

Note: Includes NGLs and condensates.
Sources: IHS Energy database; IEA analysis.

The average quality of the crude oil produced in MENA is expected to deteriorate slightly over the projection period (Figure 4.16). The API gravity of MENA production has averaged 33.2° over the past 35 years and stood at 33.1° in 2004. It is projected to drop to 32.3° in 2030 in the Reference Scenario, averaging 33.0° over 2005-2030. The decline will be more marked after 2015, when more oil will come from the Middle East which has a lower average API gravity compared to that produced in North Africa. The average sulphur content of MENA crude oil is expected to increase too. These developments will increase the challenge to refiners, who will have to meet a continuing increase in the share of light products in total oil demand and meet increasingly stringent rules on sulphur in refined products.

Box 4.4: **Field Production Analysis Methodology**

Our projections of MENA crude oil production are underpinned by an extensive bottom-up, field-by-field analysis of currently producing fields and discovered fields awaiting development. We compiled detailed information on proven and probable reserves, initial oil in place, production history and oil quality for all fields containing proven and probable (2P) oil and gas reserves of at least 500 million barrels of oil equivalent. The primary source was the IHS Energy database. Additional information was obtained from a number of other sources, including international oilfield service companies, national and international oil companies, consultants and the IEA's own databases. Despite the difficulties, we believe that our analysis is based on the best data available.

The methodology for projecting oil production follows three steps:

1. Production from *currently producing fields* is projected on the basis of each field's current reserves, cumulative production to date, historical production trends, the age of the reservoir and current and expected future decline rates. Estimates of natural decline rates, based on published data and information obtained from oil service companies, vary from around 5% to 10% per year. This analysis takes into account and assesses all current and planned development projects. In general, the approach adopted is conservative about their impact on future production.
2. Production from *fields awaiting development* is projected on the basis of our assessment of official plans and forecasts, reserves, geography and technical factors.
3. Production from *additions to reserves and new discoveries* is then added. Production from additions to reserves is projected using a field-specific reserves-growth factor, based on the age and geological structure of known reservoirs, the amount of oil initially in place, the current recovery rate and historical trends in reserve additions. Where existing reserves are judged insufficient to meet future production and the potential for new discoveries is significant, production from new discoveries is projected on the basis of USGS data (mean values) on undiscovered resources and estimated development costs.

The final results were calibrated to the results of the top-down calculation of the call on MENA supply – the difference between world oil demand and non-MENA oil supply.

Figure 4.16: **MENA Average Crude Oil Quality**

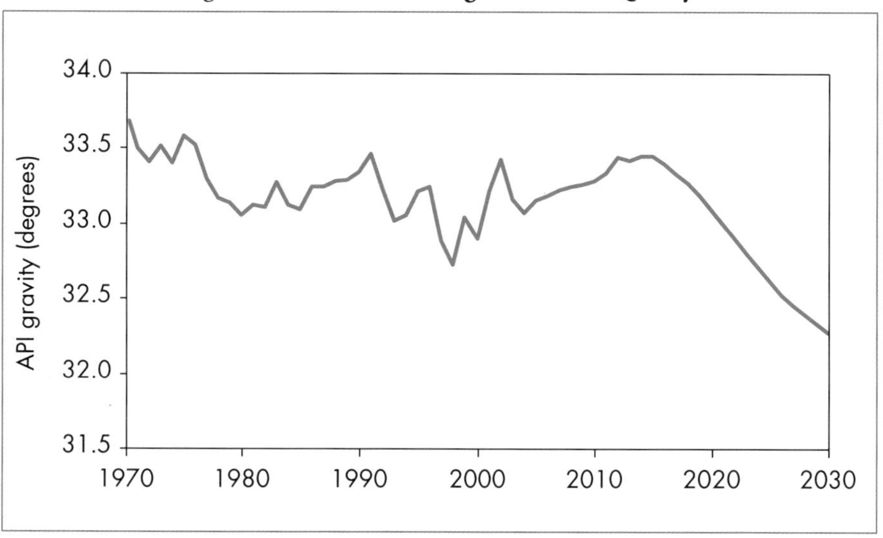

Sources: IHS Energy database; *Oil and Gas Journal* (2004); IEA analysis.

The production of non-conventional oil, other than gas-to-liquids (GTLs), is not expected to play a major role in the MENA region before 2030, because development will continue to focus almost exclusively on the region's huge resources of low-cost conventional oil. However, non-conventional resources are significant in some countries and could contribute in a small way to output growth after 2030. Total MENA non-conventional resources in place are estimated to amount to around 100 billion barrels and are mainly in the form of oil shales (WEC, 2004). The amount of oil that could be recovered economically is very uncertain. Assuming a recovery rate of 10%, a total of 10 billion barrels could be produced eventually. Although this is small compared to proven reserves of conventional oil of 784 billion barrels, some of these non-conventional resources are located in countries, such as Jordan[10] or Morocco, which have virtually no conventional reserves. These countries clearly have a stronger incentive to develop their non-conventional resources than those countries with large conventional reserves. MENA gas-to-liquids production is expected to make a growing contribution to the region's oil output over the projection period, reaching about 100 kb/d in 2010 and 765 kb/d in 2030. A 34-kb/d plant in Qatar, the first to be built in a MENA country, is scheduled to come on stream in 2006.

10. Jordan's non-conventional reserves, in oil shales, are estimated at 5 billion tonnes (WEC, 2004).

Summary of Country-by-Country Trends

Saudi Arabia

Saudi Arabia is the world's largest oil producer. Its proven reserves of 262 billion barrels[11] are equal to 69 years of production at current rates. More than half of Saudi Arabia's reserves are concentrated in just eight fields. The world's largest field, Ghawar, has estimated remaining reserves of 70 billion barrels and accounts for about half of Saudi Arabia's total oil production capacity. Saudi Arabia is implementing projects that will raise its total production capacity (excluding NGLs) to 12.5 mb/d by 2009.

Saudi Arabia currently has 10.5 to 11 mb/d of sustainable production capacity. There are around 20 fields in production, including a number of super-giant fields. Seven of these have a combined capacity of some 8 mb/d: Ghawar, Abqaiq, Shaybah, Safaniyah, Zuluf, Berri and Marjan. The Ghawar field alone contributed 5.2 mb/d in 2004. Over 60% of Ghawar's 115 billion barrels of proven reserves have already been extracted. As the field matures, costs will undoubtedly increase and sustaining production will become more difficult. The water cut – the share of water in the liquids extracted – has been a problem in recent years, but efforts to reduce it by drilling new wells appear to have been successful.

Maintaining and expanding Saudi capacity will require more investment in the future. Saudi Aramco, the national oil company, has estimated that the natural rate of decline at existing fields will be of the order of 6% per year over the next five years, so that some 600 kb/d of capacity will have to be replaced each year just to maintain the present level of capacity. This will be achieved mostly through enhanced development of existing fields, which have been managed very conservatively over several decades in order to extend their plateau production for as long as possible. But as these mature fields age, their production will decline slowly and new fields will have to contribute an increasing share of production. There has been little exploration effort in recent years and parts of the country are unexplored, including the region close to the border with Iraq, the Red Sea and the Rub Al-Khali in the south-east. At present, 70 fields await development. The Saudi Oil Minister has indicated that production of crude oil could be increased to 15 mb/d and sustained at that level for at least 50 years. Saudi Aramco believes that future investments can be financed solely out of its own cash flow.

Saudi Arabia will continue to play a central role in OPEC and in balancing the world market, not only because of the size of its production but also because of

11. On 1 January 2005. The source of reserves data for Saudi Arabia and the other countries discussed in this section is the *Oil and Gas Journal* (20 December 2004).

its spare capacity. In mid-2005, it was the only country in the world with an appreciable amount of sustainable capacity in reserve. Much of this capacity is in three offshore fields – Safaniyah, Zuluf and Marjan – all of which produce medium or heavy crude oil. Saudi Arabia is expected to remain the primary source of spare capacity worldwide. The country's official policy is to maintain 1.5 to 2 mb/d of spare capacity for the foreseeable future. In the Reference Scenario, Saudi oil production is projected to increase from 10.4 mb/d in 2004 to 11.9 mb/d in 2010 and to 18.2 mb/d in 2030.

Iraq

The near-term prospects for oil production in Iraq remain very uncertain. Output rebounded from the low reached immediately after the 2003 invasion, but fell back from a post-war high of 2.4 mb/d in March 2004 to only 1.8 mb/d in May 2005 because of sabotage to oil pipelines and production facilities. The Iraqi authorities aim to increase production to 2.5 mb/d by late 2005, but this will hinge on the completion of repairs to pipelines and other facilities as well as rehabilitation work on oil wells, separation plants and water treatment facilities and improved security. In the longer term, further significant increases in production capacity will require tens of billions of dollars of investment in adding new wells at existing fields and in developing new fields.

Iraqi production has always been low relative to the size of the country's reserves. Iraq's proven oil reserves are estimated at 115 billion barrels – the third-highest in the MENA region after Saudi Arabia's and Iran's. Some analysts estimate that exploration in the largely unexplored Western Desert could lift proven Iraqi reserves to 180 billion barrels. Many fields are thought to be relatively easy and cheap to develop and production capacity could be expanded substantially. But how quickly these reserves can be tapped hinges on domestic security, as well as on the legal and commercial framework that will emerge. In the Reference Scenario, oil production is projected to recover to 3.2 mb/d in 2010 and to rise to 7.9 mb/d in 2030, assuming security is gradually restored and reserves are opened up progressively. Its share in MENA total production will increase more than that of any other country (Figure 4.17).

Iran

Crude oil production potential in Iran is very large. Proven oil reserves are 126 billion barrels, equal to 83 years of production at current rates. In addition to recently discovered fields in the Azadegan and Bushehr regions, there is considerable potential for increasing offshore production in the Persian Gulf and Caspian Sea. Iranian output increased to 4.1 mb/d in 2004, making Iran the second-largest oil producer in OPEC and the fourth-largest exporter in the

Figure 4.17: **MENA Crude Oil Production by Country in the Reference Scenario**

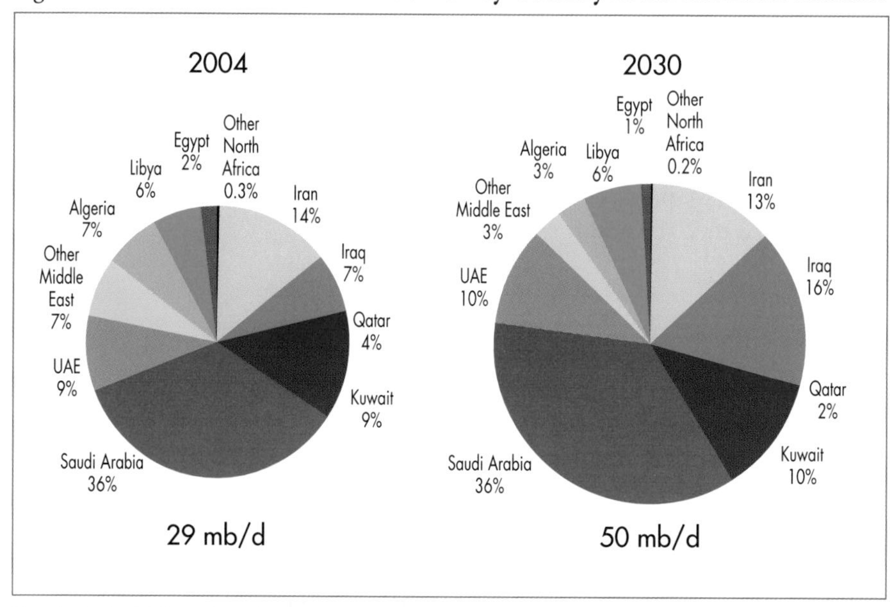

Note: Includes NGLs and condensates.

world. But Iranian production remains far below its historical peak of 5.6 mb/d in 1974. Current sustainable crude oil (excluding NGLs) capacity is estimated at 4.1 mb/d. Maintaining this capacity will require large investments in existing fields, where decline rates are thought to be relatively high. Iranian production is expected to rise to 4.5 mb/d in 2010 and 6.8 mb/d in 2030 in the Reference Scenario. Until 2010, exports will grow less rapidly, because much of the increase in output will go to meeting rising domestic needs. After 2010, exports grow faster as production outpaces demand.

United Arab Emirates

The UAE has proven crude oil (including condensates) reserves of 98 billion barrels, nearly 8% of the world total. Crude oil (excluding NGLs) production in 2004 was 2.35 mb/d, some 200 kb/d below the Emirates' total production capacity. Only Saudi Arabia has more spare capacity than the UAE at present. The UAE plans to increase production capacity to 3 mb/d by the end of 2006. Several projects are under way or are planned to expand capacity at existing oilfields. Drawing on its large reserves, UAE output in the Reference Scenario is expected to climb to 3.2 mb/d in 2010 and 5.1 mb/d in 2030. Most of the increase will come from Abu Dhabi.

Kuwait

Kuwait's crude oil production (including NGLs and condensates) averaged 2.5 mb/d in 2004. The country is the ninth-largest exporter of crude oil in the world and has ambitions to expand output to 4 mb/d by 2020. Most of the planned expansion projects involve fields in the north of the country, including Abdali, Ratqa, Rawdhatain, and Sabriyah, under the programme "Project Kuwait". The total cost of the project is estimated at about $8.5 billion. However, its implementation has been delayed, because of opposition among some members of parliament to the government's plan to involve international oil companies. In the Reference Scenario, Kuwait's total oil production is projected to reach 2.9 mb/d in 2010, 3.8 mb/d in 2020 and 4.9 mb/d in 2030.

Qatar

Qatar has 15 billion barrels of proven oil reserves, two-thirds of which are in the form of condensates (concentrated in the North Field). Around 90% of the country's 1 mb/d of oil production (including NGLs and condensates) was exported in 2004. In the Reference Scenario, oil production is projected to rise to 1.25 mb/d in 2030, with condensates making up most of the increase.

Libya

Prospects for oil production in Libya have improved considerably with the lifting of UN and US sanctions in 2004, as well as recent moves to liberalise the upstream industry. The government aims to boost production from 1.6 mb/d in 2004 to 2 mb/d by 2010 and 3 mb/d by 2015, bringing output back to the level it reached in the 1970s. The country's proven oil reserves, at 39 billion barrels, are the largest in Africa. A dozen fields hold more than 1 billion barrels each. Libya also has substantial undiscovered resources. In the Reference Scenario, Libya's production is projected to reach 1.9 mb/d in 2010 and 3.1 mb/d in 2030.

Algeria

Unlike most other major MENA countries, Algeria's oil sector has benefited from large amounts of foreign investment. Nonetheless, Sonatrach – the national oil company – still produces more than half of Algeria's output, which amounted to 1.9 mb/d in 2004. A quarter of this production was made up of NGLs. Total oil production is projected, in the Reference Scenario, to level off at about 2.2 mb/d in 2010. Enhanced oil recovery techniques, especially at the Hassi Messaoud field, where the recovery factor is thought to be as low as 20%, will be increasingly called upon to maintain production levels. Even so, overall output is projected to decline soon after 2010 in the Reference Scenario, falling to 1.6 mb/d in 2030 as reserves dwindle.

Egypt

Egypt has an estimated 3.7 billion barrels of proven oil reserves, mainly concentrated in the Gulf of Suez and Nile Delta areas. Egyptian oil production peaked over a decade ago and has since been in decline. In our Reference Scenario, production is projected to fall from 0.7 mb/d in 2010 to 0.5 mb/d 2030. Egypt, currently a minor oil exporter, is expected to become a net oil importer by 2015.

Other Countries

In the Reference Scenario, aggregate production in other MENA countries is projected to dip from 2.0 mb/d in 2004 to 1.4 mb/d in 2030. Their share of total MENA output will decline from 7% to 3%. Two-thirds of total production in this group of countries currently comes from Oman and Syria (Table 4.5). Together with Yemen, they are the only countries with significant reserves. They are expected to continue to dominate the production of this group of countries through the projection period. Israel, Jordan and Morocco have negligible reserves.

Table 4.5: **Other Oil Producers in MENA**

	Proven reserves end-2004		**Production 2004**	
	(million barrels)	**Share of MENA** (%)	(kb/d)	**Share of MENA** (%)
Other Middle East				
Bahrain	125	0.02	200	0.69
Israel	2	0.00	0	0.00
Jordan	1	0.00	0	0.00
Oman	5 506	0.70	786	2.71
Syria	2 500	0.32	505	1.74
Yemen	4 000	0.51	419	1.45
Other North Africa				
Morocco	2	0.00	5	0.02
Tunisia	308	0.04	68	0.23
Total other MENA	**12 443**	**1.59**	**1 982**	**6.84**

Note: Includes NGLs and condensates.

Sources: *Oil and Gas Journal* (20 December 2004); *IEA Oil Market Report (*September 2005); IEA analysis.

Oil Refining

In 2004, MENA had 8.8 mb/d of crude oil distillation capacity – 11% of the world total (Table 4.6). Production of petroleum products[12], at 10.6 mb/d, exceeds domestic demand by about two-thirds, with product exports running at about 4 mb/d in 2004. Most capacity, amounting to 7 mb/d at end 2004, is in the Middle East. Several countries are pursuing plans rapidly to expand their presence in refining in response to the worldwide increase in demand and shift to lighter and cleaner products, especially middle distillates. Middle East refiners, notably in Saudi Arabia, are aiming to increase their output of middle distillates and are embarking on major expansion and upgrading programmes. The Middle East is one of the few regions with significant spare export capacity at present: strong demand has stretched middle distillate and gasoline capacity to its limit almost everywhere else in the world. Strong demand growth and a tight refining market are expected to keep refining margins attractive for another few years.

4

Table 4.6: **MENA Crude Oil Distillation Capacity in the Reference Scenario** (mb/d)

	2004	2010	2020	2030	2004-2030*
Middle East	**7.0**	**8.8**	**10.7**	**12.9**	**2.4**
Iran	1.5	1.7	2.2	2.6	2.1
Iraq	0.6	0.8	1.0	1.2	2.6
Kuwait	0.9	1.3	1.4	1.4	1.8
Qatar	0.1	0.2	0.3	0.3	3.3
Saudi Arabia	2.1	2.6	3.4	4.5	3.0
UAE	0.7	0.8	0.9	1.1	1.7
Other Middle East	1.1	1.4	1.6	1.8	2.1
North Africa	**1.8**	**2.1**	**2.6**	**3.1**	**2.1**
Algeria	0.5	0.7	0.9	1.0	3.2
Egypt	0.8	0.8	0.9	1.1	1.6
Libya	0.4	0.4	0.6	0.7	2.2
Other North Africa	0.2	0.2	0.2	0.3	1.2
Total	**8.8**	**10.9**	**13.4**	**16.0**	**2.3**

* Average annual growth rate.

12. Petroleum products include liquefied petroleum gas and other products derived from NGLs and output from gas-to-liquids plants. Refinery output accounted for 8.1 mb/d, or 77% of total MENA output of oil products in 2004.

To secure an outlet for their predominantly sour crude oils, Saudi Arabia and Kuwait are planning refinery expansions at home, as well as in major growth markets such as China, India and the United States. These plans will add over 2 mb/d to world distillation capacity by 2010, of which 1.8 mb/d will be located in the Middle East itself. The biggest gross additions are expected to occur in Saudi Arabia, with the construction at Yanbu (400 kb/d) and in Kuwait (600 kb/d). Oman is also planning to add 116 kb/d, Syria 140 kb/d and Qatar 146 kb/d. Refining capacity additions beyond 2010 are assumed to be driven largely by trends in domestic demand.

Figure 4.18: **Additions to MENA Crude Oil Distillation Capacity in the Reference Scenario**

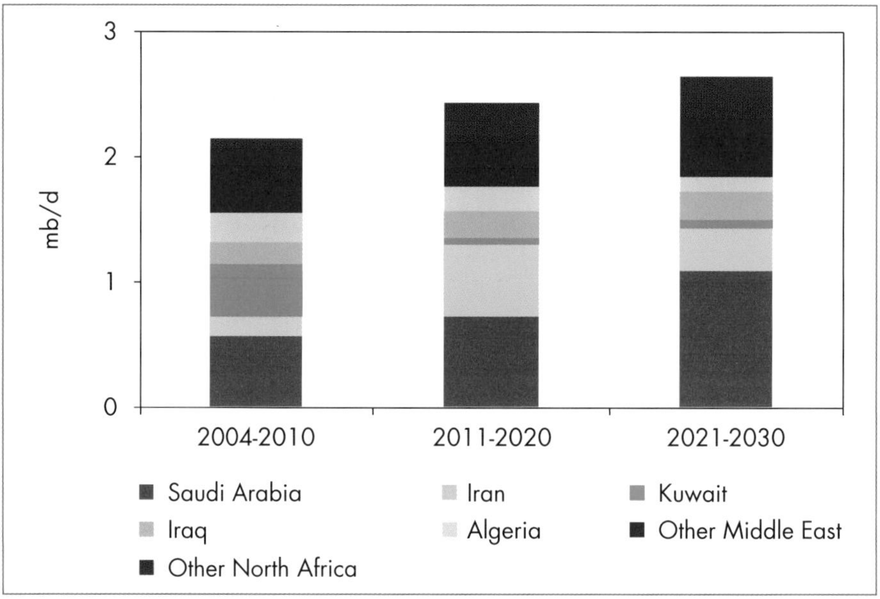

Note: Other Middle East includes Qatar and the UAE in addition to Bahrain, Israel, Jordan, Lebanon, Oman, Syria and Yemen.
Other North Africa includes Egypt and Libya in addition to Morocco and Tunisia.

Middle East refinery expansions over the past 10 years have largely focused on condensate splitters to take advantage of the readily available supplies. The lighter feedstock used by these units has helped refiners to increase their light product output without having to invest heavily in expensive hydro-cracking or desulphurisation units. But plans are in hand for secondary desulphurisation units to be built. Saudi Arabia will add 137 kb/d of hydro-treating capacity in 2006 at its Yanbu and Riyadh plants, while Iran is planning to add 142 kb/d. Kuwait plans to increase its capacity for producing low-sulphur fuels. Most of these expansions are aimed at meeting rapid growth in domestic demand for oil products, particularly gasoline. In recent years, this has encouraged the building of additional fluidised catalytic cracking capacity in Qatar, Oman and Iran. Iran's expansions will mostly serve to supply its domestic market. However, Saudi Arabia and Kuwait plan to boost their exports of refined products.

North Africa, with 1.8 mb/d of refining capacity in 2004, barely meets its domestic product demand. Oil demand growth of 2.4% per year over the projection period is expected to lead to an increase in domestic refining capacity, to about 3.1 mb/d in 2030. Out of all the North African countries, Algeria is expected to see the largest increase in capacity, not only to meet domestic demand but also to export.

In total, MENA refining capacity is projected in the Reference Scenario to reach 10.9 mb/d in 2010 and 16 mb/d in 2030 (Table 4.6). Middle Eastern countries will account for over 80% of the additions to MENA capacity (Figure 4.18). Middle East refining capacity is projected to reach 8.8 mb/d in 2010 and 12.9 mb/d in 2030. North African capacity will increase to 2.1 mb/d in 2010 and 3.1 mb/d in 2030.

Oil Exports

MENA countries as a group will play an increasing role in growing international oil trade over the *Outlook* period. World inter-regional trade is projected to grow from 36 mb/d in 2004 to 43 mb/d in 2010 and 60 mb/d in 2030. MENA's share of global exports is projected to rise from 62% to 64%. Net oil exports from MENA countries to other regions will edge up from 22 mb/d in 2004 to 25 mb/d in 2010 and then surge to 39 mb/d in 2030. The region will increase its exports to all the major consuming regions (Figure 4.19). Exports to developing Asian countries will increase the most.

All the net oil-importing regions – the three OECD regions and developing Asia – will become even more dependent on both oil imports in total and imports from MENA, in particular, over the projection period. The OECD Pacific region will remain both the most dependent on imports and the most dependent on MENA. About 90% of its imports will come from the Middle East in 2030, compared with 83% in 2004. North America will also rely more and more on MENA oil. It will even become the biggest importer from MENA countries in 2030, when its MENA imports will hit 11 mb/d – a full one-quarter of total MENA exports. However, China will see the biggest percentage increase in MENA imports, which will be multiplied by almost four over the projection period. China, which only became a net oil importer in 1993, will have to import three-quarters of its oil needs in 2030. Most of this oil will come from the Middle East. Saudi Arabia will retain its position as the largest oil exporter in MENA in 2030 (Table 4.7 and Figure 4.20). Its exports

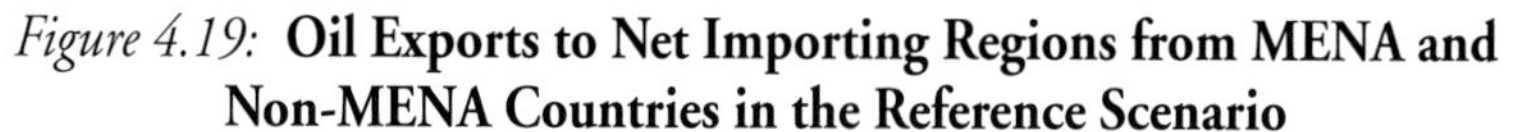
Figure 4.19: **Oil Exports to Net Importing Regions from MENA and Non-MENA Countries in the Reference Scenario**

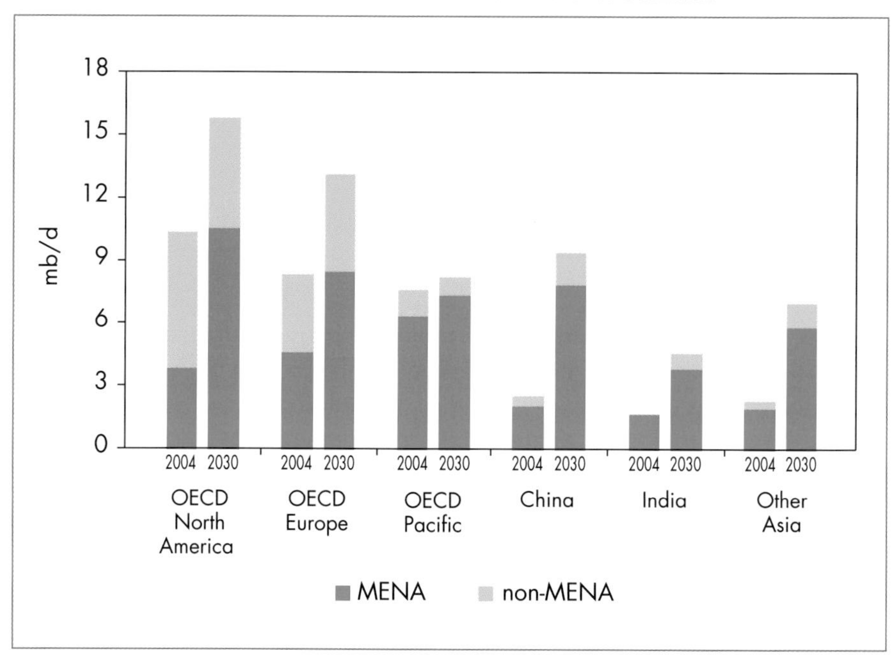

Table 4.7: **MENA Net Oil Exports in the Reference Scenario** (mb/d)

	2004	2010	2020	2030	2004-2030*
Middle East	**19.3**	**21.8**	**28.7**	**36.0**	**2.4%**
Iran	2.7	2.8	3.6	4.4	1.9%
Iraq	1.4	2.5	4.6	6.9	6.2%
Kuwait	2.2	2.5	3.4	4.4	2.8%
Qatar	0.9	1.0	1.0	0.9	0.0%
Saudi Arabia	8.3	9.3	12.1	14.4	2.1%
United Arab Emirates	2.5	2.9	3.7	4.7	2.4%
Other Middle East	1.2	0.8	0.4	0.1	-8.3%
North Africa	**3.0**	**3.2**	**3.0**	**2.7**	**-0.4%**
Algeria	1.7	1.8	1.5	1.1	-1.7%
Egypt	0.2	0.1	-0.2	-0.5	-
Libya	1.4	1.5	2.1	2.5	2.5%
Other North Africa	-0.2	-0.2	-0.3	-0.4	3.1%
Total MENA	**22.3**	**25.0**	**31.8**	**38.7**	**2.1%**

* Average annual growth rate.

Note: Negative numbers correspond to net imports.

Figure 4.20: **MENA Oil Exports by Country in the Reference Scenario** (mb/d)

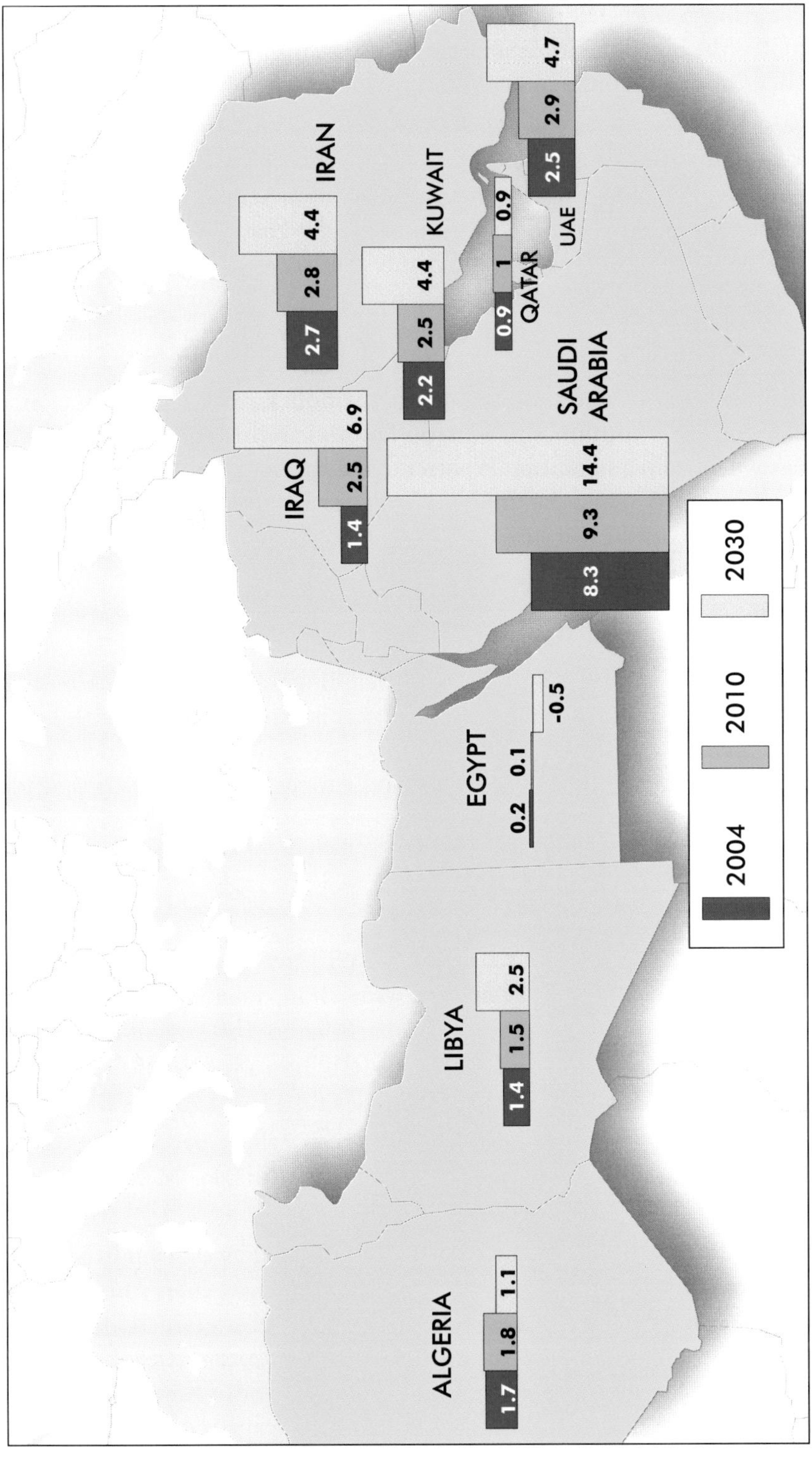

will grow in the Reference Scenario from 8.3 mb/d in 2004 to 14.4 mb/d in 2030 – the biggest volume increase of any country in the world. Iraq's exports will grow most rapidly in percentage terms.

Most MENA exports will still be crude oil in 2030, but refined products will account for a growing share. The Middle East is already the largest net exporter of oil products in the world. In 2004, it accounted for around half total inter-regional product trade. By 2030, the Middle East's share will grow to two-thirds, with volumes rising to 6.5 mb/d in the Reference Scenario. Almost three-quarters of this, about 4 mb/d, will come from Saudi Arabia and Kuwait alone. The Middle East region is already a major supplier of jet fuel and other middle distillates to Europe and heavy fuel oil to Asia. The United States will emerge as an important new market for Middle East refined-product exports, as US demand continues to outstrip domestic refining capacity. The Middle East will remain a significant exporter to the Far East especially providing heavy fuels; but as demand for middle distillates rises in OECD markets, Saudi Arabia and Kuwait will supply the changing needs.

Algeria and Libya will remain key suppliers of high-quality products to European markets. Algeria plans to boost its refined-product exports. In the Reference Scenario, Algerian exports are projected to grow by 52% from 2004 to 2030, reaching 890 kb/d in 2030 (including more than 500 kb/d of NGLs and the output from a new GTL plant that is expected to be built). Libya has no clear plans to increase its product exports, but is expected to continue exporting around 50 kb/d of refined products per year over the *Outlook* period.

Investment and Financing

Projected Investment Needs

The projected growth in MENA crude oil production and refinery output implies a need for $614 billion (in year-2004 dollars) of capital spending between now and 2030 in the Reference Scenario. Investment flows will need to rise substantially from an average of $16 billion a year in the current decade (in year-2004 dollars) to $28 billion a year in the last decade of the projection period. The increase in the investment needs will be particularly important after 2010, reflecting the increase in the call on MENA oil supply (Figure 4.21).

Middle Eastern countries will absorb most of the oil investment in MENA. Oil will account for more than 40% of total Middle East energy investment and one-third in North Africa (Figure 4.22). While large in dollar terms, oil investment needs will amount to less than 6% of projected oil revenues and about 1% of GDP over the entire projection period. Thus, financing should not, in principle, be a major obstacle to industry expansion in most countries.

Figure 4.21: **MENA Oil Investment in the Reference Scenario**

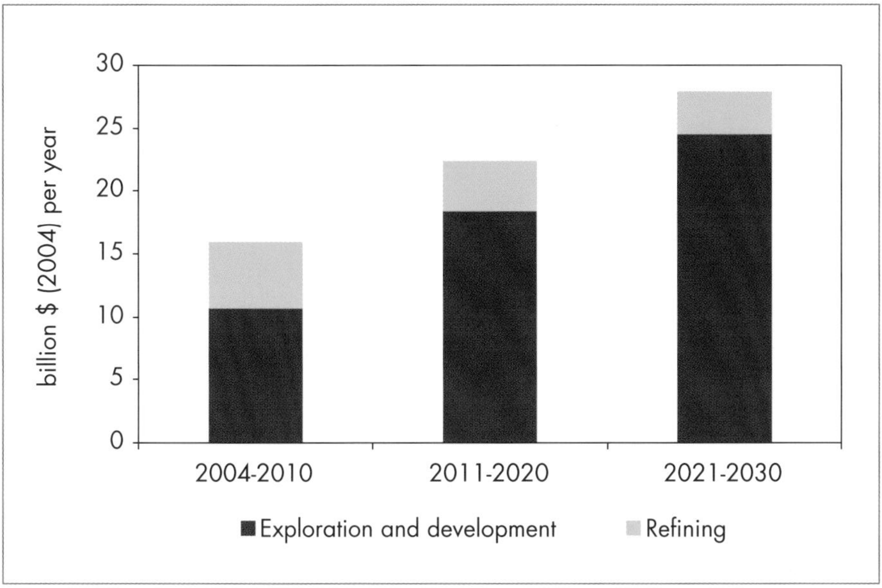

Investment needs will be largest in Saudi Arabia, at $174 billion in total or $6 billion a year over the projection period in the Reference Scenario (Figure 4.22). Other big producers – Iran, Iraq and Kuwait – follow. Qatar will have important investment needs, totalling $50 billion over 2004-2030. This reflects high exploration and development costs compared with most other MENA countries and large investments in GTL capacity.

Upstream Investment

Exploration and development will account for more than three-quarters of total investment. Upstream investment needs will be highest in the last decade, when more new fields will need to be brought on stream to replace ageing giant fields. A small but growing share of upstream investment will go to exploration. Exploration and development costs in the Middle East will remain the lowest in the world and well below assumed price levels (see the earlier section on Production Costs). It is assumed that costs for new supplies will grow by about 50% over the projection period. However, the increase in investment needs is mainly linked to the increase in capacity and assumed natural decline rates: the higher the rate at which production declines naturally in the absence of new investment, the more investment is needed to develop reserves and make new discoveries.

Decline rates in the MENA region are among the lowest in the world. They are assumed to vary over time and range from 5-6% in Iraq and Saudi Arabia

Figure 4.22: **Cumulative MENA Oil Investment by Country in the Reference Scenario, 2004-2030**

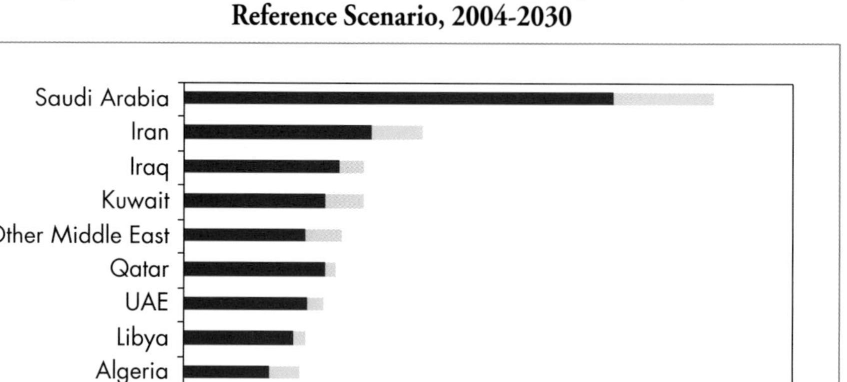

to 10% per year for smaller producers, such as Egypt or Oman. The average rate for the region as a whole is about 7%, which is low compared to rates of more than 10% seen in more mature regions such as the North Sea and North America. About 78 mb/d of new capacity will be needed in MENA countries over the projection period – roughly two-thirds to replace the capacity lost through natural declines in production and one-third to meet the projected increase in the call on MENA oil.

Downstream Investment

Cumulative investment in MENA refineries over the *Outlook* period is projected to amount to $110 billion, or $4 billion per year in the Reference Scenario. This represents nearly a quarter of total world refinery investments. Saudi Arabia alone will account for 30% of this investment and over 37% of Middle East investment (Figure 4.23). In contrast to the upstream sector, spending on refineries will rise most strongly in the first decade of the projection period. From an average of $1.7 billion per year in the last ten years, MENA investment will average $5.3 billion per year up to 2010 to $3.4 billion per year in the 2020s. For the rest of the current decade, most investment will be needed to build new primary distillation capacity, as current capacity will be absorbed rapidly by rising domestic demand. In the second and third decades, more of the investment in refining will provide for exports.

Figure 4.23: **MENA Refinery Investment by Country in the Reference Scenario, 2004-2030**

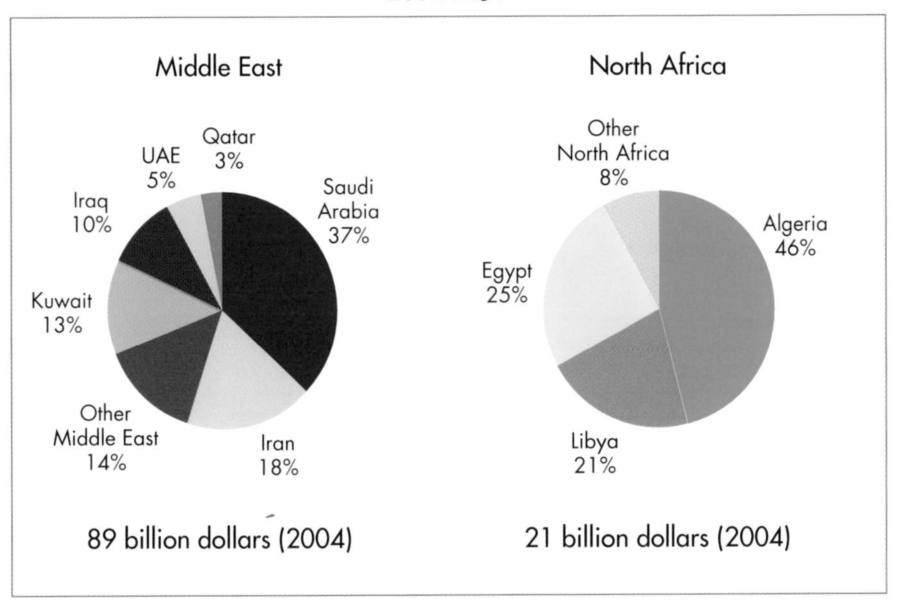

More than 70% of the MENA investment will go towards financing capacity additions (Table 4.8). In the Middle East where capacity is set to increase by more than 80% between 2004 and 2030, investment in capacity additions will total $62 billion. Over the rest of the current decade, some $21 billion – nearly a quarter of world investment in new capacity – will be spent in the Middle East, mostly in major new refineries in Saudi Arabia ($7 billion) and Kuwait ($5 billion). Algeria will also need to invest, mainly in capacity additions, with 87% of its refinery investment needed for additions and only 13% for new conversion facilities to produce products better suited to the domestic market. This is also the case for the net product-importing countries: Iran, Egypt and Morocco. But conversion will also be necessary in exporting countries to enable them to comply with an increase in demand for middle distillates and with tighter environmental standards in OECD countries. This is especially true for the Middle Eastern countries, whose crude oil will become heavier and more sour. The bulk of the $26 billion of capital needed for conversion capacity in these countries will have to be invested in the next 10 years.

Transportation

The volume of international trade in oil and refined products will increase by over two-thirds between 2004 and 2030. Considerable investment to transport crude oil and refined products to market via tankers and pipelines will be

Table 4.8: **MENA Net Crude Oil Distillation Capacity Additions and Total Refinery Investment Needs in the Reference Scenario, 2004-2030**

	Capacity additions (mb/d)	**Investment** (billion dollars in year-2004 dollars)		
		Additions	**Conversions**	**Total**
Middle East	**5.9**	**62**	**26**	**89**
Iran	1.1	11	5	16
Iraq	0.6	6	2	8
Kuwait	0.5	6	6	12
Qatar	0.2	2	1	3
Saudi Arabia	2.4	26	7	33
UAE	0.4	3	2	5
Other Middle East	0.8	8	4	12
North Africa	**1.3**	**16**	**6**	**21**
Algeria	0.6	9	1	10
Egypt	0.4	3	2	5
Libya	0.3	3	1	4
Other North Africa	0.1	1	1	2
Total MENA	**7.2**	**78**	**32**	**110**

necessary. Because of lengthening supply chains resulting from increased production in the Middle East, which is far from the main consuming centres, an increasing share of trade will be waterborne. As a consequence, most of the investment will go towards expanding the capacity of the oil-tanker fleet. Cumulative global investment in the oil-tanker sector and in pipelines will amount to $252 billion over the period 2004-2030, or an average of $9.3 billion per annum. This represents 8% of total oil-sector investment. A very small proportion of this investment will be made within the MENA region. Most oil tankers are built in Asian shipyards, mainly in South Korea and Japan. Most of the very large oil-pipeline projects that are currently under construction or being planned are in countries where oil reserves are landlocked, such as Azerbaijan and Kazakhstan.

Sources of Finance and Investment Uncertainty

The bulk of oil investment in MENA countries is expected to be financed out of the internal cash flows of the national and international companies operating in the region. National oil companies dominate the industry in the region and are likely to continue to be responsible for most of the investment over the *Outlook* period (Box 4.5). The governments that own them will determine how much of the revenue they generate will be allocated to investment programmes and how much will be used for other purposes. Budgetary pressures could limit the amount of their earnings that the national companies are allowed to retain for investment purposes and, therefore, increase their need to borrow. Some governments may decide to give priority to using oil revenues to pay for non-oil infrastructure projects or for current spending on various national programmes. Revenues are more likely to be used for such purposes in countries where the national oil company is a department within the government. Lower prices than assumed in our Reference Scenario would have a disproportionately pronounced impact on oil company cash flows (because of fixed costs) and, therefore, the extent to which they would need to rely on debt financing.

Box 4.5: **National Oil Companies in MENA Countries**

National oil companies (NOCs) play a major role in MENA countries' oil and gas sectors. In most of the largest oil-producing countries, the government acquired ownership of its hydrocarbon resources through nationalisation in the 1970s. NOCs were subsequently established to take responsibility for the development and the marketing of those resources. Saudi Aramco and the National Iranian Oil Company – both NOCs – are the two largest oil companies in the world, ranked by production and reserves. Most other NOCs in the MENA region are ranked among the world's top 20 companies.

NOCs in all major oil-producing MENA countries are 100%-owned by the government. The role of NOCs has evolved significantly since nationalisation. In addition to the development of hydrocarbon resources, the governments of many MENA countries expect NOCs to contribute to the implementation of other national policies aimed at promoting socio-economic development. This includes building non-oil infrastructure, creating opportunities for the private sector and providing employment for nationals. The balance between these two roles varies across countries and has been changing over time. In most MENA countries, the NOC has become more focused on hydrocarbon-related activities.[13]

The technical and managerial capabilities of NOCs in the region also vary considerably. Saudi Aramco maintains high technical and managerial standards and has a relatively large degree of autonomy in its operations. The degree of autonomy in financial management affects the efficiency of investment. NOCs with little financial autonomy may find it difficult to carry out investments in a timely fashion, because of constraints resulting from the government's budget-approval process.

13. Marcel (2005a and 2005b).

In some MENA countries, government revenue needs for non-oil expenditure and policies on private and foreign investment may constrain capital flows to the sector. Saudi Arabia and Kuwait are the only MENA countries that do not allow companies other than the national oil company to have any direct involvement in the development or production of oil resources, although foreign companies operate in both countries under technical services contracts. Since 1980, when the nationalisation of Aramco was completed, all upstream operations in Saudi Arabia have been managed by Saudi Aramco on behalf of the government (with the exception of the Neutral Zone, where international oil companies are allowed to operate). The Saudi government recently signed four deals with foreign companies to explore for and develop gas, but the upstream oil sector remains off limits to foreign companies. Kuwait, however, is planning to invite foreign companies to invest in its upstream oil and gas sectors.

There are restrictions on where foreign companies are allowed to operate and the nature of their involvement in other MENA countries. In most cases, foreign investment is restricted to either production-sharing or to buy-back deals, under which ownership of the reserves remains with the national oil company:

- *Iran:* Foreign companies are allowed to invest in Iran only under buy-back contracts – a special form of fee-for-services deal. US sanctions effectively block US companies' involvement in Iran.
- *Iraq:* Opportunities for foreign companies to invest in Iraq are expected to emerge once a new constitution and new hydrocarbons law are adopted, possibly in 2006. The status of Development and Production Contracts negotiated with foreign companies before the 2003 invasion is uncertain.
- *Qatar:* International oil companies participate in joint-venture projects under production-sharing contracts.
- *UAE:* Exploration and production operations are generally governed by concession agreements with international oil companies, usually in partnership with the national oil companies.
- *Algeria:* A new hydrocarbon law, adopted in 2005, opens the way for increased foreign participation in the upstream and downstream sectors. The national oil company, Sonatrach, will no longer be responsible for government oil and gas policy and the existing production-sharing regime will be fully replaced by a tax and royalty system.
- *Egypt:* Foreign investment is governed by standard production-sharing contracts.
- *Libya:* Foreign companies can now invest under concession agreements as well as exploration and production-sharing agreements (EPSA), following the lifting of UN sanctions in 2003 and of US sanctions in 2004.

In countries that permit foreign investment, the extent to which international oil companies actually invest in MENA countries will depend largely on the commercial and fiscal terms on offer. Future investment could be delayed or blocked by disagreements between host countries seeking the largest possible share of upstream rent and investors seeking returns that are high enough to compensate for the risk involved and competitive with the returns in other producing countries. At present, the rate of government take (taxes and royalties as a share of profits) varies considerably among MENA countries, as well as within countries, according to the maturity of the upstream sector, short-term economic and political factors and investment risk (Table 4.9). Most major international oil companies and a growing number of national oil companies are interested in gaining access to MENA's low-cost oil reserves, but will only strike deals if the commercial and fiscal terms are good enough. Oil-service companies and smaller oil companies already play an important role in assisting national companies to develop reserves, without taking equity stakes. This is starting to happen in Iraq, where the international majors are reluctant to invest at present because of insecurity and legal uncertainty. Foreign national oil companies are also seeking opportunities to participate in the development in MENA countries in partnership with the host company.

Where capital availability is not an obstacle, government policies on long-term resource development and production determine how much capital is actually invested in the oil industry in MENA countries and, therefore, how much additional supply will be forthcoming. A Deferred Investment Scenario analyses the impact of lower investment than projected in the Reference Scenario (see Chapter 7).

Table 4.9: **Licensing and Fiscal Terms in MENA Countries***

Country	Type of contract	General terms	Profit-sharing**	Maximum cost recovery***
Algeria	PSC	A new royalty/tax regime is being introduced under the 2005 Hydrocarbons Law. Acreage will be offered on the basis of transparent bidding rounds.	60%-85%	No limit
Egypt	PSC	50:50 joint-venture operating company (contractor/NOC). Royalty rate is 10% of revenue.	75%-85%	40%
Iran	Buy-back	Contractors fund and implement E&P on behalf of NIOC, providing capital, and bearing all risks. Allowed rate of return is 15%. No royalty is payable.		Competitive bidding
Iraq	PSC/DPC	PSCs guarantee state participation through a carried 25% interest. DPCs specify a 10% carried interest for state. Remuneration fee (after cost recovery or start of production) is up to 10% of production. Status of contracts signed before 2003 is uncertain.	91%	50%
Kuwait	OSA	With OSA yet to receive official approval, it is expected that the state will not take an equity position. No royalty will be levied. IOC consortia to fund 100% of the required capital.	n.a.	Expected to be biddable
Libya	Concession	NOC has a minimum 51% stake in these licences. Contractors pay 100% of all capital agreement and operating costs. Royalty is at 16.67% of the value of the recovered crude.	n.a.	No limit
	EPSA	NOC has a 50%-75% interest in any field, paying 50% of the capital costs and a share of operating costs.	Up to 90%	35%

Table 4.9: **Licensing and Fiscal Terms in MENA Countries*** *(continued)*

Country	Type of contract	General terms	Profit-sharing**	Maximum cost recovery***
Qatar	PSC	The state does not have the right to back-in to contracts/licences.	35%-90%	65%
Saudi Arabia	Onshore concession contract	Apart from the recent award of gas exploration contacts, the only active contract in Saudi Arabia remains the concession agreement in the onshore Neutral Zone, where the kingdom has no equity. Royalty is 20%.	80% income tax	No limit
UAE	Concession agreement	Direct state participation in upstream only in Abu Dhabi and Sharjah. Royalties are negotiable. In Abu Dhabi, some contracts operate with flat rates of return of between 12.5% and 20%. Royalty and tax calculations are commonly based on the posted price for the crude.	55%-85% (tax rate)	No limit

* Figures shown are indicative. ** NOC/state share after cost recovery. *** As % of revenues.
Note: PSC = production-sharing contract; DPC = development and production contract; OSA = operating service agreement; EPSA = exploration and production-sharing agreement.
Source: IEA database.

CHAPTER 5

NATURAL GAS OUTLOOK IN THE MIDDLE EAST AND NORTH AFRICA

HIGHLIGHTS

- Aggregate natural gas demand in MENA countries is projected to nearly treble over 2003-2030 in the Reference Scenario, reaching 767 bcm. Most of the increase will come from power generation, but industrial use of gas – mainly for petrochemicals and fertilizer – will increase rapidly in some countries. Gas-to-liquids will absorb an increasing share of gas supply in some countries, notably Qatar.
- Marketed gas production will grow even more strongly than demand, and faster than oil, underpinned by large reserves in several countries. MENA proven natural gas reserves stood at 81 trillion cubic metres at the end of 2004, equal to 45% of total world reserves. Just two countries, Iran and Qatar, hold around two-thirds of MENA reserves and one-third of global reserves.
- MENA gas output will surge from 385 bcm in 2003 to 1 210 bcm in 2030. The biggest increases will occur in Qatar, Iran, Algeria and Saudi Arabia. The increase in MENA production in volume terms will be greater than that of any other major world region. In the Deferred Investment Scenario, MENA gas output is 20% lower in 2030 than in the Reference Scenario.
- A third of MENA production by 2010 will come from just one super-giant field: North Field/South Pars, shared by Qatar and Iran. Hassi R'Mel in Algeria will contribute another 13%. No new gas will need to be discovered to attain projected output to 2030. The share of non-associated gas in total output will rise substantially.
- The call on MENA gas supply will increase rapidly over the projection period, a result of strong global demand and dwindling output in many non-MENA regions. Net exports from MENA countries to other regions are projected to climb from 97 bcm in 2003 to 188 bcm in 2010 and 444 bcm in 2030. Most of the increase will be in the form of LNG.
- There will be a marked shift in the balance of Middle East exports from eastern to western markets. The bulk of incremental Middle East gas exports will go to Europe, which will remain the primary destination for North African gas exports. Qatar will overtake Algeria as the largest gas exporter in the MENA region by around 2010.
- Expanding gas-supply infrastructure will entail cumulative investment of $436 billion over 2004-2030, or $16 billion per year. Just under two-thirds of this investment will be made in upstream.

Natural Gas Demand[1]

Aggregate consumption of natural gas in MENA countries in the Reference Scenario is projected to increase by more than any other primary fuel, almost trebling to 767 bcm (639 Mtoe) in 2030 (Table 5.1). Demand will grow at an average annual rate of 3.7%, compared with 9.5% over the past three decades. The slow-down in demand growth mainly reflects the maturing of the gas industry in many countries.[2] The share of MENA in total world primary natural gas demand is projected to increase from 11% in 2003 to 16% in 2030.

Table 5.1: **MENA Primary Natural Gas Demand in the Reference Scenario** (bcm)

	2003	2010	2020	2030	2003-2030*
Middle East	**226**	**324**	**507**	**615**	**3.8%**
Iran	79.8	103.9	146.5	183.1	3.1%
Iraq	1.6	3.2	8.2	15.3	8.8%
Kuwait	9.6	11.8	20.6	26.2	3.8%
Qatar	13.7	36.9	93.4	103.4	7.8%
Saudi Arabia	60.1	86.3	129.3	155.4	3.6%
UAE	36.9	52.7	70.7	83.6	3.1%
Other Middle East	24.1	28.7	38.4	48.4	2.6%
North Africa	**62**	**85**	**121**	**152**	**3.4%**
Algeria	24.0	31.1	45.3	54.0	3.0%
Egypt	28.7	38.8	51.7	63.8	3.0%
Libya	5.7	10.1	16.1	23.1	5.4%
Other North Africa	3.5	4.9	8.1	10.7	4.2%
MENA	**288**	**408**	**628**	**767**	**3.7%**
World	**2 709**	**3 215**	**4 061**	**4 789**	**2.1%**

* Average annual growth rate.

The biggest increases, in volume terms, will occur in Iran and Saudi Arabia, already the dominant gas markets in the region, and Qatar. Together, their share in the region's gas use will increase, from 53% in 2003 to 58% in 2030.

1. All the primary natural gas demand numbers cited in this chapter include gas used both as a feedstock and as an energy input to gas-to-liquids plants. For methodological reasons, the energy balances in the tables shown in Annex A take into account only the energy input.
2. The share of natural gas in total primary energy demand in MENA increased from less than one-fifth in 1971 to 42% in 2003. The rate of growth was high, but demand started from a very low base. The scope for further switching to gas, while still significant, is less than in the past.

Qatar and Iraq will register the fastest *rates* of increase. Most of the increase in Qatari gas demand will be for gas-to-liquids (GTL) production.

In most countries, growth in demand over the projection period will come predominantly from the power-generation and water-desalination sector (Figure 5.1), which will account for about half of the increase. As a result, the sector's share in regional gas demand will rise from 44% in 2003 to 46% in 2030. Gas is expected to be the preferred fuel for new power stations in those countries with ample gas resources or access to supplies from neighbouring countries. Most new plants will use combined-cycle gas-turbine (CCGT) technology, whose capital costs are lower and construction lead-times shorter than those for oil-fired plants, which provide the bulk of power supplies in the region today. Gas-fired CCGTs are also less polluting (see Chapter 6). Moreover, most oil-producing countries aim to replace oil with gas, or at least limit the increase in oil-fired generation, in order to free up more oil for export.

Figure 5.1: **MENA Demand for Natural Gas by Sector in the Reference Scenario**

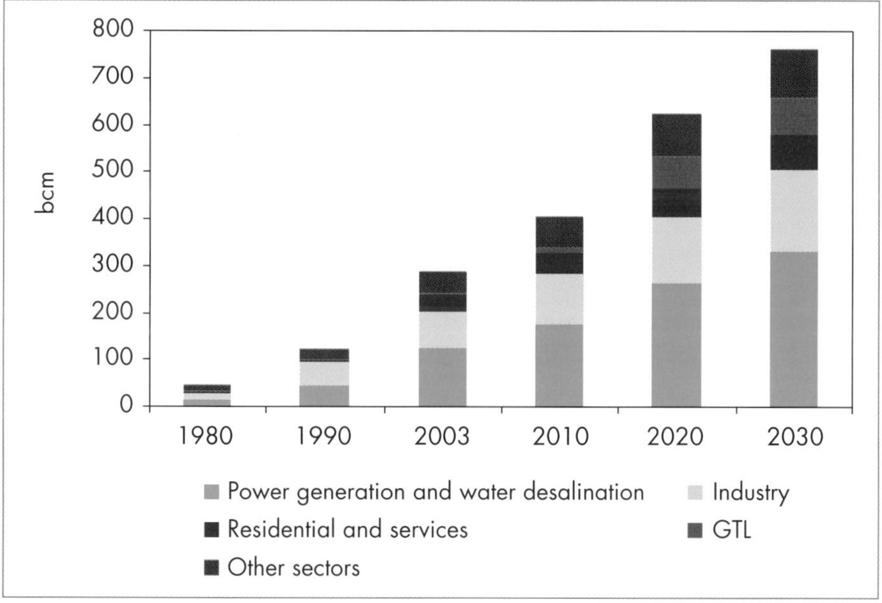

The share of natural gas as a fuel for power generation is projected to grow in most MENA countries between 2003 and 2030 (Figure 5.2). Qatar will remain wholly dependent on power generated from gas. Electricity output from gas-fired stations is projected to grow even more rapidly than gas inputs to power plants because of continuing improvements in the thermal efficiency of gas-fired stations, largely resulting from the more widespread use of CCGTs.

Figure 5.2: **Share of Natural Gas in MENA Power and Water Production in the Reference Scenario**

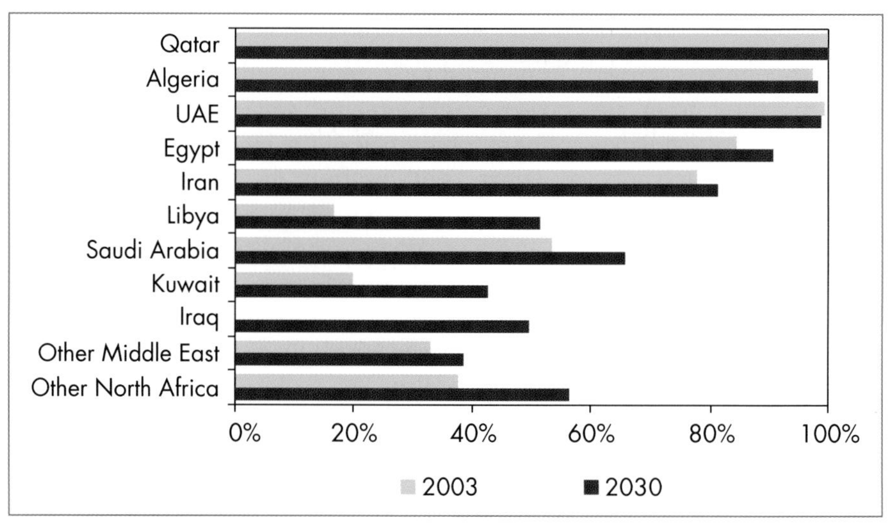

GTLs will absorb an increasing volume of gas supply in some countries, notably Qatar. The country's first plant, Oryx, is under construction and is expected to be commissioned in 2006. It will have a capacity of 34 kb/d and will take gas from the huge North Field. The plant is a joint venture between Qatar Petroleum (51%) and South Africa's Sasol and Chevron (49%). Two other projects – Shell's 140 kb/d Pearl project, and ExxonMobil's 154 kb/d plant – are at an advanced planning stage. If these projects proceed to schedule, total Qatari gas-to-liquids capacity will reach 104 kb/d in 2009, when Phase 1 of the Pearl project is commissioned, and 330 kb/d in 2011, when Pearl Phase 2 and the ExxonMobil plants are completed. These plants, when operating at full capacity, will together consume about 33 bcm a year of gas – equal to Qatar's entire output of natural gas in 2003. The Qatari government, which wants to allow more time for a careful evaluation of the impact of a rapid expansion of production on reservoir performance and the recovery rate of the North Field, has decided to slow down the approval process for three other projects – a second phase of the Oryx project and new projects being developed by Marathon and ConocoPhillips. Spiralling engineering, procurement and construction costs, and concerns about the socio-economic impact of the planned flurry of large-scale construction projects, are thought to have contributed to this decision.

Qatari consumption of gas for GTL production is expected to reach 10 bcm in 2010 and 63 bcm in 2030, equivalent to a production of about 100 kb/d in 2010 and more than 600 kb/d in 2030. In Algeria, one 36-kb/d GTL plant, consuming 3.1 bcm of gas per year, is expected to come on stream some time around 2010. GTL plants are also expected to be built later in the projection period in Iran, where gas use for this purpose is expected to amount to 12.5 bcm in 2030 (producing 120 kb/d). The overall energy transformation sector, which includes GTLs, liquefied natural gas production and oil and gas extraction, will in 2030 account for 19% of total primary gas demand in MENA as a whole, an increase from 16% in 2003.

5

Box 5.1: **Global Prospects for Gas-to-Liquids Technology**

Gas-to-liquids (GTL) plants are emerging as a major new outlet for natural gas, making use of low-cost reserves, located far from traditional markets, to produce oil products. There has been a surge in interest in such projects in recent years due to higher oil prices and the technological advances that have greatly reduced production costs. Holders of gas reserves that cannot be transported economically to market by pipeline have the option of turning to GTL as an alternative or a complement to liquefied natural gas (LNG). In practice, the choice between GTL and LNG is driven mainly by financial considerations: but GTL can also help to diversify an oil company's activities and reduce its overall portfolio risk.

Fischer-Tropsch technology is used in all GTL plants now in operation, under construction or planned. This converts natural gas into synthesis gas (syngas) and then, through catalytic reforming or synthesis, into very clean conventional oil products. The main fuel produced in most plants is diesel. The production process is very energy-intensive: about 45% of the gas supplied to GTL plants is currently consumed in the conversion process. The share of energy used in conversion is, however, assumed to drop by 2030 to around 35%, as a result of efficiency improvements.

Global demand for gas from GTL producers is projected to surge from just 3.5 bcm in 2003, corresponding to about 30 kb/d of refined-product output, to about 30 bcm in 2010 (300 kb/d) and around 220 bcm (2.3 mb/d) in 2030. The MENA region is projected to make up 0.76 mb/d, or around 35%, of the global output of GTLs in 2030.

The rate of construction of GTL plants is very hard to predict and our projections should be treated with caution. Further technology improvements could reduce the energy intensity of GTL processes; and rising demand for sulphur-free diesel oil could significantly boost the value of GTL output. On the other hand, further declines in LNG supply costs and an increase in the export price of gas relative to oil could reduce the attraction of GTLs. Major improvements in conversion efficiency will be crucial to the prospects for GTLs.

Industrial demand for gas in the MENA region is projected to grow at an average annual rate of 3% over the *Outlook* period, rising from 78 bcm in 2003 to 174 bcm in 2030. Industry will remain by far the largest end-consumer of gas. Much of the increase in industrial gas use will be in petrochemicals. Large expansions in capacity are planned in several MENA countries, notably Saudi Arabia, Iran and Algeria. These expansions will benefit from the low cost of gas feedstock in these countries.

Demand for gas in other end-use sectors – mainly residential and services – will remain relatively modest. In most MENA countries, there is little scope for establishing and extending local distribution networks because of the hot climate, which limits space heating needs. Gas load for cooking and water heating – the main uses of gas in these sectors – is rarely big enough to justify the high cost of building local distribution networks. In most cases, liquefied petroleum gas and electricity are more economic. The only major gas networks that exist today are in northern Iranian cities. There are small grids in Algeria, Egypt, Kuwait, Oman, Syria and Tunisia, but consumption in the residential and services sectors in those countries is very small. In 2003, Iran accounted for 85% of that demand. Total MENA gas demand for residential and services will grow by 3% per year over 2003-2030.

Natural Gas Supply

Natural Gas Reserves and Resources

MENA proven natural gas reserves stood at 81 trillion cubic metres at the end of 2004, according to Cedigaz, an international centre for gas information. This is equal to 45% of total world reserves (Table 5.2). Two countries – Iran and Qatar – account for around two-thirds of MENA reserves. The bulk of their gas reserves are in a single field in the Persian Gulf, known as South Pars in Iran and the North Field in Qatar, which straddles their maritime border. Its remaining reserves are put at 13 tcm for South Pars and 28 tcm for the North Field. Saudi Arabia, with 7 tcm, and the UAE, with 6 tcm, are the only other MENA countries with more than 5 tcm of reserves. In aggregate, MENA's reserves-to-production ratio is very high compared to the rest of the world. At present rates of production, MENA reserves would last 211 years, compared to 66 years for the world as a whole.

The bulk of proven gas reserves in the MENA region is non-associated. This is largely because the South Pars/North Field reserves are classified as non-associated, although the field contains large volumes of condensate. Outside Iran and Qatar, most reserves are associated with oil.[3] This reflects the fact that

3. Saudi Arabia is thought to have significant amounts of non-associated gas in Khuff reservoirs, but the share of this gas in official estimates of total proven gas reserves is not known.

Table 5.2: **MENA Proven Natural Gas Reserves by Country**

	Reserves, end-2004 (tcm)	**Share of world reserves**	**Share of world production**	**R/P ratio*** (years)
Middle East	**73.3**	**41%**	**9.5%**	**283**
Iran	28.2	16%	2.9%	364
Iraq	3.1	2%	0.1%	1 974
Kuwait	1.6	1%	0.4%	163
Qatar	25.8	14%	1.2%	776
Saudi Arabia	6.7	4%	2.2%	111
UAE	6.1	3%	1.6%	140
Other Middle East	1.9	1%	1.2%	57
North Africa	**8.0**	**4%**	**4.6%**	**64**
Algeria	4.6	3%	3.2%	52
Egypt	1.9	1%	1.1%	65
Libya	1.5	1%	0.2%	227
Other North Africa	0.1	0%	0.1%	32
MENA	**81.3**	**45%**	**14.2%**	**211**
World	**180.0**	**100%**	**100%**	**66**

* Reserves-to-production ratio.
Sources: Cedigaz (2005); IEA databases.

there has been little exploration for non-associated gas: most gas has been discovered when looking for oil.

Proven gas reserves have risen sharply since the 1970s, as more gas has been found and estimates of reserves in discovered fields have been upgraded in response to better appraisal and improved expected recovery rates. Reserves are now more than twice the level of 15 years ago (Figure 5.3). Most of the recent growth in reserves has come from Qatar's North Field. Most other countries have not adjusted their reserve estimates since the 1990s, or have done so marginally. As with oilfields, new gas fields that have been discovered in the last few years are, on average, smaller than in the past.

Proven reserves are a poor indicator of MENA's ultimate production potential. Some parts of the region have hardly been explored, and the exploration that has been undertaken has focused on oil. Proven and probable (2P) reserves are

4. The USGS has only assessed resources in conventional petroleum systems, but identified 24 continuous (non-conventional) systems, including coal-bed methane and basin-centre (shale) gas. One gas-shale field, Risha in Jordan, has been in production since 1989. The USGS believes that non-conventional gas resources in MENA are large and cover a wide area.

Figure 5.3: **MENA Proven Natural Gas Reserves**

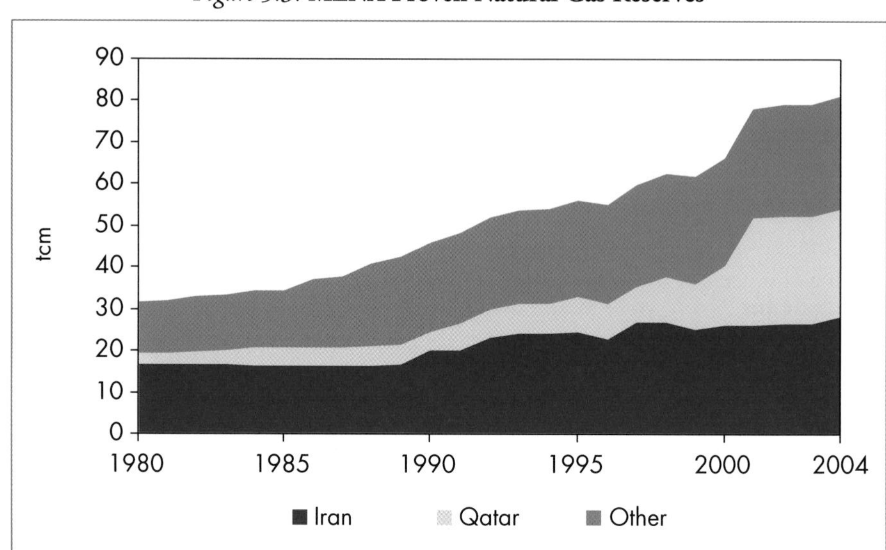

Source: Cedigaz (2005).

estimated by IHS Energy at 85.5 tcm on 1 January 2005. According to the US Geological Survey, undiscovered gas resources[4] in MENA, including reserves growth, are estimated (in a mean probability case) at 39 tcm, of which 25% is associated with oil. Remaining *discovered* recoverable resources, a category that includes proven, probable and possible reserves, are estimated at 82 tcm, according to Cedigaz. Cumulative production to 2003 was 5.7 tcm. Adding up these four categories yields an estimate of ultimately recoverable gas resources of 126 tcm, of which 5% has already been produced (Table 5.3).

Table 5.3: **Estimates of Ultimately Recoverable Natural Gas Resources in MENA** (tcm)

Category	Middle East	North Africa	MENA
Undiscovered (mean)	36.4	2.8	39.2
Remaining reserves	73.3	8.0	81.3
Cumulative production	3.8	2.0	5.7
Total ultimately recoverable resources	113.5	12.7	126.2
Remaining ultimately recoverable resources	**109.7**	**10.8**	**120.5**

Sources: USGS (2000); Cedigaz (2005).

Natural Gas Production

In the Reference Scenario, the volume of MENA gas production that is marketed[5] is projected to continue to grow rapidly over the projection period, at an average rate of 4.3% per year. Output will rise from 385 bcm in 2003 to some 600 bcm in 2010 and 1 210 bcm in 2030. More than two-thirds of this increase will come from the Middle East alone, which will see the largest growth of any world region.

The biggest increases in production will occur in Qatar, Iran, Algeria and Saudi Arabia (Figure 5.4 and Table 5.4). Qatar and Iran will overtake Algeria as the largest producers in the region by 2010. The biggest increase in volume terms will occur in Qatar. In some cases, notably Iran, gross production will increase considerably more than marketed production because of the scale of reinjection needs to boost oil production. In Iraq, some of the increase in production for reinjection is expected to be offset by reduced flaring. Abu Dhabi and Algeria are also implementing programmes to reduce gas flaring.

Figure 5.4: **MENA Natural Gas Production by Country in the Reference Scenario**

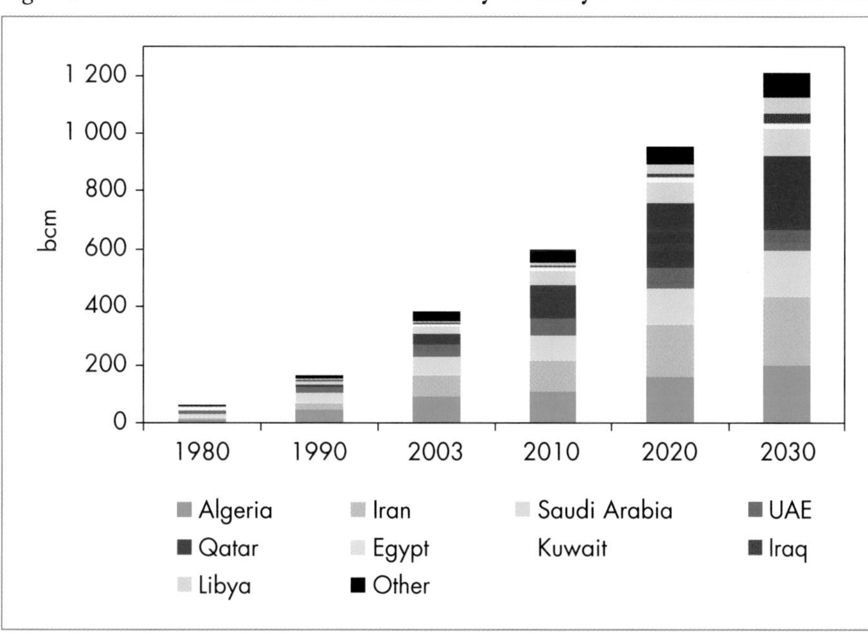

5. In many countries, a part of gross domestic production is reinjected or flared. All the figures on gas production cited in this chapter refer to marketed production only.

Table 5.4: **MENA Natural Gas Production in the Reference Scenario** (bcm)

	2003	2010	2020	2030	2003-2030*
Middle East	**259**	**425**	**692**	**860**	**4.5%**
Iran	78	109	177	240	4.3%
Iraq	2	4	15	32	11.8%
Kuwait	10	11	18	21	2.9%
Qatar	33	115	220	255	7.8%
Saudi Arabia	60	86	129	155	3.6%
UAE	44	59	72	75	2.0%
Other Middle East	33	41	61	82	3.4%
North Africa	**125**	**171**	**264**	**352**	**3.9%**
Algeria	88	107	160	198	3.0%
Egypt	29	49	71	92	4.4%
Libya	6	12	29	57	8.4%
Other North Africa	2	3	4	5	3.1%
MENA	**385**	**596**	**956**	**1 211**	**4.3%**
World	**2 717**	**3 215**	**4 061**	**4 789**	**2.1%**

* Average annual growth rate.

Part of the increase in capacity required to offset declining output from fields already in production will come from further development of those fields. Additional capacity will come from the deployment of enhanced oil recovery techniques, which will yield additional associated gas. But a growing share of production will come from discovered fields that have yet to be developed and new fields that have yet to be found. Exploration for non-associated gas is expected to be stepped up in several countries. Figure 5.5 summarises the breakdown of future production according to these categories.[6] By 2030, less than half of MENA gas production will be sourced from fields currently in production. No new gas will need to be discovered to attain projected output to 2030. The share of output from new developments – 53% in 2030 – is significantly higher than for oil, mainly because the projected growth in MENA gas output is higher. Details of gas-production profiles for major fields can be found in the individual country chapters.

6. See Chapter 4 (oil) for a description of the methodology used to derive projections of production by type of reserve.

Figure 5.5: **MENA Natural Gas Production by Field in the Reference Scenario**

bcm
1 400
1 200
1 000
800
600
400
200
0
2004 2006 2008 2010 2012 2014 2016 2018 2020 2022 2024 2026 2028 2030

Hassi R'Mel
North Field
South Pars
Other currently producing fields
New developments

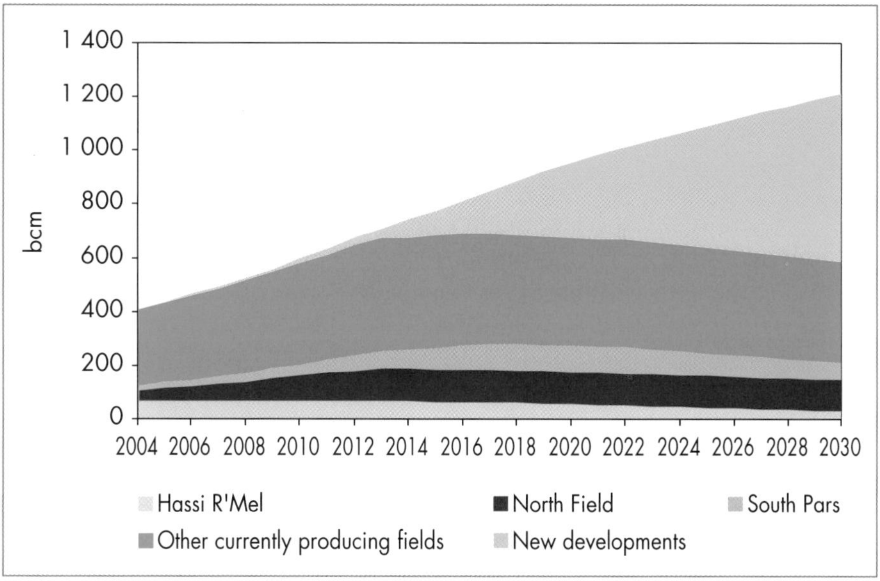

Natural Gas Trade

MENA Natural Gas Export Flows

The projections of gas production and demand in each region, which take account of new historical data and revised assumptions concerning MENA population and energy prices have been updated since *WEO-2004* and can be found in Chapter 1. In the Reference Scenario, as with oil, MENA countries are expected to fill the gap between the total global gas demand and production in all other *WEO* regions. Transporting gas from the Middle East to the main importing markets is generally more expensive than transportation from other exporting regions, because of the longer distances involved. So the Middle East is likely to remain the marginal exporter to regions short of gas. Our gas-price assumptions reflect our judgment of the price paths that would be needed to balance global gas demand and non-MENA supply, taking account of oil prices. In other words, we estimate that MENA as a whole is capable, at the prices assumed, of supplying the required gas, at a profit, to those regions facing a deficit in supply.

Our gas projections in the Reference Scenario imply a major expansion of inter-regional trade, as a result of the growing geographical mismatch between the location of resources and that of demand. MENA is expected to play an increasingly important role in meeting the growth in gas demand in importing regions. Net exports from MENA countries to other regions are projected to climb from 97 bcm in 2003 to 188 bcm in 2010 and 444 bcm in 2030.

Our near-term projections of MENA exports to importing regions are derived largely from an assessment of current long-term contracts and projects under development, under which most gas will be supplied. The bulk of projected exports for 2010 have already been contracted, from LNG facilities or through pipelines that are in operation or under construction. These projections can, therefore, be considered relatively certain. However, delays in completing projects and under- or over-lifting of contracted volumes, due to higher or lower demand than expected, might cause actual trade flows to deviate from the projected volumes. Further ahead, additional export flows are modelled on the basis of the netback value of gas from MENA countries, taking account of the cost of transportation, compared with that of gas from competing exporting regions. Consequently, most North African exports are projected to go to Europe and, to a lesser extent, to the United States. Middle East exports will serve markets in the Asia-Pacific, Europe and the United States.

A growing share of MENA exports will be in the form of LNG. Altogether, the share of LNG in total MENA exports is projected to grow from around one-third in 2003 to more than 60% in 2030. Most Middle East exports will be as LNG, though some pipelines are expected to be built to Europe and South Asia after 2010. North African gas will continue to be exported both by pipeline and as LNG.

Table 5.5 details LNG plants that are already in operation, that are being built and that are expected to be completed in MENA countries by 2010. New export pipeline capacity will also boost exports from Algeria to Europe by 2010. A new 10-bcm/year sub-sea pipeline from Algeria to Italy, via Sardinia, is planned. And a second pipeline from Algeria to Spain, known as Medgaz, with an initial capacity of 4 bcm/year, is also expected to be commissioned in 2008. Algeria already exports through two pipeline systems; one to Italy, which is being expanded, and a second to Spain.

Most gas exports will continue to be traded under long-term contracts, but spot transactions are expected to grow. Several LNG carriers being built today are not tied to long-term contracts and some older vessels are likely to become available for shipping spot traded volumes when the contracts to which they are currently assigned expire. As a result, some volumes of gas may, at various times, be traded on the spot market for delivery to various markets, according to short-term arbitrage opportunities.

There will be a radical shift in the balance of MENA exports between eastern and western markets (Figure 5.6):

- At present, most *Middle East* gas exports move eastwards. This situation could be reversed as early as 2010. Gas exports to Europe will increase most in volume terms, from 2 bcm in 2003 to 35 bcm in 2010 and 117 bcm in 2030, at which time Europe will become the largest market for Middle East

Table 5.5: **MENA LNG Existing and Planned Capacity**

	Project	Current status	Date of start-up	Number of trains	Capacity in 2010 (bcm/year)
Algeria	Arzew GL 1-4Z	In operation	1964	15	22.8
	Skikda GL 1K	In partial operation	1972	3+1	4+5.5
	Gassi Touil	Planned	2007/2008	1	5.5
	Sub-total			*20*	*37.8*
Egypt	Egyptian LNG Idku	In partial operation	2005	2	9.8
	SEGAS Damietta	In partial operation	2005	2	6.5
	Sub-total			*4*	*16.3*
Libya	Marsa-el-Brega	In operation		1	0.8
Oman	Qalhat	In operation	2000	3	10
Qatar	Qatargas	In operation	1997	1	12.9
	Qatargas II	Under construction	2008/2009	2	21.2
	Qatargas 3	Planned	2009/2010	1	10.6
	RasGas	In operation	1999	2	9
	RasGas-II	In partial operation	2004	3	19.5
	RasGas-III	Under construction	2008/09	2	21.2
	Sub-total			*11*	*94.4*
UAE	Adgas (1 and 2)	In operation	1977	3	7.8
Yemen	Yemen LNG (Balhaf)	Under construction	2008	2	9.1
Total				**44**	**176.2**

Source: IEA databases.

exports, just ahead of Asia. Exports to Asia, the main market for Middle East gas at present, will grow more slowly, from 50 bcm in 2003 to 106 bcm in 2030.

- Europe will remain the primary destination for *North African* gas exports, absorbing most of the volume increase. Exports to OECD Europe are projected to rise from 61 bcm in 2003 to 83 bcm in 2010 and 170 bcm in 2030.

MENA exports of LNG to the United States will grow substantially, from 2 bcm in 2003 to 51 bcm in 2030. As a share of total US gas demand, however, MENA gas will account for only a small share. Among MENA countries, Qatar will be the largest gas exporter in 2030 (Figure 5.7). Its exports will grow from 19 bcm in 2003 to 78 bcm in 2010 and 152 bcm in 2030. Qatar will also register the biggest increase in exports. Egypt, which began exporting LNG in

Figure 5.6: **MENA Natural Gas Exports by Destination, 2010 and 2030** (bcm)

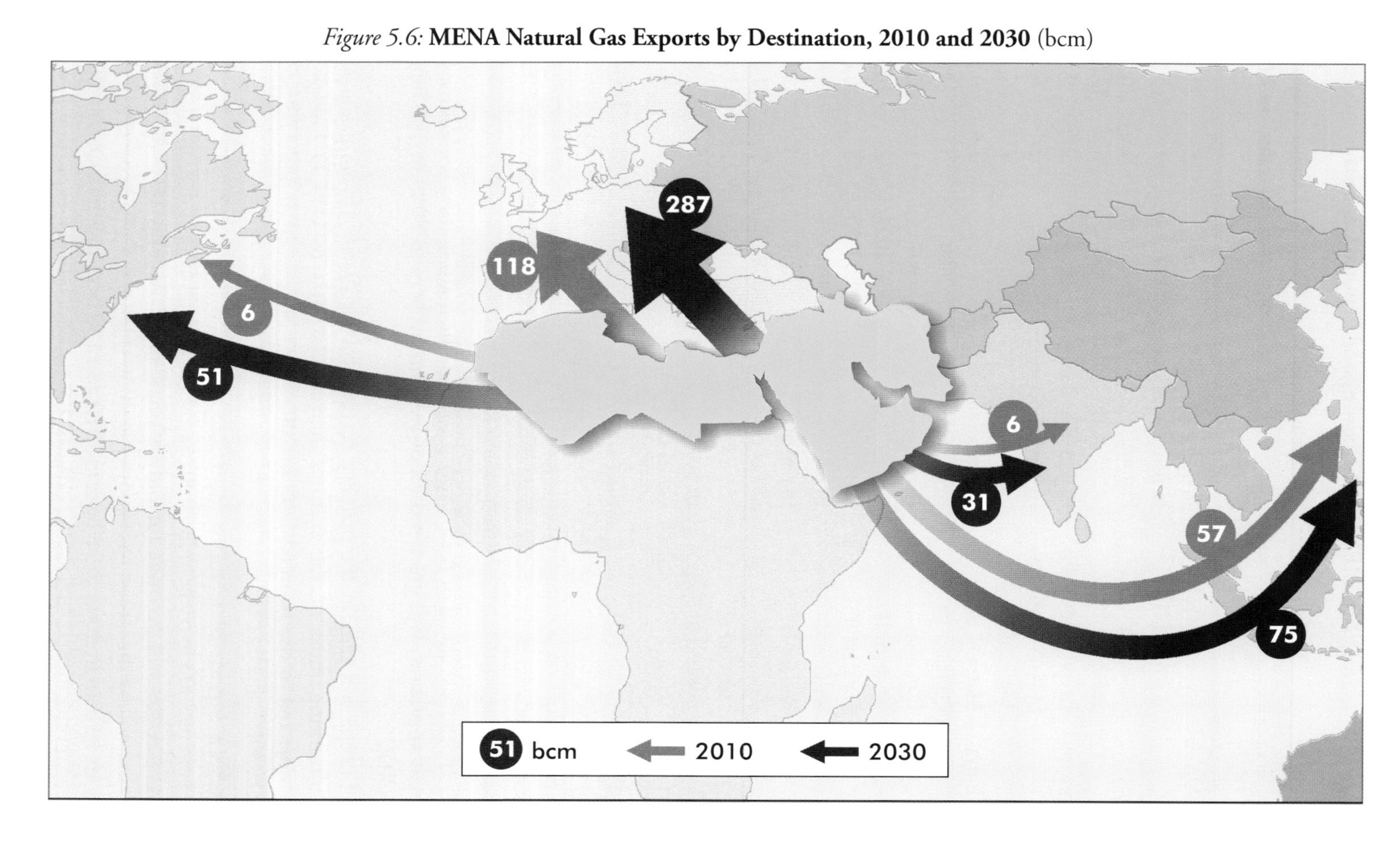

Figure 5.7: **MENA Net Natural Gas Exports in the Reference Scenario**

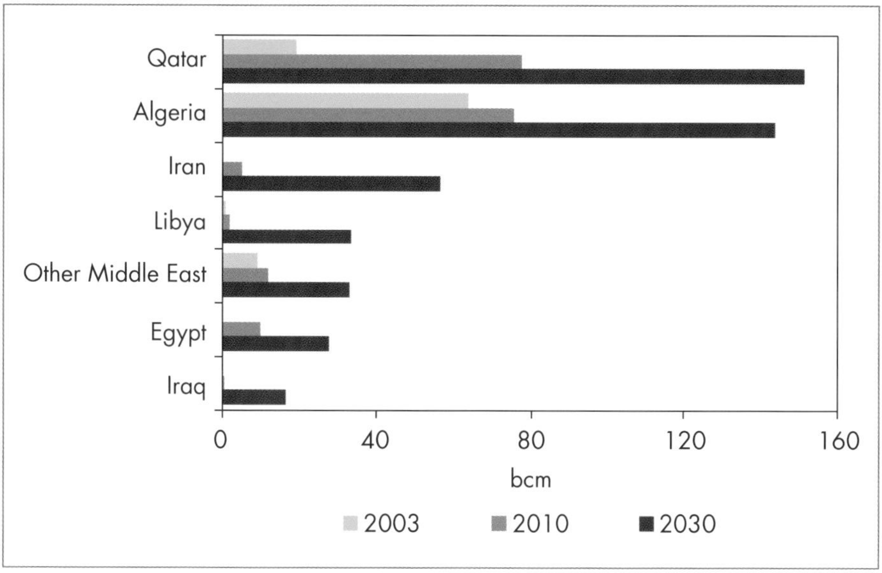

Table 5.6: **Net Exports as Share of Natural Gas Production by MENA Country in the Reference Scenario (%)**

	2003	2010	2020	2030
Middle East	**12.9**	**23.9**	**26.7**	**28.4**
Iran	(3.0)	4.7	17.3	23.7
Iraq	0.0	15.7	46.1	52.3
Kuwait	0.0	(5.3)	(17.4)	(27.8)
Qatar	58.6	67.9	57.5	59.4
Saudi Arabia	0.0	0.0	0.0	0.0
UAE	16.0	11.0	1.7	(11.1)
Other Middle East	27.9	29.6	36.8	40.7
North Africa	**50.6**	**50.4**	**54.1**	**56.9**
Algeria	72.7	70.9	71.6	72.7
Egypt	0.0	20.8	26.9	30.6
Libya	11.7	16.5	45.4	59.4
Other North Africa	(53.6)	(68.4)	(97.4)	(105.1)
MENA	**25.2**	**31.5**	**34.3**	**36.7**

Note: Numbers in parenthesis indicate net importers.

2005, will also see its exports grow markedly, to 10 bcm in 2010 and 28 bcm in 2030. Iran will emerge as a gas exporter, as will Iraq, although to a lesser extent. The share of exports in total gas production will climb from 25% in 2003 to 37% in 2030 (Table 5.6). The share will remain highest in Algeria, at 73%.

Intra-Regional Natural Gas Trade

Growing exports from MENA countries to other regions will be accompanied by an increase in trade *between* MENA countries. Several countries in the region face a growing deficit in indigenous gas supplies and need to import gas to meet power generation and industrial needs. The Dolphin Project, involving the construction of pipelines from Qatar to the UAE and Oman, is due to be commissioned in 2006 (see Chapter 17). The UAE will initially import about 20 bcm per year from Qatar through a 370-km pipeline, with first gas expected by the end of 2006. A plan to increase imports to around 30 bcm per year is under discussion. Iraq is expected to export small quantities of gas to Kuwait, which does not produce enough gas to fuel its power stations. Kuwait is also seeking to import gas from other neighbouring countries.

In North Africa, Algeria is expected to expand its exports to Tunisia, which are currently running at just over 1 bcm per year. In addition, there are plans for Egypt to pipe gas from the Nile Delta, overland across the Sinai Peninsula or via a sub-sea pipeline, to Israel. However, negotiations have been hindered by regional politics and disagreements over financing and pricing. Additional sales to the Gaza strip have also been mooted. A pipeline from Egypt to Jordan began commercial operation in July 2003. There are plans to extend the pipeline to Syria and Lebanon and, possibly, Turkey. It is expected that the pipeline could eventually deliver a total of 3 bcm per year.

Investment and Financing

Fulfilling these gas-supply projections for MENA countries will entail cumulative investment of $436 billion over 2004-2030, or $16 billion per year. Over 60% of this investment – $270 billion – will be needed in the upstream sector, including to replace existing capacity investment. The rest will go to liquefaction plants, transmission and distribution networks and storage facilities.

Qatar, where production is projected to grow the most, will have by far the largest investment needs (Figure 5.8). Almost all the additional Qatari output will be converted to LNG and oil products in GTL plants.[7] Upstream development costs, including processing, will account for 65% of the

7. GTL investment costs are included in the total for oil.

$100 billion that Qatar will need to spend on its gas industry. The cost of incremental capacity is high relative to some other Middle Eastern countries, but will be mostly recouped from the sale of condensates and NGLs produced together with the gas. LNG will account for the bulk of the remaining $34 billion. Iran, Algeria and Saudi Arabia have the highest capital needs for gas, after Qatar.

Figure 5.8: **Cumulative Investment in Natural Gas Supply Infrastructure in MENA Countries in the Reference Scenario, 2004-2030**

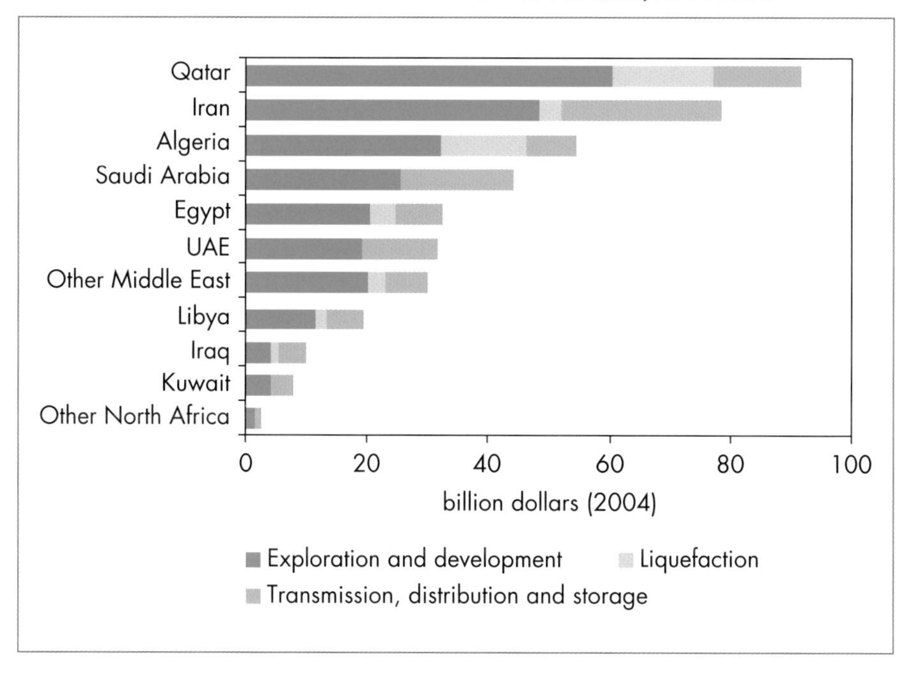

Despite political tensions in the region, which could undermine credit ratings for new projects in certain parts of the region and raise the cost of capital, this investment is expected to be forthcoming. Most MENA countries, particularly the Gulf states, enjoy good risk ratings (Table 5.7). Project risks in the region vary according to geopolitical and technical factors: cross-border pipeline projects are considered highly risky in view of regional political tensions. For this reason, the probability and timing of the construction of export pipelines to Europe and India are very uncertain. LNG will undoubtedly remain easier to finance than pipeline projects.

Table 5.7: **Sovereign Ratings of MENA Countries*, July 2005**

	Moody's	Standard & Poor's	Fitch
Bahrain	Baa1	A-	A -
Egypt	Ba1	BB+	BB+
Iran	Not rated	Not rated	BB-
Israel	A2	A-	A-
Jordan	Ba2	BBB	Not rated
Kuwait	A2	A+	AA-
Lebanon	B3	B-	B-
Morocco	Ba1	BBB	Not rated
Oman	Baa2	A-	Not rated
Qatar	A1	A+	Not rated
Saudi Arabia	Baa2	A+	A
Tunisia	Baa2	A	BBB
UAE	A1	Not rated	Not rated

* Long-term foreign currency debt.
Note: Ratings for Algeria, Yemen, Iraq, Libya and Syria not available.
Source: Company reports.

Most projects will continue to be funded out of a mixture of retained earnings, state budget allocations and, in the case of most export-oriented projects, project finance and/or international bond issues. Qatar has been very successful in raising finance through bonds for its LNG projects. However, the ability of some states to finance growing capital needs for new projects might be constrained in the future by budget deficits and competing demands for state financial resources, especially if oil prices were to fall sharply. National oil companies still dominate the gas industry in most of the major producing countries and they might have to rely more on project finance and borrowing from international capital markets.

An increasing proportion of gas investment, as for oil, is expected to come from private sources. Most countries in the region allow foreign companies to play some role in new gas projects, often in partnership with the NOC. This approach facilitates access both to capital and to the technological and project-management expertise of foreign companies. The approach to market opening and the pace of negotiations vary with political and cultural factors. Some countries, such as Qatar and the UAE, have a long history of co-operation with international oil companies in gas projects. Iran and Saudi Arabia have only recently opened their upstream gas sectors to foreign participation.

Pricing terms, including the amount of government tax-take, are a key factor for all gas projects. The rent available on gas projects is smaller than for most upstream oil investments. This makes gas investments in the region highly sensitive to oil and gas prices. Gas projects may, in some cases, struggle to compete for capital against oil projects, which typically yield higher revenues to the host countries. New gas projects, especially cross-border pipelines and ventures involving foreign investors, may be delayed by protracted negotiations over investment terms and intergovernmental agreements. The required investment in MENA gas projects would be affected by any shortfall in investment in MENA oil projects. The implications are examined in the Deferred Investment Scenario (Chapter 7).

CHAPTER 6

ELECTRICITY AND WATER OUTLOOK IN THE MIDDLE EAST AND NORTH AFRICA

HIGHLIGHTS

- To meet rapidly expanding electricity demand, MENA electricity generation is projected to increase by 3.4% per year on average in 2003-2030, reaching 1 800 TWh. The region will need some 300 GW of new generating capacity, about 6% of the world total.
- The share of gas in electricity generation will rise from 56% now to 58% in 2010 and to 69% in 2030 in the Reference Scenario. The share of oil will fall, but in some countries, notably Kuwait, Saudi Arabia, Iraq and Libya, it will still be fairly high in 2030.
- The power sector in most MENA countries is not commercially viable. Under-pricing is particularly marked in Iran, Egypt and the Gulf countries. Revenue collection is very poor because of low retail prices and illegal connections. Power companies rely on government budgets to cover their costs and finance investment. Price reforms are essential.
- Investment needs for power generation, transmission and distribution in MENA countries are projected to amount to $458 billion (in year-2004 dollars) – nearly as high as the upstream investment in the oil sector. Power generation will take $203 billion, while networks will need $255 billion. Saudi Arabia's power sector will need the largest investment, some $110 billion.
- In the absence of market reforms, these investments will remain a burden on government budgets. Investing in the demand side may be more cost-effective than investing in new power plants and networks. Pricing reforms would cut demand by encouraging the purchase of more efficient appliances.
- Desalination plants will be increasingly relied upon to meet freshwater needs, especially in Saudi Arabia, the UAE, Kuwait, Qatar, Algeria and Libya. Energy use in such plants will account for more than a quarter of the total increase in fuel use in the power and water sector in these countries. Desalination capacity in these countries will more than triple over 2003-2030, requiring investment of $39 billion. More than half of new power-generation capacity will be in combined water-and-power (CWP) plants.

Introduction

On the assumptions in the Reference Scenario, MENA electricity use will grow by two-and-a-half times in 2003-2030. Installed generating capacity will grow from 178 GW in 2003 to 447 GW in 2030. Gas-fired plants will provide almost 70% of the new capacity. The investment required in the sector is equivalent to 1% of GDP. Over half of this is required for transmission and distribution.

Water supply is a critical issue for the region and more than a quarter of the increase in energy consumption in the power and water sector will be in new water desalination plants. For this reason, this chapter concludes with a section examining water demand and supply, looking particularly at the prospects in the six MENA countries which are most water-stressed.

Electricity Demand[1]

Demand for electricity in the countries of the MENA region (as measured by generating output) is projected to increase from 724 TWh in 2003 to 1 799 TWh in 2030, a rate of 3.4% per year. The share of MENA in global supply will increase from 4.3% to almost 6%. Estimated demand in 2004 was 770 TWh.

Table 6.1: **MENA Electrification Rates, 2002**
(% of population with access to electricity)

Kuwait	100	Lebanon	96
Israel	100	Qatar	96
Bahrain	100	Jordan	95
Libya	100	Iraq	95
Iran	99	Tunisia	95
Algeria	99	Oman	95
Saudi Arabia	98	Syria	87
Egypt	98	Morocco	77
United Arab Emirates	97	Yemen	50

Source: IEA (2004).

1. There is a particular problem of definition in discussing MENA electricity supply. Conversion losses always create a disparity between primary energy demand for electricity generation and electricity output; and transmission and distribution losses reduce final demand below the level of output from generation. However, losses in MENA countries before the supply reaches the final customer, whether for technical or other reasons, are particularly high. This makes it necessary sometimes, in discussing electricity demand in MENA countries, to discuss output from generation rather than measured final demand.

Figure 6.1: **Per Capita Electricity Demand in MENA Countries and the OECD, 2003**

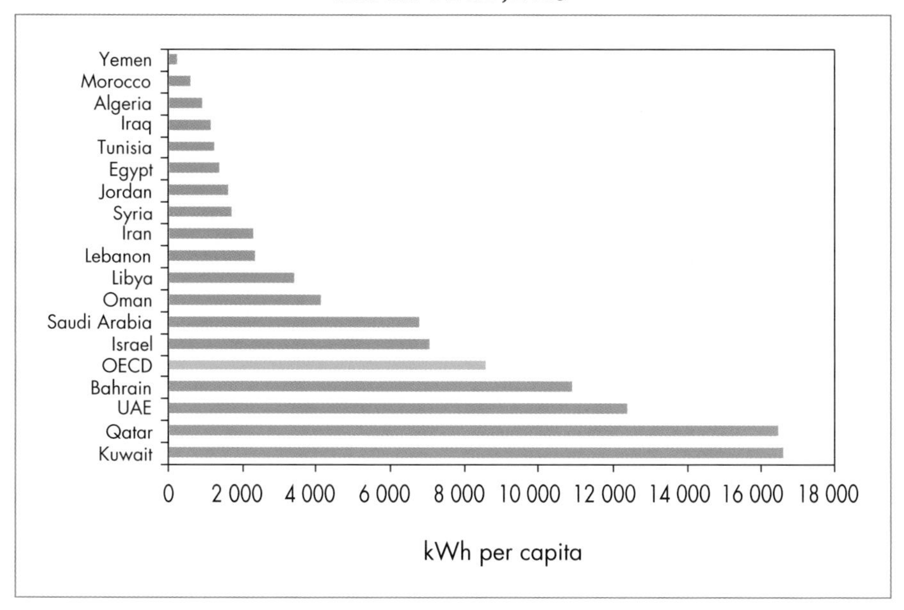

Figure 6.2: **Electricity Losses in MENA and the OECD, 2003**

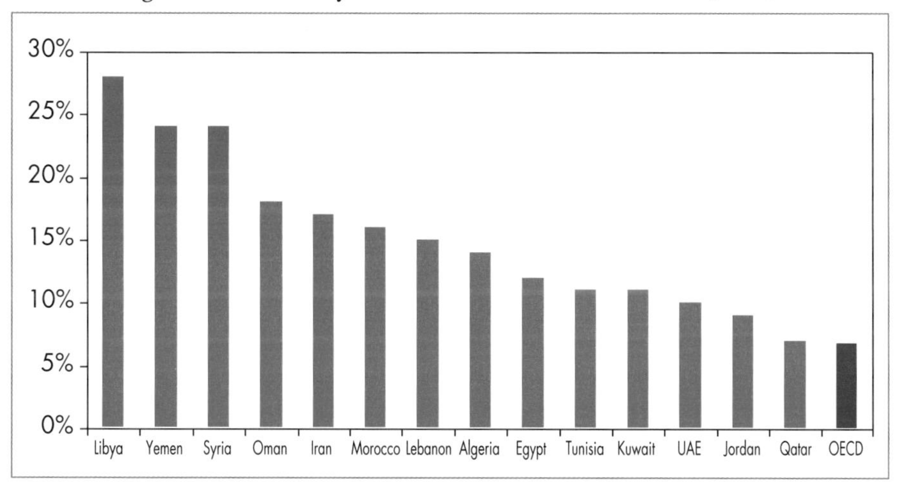

Per capita electricity demand in the Gulf countries was already high in 2003, in some cases higher than in the OECD (Figure 6.1). Kuwait then had the fourth-largest per capita consumption in the world (after Iceland, Norway and Canada) and Qatar the fifth. By contrast, the countries of North Africa, except Libya, were at the low end of the range. Yemen, the poorest country in

MENA, has the lowest per capita electricity demand. The region as a whole already enjoyed some of the highest electrification rates outside the OECD (Table 6.1).

Demand for electricity in the region grew rapidly in recent years. The average growth rate in the period 1998-2003 was 6.3% per year, very close to the growth rate across all developing countries in the same period. In some

Box 6.1: **Network Losses in MENA**

Technical and non-technical losses are high almost everywhere in MENA. In general, networks are well maintained but because of the climate, there are some problems which affect the performance of networks, such as sandstorms that stress transmission lines. Further, because of the high temperatures in the summer, networks often operate in extreme conditions.

The main cause of high losses is, however, illegal connections. Some countries also include unpaid electricity in their losses. Figure 6.2 clearly illustrates the problem. The share of transmission and distribution losses across all OECD countries was 7% in 2003. This share is higher in all MENA countries.[2] One extreme case – although not the only one – is that of Libya, where over a quarter of electricity produced is considered as lost in the system. Some sources suggest that about 60% of Libyans do not pay their electricity bills, although the residential electricity price is one of the lowest in the region. Often, the largest share of receivables does not come from residential consumers but from the public sector, where utilities may face difficulties in getting paid from public consumers such as the army or some government institutions.

Despite the significance of the problem, few countries are making efforts to remedy the situation because of strong resistance from citizens and the potentially high political cost. Better subsidy targeting will, therefore, be essential if price reform is to be politically feasible. Algeria is one of the countries trying to address the non-payment problem. SONELGAZ, the electricity company, has taken action against fraud and retrocession and is trying to improve electronic metering.[3] The rate of losses, however, remains high, at around 14% in 2003.

2. Because of data problems, some countries appear to have extremely low losses, which is not the case. These countries have been omitted from the chart.

3. Ministry of Energy and Mines (2004). Retrocession is a practice where a household connects to a neighbour's electricity supply, sharing the electricity bill and avoiding the fixed charges or the trouble of obtaining the connection.

Figure 6.3: **Average Annual Growth Rates in Electricity Demand, 1998-2003**

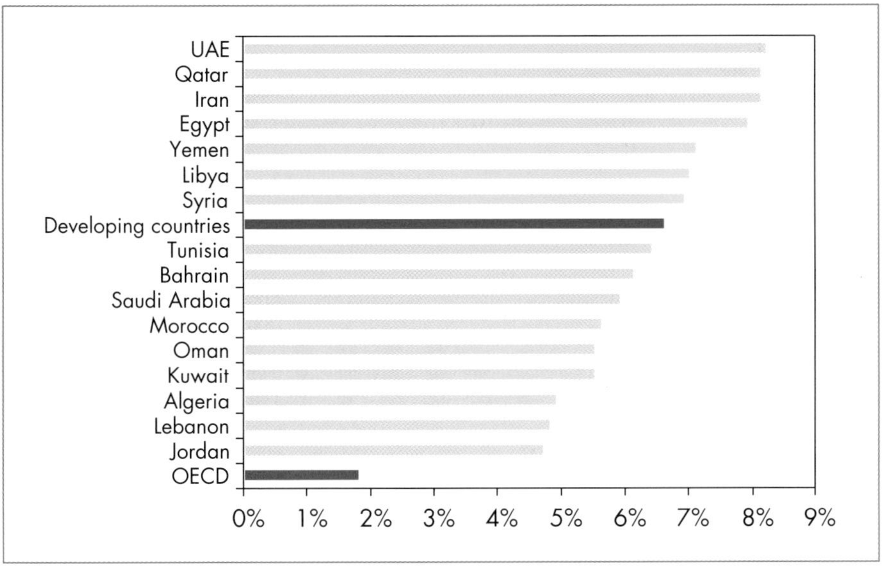

UAE
Qatar
Iran
Egypt
Yemen
Libya
Syria
Developing countries
Tunisia
Bahrain
Saudi Arabia
Morocco
Oman
Kuwait
Algeria
Lebanon
Jordan
OECD
0% 1% 2% 3% 4% 5% 6% 7% 8% 9%

countries, such as the UAE, Iran, Qatar and Egypt, electricity demand grew at rates as high as 8% per year (Figure 6.3). In Iraq (not shown in the chart), exceptionally, electricity demand decreased in 2003 because of the conflict.

One reason for this has been that electricity prices in most MENA countries are far below OECD prices and often they do not even cover the long-run marginal cost of supply (Figure 6.4). This is because electricity is considered as a service that the government provides to its citizens and subsidies are a way to distribute the oil rent in resource-rich countries. For example, electricity is free to Qatari nationals. In summer 2005, Kuwait's parliament approved unanimously a law waiving 2 000 Kuwaiti dinars (about $6 800 in current dollars) worth of electricity bills for each Kuwaiti family. The total cost to the government is expected to be around $1 billion.

Fully cost-reflective electricity prices would not be much lower than in OECD countries, because a large part of the electricity price is related to the investment cost. Though operating costs are low in MENA countries, mainly because fuel, accounted for at prices below the opportunity cost, is cheap, charges related to generation, transmission and distribution infrastructure should be close to those of the OECD. For example, distribution network charges in Europe range between 3 and 5 US cents per kWh in most areas. These charges are not reflected in electricity prices in most countries in the

Figure 6.4: **Residential Electricity Prices in MENA and the OECD, 2003**

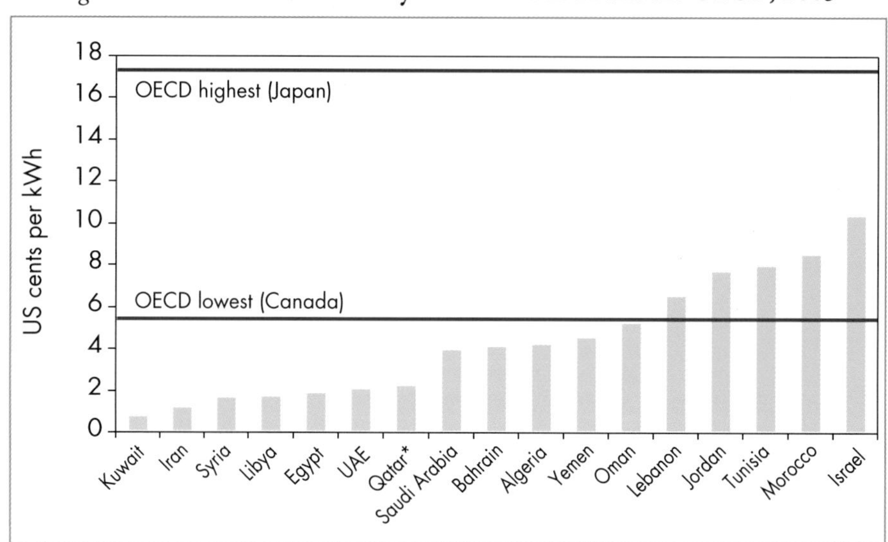

*The price shown is for expatriates.
Sources: Arab Union of Producers, Transporters and Distributors of Electricity (AUPTDE); national power companies; IEA estimates.

MENA region. The projections assume only slow progress in tackling this issue.

In the period to 2010, electricity demand is projected to increase by 5.1% per year. This growth is underpinned by current high oil revenues and the resulting economic boom, which on the one hand allows consumers to purchase more appliances and on the other hand facilitates investment in new power plants.

Saudi Arabia and the UAE are expected to see the highest growth rates in this decade. Although per capita electricity demand is already very high in the UAE – over 12 000 kWh per person in 2003 – demand for electricity will continue to grow. This is the combined result of high oil revenues and the current construction boom, particularly in the services sector.

Longer term, Iraq, Libya and Saudi Arabia are expected to have the highest growth rates. Iraq's electricity demand is projected to rise gradually as the country recovers politically and economically and new capacity is commissioned. Given the current difficulties and instability facing the country, this projection is highly uncertain. Libyan demand is also expected to grow rapidly, following the removal of sanctions. In Saudi Arabia, demand will grow fast in response to rapidly growing population.

Figure 6.5: **Electricity Demand in MENA Countries**

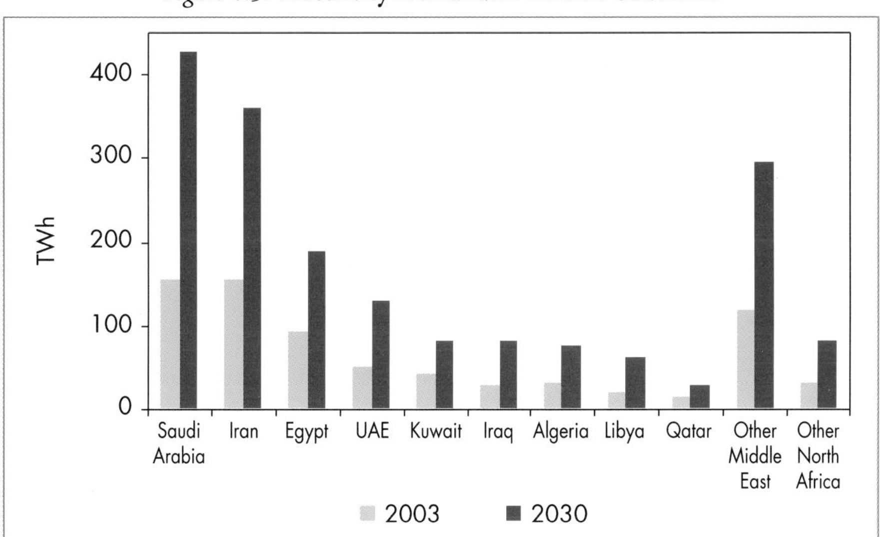

The projected increases in electricity demand will bring some changes in the relative weights of the electricity sectors of the countries in the region. Saudi Arabia's and Iran's electricity demand were almost equal in 2003. But Saudi Arabia's electricity demand is projected to increase much faster than Iran's, largely because of its rapid population increase. Consequently, Saudi Arabia will have the largest electricity market in MENA in 2030 (Figure 6.5). Its annual production level is expected to hit 426 TWh in 2030, more than the current level in the United Kingdom.[4] The country's electricity sector is likely to maintain its position as one of the ten largest outside the OECD.

The strong growth in electricity demand has, hitherto, come mainly from the residential and services sectors. In Saudi Arabia, the residential sector currently accounts for 56% of total electricity consumption (mainly because of air-conditioning). The services sector is very strong in the United Arab Emirates, accounting for 43% of the total. The share of industry is high in just a few countries, such as Morocco, where industry consumed almost half of total electricity in 2003.

4. In terms of per capita consumption, Saudi Arabia and the UK are now at about the same level.

Over the projection period, the residential sector will grow the most rapidly, increasing its share in electricity consumption from 41% now to 44% in 2030. Demand for air-conditioning will contribute substantially to this increase (Box 6.2). Industry is projected to maintain its current share of about 25%, while the share of the services sector will drop slightly, from 29% to 28%.

Box 6.2: **Air-Conditioning in MENA**

Among various uses of electricity, cooling contributes substantially to demand for electricity in the summer when temperatures and humidity are high. Air-conditioning is widespread in the Gulf and rising in other MENA countries and the severe conditions in which these systems operate make their efficiency deteriorate quickly. Figure 6.6 shows that the efficiency of chillers used in commercial buildings in MENA is below the world average. This is also the case for smaller size systems used in households. Low electricity prices give no incentive to consumers to purchase more efficient appliances and building efficiency standards are, in general, absent in MENA. District cooling can improve the efficiency of the system, but its use is very limited now and it will remain so in the absence of price signals.

Figure 6.6: **Comparison of Chiller Efficiency**

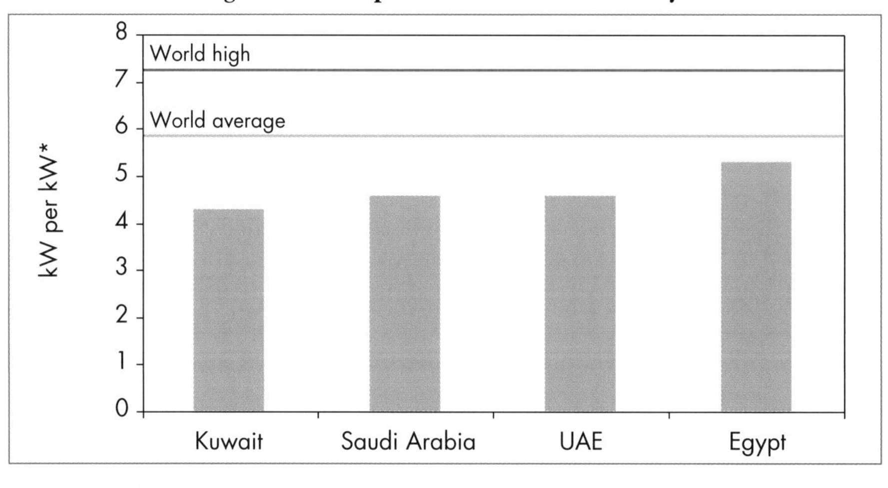

* The energy efficiency ratio of a cooling system is the ratio of the appliance's cooling capacity divided by the watts of power consumed at a specific outdoor temperature and is commonly measured in kW per kW.
Note: Chillers are large air-conditioners used in buildings.
Source: UNEP (2004).

Electricity Supply

Power Generation Capacity

Installed capacity in MENA is projected to increase from 178 GW in 2003 to 447 GW in 2030. During that period, some 18% (32 GW) of existing power plants will be retired. Total gross capacity additions in Middle East and North Africa will be 301 GW, which is roughly 6% of global capacity additions and 12% of the additions needed outside the OECD in the period to 2030. The region will need 78 GW in the period to 2010, 105 GW in the period 2011-2020 and 118 GW in the last decade. The largest increases will be in Saudi Arabia and Iran (Figure 6.7). These two countries represent over 40% of the requirements.

6

Figure 6.7: **MENA Power-Generation Capacity Additions by Country, 2004-2030** (GW)

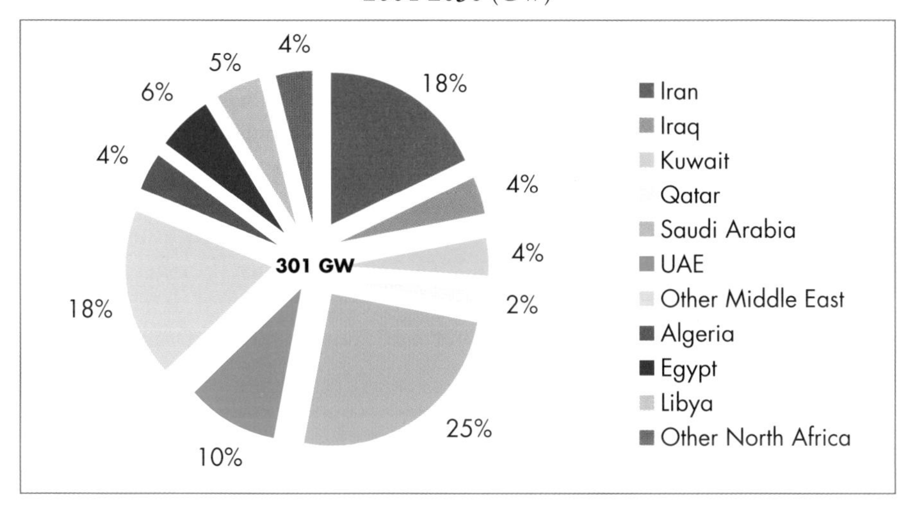

Installed capacity in MENA had reached 178 GW in 2003, less than 5% of the world total. Iran and Saudi Arabia, with 33 GW and 35 GW respectively, had the largest installed capacities in the region (Figure 6.8). Installed capacity was estimated to have reached 186 GW in 2004.

Generating capacity in the Gulf area has to be planned to cover peak demand for power, which more than doubles in the summer because of the air-conditioning load. In Abu Dhabi, for example, the winter peak (that usually occurs in January/February) was about 2 100 MW in 2004 but reached 4 300 MW in the summer (peak demand occurs in July/August). The seasonal variation is even more pronounced in some other Gulf countries. This uneven demand pattern means that the load factors of power plants in these countries

Figure 6.8: **MENA Installed Power-Generation Capacity by Country, 2003**

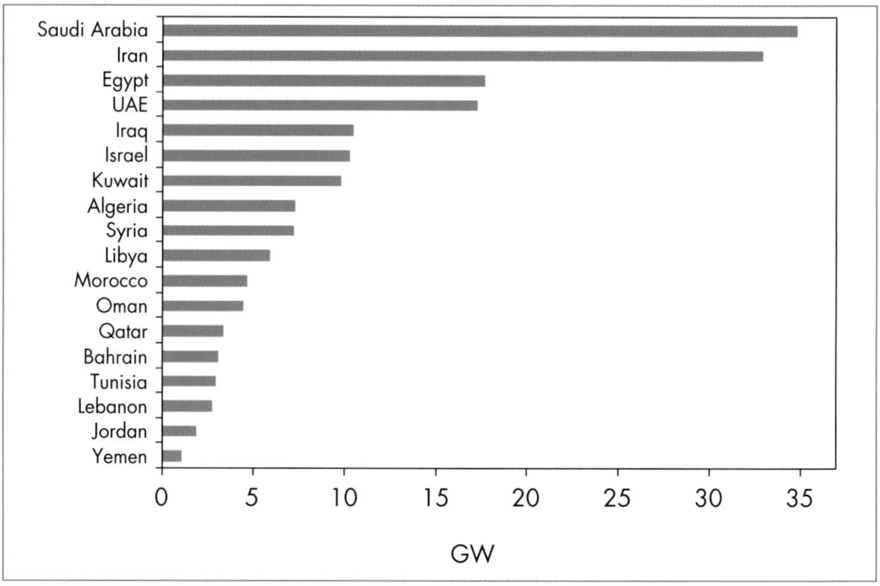

Source: Platt's (2003).

are low (Figure 6.9).[5] As a result of the high demand in the summer, MENA power plants are built to operate mostly in the summer. Further, the high summer load makes it more difficult to match electricity and desalination in countries that operate combined water and power facilities, as the water

Figure 6.9: **Load Factors in MENA, 2003**

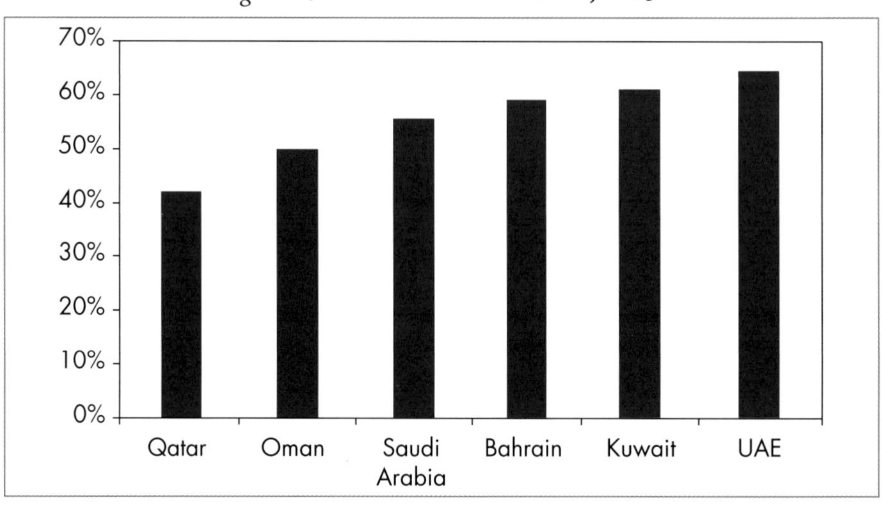

Sources: AUPTDE (2004) and IEA databases.

5. The load factor is the ratio of average load to peak load. The higher it is the better.

demand profile is much flatter throughout the year and storing excess water produced during the summer for use in the winter is not economically viable.

Although a number of countries in the region, particularly in the Gulf area, have state-of-the-art power plants, their reliance on open-cycle gas turbines (OCGTs) and boilers has kept power generation efficiency below the levels observed in OECD countries. On average, the efficiency of gas-fired power plants in MENA is 33% compared with 43% across the OECD region. Similarly, the efficiency of oil-fired generation is 34%, against 42% in the OECD. Besides the technology, the use of combined water and electricity production in the Gulf countries requires some additional fuel and this is another factor reducing the apparent electrical efficiency of power plants.

Combined-cycle gas turbines (CCGTs), which are generally considered the most economic way to produce electricity in OECD countries (see discussion of power plant economics below), make up a small share of MENA capacity (Figure 6.10). OECD countries have built about 250 GW since 1990. During this period, MENA countries built mostly OCGTs, attracted by their lower capital costs than CCGTs despite their lower efficiency. This was the most economic option for MENA, because all countries needed to make large investments and because the abundance of cheap fuel made operating efficiency less of a consideration. Another reason why OCGT has been the technology of choice is the short construction time, which enables rising electricity demand to be met rapidly.

Figure 6.10: **Natural Gas Use in Power Generation and Role of CCGTs in MENA and the OECD, 2003**

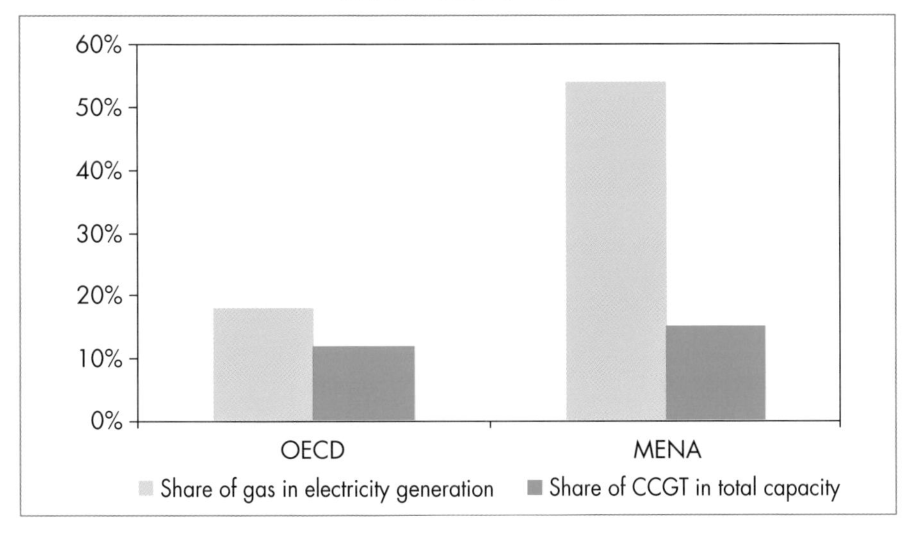

The high temperatures and humidity which characterise the climate of most MENA countries affect the performance of power plants. These two factors reduce the effective capacity of power plants: more MW of capacity must be installed to obtain the same amount of electricity as in other parts of the world with temperate climates. Fuel efficiency is also negatively affected.

Over the projection period, gas-fired power plants will meet over two-thirds of new capacity needs in the region (Table 6.2). Most of the new gas-fired power plants will be CCGTs (137 GW, roughly two-thirds of total gas-fired capacity). The use of open-cycle gas turbines will be increasingly concentrated on meeting peak load requirements, although some will be used for base- to mid-load electricity generation as well. Altogether, open-cycle gas turbine capacity additions are projected to be of the order of 54 GW. Some new combined water and power (CWP) facilities may also use OCGTs. The remaining gas-fired capacity, some 14 GW, is expected to be in steam boilers. This technology is now becoming obsolete everywhere in the world. Some heat-recovery boilers may be constructed where flexibility to burn gas or crude is needed or to repower existing open-cycle gas turbines, improving the efficiency of the power plant at a lower cost than building a new CCGT power plant.

Table 6.2: **MENA Capacity Additions by Fuel, 2004-2030**

	GW additions	**Share**
Coal	8	3%
Oil	63	21%
Gas	205	68%
of which:		
CCGT	*137*	*46%*
OCGT	*54*	*18%*
Nuclear	0.9	0.3%
Hydro	13	4%
Other renewables	10	3%
of which:		
Wind power	*6*	*2%*
Solar power	*2*	*1%*
Total	**301**	**100%**

Oil-fired capacity additions represent 63 GW, or a little more than 20% of the total. New oil-fired power plants will be in the form of open-cycle turbines and diesel engines, or boilers using fuel oil or even crude (mainly in Saudi Arabia). New hydropower capacity will reach 13 GW, while 10 GW will come from other renewable energy sources (including over 6 GW of wind farms and about 2 GW of solar). Coal-fired capacity increases will amount to 8 GW, while Iran's nuclear power plant will add almost 1 GW of new capacity.

Power Plant Economics

Until the late 1980s, most electricity generation in MENA was by firing crude oil or oil products, particularly heavy fuel oil (sometimes even in open-cycle gas turbines). In recent years, there has been a widespread switch to firing natural gas, where this is available.

Gas-fired electricity generation is now the most competitive way to produce electricity in MENA. While both oil and gas are extremely cheap to produce, oil has a particularly high opportunity cost as an export commodity. Figure 6.11 shows indicative base-load electricity generating costs for three technologies used in MENA. The fuel costs in this example are assumed to be \$4 per barrel for oil and \$0.50 per MBtu for gas.[6] Electricity generated by oil boilers is more expensive than gas because of the higher investment required, higher operating costs and lower fuel efficiency.

Figure 6.11: **Average MENA Electricity Generation Costs by Technology**

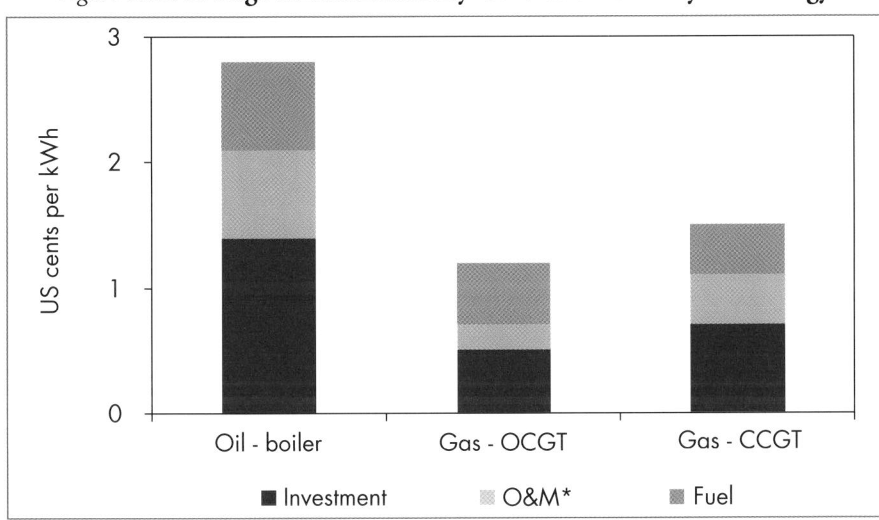

* Operation and maintenance costs.

If prices of fuel inputs were adjusted to take account of international market values, natural gas would become even more economic (Figure 6.12). Assuming an opportunity cost of \$23 per barrel for oil and \$1.50 per MBtu for gas, the cost of a kWh of electricity based on oil is up to four times higher than a kWh of electricity based on gas.[7] In other words, it is much more attractive to use gas for domestic power generation, since oil brings greater export earnings. This largely explains policies to switch from oil to gas.

6. The assumed price for gas is taken as an average across the region. This price is higher in some countries, reaching \$1 per MBtu. Even at this higher price, gas-fired electricity is cheaper than oil.
7. Average IEA import price over the past five years, exclusive of transport costs (assumed to be \$6 per barrel). Gas prices are for 2004.

Figure 6.12: **Cost of Electricity Generation Based on Fuel-Export Values**

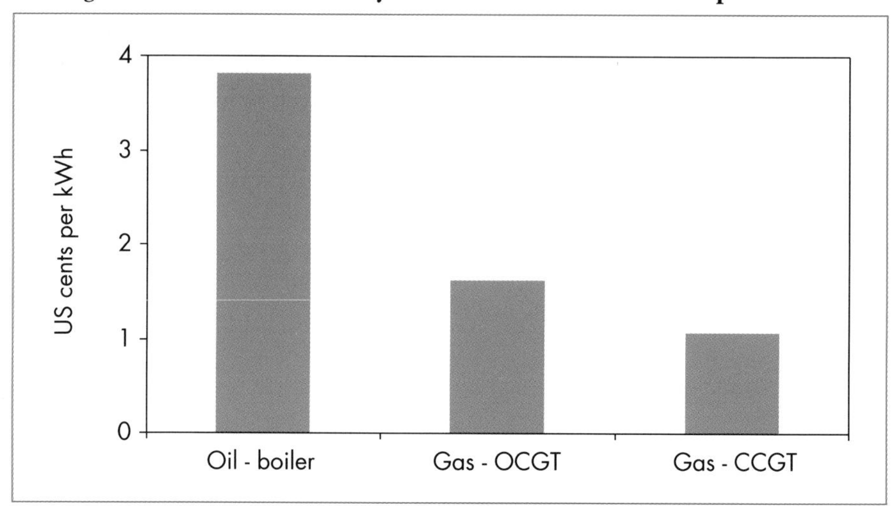

Note: Refers to the fuel component of electricity-generating costs only.

The opportunity cost of gas is now low because of the abundance of reserves and because the export potential is still small. To date, this has resulted in countries building open-cycle gas turbines instead of combined-cycle power plants, mainly because of their lower investment cost. With demand for gas from outside MENA rising, the opportunity cost of gas is expected to increase to between $1 and $1.50 per MBtu or more, making CCGTs more competitive (Figure 6.13).[8] Economic pricing of gas, which is now subsidised or available free in most gas-producing countries in MENA, will also encourage the use of CCGTs.

Several countries in the region, particularly in the Gulf, are already switching to CCGTs, so as to co-generate steam for desalination. Figure 6.14 compares the additional investment cost for a CCGT with the potential earnings if the fuel saved (because of the higher efficiency of a CCGT plant) is exported. The total undiscounted value of this fuel over the lifetime of the power plant is at least six times higher than the additional initial investment required to build a CCGT power plant.

In the Gulf region, an additional incentive for investing in CCGT is the generation of steam for combined electricity and water production. Co-generation is most important for the countries of the Gulf and is growing in importance in other MENA countries suffering from water shortages.

8. These figures also reflect the market price of gas in MENA countries, where data are available.

Figure 6.13: **CCGT and OCGT Electricity Generating Costs**

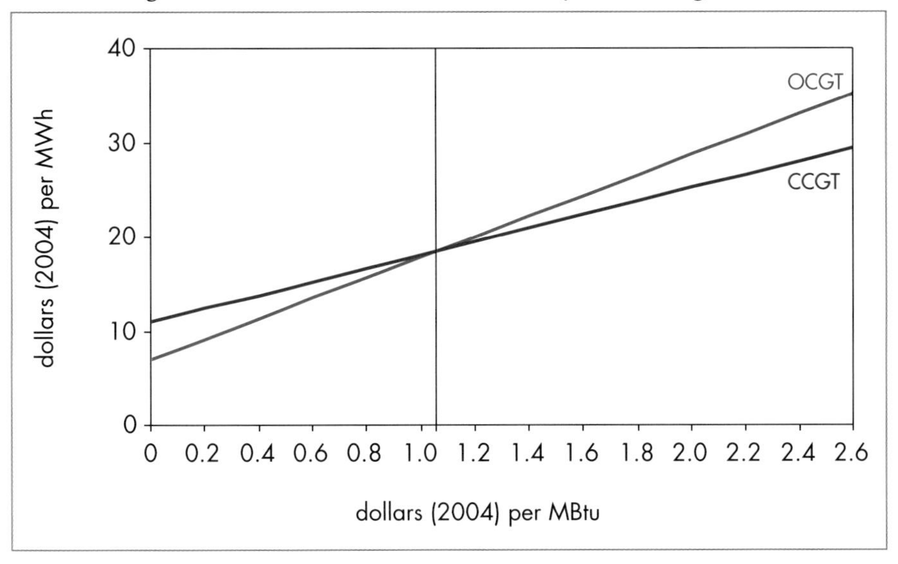

Figure 6.14: **Additional Investment for CCGT versus the Value of Fuel Savings**

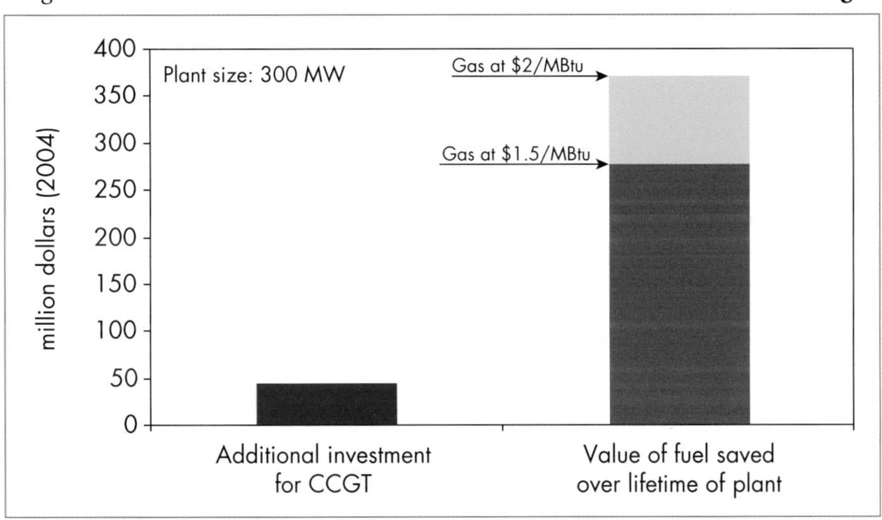

Note: The following assumptions have been used: CCGT has a capital cost of $500 per kW and efficiency of 52%; OCGT has a capital cost of $350 per kW, efficiency of 34% and plant lifetime of 25 years. The additional value of steam for desalination is not taken into account.

Another reason for using CCGTs instead of OCGTs is the reduced environmental impact of the former because of their higher efficiency. Since many power stations are built around large cities, the pressure to reduce NO_x

emissions will increase. In many OECD countries it is now very difficult to build OCGTs near towns because they do not meet environmental standards. Carbon dioxide emissions from OCGTs are also higher. Although MENA countries have no obligation to reduce such emissions under the Kyoto Protocol, projects based on the Protocol's clean development mechanism (CDM) would favour CCGTs over OCGTs. For those countries seeking to borrow money in international markets, pressure to invest in CCGTs may also come from international lending institutions, which tend to favour cleaner technologies.

Electricity Generation Mix

The MENA region will continue to rely predominantly on hydrocarbons to produce electricity. The share of hydrocarbons in total electricity generation will remain at almost 90% throughout the projection period, but the oil and gas mix will change. The region will continue to move away from oil in power generation. The share of oil in total electricity generation was maintained above 50% until 1990 and then declined faster in the 1990s, falling to about a third of the total by 2003. It is projected to drop to 20% by 2030. In absolute terms, however, oil-fired electricity generation is projected to increase from 239 TWh in 2003 to 366 TWh in 2030. Oil use includes heavy fuel oil, diesel and crude. Some countries may continue to use heavy fuel oil that they cannot export because of its quality. Crude oil will continue to be used, particularly in Saudi Arabia, although much less than in the past. Building crude oil-fired power stations may pose specific challenges, since they are based on a technology for which global demand is low, with fewer experienced suppliers and limited product development. Total oil consumption in power stations is projected to increase from 60 Mtoe in 2003 to 92 Mtoe in 2030, approximately equivalent to 1.2 mb/d and 1.9 mb/d.

Gas-fired electricity generation is expected to increase substantially at the expense of oil, the gas share in the electricity mix rising from 56% in 2003 to 69% by 2030 (Table 6.3 and Figure 6.15). Natural gas consumption for electricity generation is projected to increase from 107 Mtoe (129 bcm) in 2003 to 277 Mtoe (335 bcm) in 2030. This amount represented 44% of MENA primary gas consumption in 2003 and will be 46% in 2030.

In 2003, natural gas was used in 14 countries in the region, its share ranging from 100% in Qatar and Bahrain to 9% in Jordan and to 0.5% in Israel. In eight out of the 18 countries in the region, gas accounted for more than two-thirds of total electricity generation (Figure 6.16). Lebanon, Yemen and Morocco used no gas. Iraq had a large gas pipeline connected to many of its power plants but it was not clear how much gas Iraq used. Problems of underinvestment, pipeline sabotage and other breakdowns often affect gas pressure and the ability to supply. For the countries that hold oil and gas reserves, the main reason for shifting to natural gas is to free up oil for export. Gas-based electricity generation is often the most competitive option.[9] Oil brings greater earnings if exported and is easier to transport. For countries without domestic gas supplies, imported natural gas is the most economic way to produce electricity.

Table 6.3: **Shares of Oil and Gas in Electricity Generation** (%)

	2003		2030	
	Oil	**Gas**	**Oil**	**Gas**
Middle East	**38**	**52**	**24**	**66**
Iran	16	77	7	80
Iraq	98	0	36	53
Kuwait	80	20	59	41
Qatar	0	100	0	100
Saudi Arabia	54	46	34	66
UAE	1	99	1	99
Other Middle East	39	27	27	41
North Africa	**15**	**68**	**9**	**78**
Algeria	2	97	1	96
Egypt	6	80	2	86
Libya	80	20	42	56
Other North Africa	17	36	8	60
MENA	**33**	**56**	**20**	**69**

While all countries in MENA seek to increase the use of gas, some are facing difficulties. The western region of Saudi Arabia, for example, has no easy access to gas supplies and building the necessary gas infrastructure will be costly. The situation could improve gradually in the future, if gas networks are developed and particularly if Saudi Arabia's new efforts to develop its non-associated gas reserves – located essentially in the Rub Al-Khali (the Empty Quarter) region – begin to bear fruit. In Kuwait, projects to switch to natural gas are facing delays. Iraq's shift to gas is also likely to be slow in the near term and is not likely to start before 2010. In Libya, the switch to natural gas will also be

9. See the previous section on power plant economics.

Figure 6.15: **MENA Electricity Generation Fuel Mix**

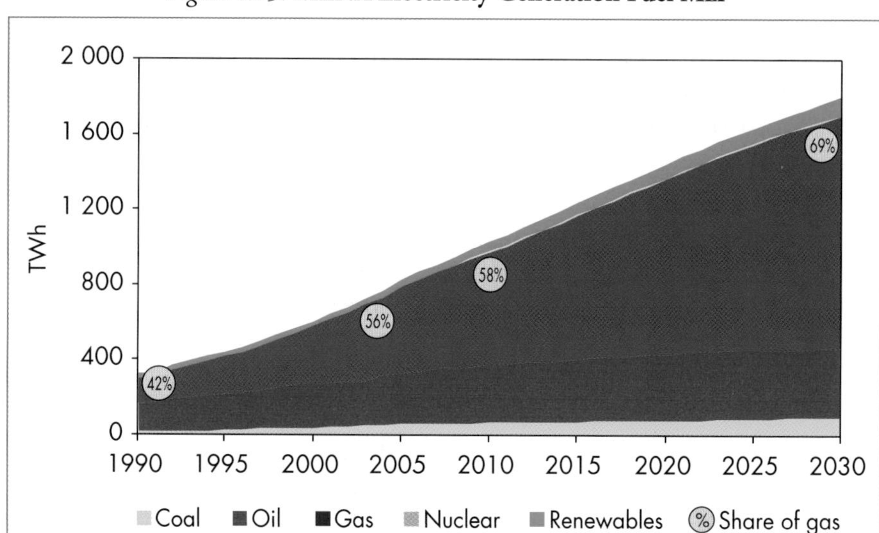

Figure 6.16: **Share of Natural Gas in MENA Electricity Generation, 2003**

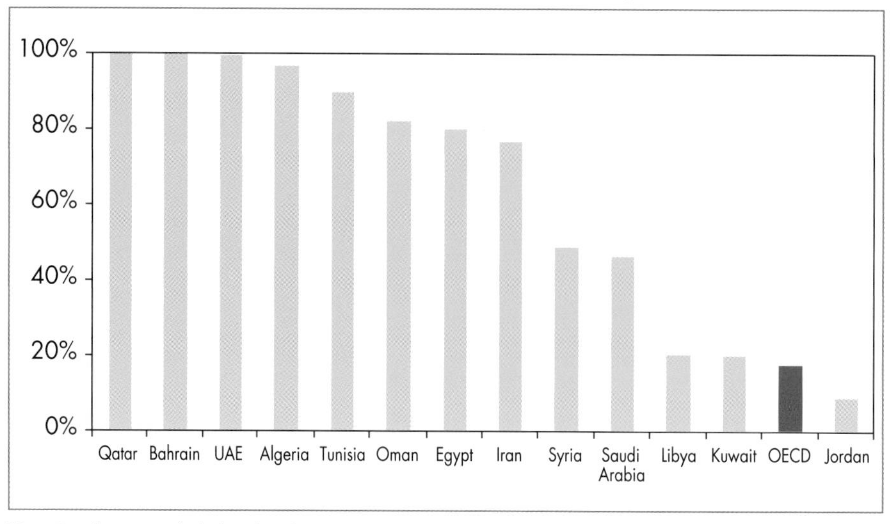

Note: Israel is not included in this chart. Its share of gas in electricity generation in 2003 was 0.5%.

somewhat slow and will depend on how fast foreign investment in gas projects comes in.

A few countries in the region today use resources other than oil and gas to produce electricity. Morocco and Israel, which do not have rich hydrocarbon reserves, rely mostly on coal to produce their electricity. Hydropower is used in eight countries in the region. Its share is fairly significant in Egypt (14% of total electricity in 2003), Lebanon (13%), Syria (9%), Morocco (8%) and Iran (7%), but very small (less than 2%) in Iraq, Algeria, Tunisia and Israel. Some countries use solar and wind power, but in very limited applications. Only in Morocco did wind power account for more than 1% of total electricity generation in 2003.

Total coal-fired generation is projected to double, from 48 TWh in 2003 to 96 TWh in 2030. Coal will continue to be used in Israel and Morocco, but it is likely to lose share in both countries, as more gas projects are developed.

There is no nuclear power in MENA now, but Iran's Bushehr power plant is expected to be the first nuclear reactor for electricity generation to come on stream in the region.[10] Nuclear power is expected to reach 6 TWh by 2010 and remain at this level to 2030. A number of countries in the region operate research nuclear reactors and some of them have in the past expressed their interest in developing nuclear power plants for electricity production. The UAE's ADWEA (Abu Dhabi Water and Electricity Authority) signed a Memorandum of Understanding in 2004 with South Korea's Ministry of Science and Technology to carry out a feasibility study about constructing a nuclear power plant. Under the Reference Scenario assumptions, such a project has not been included in our projections.

Hydropower will increase at 3.1% per year, maintaining its current share of 4% in the electricity generation mix. The largest increase will be in Iran, where several projects are underway. But the region's hydro potential is rather small, and consequently, hydropower development is projected to slow over time.

The use of non-hydro renewables is expected to remain limited, although their share will rise from 0.1% now to 1.5% of total generation output in 2030. Wind and solar will account for the majority of this increase. Wind power is projected to increase from less than 1 TWh now to 16 TWh in 2030. Several countries in the region, including Morocco, Tunisia, Egypt, Iran, Israel, Jordan and Syria, have developed wind farms. The region's overall wind potential is, however, somewhat limited. The solar potential, on the other hand, is exceptional. Iran and most countries in North Africa have solar thermal projects under development. Most of these projects are combined gas and solar facilities. Israel has several solar demonstration projects. Photovoltaics are used in rural areas for electrification. However, the use of solar energy in the region

10. See Chapter 11 for a discussion about Iran's power sector.

is likely to remain limited by 2030 under the Reference Scenario assumptions, reaching 5 TWh. The main reason for this is the high cost of such power plants. Box 6.3 analyses MENA's solar power prospects.

Box 6.3: Solar Power Prospects in MENA

Global solar energy potential is estimated to be between 1 575 and 50 000 exajoules (EJ) per year, depending on land availability assumptions. This is between 3.5 and 110 times higher than the world's current energy consumption.[11] About a quarter of this potential is concentrated in the MENA region.

The economic potential is, however, much lower at present because of the high costs of producing electricity from solar thermal or photovoltaics (PV) power plants compared with conventional alternatives and, notably, natural gas. Electricity from solar thermal power plants is cheaper than PV but still about three times higher now than the generating cost of a CCGT power plant. The capital cost of solar thermal is expected to fall to around $1 250 per kW in 2030 in the World Alternative Policy Scenario – about 40% less than now – and the generating cost to about 5 cents per kWh. Figure 6.17 compares the generating costs of solar thermal and CCGT. It shows that solar thermal will still not be competitive in 2030 for gas prices under $6.5 per MBtu. This price is much higher than the international price of gas in 2030 in the Alternative Policy Scenario, but is in line with the price level reached in the Deferred Investment Scenario.

In this context, large increases in solar thermal in MENA can happen only if these countries seek to diversify their electricity mix or if they can generate more income by exporting gas instead of using it to produce electricity. Increases in solar are also possible if MENA countries participate in global emission reduction projects (for example, an international emissions trading scheme), since solar thermal electricity is much cheaper in MENA than in other world regions (for example it is a third less than in Europe). This, however, will require new policies within and outside MENA. Among various policies, the clean development mechanism under the Kyoto Protocol can provide substantial opportunities for investment in such projects.

11. IPCC (2001).

Figure 6.17: **Solar Thermal and CCGT Electricity Generating Costs in 2030**

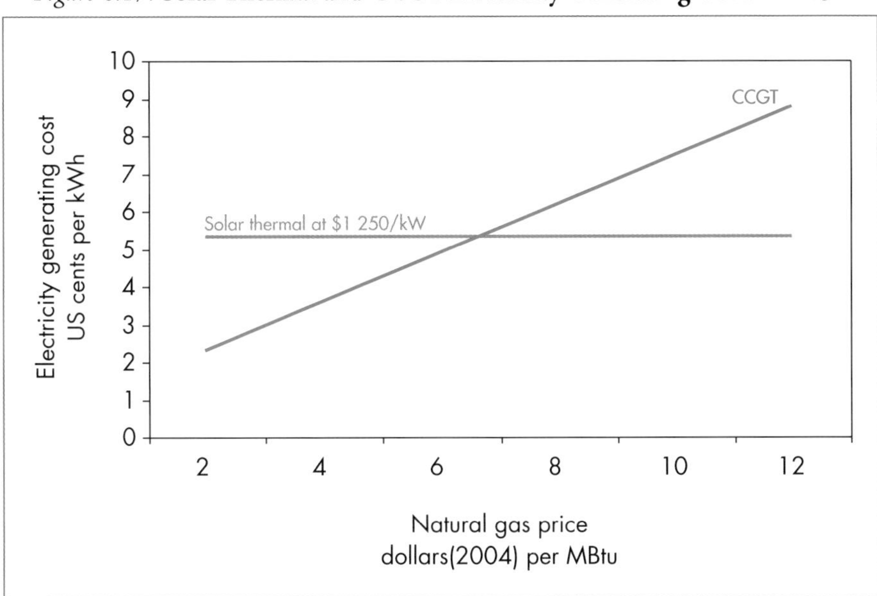

6

Electricity Investment

The total investment needed in the region's power sector is expected to reach $458 billion (in year-2004 dollars), nearly as high as the investment in upstream oil. In Iran, investment in the electricity sector will be higher than investment in oil or gas and in Egypt about the same as investment in the gas sector. On average, power sector investment makes up more than 30% of energy-related investments, about 5% of total domestic investment and nearly 1% of GDP.[12]

Investment in the period to 2010 is expected to reach $19 billion per year and then fall to about $16 billion per year in the rest of the projection period (Figure 6.18). During the first period, expenditure in electricity infrastructure will also be the highest relative to GDP, at 1.5%. In Iran, this share will be the highest, reaching 2.2%, indicating that this country may face a challenge in attracting the necessary investments.

12. Domestic investment refers to gross capital formation and is assumed to be around 20% of GDP (median across MENA over the past ten years).

Figure 6.18: **MENA Average Annual Electricity Investment**

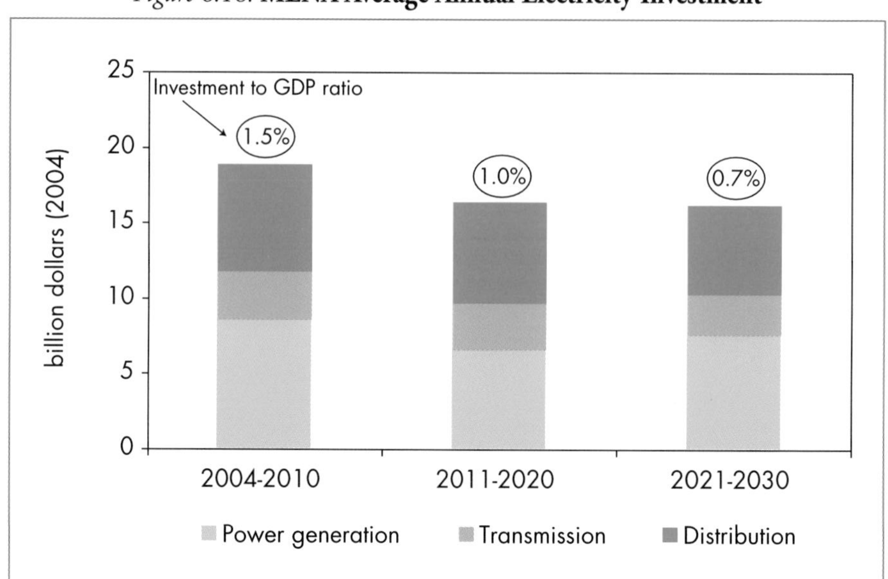

About 80% of the total investment will be in the Middle East and 20% in North Africa. Investment in Saudi Arabia's power sector will reach $110 billion, the largest in the region, while Iran will need some $92 billion (Figure 6.19). Egypt and the UAE will need about $35 billion each. The countries grouped together as other Middle East and other North Africa will need almost $100 billion.

Figure 6.19: **MENA Cumulative Electricity Investment by Country, 2004-2030**

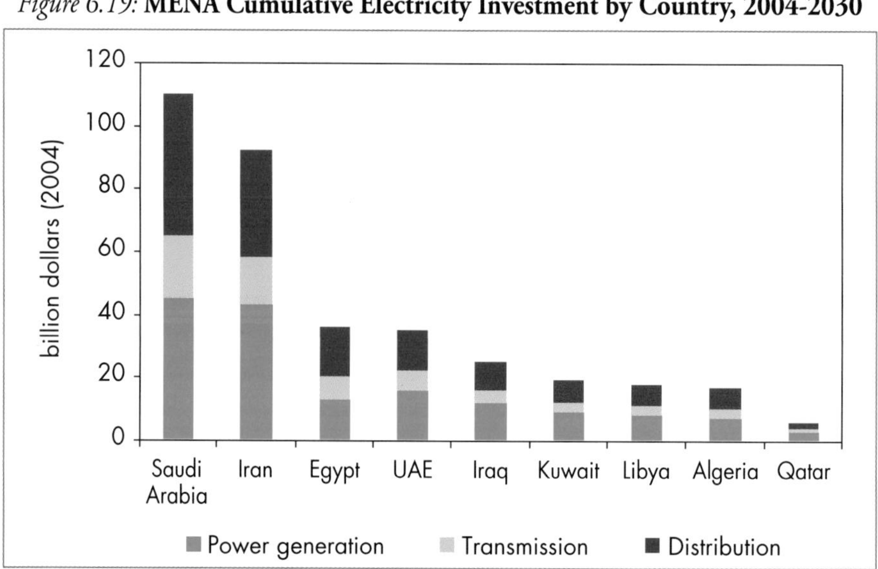

Investment in power generation is expected to reach $203 billion, including some $26 billion in refurbishment. Investment in electricity transmission and distribution networks will be higher, at $255 billion or 56% of the total. Distribution networks alone will cost $176 billion, more than twice as much as transmission. On average, one kilowatt of new capacity added in MENA will cost almost $600[13] and the accompanying investment in networks will be almost $850 per kW. Table 6.4 summarises the capacity additions and investment needs by country. Although a large number of projects are underway in MENA, a shortage of EPC (engineering, procurement and construction) contractors has been observed in some cases over the past few years. This shortage is pushing up the cost of projects but it is not likely to be a problem in the long term.

6

Table 6.4: **MENA Capacity Additions and Investment, 2004-2030**

	Capacity additions (GW)		**Investment** ($ billion)		
		Total	**Power generation**	**Transmission**	**Distribution**
Middle East	**242**	**368**	**166**	**63**	**139**
Iran	54	92	43	15	34
Iraq	12	26	12	4	9
Kuwait	13	19	9	3	7
Qatar	5	6	3	1	2
Saudi Arabia	75	110	45	20	45
UAE	31	35	16	6	13
Other Middle East	53	80	38	13	29
North Africa	**59**	**90**	**37**	**17**	**37**
Algeria	12	17	7	3	7
Egypt	19	36	13	7	16
Libya	15	18	8	3	7
Other North Africa	12	19	9	3	7
MENA	**301**	**458**	**203**	**79**	**176**

The investment estimates in this *Outlook* do not take explicitly into account the cost of future interconnections. This cost is not expected to change significantly the investment figures given here, which are hundreds of billions

13. This figure is the result of the projected technology mix with different capital costs by technology.

of dollars. Any additional increase in transmission investment because of interconnections is expected to be offset to a large extent by reduced investment in power generation (Box 6.4).

Box 6.4: **Regional Interconnections in MENA**

Several interconnections exist between MENA countries, as well as with countries outside the region, but total exchanges are rather limited. Most interconnections are between North African countries, where the first links were developed a few decades ago. Transmission lines link Iran with Turkey, Azerbaijan, Turkmenistan and Iraq. The EIJLST project links the grids of Egypt, Iraq, Jordan, Lebanon, Syria and Turkey. Interconnections are absent in the Gulf area. Two major developments could significantly improve interconnections in MENA.

The *GCC Grid* (Gulf Cooperation Council Grid) will link the power grids of the six Gulf states (Bahrain, Kuwait, Oman, Qatar, Saudi Arabia and the UAE). The project has three phases and is expected to improve security of supply in the Gulf area:

- Phase I: Interconnection of Kuwait, Saudi Arabia, Bahrain and Qatar. This system is the GCC North Grid.
- Phase II: Interconnection of the independent systems in the UAE as well as Oman. This is the GCC South Grid.
- Phase III: Interconnection of the GCC South Grid with the North Grid. This phase will complete the interconnection of the six Gulf states.

This project has for a long time been on the agenda of the Gulf countries, but was delayed mainly because of political differences. Contracts for the first phase were awarded in October 2005. The completion of the first phase, scheduled for 2008, is expected to cost over a billion dollars, but it will save investment in power generation since interconnections are expected to reduce the requirements for spinning and stand-by reserves as well as simultaneous combined peak demand. At the moment, there is no timing for the second phase, but the third phase is scheduled for 2010. Consequently, the second phase is likely to be completed between 2008 and 2010.

The *MEDRING* (Mediterranean Ring) project will link the power grids of Spain, France, Italy, Greece, Turkey, Jordan, Egypt, Syria, Algeria, Tunisia. The project, which will involve over 10 000 km of transmission lines, will facilitate power exchanges between countries around the Mediterranean. Countries in the south and east Mediterranean, in particular, hope to become competitive exporters of cheap electricity. While this is an ongoing project (Morocco, Algeria and Tunisia are already connected to Europe through a submarine cable between Morocco and Spain), there still remain several impediments, such as operational questions and compliance with the European transmission co-ordinator, UCTE (Union for the Coordination of Transmission of Electricity) guidelines, political sensitivities, competition with other countries which potentially supply electricity to Europe (for example Russia) and concerns in Europe about over-reliance on imports from distant countries. If these obstacles are overcome, MEDRING could be operational in 2010. However, because of the large distances involved and the limitations of current transmission technology, the exchanges are not likely to be very large. Further, some of the countries involved are finding it already difficult to build enough power plants to meet domestic demand, let alone for export.

In MENA countries, the electricity sector has typically been exclusively publicly owned, controlled by a ministry or a vertically-integrated power company (Table 6.5).[14] Some industrial companies (mainly the oil and gas industry) own power plants in most MENA countries, but their share in total capacity is, on average, small. The share in total electricity generation is even smaller, as they prefer to take subsidised electricity from the grid. In Saudi Arabia, where electricity tariffs have been increased, the share of autoproducer electricity is one of the highest in MENA, reaching 16% in 2003.

The electricity supply industry in most MENA countries is not commercially viable. This is because power companies do not have sufficient revenues to cover their expenses and to invest in new projects, resulting from heavily subsidised electricity prices, the non-payment of electricity bills and illegal connections.

Power companies accordingly have to rely on government budgets to carry out their investment programmes. To alleviate this burden, some countries, particularly in the Gulf region, turned to the private sector in the late 1990s, adopting the single-buyer model. This process was initially slow in most countries. Recently, however, despite regulatory frameworks not characterised by independence, transparency and accountability (World Bank, 2003), there has been a surge of new independent power producer (IPP) projects and independent water and power producer (IWPP) projects, particularly in the

14. See the country chapters for more details about the structure of the power sector in each country.

Table 6.5: **Main Electricity Suppliers in MENA Countries**

	Main electricity supplier	Remarks
Algeria	SONELGAZ	Vertically-integrated
Bahrain	Ministry of Electricity and Water	Ministry will be single buyer of output from first IPP* (awarded in 2004)
Egypt	Egyptian Electricity Holding Company (EEHC)	Joint-stock company, controls generation, distribution and transmission through subsidiaries Single buyer of IPP output
Iran	Tavanir	Controls generation, distribution and transmission through subsidiaries
Iraq	Ministry of Electricity	Established in 2003, replacing the Commission of Electricity
Israel	Israel Electric Corporation	Vertically-integrated
Jordan	Central Electricity Generation Company (CEGCO) for generation	Unbundled activities, remaining government-owned; CEGCO and EDCO will eventually be privatised
	National Electric Power Company (NEPCO) for transmission	NEPCO is the single buyer of electricity produced by CEGCO and IPPs
	3 distribution companies: EDCO, JEPCO and IDECO	
Kuwait	Ministry of Electricity and Water	Vertically-integrated
Lebanon	Electricité du Liban (EDL)	Vertically-integrated
Libya	GECOL	Vertically-integrated
Morocco	Office National de l'Electricité (ONE)	Distribution in rural areas is controlled by ONE, while in urban areas by the Ministry of Interior
Oman	Ministry of Housing, Electricity and Water	Privatisation underway following 2004 power privatisation law Newly established Power and Water Procurement Company single buyer of IPP/IWPP** output

Table 6.5: **Main Electricity Suppliers in MENA Countries** *(continued)*

	Main electricity supplier	Remarks
Qatar	Qatar Electricity & Water Corporation (QEWC) for generation Qatar General Electricity & Water Corporation (QGEWC) for transmission and distribution	Partial unbundling QGEWC single buyer of IPP output
Saudi Arabia	Saudi Electricity Company (SEC)	Vertically-integrated Water and electricity facilities are controlled by the Saline Water
Syria	Public Establishment for Generation and Transmission of Electricity Public Establishment for Distribution of Electricity	Conversion Corporation (SWCC) Partial unbundling with the separation of distribution
Tunisia	Société Tunisienne de l'Electricité et du Gaz (STEG)	Vertically-integrated
UAE Abu Dhabi	Abu Dhabi Water and Electricity Company (ADWEC) TRANSCO (Transmission Company) Abu Dhabi Distribution Company (ADDC) Al-Ain Distribution Company (AADC)	Abu Dhabi Water and Electricity Authority (ADWEA) is the state holding company for ADWEC, TRANSCO, ADDC and AADC
UAE Dubai	Dubai Electricity and Water Authority (DEWA)	Vertically-integrated
UAE Sharjah	Sharjah Electricity and Water Authority (SEWA)	Vertically-integrated
UAE Other	Federal Electricity and Water Authority (FEWA)	Vertically-integrated
Yemen	Public Electricity Corporation	Vertically-integrated

* IPP: independent power producer.

** IWPP: independent water and power producer.

Gulf area. Iran has also started developing projects on a build-operate-own (BOO) and build-operate-transfer (BOT) basis. Countries in North Africa are seeking to increase or to introduce private investment in their power sectors. The exception here is Egypt, which, having led the development of two large IPPs, is turning again to the public sector, reflecting the current availability of public investment funds.

IPP projects carry certain advantages, notably they are often built faster and can result in lower production costs.[15] IPPs are also the easiest way to introduce competition. But they have considerable drawbacks (World Bank, 2001). IPPs in MENA are requesting sovereign governmental guarantees and high returns. Rich MENA countries are offering 8%-12% rates of return, while investors are pressing for 15% to 18%. This means that these projects may turn out to be very expensive, as has been the case in other countries in the world. A typical structure of an IPP project in MENA is as follows (Khatib, 2005):

- 80:20 debt/equity ratio.
- Ownership: 20 years.
- Regulated power-purchase agreement.
- Equity from developers (possibly with government and utility participation).
- Debt from local and/or international financial institutions.

The long-term impact of recent reforms aimed at attracting private investment remains unclear but it seems likely that governments will continue to provide the bulk of financing or the guarantees for future projects. Loans will come principally from Arab development banks (such as the Arab Fund, which loaned $654 million, over 60% of all loans, to the MENA power sector in 2003) or international banks (including the World Bank and the European Investment Bank). The poorer members of the region may also rely on development agencies.

MENA countries have given little consideration to demand-side measures, such as more efficient equipment in industry and buildings. These measures can be cost-effective and can substantially reduce energy consumption and the accompanying need for investment in power supply (Box 6.5).

Since 2003, high oil prices have created an abundance of liquidity in these countries and many regional banks are more than willing to finance new projects, although some of the large projects are becoming more difficult to

15. In MENA countries, the electricity and water sectors are often used as sectors to place nationals in, which leads to gross overstaffing. IPPs can operate with fewer staff, which helps keep generation costs low.

Box 6.5: Supply and Demand Side Investments in the World Alternative Policy Scenario

The World Alternative Policy Scenario (WAPS), described in Chapter 8, analyses the impact of a more efficient and more sustainable energy future compared with the Reference Scenario.[16] WAPS shows that investing in energy efficiency requires increased capital expenditure on end-use equipment, but this is offset by lower investment on the supply side.

World electricity generation in WAPS in 2030 is 13% lower than in the Reference Scenario because of policies to improve end-use efficiency. In the residential sector, these policies include measures relating to lighting, electric appliances, space heating, water heating, cooking and air-conditioning. In the services sector, they include lighting, space heating, air-conditioning and ventilation.

Investment by final consumers in more efficient electrical equipment in the industrial, residential and commercial sectors is more than $600 billion *higher* in the Alternative Scenario. The capital costs of more efficient and cleaner end-use appliances are generally higher, but the result of such investment is to drive down energy demand, thereby reducing consumption, global energy bills and investment requirements for energy-supply infrastructure. Savings in electricity-supply investment amount to $1.4 trillion and they more than offset the increase in demand-side investments.

MENA countries have rapidly growing supply-side investment needs. At the same time, they have an enormous potential to save energy. Promoting end-use efficiency policies may be a cost-effective way to reduce investment in power infrastructure.

finance because of their complexity. At the same time, the favourable investment climate has delayed the need for deeper reforms, which would involve making the electricity sector commercially viable, starting with reforms in the distribution sector. Such reforms are difficult and take considerable time, at least five years and more likely ten (IEA, 2003). The current surge of liquidity will not be indefinite and MENA countries need to plan for future needs if they do not want to face the same difficulties as they did in the 1990s.

16. WAPS was first developed in *WEO-2004*.

Water Desalination

Overview

Water desalination is used around the world where sources of freshwater are limited. It is an energy-intensive process and will, accordingly, account for a growing proportion of energy use in the MENA region as the importance of desalinated water continues to grow over the projection period. Figure 6.20 shows global renewable water resources per capita on a WEO regional basis. The Middle East is the most water-stressed region in the world.

Desalination is the process of removing dissolved solids from sea-water, groundwater or waste water. The World Health Organization recommends that water for human consumption should have total dissolved solids (TDS) of less than 500 parts per million (ppm). Sea-water typically has salinity of about 35 000 ppm or more while brackish water has salinity of about 1 500 ppm.

The MENA region accounted for over half of global desalination capacity in 2003 (Table 6.6). Global planned additions to desalination capacity from 2004 to 2013 amount to 8 300 million cubic metres, an increase of about 75% over current capacity. The MENA region will account for almost 70% of these capacity additions. In that time-scale, desalination capacity will double in the region, from about 5 700 million cubic metres per year in 2003 to more than 11 400 mcm per year in 2013.

The cost of producing desalinated water is fairly high and depends on the level of salinity and on energy costs. Desalination costs are going down and further reductions are likely in the future along with technological and energy efficiency improvements.

The most widely used desalination processes are thermal separation through distillation and membrane separation through reverse osmosis (RO). There are three predominant types of distillation processes – multistage flash (MSF), multiple effect distillation (MED) and mechanical vapour compression (MVC). The MSF and RO processes dominate the global market for both brackish water and sea-water desalination, with a share of more than 80%.

Distillation plants account for about 40% of global capacity. Another 47% of global desalination capacity is from the reverse osmosis process. RO is widely used in sea-water, brackish water and water reclamation projects and is the fastest growing segment of the desalination market. In the Middle East, MSF plants account for 68% of desalination capacity, while in North Africa, MSF and RO plants capture about 40% each of the desalination market.

A rule of thumb for capital investment costs is about $1 000 per cubic metre per day of capacity installed for both distillation and membrane processes (Table 6.7). For distillation units, the investment required for the co-generation plant can more than double the overall investment requirements.

Figure 6.20: **Global per Capita Renewable Water Resources**

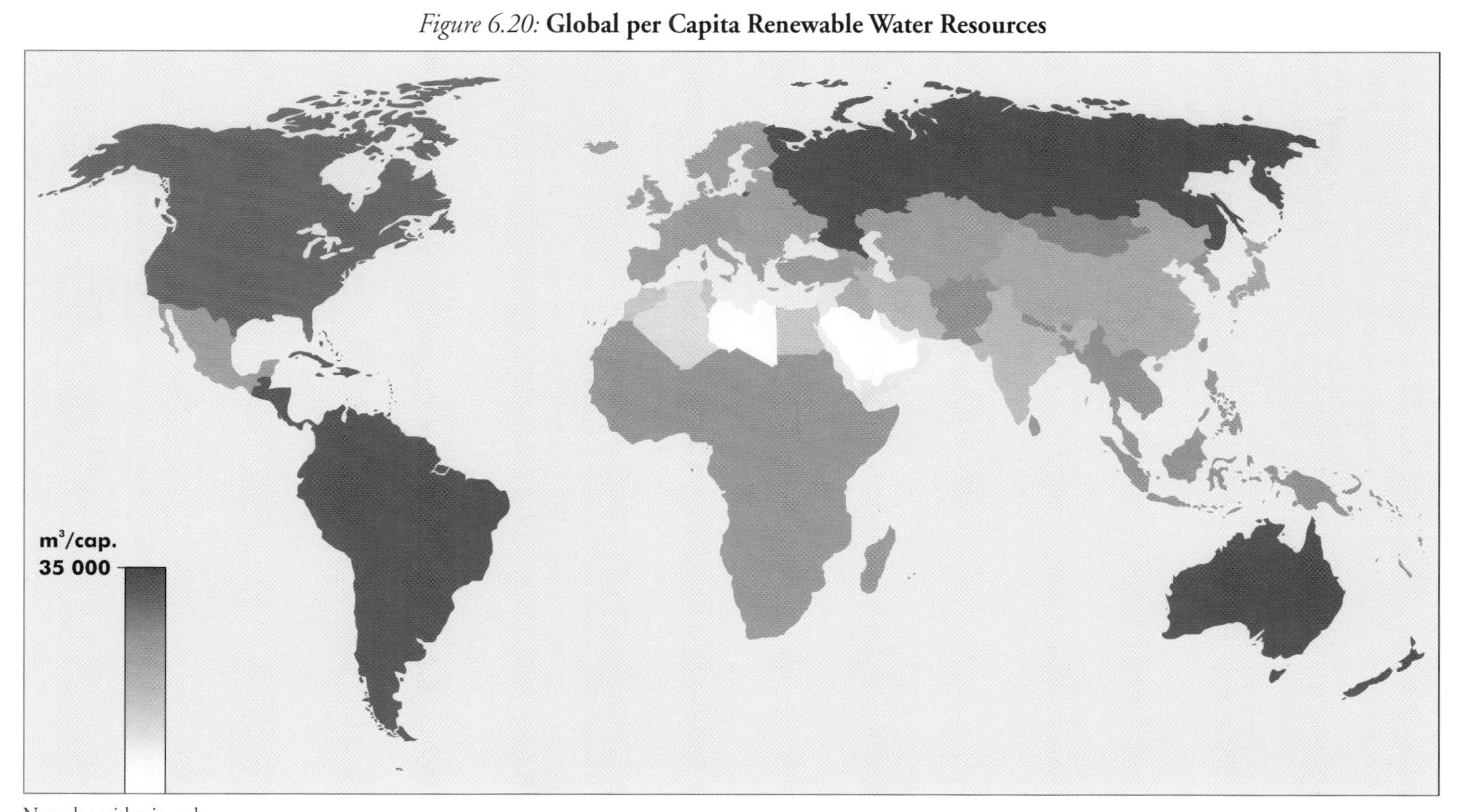

Note: logarithmic scale.
Source: FAO Aquastat database.

Table 6.6: **Global Water Desalination Capacity, 2003**

Country/region	**Desalination capacity** (million cubic metres/year)	**Share in global capacity** (%)
OECD North America	2 067	18%
OECD Pacific	753	7%
OECD Europe	1 399	12%
Transition economies	264	2%
China	191	2%
India	164	1%
Other Asia	310	3%
Latin America	323	3%
MENA	**5 692**	**51%**
North Africa	*572*	*5%*
Middle East	*5 120*	*46%*
Other Africa	67	1%
Total	**11 229**	**100%**

Source: IEA analysis based on Wangnick (2004).

The main environmental impacts arising from the desalination process are brine concentrates and discharges of added chemicals. These effluents can harm coastal and marine ecosystems or pollute aquifers, wadi flows and soils. In the Persian Gulf, the increased amount of brine discharge could lead to a build-up of salt concentrations given the enclosed nature of the sea.

Table 6.7: **Capital Costs for Desalination Processes** ($ per cubic metre per day)

Process	**2000**	**2005**
MSF	1050 - 3150	1000 – 3400
MED	925 - 2120	1000
MVC	1580 - 3170	1000 - 1650
ED	637	n.a.
RO (sea-water)	925 - 2100	800 - 1200
RO (brackish water)	n.a.	500

Note: MSF – multistage flash; MED – multiple effect distillation; MVC – mechanical vapour compression, ED – electrodialysis; RO – reverse osmosis; n.a. – not available.
Source: Middle East Desalination Research Centre, various journals.

Box 6.6: **Energy Requirements and Economics of Desalination Systems**

The energy requirements for the reverse osmosis (RO) process, mostly electricity to operate a pump, are less than for distillation processes. Depending on the salinity, for one cubic metre of freshwater produced through RO, 0.5 kWh to 1.5 kWh is required to purify brackish water, and 3 kWh to 8 kWh to purify sea-water. The electricity required for distillation systems is about 2 kWh to 4 kWh per cubic metre, but the thermal energy requirements are much higher. As the distillation process is highly energy-intensive, requiring a lot of low-grade steam, co-generation plants are often used in order to utilise low-grade steam from power plants. These dual-purpose, combined water and power (CWP) plants reduce plant construction and permitting costs and are more efficient than single-purpose plants.

RO system capital costs tend to be lower than those of evaporative processes. Membrane processes are also more economical for use with brackish and less saline raw water. Poor water quality, however, tends to favour evaporative processes because pre-treatment requirements for thermal processes are much lower than those of RO. The distillation process can better deal with more saline water than membrane processes. Pre-treatment, membrane and energy recovery technologies, however, have improved considerably and the use of RO is likely to expand in the future.

The distillation and membrane systems can be combined in a hybrid desalination system. The hybrid concepts have been proposed for over a decade, but today there are very few applications of the systems. These systems are estimated to offer considerable cost savings over current plants. They are also expected to reduce installed capacity requirements because they allow for more flexibility between water and electricity production. With a hybrid system, water production can be maintained efficiently in winter time when electricity demand is low.

MENA Water Demand and Supply[17]

The MENA region has been facing dwindling water supplies and growing demand for many years, as a result of high population growth, improvements in the standard of living, industrial development in urban centres and efforts to increase food self-sufficiency. The population in several countries in the

17. The projections for desalinated-water demand are for the six most water-stressed countries, Saudi Arabia, the UAE, Kuwait, Qatar, Algeria and Libya. The share of desalinated water in total water demand is highest for these countries. More details can be found in the country chapters.

region is expected to grow by about 2% or more per year over the period 2003-2030. The population of the region's cities is growing even faster, increasing the need for other water options, like sea-water desalination and waste water reclamation.[18]

Desalination is used extensively in the Gulf countries, where natural renewable sources of water are too low to meet demand. There is already considerable desalination capacity in Saudi Arabia, the UAE, Kuwait, Qatar, Bahrain and Oman. There is also reasonably high interest in the technology in Algeria and Libya.

The Persian Gulf has the highest salinity of any body of water, ranging from 45 000 to 50 000 ppm, compared with salinity in the Mediterranean Sea of about 40 000 ppm and average ocean salinity of 35 000 ppm.[19] This is one of the main reasons for the high share of distillation units in the Gulf countries.

This section focuses primarily on residential and industry water demand because these two sectors are the main consumers of desalinated water. In countries that have little agriculture or industry, such as Kuwait, most water is used in households. Household demand is affected by a variety of factors, such as per capita income, household size, distance from a source of water and consumption patterns. Some industries that use large amounts of water are food, paper, chemicals, refined petroleum and primary metals. The petrochemical industry is considered the most water-intensive in the Middle East. Industry's share of water use is quite low in most countries.

The agricultural sector's share of water use in most MENA countries is more than 70%, largely for irrigation of crops. Irrigation water is mainly supplied by groundwater and by renewable surface water. Rapid expansion of irrigated areas has resulted in substantial groundwater extraction. Many countries have a policy goal of self-sufficiency in agricultural consumption and the pressure to supply potable water is intense.

Due to substantial losses in the distribution network, unaccounted-for water represents about 30% to 40% across the MENA region. Efficiency in the water system is also low because of poor maintenance of water reservoirs and frequent theft of water. Technical improvements to the water infrastructure are needed to reduce losses. Repairing leaking distribution systems and water pipes,

18. The waste-water infrastructure of the MENA region is significantly underdeveloped. Less than half of households are connected to a sewage system, low by international standards. Only a small proportion of the waste water is treated.
19. Pankratz and Tonner (2003).

expanding central sewage systems, metering water connections and rationing water use could all be important. Improving the efficiency of water use in the agricultural sector is also necessary. But the most cost-effective means of achieving greater efficiency in water use is effective water pricing. Water supply is heavily subsidised in MENA countries (Table 6.8).

Table 6.8: **Average Water Tariffs in Selected Countries** ($ per cubic metre)

Saudi Arabia	0.03
Kuwait	0.65
Qatar	1.20
United States	1.30
France	3.15

Sources: Global Water Intelligence (2005); OECD (2003).

The rate of depletion of surface water and renewable groundwater in the Middle East is phenomenal and water scarcity is a major issue for Kuwait, Qatar, the UAE, Libya, Algeria and Saudi Arabia. These six countries are among those with the lowest natural renewable water resources in the world. Groundwater depletion across the Middle East is such that salinity is a problem in many coastal areas. All countries will face a reduction in the availability of potable groundwater over the coming decades.

In 2003, total installed desalination capacity in the MENA region was 5.7 billion cubic metres per year. Saudi Arabia accounted for about 40% of total capacity. Water production from desalination plants was nearly 3.7 billion cubic metres in 2003 in Saudi Arabia, the UAE, Kuwait, Qatar, Algeria and Libya, meeting some 10% of total water demand in these countries.

In general, countries with small-scale desalination plants and with relatively high natural freshwater availability favour the smaller, more cost-effective RO plants. Large-scale sea-water desalination plants in the Middle East tend to use the MSF process (Table 6.9).

Total water demand (in agriculture, industry and residential) in the six countries analysed is projected to rise to 44 billion cubic metres in 2030, from 37 billion cubic metres in 2003, representing growth of 0.6% per year over the projection period. Residential water demand, however, is projected to grow by 2.5% per year. Rising demand in this sector will be the driving force behind expected growth in desalination capacity. Water from desalination plants will rise to almost 12 billion cubic metres, increasing the share of desalinated water in total water demand from 10% in 2003 to 27% in 2030.

Table 6.9: **Water Demand and Desalination Capacity in Selected MENA Countries, 2003**

	Water demand (mcm)	Water demand per capita (litres/day/capita)	Water supply from desalination (%)	Reverse osmosis capacity		Distillation capacity		Total capacity (mcm)
				(mcm)	(%)	(mcm)	(%)	
Saudi Arabia	22 484	2 734	8	780	35	1 427	65	2 207
UAE	2 694	1 843	42	75	5	1 390	95	1 465
Kuwait	679	775	64	62	11	519	89	582
Libya	4 867	2 401	2	71	26	201	74	272
Qatar	375	1 408	38	5	3	201	97	206
Algeria	6 244	537	2	67	54	58	46	125
Iran	92 000	3 798	0.2	58	32	124	68	182
Egypt	73 533	2 975	0.1	92	86	15	14	107
Iraq	43 208	4 768	0.2	82	95	4	5	86

Note: Due to the lack of accurate data by sector and by country, water consumption for 2003 is estimated for some countries, on the basis of previous years.

The projections are based on assumptions for growth in population and per capita income and efficiency improvements. Population and GDP assumptions are in Tables 1.2 and 1.4. Efficiency improvements in water losses are assumed to average 1.5% per year over the projection period, reflecting government policies to enforce strict conservation measures and price mechanisms to reduce water demand.

Figure 6.21 shows the evolution of water demand in the residential and industry sectors in the six most water-stressed MENA countries. The graph shows the rising share of desalinated water supplies in residential and industrial water demand. Total water supply in these sectors will increase to 13.3 billion cubic metres in 2030, an annual increase of 2.4% from the 2003 level of 7 billion cubic metres. The supply of water from desalination plants in these countries is expected to grow by 4.5% per year.

One of the main uncertainties for water supply over the next thirty years is the rate of depletion of the non-renewable groundwater resources. Non-renewable water reservoirs in Saudi Arabia and Algeria are assumed to deplete at about 0.5% per year. Libya and Kuwait's resources are assumed to deplete at a slightly faster rate over the *Outlook* period. Faster rates still are assumed for the UAE and Qatar, at around 1%. Trade in water resources is not included in the projections, but it is likely to remain limited, given current policies.

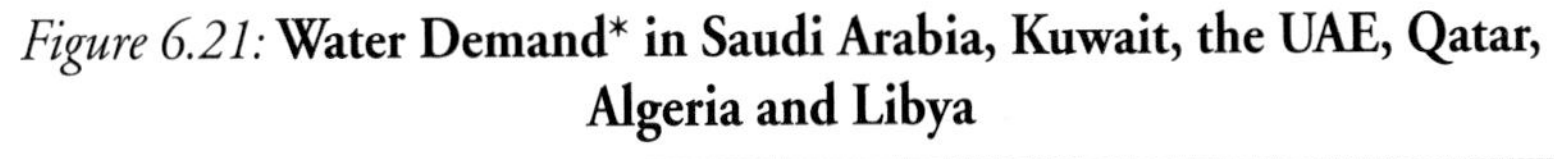

Figure 6.21: **Water Demand* in Saudi Arabia, Kuwait, the UAE, Qatar, Algeria and Libya**

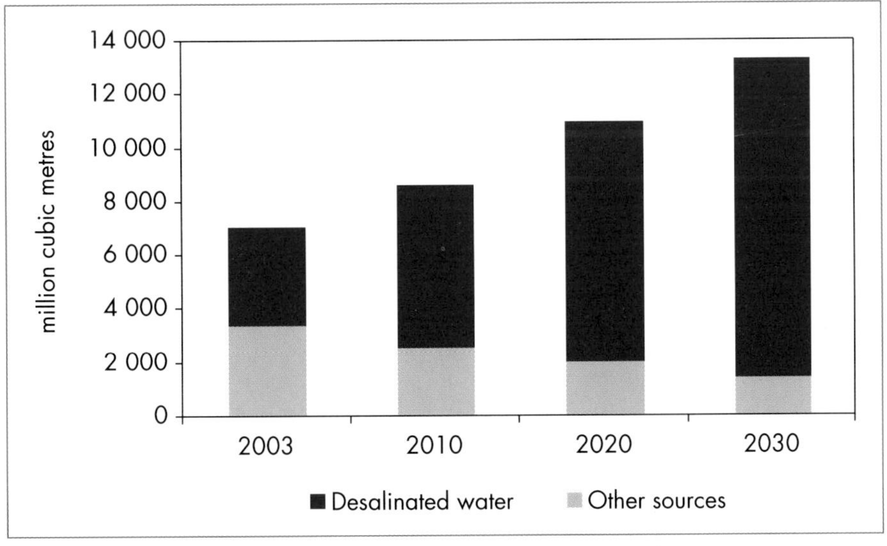

* Projections do not include water demand in the agricultural sector.

Water Desalination Production and Capacity

In the Reference Scenario, production of desalinated water in the six water-stressed countries is expected to grow from 3.7 billion cubic metres in 2003 to 11.9 billion cubic metres in 2030, growth of 4.5% per annum (Table 6.10). Over the projection period, desalination capacity doubles in the UAE, Kuwait and Qatar, and triples in Saudi Arabia and Libya. Desalination capacity will increase substantially in Algeria.

Table 6.10: **Projected Desalination Capacity in Selected MENA Countries**
(million cubic metres per year)

	2003	2010	2020	2030
Saudi Arabia	2 207	3 523	5 593	7 794
UAE	1 465	2 482	2 684	2 948
Kuwait	582	934	1 006	1 088
Libya	272	465	532	772
Qatar	206	282	336	401
Algeria	125	542	1 307	2 004
Total	**4 857**	**8 227**	**11 458**	**15 007**

The rate of growth in water production from the RO process, at 6.7% per year, is much faster than water production from distillation plants (3.7%). Technological improvements, as well as increasing cost-effectiveness, will favour the RO process, particularly in the long term.[20] The share of RO in total desalination capacity is expected to increase from 22% in 2003 to 27% in 2010 and to 37% in 2030.

The percentage of new CWP plants in total power generation plant additions in 2030 is illustrated in Figure 6.22. New CWP capacity will account for more than half of capacity additions in the power sector in Saudi Arabia, the UAE, Kuwait and Qatar, for almost half in Algeria and for one-third in Libya.

20. In the past, RO plants had operational problems in the MENA region due to higher salinity and more abundant marine life in the Persian Gulf. But the industry has gained experience with design and operation of pre-treatment systems, and RO technology will increasingly be perceived as the cost-effective and competitive option.

Figure 6.22: **Share of CWP in Power-Generation Capacity Additions in Selected MENA Countries, 2004-2030**

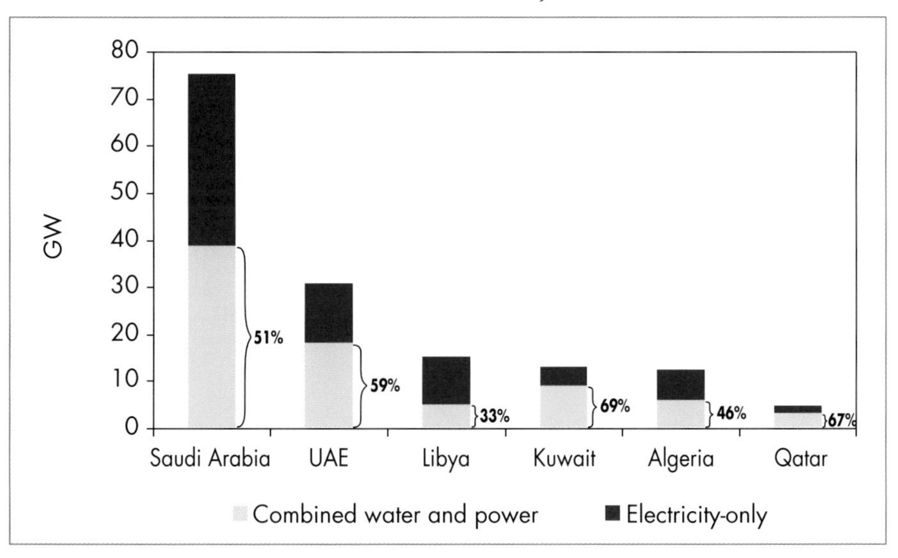

Primary energy requirements[21] for water desalination will increase from 26 Mtoe in 2003 to 61 Mtoe in 2030, growth of 3.2% per year. In 2030, the energy required to meet expected water demand will account for 10% of total primary energy demand in the six most water-stressed countries, about the same share as today. The additional fuel requirements for water desalination will account for more than a quarter of the total increase in fuel requirements in the power and water sector over the projection period.

Electricity consumption in new desalination plants will be 3.2 Mtoe (37 TWh) in 2030, or 6% of total electricity consumption. The electricity needs for new distillation plants will be 1.3 Mtoe, or 42%. Electricity required in new reverse osmosis plants will be 1.9 Mtoe.

Table 6.11 shows the total fuel requirements for steam and for the electricity consumption in the desalination process. Energy demand to purify water in Saudi Arabia will account for over half of the total requirements for desalination in the six most water-stressed countries.

The predominant fuels used for electricity and water production are oil and natural gas. Gas will be the preferred fuel for water desalination in the Middle

21. The primary energy requirements include both the fuel used to generate the electricity consumed by the desalination plants and the fuel requirements for the steam production. For a detailed description of the fuel-allocation methodology, see the discussion of the World Energy Model on www.worldenergyoutlook.com.

Table 6.11: **Energy Use for Water Desalination in Selected MENA Countries**
(Mtoe)

	2003	2010	2020	2030
Algeria	0	1	3	4
Kuwait	3	4	4	5
Libya	1	1	2	3
Qatar	1	2	2	2
Saudi Arabia	11	17	24	31
UAE	9	13	15	16
Total	**26**	**38**	**50**	**61**

East over the next 25 years, as it is more efficient, less costly and more environmentally acceptable than oil. Gas will gain share in countries that currently use oil for CWP.

Water Desalination Investment Needs and Financing

Table 6.12 shows investment needs for additional desalination capacity in the water sector over the projection period. The figures refer to the investment needs for desalination plants, excluding the power generating capacity. A total investment of $39 billion will be needed over the projection period in the six water-stressed countries in order to add almost 14 billion cubic metres of desalination capacity. The total investment for desalination plants accounts for 30% of the total investment in the power and water generation sector. Required investments in Saudi Arabia overwhelm the other countries, accounting for more than half of the requirements in the six most water-stressed countries in 2030.

Table 6.12: **Cumulative Investment in Water Desalination in Selected MENA Countries** ($ billion in year-2004 prices)

	2004-2010	2011-2020	2021-2030	2004-2030
Algeria	1.3	2.2	1.9	5.4
Kuwait	1.4	1.2	0.3	2.9
Libya	0.8	0.5	0.8	2.1
Qatar	0.2	0.5	0.2	0.9
Saudi Arabia	4.4	8.9	7.6	20.9
UAE	3.2	1.5	1.7	6.3
Total	**11.3**	**14.8**	**12.5**	**38.6**

DEFERRED INVESTMENT SCENARIO

HIGHLIGHTS

- The Deferred Investment Scenario analyses how global energy markets might evolve if upstream oil investment in each MENA country were to increase much more slowly over the projection period than in the Reference Scenario. This may reflect government decisions to limit budget allocations to the industry or constraints on the industry's ability or willingness to invest in upstream projects.
- The resulting international crude oil price is $13 higher in 2030, or $21 in money of the day – an increase of almost one-third. The price gap between the two scenarios widens gradually over time as MENA output is increasingly held back by lower investment and non-MENA supply struggles to meet rising global demand. Natural gas prices also rise, broadly in line with oil prices. Coal prices rise, but less quickly.
- World energy demand is reduced by around 900 Mtoe, or 6%, in 2030 compared with the Reference Scenario, as a result of higher prices and lower GDP growth. Global oil demand, at 105 mb/d in 2030, is 10 mb/d lower than in the Reference Scenario. Primary energy demand growth in MENA countries slows with lower oil and gas revenues and higher prices, but less than in non-MENA regions.
- MENA oil production falls by 15 mb/d, or 30%, in 2030 compared with the Reference Scenario. MENA's share of world oil production drops from 35% in 2004 to 33% in 2030 (it *increases* to 44% in the Reference Scenario). As a result, MENA oil exports are considerably lower – by almost 15 mb/d, or 38%, in 2030. Non-MENA oil production is 5 mb/d *higher* in 2030.
- Natural gas production in MENA countries and in all other regions also falls significantly, due to lower associated gas output and lower global demand. MENA gas exports fall by 46% in 2030, with Qatar's falling most in absolute terms.
- The cumulative value of MENA oil and gas export revenues over 2004-2030 is $1 000 billion lower than in the Reference Scenario, because lower export volumes more than outweigh the increase in prices. This is almost five times more than the reduction in investment. Revenues also fall in net present value terms. The fall in revenues is proportionately higher for gas than for oil.

Background and Approach

Why a Deferred Investment Scenario?

In the Reference Scenario, MENA oil and gas production is projected to rise strongly over the *Outlook* period in order to satisfy escalating global demand. Crude oil production increases from 29 mb/d in 2004 to 50 mb/d in 2030, while gas production grows from 385 bcm to 1 211 bcm in 2003-2030. Achieving these production levels will call for cumulative investment over 2004-2030 of $614 billion, or $23 billion per year on average for oil and $436 billion ($16 billion a year) for gas. For both fuels, supply from MENA countries is assumed to meet the portion of global supply that is not met by non-MENA producers.

Projected non-MENA supply is based on assumptions about international oil and gas prices. Those assumptions reflect our judgment of the prices that will be required to encourage sufficient investment in overall supply of both oil and gas to meet projected world demand over the *Outlook* period. Underpinning these figures is an assessment of the factors other than price that will drive future investment in supply capacity for a given level of demand. These include developments in technology and supply costs, as well as the strategic interests and production policies of producer countries. Of course, the demand numbers in the Reference Scenario are themselves based on assumptions, which could be invalidated. The impact of new government policies is developed further in a World Alternative Policy Scenario in Chapter 8.

History teaches us that the trajectory of energy prices that balance supply and demand is neither stable nor easily predictable. Investment flows to the oil and gas sector and oil and gas prices have been highly volatile for much of the time since petroleum production began. So it is quite possible that both investment and prices could deviate by a wide margin and for prolonged periods from the paths envisaged in our Reference Scenario. In recent years, investment in crude oil production and refining capacity have lagged the rise in demand, causing prices to rise. Our projections call for a significant increase in global upstream and downstream investment, but that investment may not be forthcoming. There would be a particular problem for the global economy if investment in the oil sector of MENA countries turns out to be markedly lower than projected, curbing gas investment as one consequence.

In recognition of these uncertainties, we have developed a Deferred Investment Scenario to analyse how global energy markets might evolve if investment in the upstream oil industry of MENA countries were *not* to grow as rapidly as called for in the Reference Scenario. The assumption adopted is that the level of upstream investment in oil remains constant as a share of GDP, based on the average for the past decade. This is a pessimistic assumption, since low prices in much of the decade to 2003 constrained investment. The same severe investment cut is not applied, initially, to investment in gas; but gas investment is cut as a secondary effect of the reduced investment in oil. Such a shortfall in investment would lead to slower development of resources and, consequently, slower growth in actual production of oil and gas in the region. To bridge the gap between MENA output and world demand, supply from non-MENA regions would have to increase and/or demand would need to fall to bring the global oil and gas markets back into balance. For this to happen, prices would need to rise. A key objective of the Deferred Investment Scenario is to assess how much global demand would fall and how much non-MENA supply would rise as a result of the higher prices brought about by a given reduction in investment, compared with the Reference Scenario.

What Might Cause Oil and Gas Investment to be Deferred?

There are several reasons why oil and gas investment might be deferred or restricted. These are discussed in each country chapter. They can be categorised according to whether the lower level of investment is a consequence of government policy or whether it stems from external factors.

MENA governments could choose deliberately to *develop production capacity more slowly* than we project in our Reference Scenario. There are several possible motivations for such a move. OPEC members among the MENA producers might judge that curbing investment and limiting capacity – if decided and implemented in a co-ordinated, formal manner – would be likely to boost net earnings from exports of oil and gas, on the basis that the increase in international market prices might be large enough to offset the revenue loss due to this fall in production compared to the Reference Scenario. On the other hand, there would always be a risk that other producers might boost their capacity more quickly, tempering the increase in prices and potentially resulting in *lower* export earnings. This is precisely what happened in the 1980s and 1990s. There would also be the risk that the price elasticity of global demand for oil and gas might turn out to be higher than expected, leaving each producer worse off than if they had pursued an expansion policy in line with the Reference Scenario. Our analysis suggests that MENA producers would, indeed, be *worse* off by deferring investment in the way assumed in the Deferred Investment Scenario.

Another reason why MENA countries might defer investment could be to preserve hydrocarbon resources and revenues for future generations. This could be a legitimate policy for a country with relatively high GDP per capita and no pressing need for additional oil revenues to fund infrastructure or social programmes, such as the UAE. Yet, holding back development of oil and gas carries the risk of driving up prices, thereby accelerating the development of alternative energy resources. It is particularly important to those countries with the largest reserves to prolong the life of the oil and gas industry as a means of sustaining in the long term the modernisation and diversification of the economy and thereby to allow future generations to share the benefits. They are well aware of the need to steer a course between pushing up oil revenues in the short to medium term and preventing prices from rising so high as to undermine the longer-term prospects for their oil and gas industries – even if there is no agreement on exactly what that course might be.

A deferral of investment may also occur as a result of *external factors that prevent producers from investing* as much in expanding capacity as they would like. A lack of capital is one such cause. This could occur for several reasons. Despite the amplitude of current financial surpluses, in countries with national oil companies financing new projects could become a problem where the national debt is high and, in changed market circumstances, there is a need to borrow large sums to finance new projects. Strong guarantees could be difficult to find or involve a high premium. Sovereign risk in some Middle Eastern countries is high, pushing up borrowing costs. A combination of these factors could delay or even prevent investments altogether.

Considerations of national sovereignty might discourage reliance on or block foreign investment. The producing countries' policies on opening up their oil industries to private and foreign investment, the legal and commercial terms on offer and the fiscal regimes will, together, have a major impact on how much external capital Middle Eastern producers will be able to secure. This could become a particularly crucial factor in the middle decade of the projection period, when real oil prices are assumed in the Reference Scenario to be at their lowest.

In many countries, education, health, defence and other sectors of the economy – including public electricity and water services – could command an increasing share of government revenues and, thus, constrain capital flows to the oil sector. The populations of all MENA countries are expected to grow rapidly in the coming decades, averaging 1.7% per year over the period 2003-2030. Even in countries that are open to foreign investment or plan to become so, thereby becoming less dependent on government revenues, the needs of an expanding population could lead governments to increase taxes and royalties on oil and gas production. This would lower the profitability of upstream projects and might deter investment.

Inadequate infrastructure to support upstream oil and gas developments – through poor planning or lack of capital to finance such related projects – could constitute another barrier to investment. If access to production sites is difficult, or roads, railways, pipelines or export facilities are not available, upstream projects could be delayed or cancelled. Operational and financial performance could also be affected by a shortage of qualified labour. There are signs that increasing upstream activity is driving up engineering, procurement and construction (EPC) costs, though this phenomenon is expected to be short-lived. Insecurity and conflict may also be a barrier to investment.

Some deferral of upstream oil and gas investment in MENA countries is clearly a strong possibility, whether due to deliberate policy decisions or to other factors, including those beyond governments' control. Some combination of such causes is possible. Our intention here is not to predict the events which might lead to lower investment, or to analyse their risk, but rather to assess the consequences of any such deferral for global energy markets. Summarised below is the approach adopted in analysing the impact of a scenario in which MENA oil and gas investment is markedly lower, together with the precise assumptions underlying that analysis.

Methodology and Assumptions

We developed a detailed methodology for evaluating the impact of lower investment that takes account of the complex inter-relationships between investment, economic growth, energy demand and supply, and international energy prices. Without attempting to define exactly how and why this might occur, for the purposes of our analysis we adopted the assumption that the proportionate share of upstream oil investment in GDP in each MENA country remains broadly constant over the projection period at the estimated level of the past ten years (1995-2004).

In the Reference Scenario, upstream oil investment is projected to average 0.52% of GDP in MENA countries as a whole between 2004 and 2030, compared to an estimated 0.38% for 1995-2004 (8% of total investment). In all countries, the share of investment goes up, though to varying degrees. As a result, applying the Deferred Investment Scenario rule of constant investment relative to GDP over the projection period results in marked differences among MENA countries. Investment is lower in all countries, but the fall is largest in percentage terms in Iraq and Libya (Table 7.1), because investment increases relative to GDP the most in the Reference Scenario. Investment falls least in Algeria, Egypt and Qatar. In aggregate, the reduction in cumulative upstream oil investment in the Deferred Investment Scenario vis-à-vis the Reference Scenario amounts to some $110 billion, or 23%, over 2004-2030. Nonetheless, upstream investment continues to grow in absolute terms. These changes should not be interpreted as a *prediction* of investment flows, nor as an indication of the relative probability that each MENA country will invest less than in the Reference Scenario.

Though the same proportionate cut has not been applied initially to gas investment, the assumption for oil has implications for gas investment, as production of gas is, in many countries, associated with the production of oil. In addition, higher oil prices resulting from lower investment and production would push up gas prices too, curbing global gas demand and the call on MENA gas supply. The net impact on gas investment has been estimated taking account of the impact of higher oil prices on gas prices and global gas demand and the share of non-associated gas in each country. In total, MENA upstream gas investment is 19% lower than in the Reference Scenario.

Table 7.1: **Cumulative Upstream Oil Investment in the Reference and Deferred Investment Scenarios, 2004-2030**

	Reference Scenario ($ billion*)	**Deferred Investment Scenario** ($ billion*)	**Difference vs. Reference Scenario**
Middle East	**409**	**313**	**–23%**
Iran	59	44	–24%
Iraq	51	30	–41%
Kuwait	47	36	–24%
Qatar	31	28	–10%
Saudi Arabia	141	115	–19%
UAE	41	30	–27%
Other Middle East	40	31	–23%
North Africa	**74**	**58**	**–22%**
Algeria	27	24	–10%
Egypt	9	8	–10%
Libya	36	24	–34%
Other North Africa	2	1	–22%
Total MENA	**484**	**371**	**–23%**

* In year-2004 dollars.

The assumptions about all other aspects of MENA government policy, about population growth and about technology are the same in this scenario as in the Reference Scenario. However, international energy prices are higher. Economic growth rates in both MENA and non-MENA countries also change, because of the impact of lower investment on the future level of oil and gas production, international energy prices and revenues. The way in which the impact of lower investment on energy prices and economic-growth rates is modelled is described below.

The methodological approach involved several steps, summarised graphically in Figure 7.1:

- *Oil and gas supply:* The impact of lower oil investment on oil production in each country is modelled separately for existing fields and for new field developments (including future reserve additions at existing fields and fields yet to be discovered). The reduction in investment is assumed to be the same for each. If, for example, the reduction in investment compared with the Reference Scenario is 10%, the reduction in supply from existing fields[1] is calculated as 10% of the difference between the production in the Reference Scenario and the production that would occur in the total absence of any investment in those fields, *i.e.* the natural decline in field production. The latter element was derived using indicative estimates – verified by country experts – of average decline rates for each MENA country, ranging from 5% to 10% per year. For new field developments, production is assumed to fall at the same rate as investment, compared with the Reference Scenario.
- *Oil price and non-MENA oil supply:* The impact on international energy prices of reduced MENA supply – a central element of our analysis – is quantified using a world oil equilibrium model. The model solves the oil price trajectory that brings non-MENA supply and global demand into equilibrium for the lower MENA supply.[2] Natural gas prices are assumed to increase broadly in the same proportion as oil prices, reducing gas demand, investment and production. Coal prices also increase, but to a lesser degree.
- *GDP growth:* Gross domestic product growth rates for each MENA country and for each year were adjusted in line with the higher price trajectories and lower oil and gas production trends, using the price/GDP model deployed in the Reference Scenario and described in Chapter 1. GDP was also adjusted for all non-MENA regions, based on the degree of energy-import dependence and the estimated short- and long-run sensitivity of GDP to oil price increases in each region.
- *Energy demand:* The World Energy Model (WEM), including the individual MENA country models, was run using the revised energy price and GDP trajectories, yielding a new set of energy-demand projections.
- *World oil and gas supply:* The new global energy-demand projections were combined with the new MENA oil and gas production trajectories to yield new profiles for aggregate non-MENA oil and gas production. It is assumed that non-MENA production rises in response to higher prices to meet the shortfall in global supply resulting from lower investment and production in MENA countries. This is different from the approach adopted in the

1. Derived from our field-by-field production analysis, described in Chapter 4.
2. A detailed description of how the model works can be found at www.worldenergyoutlook.org.

Reference Scenario, where MENA countries are assumed to be the residual supplier to the world market. Aggregate non-MENA supply is broken down by region, using an algorithm that takes account of ultimately recoverable resources.

- *Global energy-market effects:* Inter-regional oil and gas trade flows, together with investment by region, were recalculated on the basis of the new projections for demand and supply in MENA countries and non-MENA regions.

Because of the feedback from oil price to energy demand, both directly and through GDP, it was necessary to run the world oil equilibtium model and the WEM iteratively in order to reach a stable equilibrium. At that equilibrium, international fossil-fuel prices and global energy demand and supply are in balance with each other.

Figure 7.1: **Modelling Approach of the Deferred Investment Scenario**

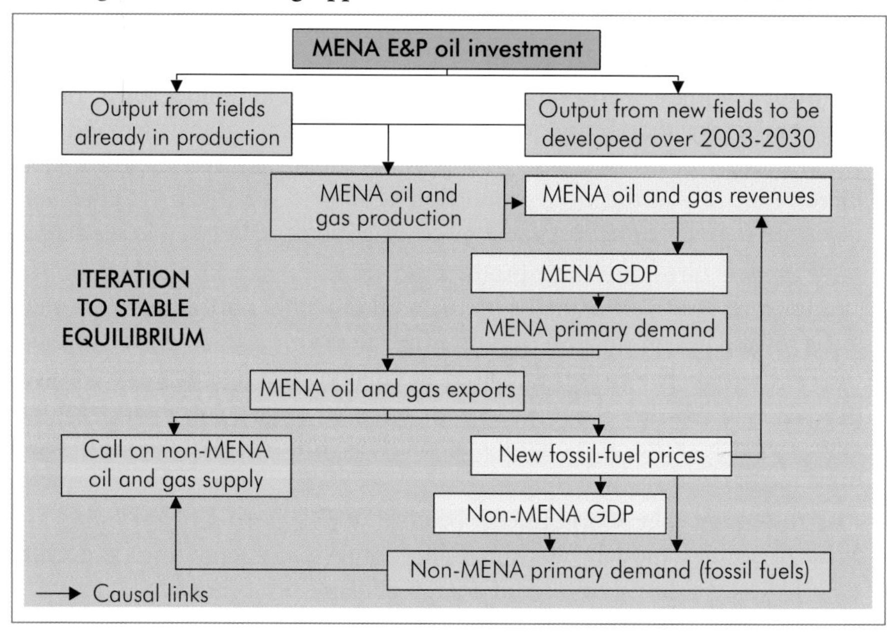

Implications for Energy Prices

The international crude oil price, for which the average IEA import price serves as a proxy, follows a higher trajectory over the projection period in the Deferred Investment Scenario compared with the Reference Scenario (Figure 7.2). On average, the price is about 20% higher over the projection period. The price gap between the two scenarios widens gradually over time, reaching about $13 in year-2004 dollars and $21 in nominal terms (assuming 2% inflation) in 2030 – an increase of 32% over the Reference Scenario. MENA output is increasingly held back by lower investment and non-MENA supply rises, but it struggles to meet rising global demand (albeit rising more slowly than in the Reference Scenario). Higher prices are needed to compensate for rising marginal supply costs.

Figure 7.2: **Average IEA Crude Oil Import Price in the Deferred Investment and Reference Scenarios**

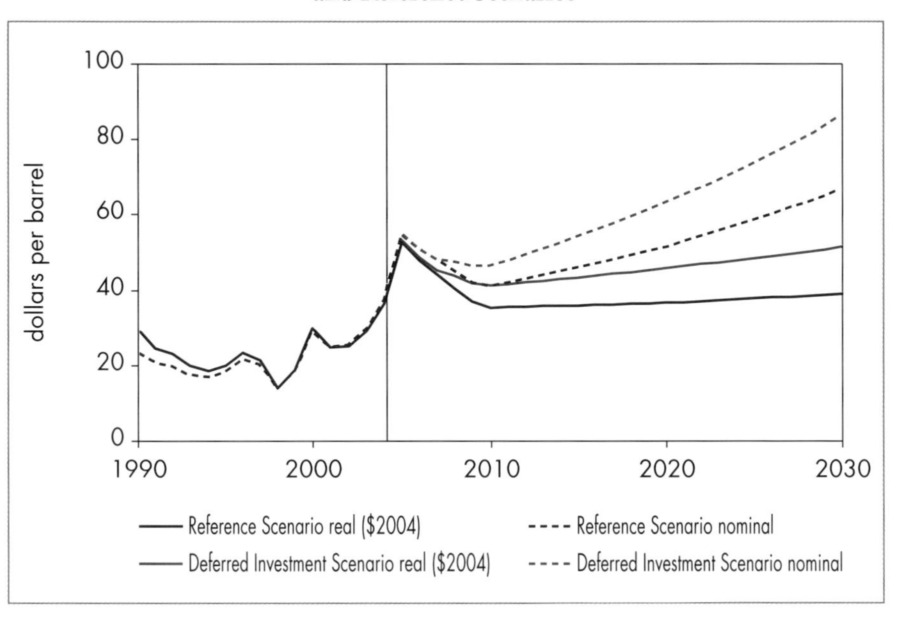

This increase in oil prices brings about an increase in natural gas prices even though liberalisation and the growing globalisation of gas markets are expected to weaken the link between oil and gas prices (Table 7.2). This results in a lower

Table 7.2: **Fossil-Fuel Prices in the Deferred Investment Scenario**
(in year-2004 dollars)

	2004	**2010**	**2020**	**2030**
IEA crude oil imports ($/barrel)	36	41	46	52
In nominal terms	*36*	*47*	*63*	*86*
Natural gas ($/MBtu):				
US imports	5.70	6.60	7.10	7.80
European imports	4.20	5.70	6.30	7.10
Japanese LNG imports	5.20	6.70	7.30	8.00
OECD steam coal imports ($/tonne)	55	53	55	57

Note: Prices in the first column are historical data. Gas prices are expressed on the basis of gross calorific value.

percentage increase in gas prices than in oil prices. The US gas import price, for example, is $0.80 per MBtu higher in 2010 and $1.60/MBtu higher in 2030 than in the Reference Scenario. The coal price also increases, by $4 per tonne in 2010 and $6 in 2030 compared with the Reference Scenario.

Implications for Energy Demand

Global Energy Demand

Energy demand in all regions is affected by changes in GDP and the increase in energy prices in the Deferred Investment Scenario. Those regions that are most dependent on oil and gas imports – notably the OECD and Asia – suffer the biggest losses in GDP. In aggregate, the average annual rate of world GDP growth (including MENA countries) is around 0.23 percentage points lower over the projection period than in the Reference Scenario – equivalent to about $3 trillion per year in 2004 prices. The differences in GDP increase over time.

The sensitivity of energy demand to higher prices in non-MENA counties varies according to the extent to which energy is subsidised and taxed. In most OECD countries, for example, demand for oil products is little affected by increases in international prices, because end-user prices are cushioned by high taxes. Taxes on oil products and other fuels are generally lower in non-OECD countries, making demand more sensitive to price changes.

As a result of higher prices and lower GDP growth, global demand for energy is reduced by around 900 Mtoe, or 5.6%, in 2030 compared with the Reference Scenario. In fact, lower energy demand in non-MENA countries is equal to more than 90% of the MENA supply reduction; lower MENA demand accounts for most of the rest. On average, global energy demand grows by 1.3% per year over the projection period, compared to 1.6% in the Reference Scenario. The biggest percentage reductions occur in China, Russia and other transition economies. A relatively large share of commercial fuels in total primary energy use in these regions, compared to other non-OECD countries, accentuate the impact of higher international fossil fuel prices on total energy demand. Demand in most developing regions – notably Africa – is less affected because traditional biomass, the price of which is unchanged, accounts for a much bigger share of the energy mix. The drop in demand in Europe and the Pacific region, despite their relatively high dependence on oil and gas imports, is smaller than that in North America. This is mainly because end-user prices are already high compared to other regions, owing to taxes. The concentration of oil use in transport, where the scope for fuel substitution is very limited, also dampens the impact of higher oil prices on demand. In these regions, hydrocarbon intensity – the amount of oil and gas used per unit of GDP – is lower than in the rest of the world.

Among the primary fuels, demand for oil falls most worldwide (Table 7.3). The average annual rate of global oil demand growth over 2003-2030 falls from 1.4% to 1.1%. By 2030, oil demand reaches 105 mb/d – some 10 mb/d, or 9%, less than in the Reference Scenario. Higher oil prices encourage faster improvements in end-use efficiency, primarily in road vehicles and aircraft. They also encourage faster deployment of alternative fuels, such as biofuels. Gas demand falls by 370 bcm, or 8%, in 2030 compared with the Reference Scenario, mainly as a result of slower growth in power-generation demand. Higher gas and electricity prices choke off electricity demand and encourage switching to other energy sources. End-users of gas also reduce their gas consumption.

Table 7.3: **World Primary Energy Demand in the Deferred Investment and Reference Scenarios, 2030**

	Reference Scenario (Mtoe)	**Deferred Investment Scenario** (Mtoe)	**Difference**
Coal	3 724	3 551	–4.6%
Oil	5 546	5 068	–8.6%
Gas	3 942	3 639	–7.7%
Nuclear power	767	772	0.7%
Hydropower	368	369	0.4%
Biomass	1 653	1 690	2.2%
Other renewables	272	278	2.3%
Total	**16 271**	**15 367**	**–5.6%**

Global coal demand is also reduced, by some 170 Mtoe, or 5%, in 2030, because of higher absolute prices and lower GDP growth, despite its improved competitiveness against gas compared with the Reference Scenario. Higher fossil fuel prices encourage energy users to switch to renewable energy sources. In 2030, hydropower supply is 0.4% higher and the use of biomass and other non-hydro renewables just over 2% higher. Consumption of traditional biomass – including wood, agricultural waste and dung – increases as households switch less rapidly to commercial fuels. Biofuels and the use of biomass for power generation also grow. Output of nuclear power is assumed to be almost 0.7% higher.

MENA Energy Demand

Primary energy demand in MENA countries as a whole grows marginally less quickly on average in the Deferred Investment Scenario, mainly because of the slower increase in economic activity, due to lower oil and gas export revenues. Energy prices are assumed to remain heavily subsidised in most MENA countries, so changes in international prices have only a limited impact on domestic prices and, therefore, demand. GDP grows slightly faster in the first half of the projection period in most MENA countries, but this is offset by reduced energy needs in the oil and gas sectors themselves. Reduced export revenues cause GDP to grow more slowly in the second half of the projection period, driving domestic demand down further. By 2030, total MENA primary energy demand is 47 Mtoe, or 4%, lower in the Deferred Investment Scenario (Figure 7.3). Oil and gas account for virtually the entire drop in demand, falling in almost equal measure in both volume and percentage terms.

Figure 7.3: **Change in MENA Primary Energy Demand in the Deferred Investment Scenario Compared with the Reference Scenario**

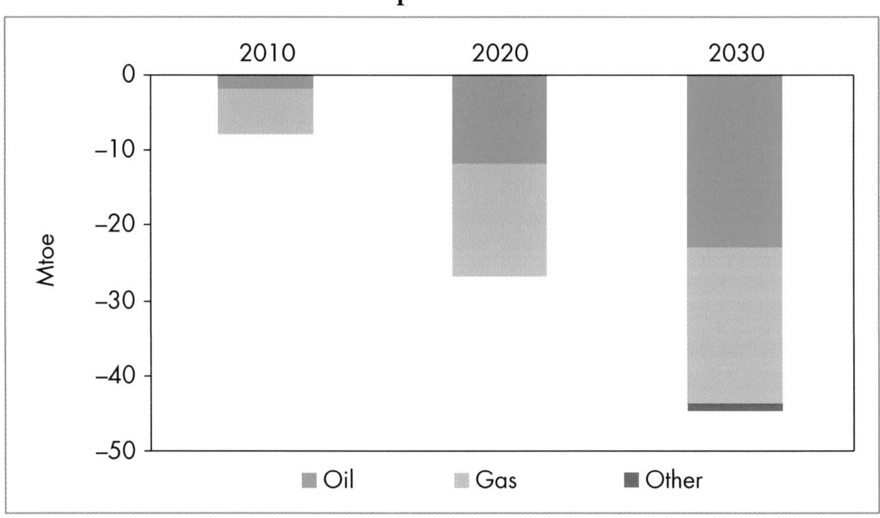

The difference in demand between the two scenarios varies markedly among MENA countries, mainly according to the weight of oil and gas revenues in GDP and the extent to which revenues fall. Demand is lower in all countries (Figure 7.4). The fall in demand is sharpest in Qatar, mainly because of reduced energy needs in gas-production and export facilities, and in Iraq, because its GDP is much lower as a result of far slower oil and gas production growth.

Figure 7.4: **Change in MENA Primary Energy Demand in the Deferred Investment Scenario Compared with the Reference Scenario by Country, 2030**

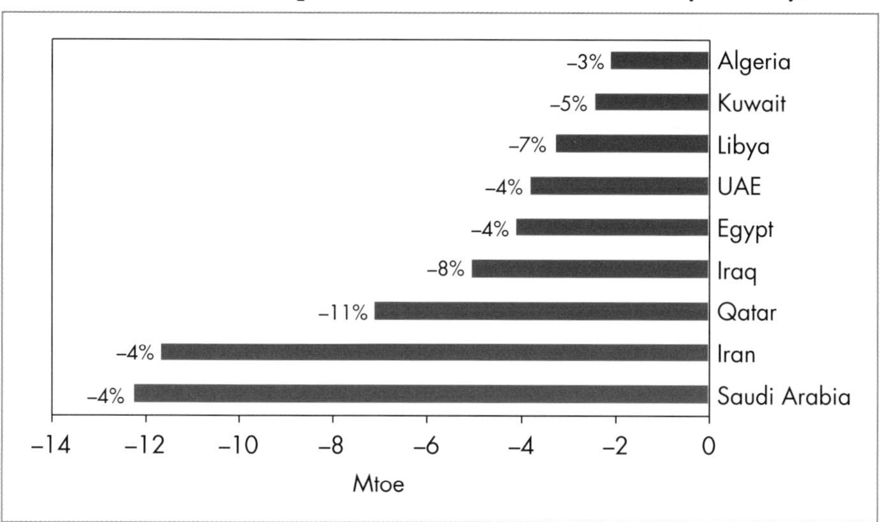

7

Oil Supply

MENA Oil Production

Crude Oil[3]

The underlying assumption that upstream oil investment over the entire projection period is constant proportionate to GDP in each MENA country at the level of the past decade results in substantially different reductions in investment and production among MENA countries. These differences largely reflect the extent to which the share of investment in GDP increases in the Reference Scenario. On average, upstream oil investment is 23% lower in the Deferred Investment Scenario, resulting in a reduction in MENA crude oil production of 7% in 2010 and 30% in 2030 compared with the Reference Scenario. The reduction in production is significantly larger than that in investment because of the *cumulative* effect of lower investment over the projection period: less investment reduces output relative to the Reference Scenario, both from fields that are already in production and from fields that will be developed in the future. For existing fields, the impact is more pronounced because of the effect of natural production decline.

3. Includes NGLs and condensates.

Total MENA crude oil production is just over 15 mb/d lower in 2030 than in the Reference Scenario. It reaches 35.3 mb/d – only 6.3 mb/d more than in 2004. The gap between the two scenarios builds up progressively over the projection period, widening markedly after 2010 (Figure 7.5). MENA's share of world oil production in the Deferred Investment Scenario falls from 35% in 2004 to about 33% in 2030. In the Reference Scenario, the share rises to 44%.

Figure 7.5: **MENA Crude Oil Production in the Reference and Deferred Investment Scenarios**

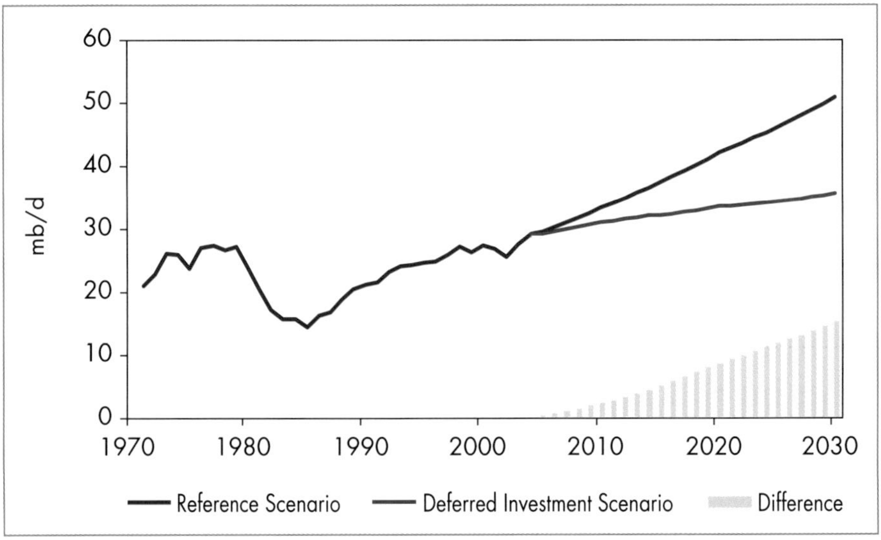

Note: Includes NGLs and condensates.

The biggest reduction in production in volume terms occurs in Saudi Arabia, where production reaches just over 14 mb/d in 2030 compared with 18.2 mb/d in the Reference Scenario (Table 7.4). This lower level of Saudi production is about 1 mb/d below the long-term sustainable output that the Saudi government believes can be achieved in the longer term based on current reserves. The difference in Saudi production between the two scenarios – just over 4 mb/d – is equal to around 27% of the total reduction in MENA oil production.

Among the main producing countries, Iraq incurs the largest percentage reduction in output – 48% in 2030 – reflecting the perpetuation of the very low level of investment in the past decade as against the expected strong increase in investment in the next 25 years in the Reference Scenario.

Table 7.4: **MENA Crude Oil Production by Country in the Deferred Investment Scenario** (mb/d)

	2004	2010	2020	2030	Difference in 2030 compared with Reference Scenario
Middle East	**24.6**	**26.3**	**29.1**	**31.5**	**–31%**
Iran	4.1	4.2	4.5	4.9	–27%
Iraq	2.0	2.4	3.3	4.1	–48%
Kuwait	2.5	2.7	3.2	3.3	–33%
Qatar	1.0	1.1	1.1	1.1	–11%
Saudi Arabia	10.4	11.2	12.8	14.1	–23%
UAE	2.7	3.0	3.2	3.4	–34%
Other Middle East	1.9	1.6	1.0	0.7	–51%
North Africa	**4.3**	**4.3**	**4.1**	**3.8**	**–26%**
Algeria	1.9	2.0	1.7	1.3	–15%
Egypt	0.7	0.6	0.5	0.4	–16%
Libya	1.6	1.6	1.8	2.1	–33%
Other North Africa	0.1	0.1	0.1	0.1	–28%
Total MENA	**29.0**	**30.7**	**33.2**	**35.3**	**–30%**

Note: Includes NGLs and condensates.

Currently producing fields generally bear the brunt of the reduction in production in the Deferred Investment Scenario. This is because of the assumption that all types of upstream projects take the same proportionate cut in investment.[4] Thus, for existing fields, there is less investment in combating the natural decline in output, as well as in drilling to raise overall output. By 2030, new fields, including fields awaiting development, reserve additions and new discoveries, account for 15 mb/d of total MENA oil production in 2030, compared with 21 mb/d in the Reference Scenario (Figure 7.6).

4. In practice, the reduction in investment may be higher or lower for new field developments, depending on costs relative to further development of existing fields.

Figure 7.6: **MENA Crude Oil Production in the Deferred Investment Scenario**

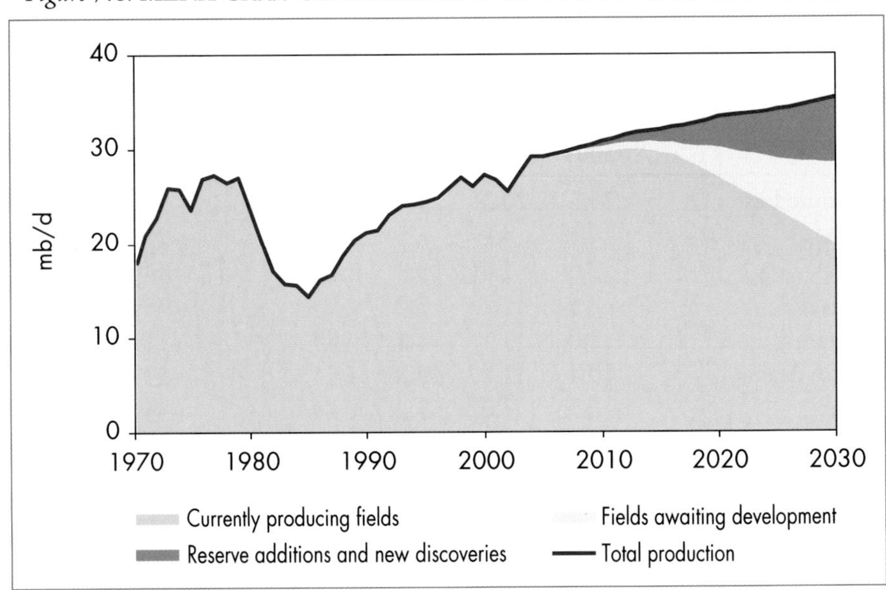

Note: Includes NGLs and condensates.

Refined Products

The fall in global oil demand in the Deferred Investment Scenario induces a corresponding fall of about 10 mb/d, or 9%, in refinery throughput and output in 2030. The largest reductions in output occur in developing countries, where oil demand falls most – notably China where installed capacity is 11% lower in 2030.

The reduction of demand resulting from the Deferred Investment Scenario will have a double impact on the capacity additions of the MENA countries. First, and most significantly, the overall drop in world refined-product demand, which reaches 10 mb/d in 2030, reduces the need for refining capacity additions in MENA, especially in Saudi Arabia, which plays a pivotal role in oil-product exports in the Reference Scenario (Figure 7.7). Secondly, lower domestic MENA demand – about 0.5 mb/d, or 4%, less in 2030 compared with the Reference Scenario – lowers the need for new refineries to produce for the domestic market. In MENA as a whole, installed distillation capacity reaches 14.1 mb/d in 2030, 12% lower than in the Reference Scenario. The fall in capacity is close to three times more than the reduction in domestic demand. About one-quarter less distillation capacity needs to be added in MENA compared to the Reference Scenario.

Figure 7.7: **Change in MENA Oil Product Demand and Distillation Capacity in 2030 in the Deferred Investment Scenario Compared with the Reference Scenario**

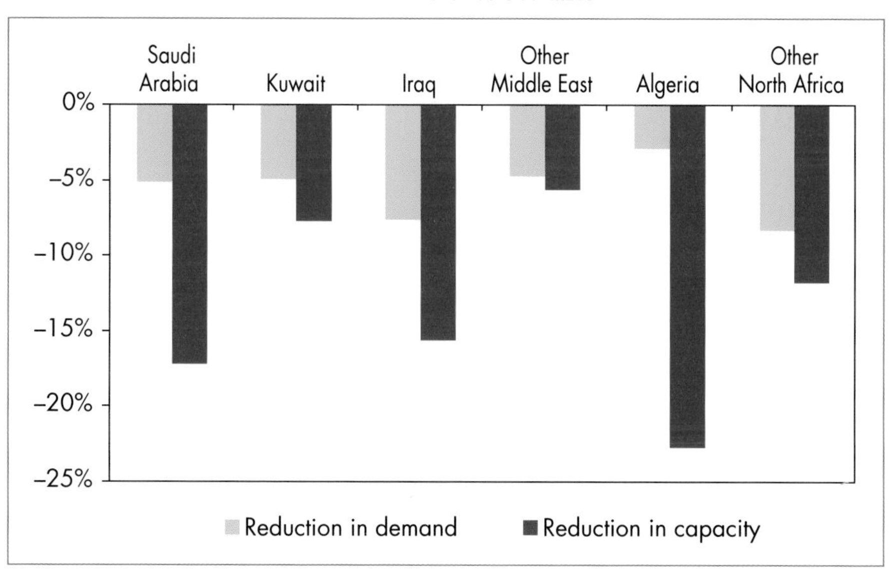

Note: Other Middle East includes Iran, Qatar and the UAE in addition to Bahrain, Israel, Jordan, Lebanon, Oman, Syria and Yemen. Other North Africa includes Egypt and Libya in addition to Morocco and Tunisia.

In the short term, the impact of deferred upstream investment in MENA is modest since domestic demand actually grows and high refining margins continue to make capacity additions for export attractive to MENA countries. Most projects that will affect capacity in 2010 have already been launched. But a 2% drop in world demand in 2010 to 91 mb/d alleviates the pressure on refining capacity. MENA refining capacity reaches 11 mb/d in 2010, a marginally lower level than in the Reference Scenario.

MENA Oil Exports

Despite the fall in domestic demand, MENA net oil exports decline even more in percentage terms than production in the Deferred Investment Scenario. In 2030, the region's exports of crude oil and products amount to 24 mb/d, almost 15 mb/d, or 38%, lower than in the Reference Scenario (Figure 7.8). The drop in exports is largest, in volume terms, in Saudi Arabia and Iraq. By 2030, total MENA oil exports are barely 2 mb/d, or 8%, higher than in 2004. Crude oil accounts for most of the reduction in exports.

Figure 7.8: **MENA Net Oil Exports in the Reference and Deferred Investment Scenarios**

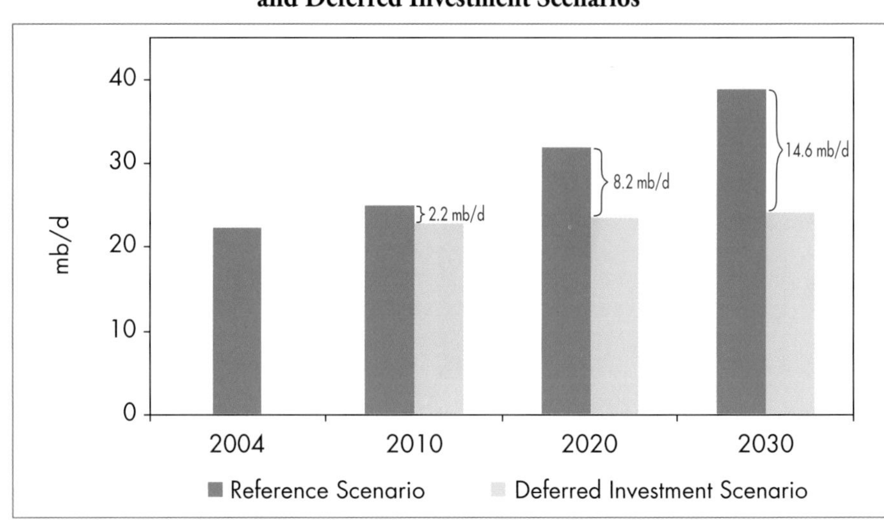

Note: Includes NGLs and condensates.

Lower world oil demand results in significantly lower exports of refined products from MENA countries after 2010 in the Deferred Investment Scenario. In total, MENA product exports reach only around 6 mb/d in 2030, or about one-fifth, lower than in the Reference Scenario.

Non-MENA Oil Production

The impact of lower MENA crude oil production on global oil supply is offset to some extent by higher non-MENA production, stimulated by higher prices. Part of the increase in non-MENA output comes from non-conventional sources – mostly tar sands in Canada and extra-heavy oil in Venezuela. By 2030, non-MENA output (including non-conventional oil) climbs to 70 mb/d – some 5 mb/d higher than in the Reference Scenario and 17 mb/d higher than in 2004 (Figure 7.9). World oil production, at 105 mb/d, is 10 mb/d lower than in the Reference Scenario (Table 7.5).

Conventional oil production increases in all non-MENA regions in the Deferred Investment Scenario. The largest increase in volume terms occurs in the developing countries (other than MENA): production there in 2030, at 29.8 mb/d, is 1.6 mb/d higher than in the Reference Scenario. Russian output is about 0.3 mb/d higher. Production in OECD is just 0.6 mb/d higher, with most of the increase in output occurring in North America. These results are sensitive to our underlying estimate of the responsiveness of non-MENA production to changes in the international oil price – the price elasticity of supply.

Figure 7.9: **World Crude Oil Production in the Reference and Deferred Investment Scenarios**

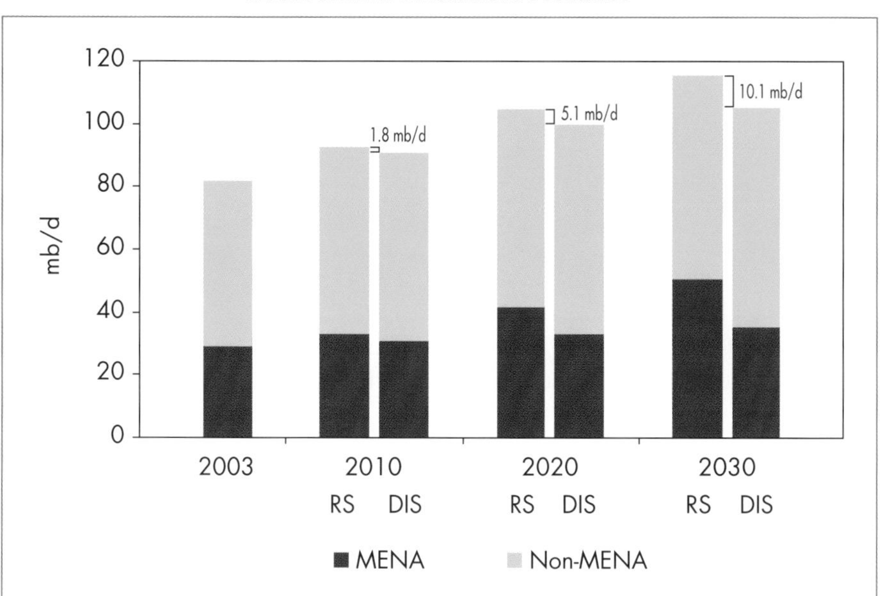

Note: Includes NGLs and condensates.

Table 7.5: **World Oil Production by Country in the Deferred Investment Scenario** (mb/d)

	2004	2010	2020	2030	Difference in 2030 compared with Reference Scenario
MENA	29.0	30.7	33.2	35.3	–15.2
OECD	20.2	19.3	16.4	14.0	0.6
Rest of the world*	30.8	37.5	43.1	44.5	3.2
Non-conventional oil	2.2	3.2	7.1	11.6	1.4
World	**82.1**	**90.6**	**99.8**	**105.3**	**–10.1**

* Including miscellaneous.

Gas Supply

The reduction in MENA oil production in the Deferred Investment Scenario leads to lower gas production in the region in two ways. First, associated gas production is reduced in line with oil output. Second, higher gas prices that result from the increase in oil prices choke off global demand and reduce the call on MENA gas exports. MENA domestic gas demand is also marginally lower at the end of the projection period, largely as a result of the impact of lower oil and gas revenues on national incomes.

In 2030, MENA gas production would be around 240 bcm, or 20%, lower than in the Reference Scenario (Table 7.6), but still 590 bcm higher than in 2003. Qatar sees the biggest reduction in output in volume terms vis-à-vis the Reference Scenario, falling from 255 bcm to over 200 bcm in 2030. As most Qatari gas is exported, lower demand for exports results in a large reduction in Qatari output. Iraqi gas output falls the most in percentage terms relative to the Reference Scenario, by 47% in 2030 – mainly due to lower oil production, with which a large share of gas production is associated. Output increases from 2 bcm in 2003 to 17 bcm in 2030 in the Deferred Investment Scenario, compared with 32 bcm in the Reference Scenario.

Table 7.6: **MENA Natural Gas Production by Country in the Deferred Investment Scenario** (bcm)

	2003	2010	2020	2030	Difference in 2030 compared with Reference Scenario
Middle East	**259**	**412**	**598**	**701**	**–159**
Iran	78	107	163	205	–35
Iraq	2	3	10	17	–15
Kuwait	10	11	15	15	–6
Qatar	33	106	172	202	–53
Saudi Arabia	60	87	127	150	–5
UAE	44	57	59	51	–24
Other Middle East	33	41	51	60	–21
North Africa	**125**	**154**	**219**	**273**	**–79**
Algeria	88	95	134	156	–42
Egypt	29	46	62	78	–13
Libya	6	10	19	33	–24
Other North Africa	2	3	4	5	0
Total MENA	**385**	**566**	**817**	**973**	**–238**

Global gas demand grows on average by 1.8% per year, compared with 2.1% in the Reference Scenario. As a result, demand is about 370 bcm lower in 2030. The biggest slow-downs in demand occur in Europe and North America (Figure 7.10), which between them account for 46% of the total difference in demand between the two scenarios. In both cases, most of the reduction in demand is due to lower final demand for electricity, which reduces gas needs in power generation, and to switching from gas to renewables, coal and nuclear in power generation as they become more competitive.

Figure 7.10: **Change in Natural Gas Demand in 2030 in the Deferred Investment Scenario versus the Reference Scenario**

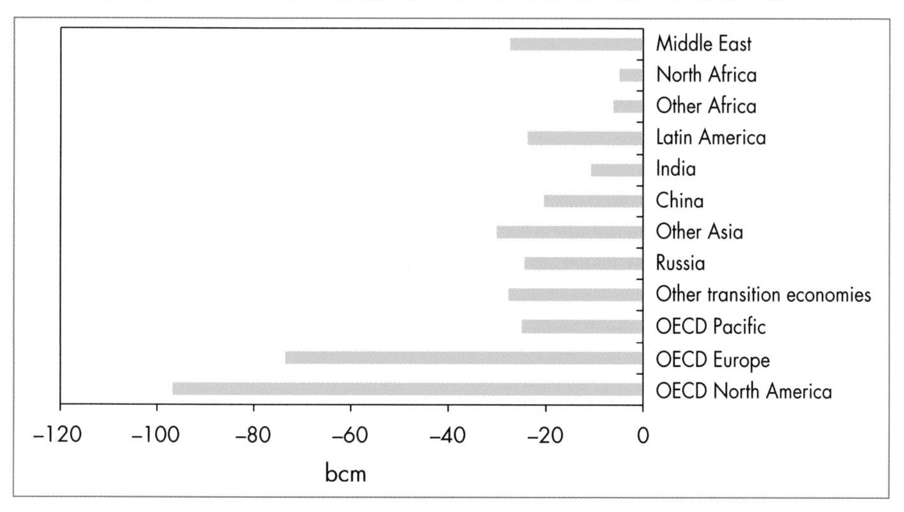

The fall in total MENA gas exports in the Deferred Investment Scenario is slightly smaller in volume terms than the fall in gas production, because lower output is partially offset by lower domestic demand. Exports are almost 210 bcm lower in 2030 compared with the Reference Scenario (Figure 7.11).

Figure 7.11: **MENA Natural Gas Exports in the Reference and Deferred Investment Scenarios**

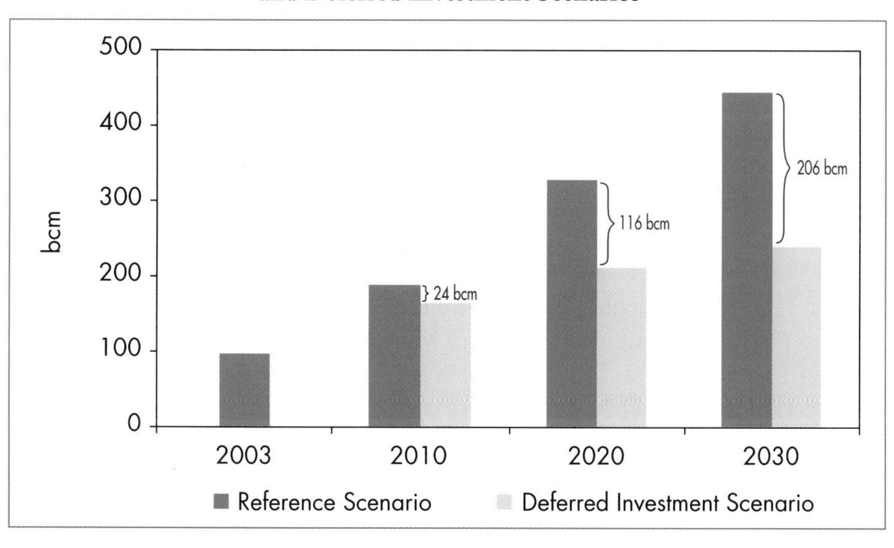

Yet, in *percentage* terms, exports fall much more – by 46% in 2030. MENA exports rise from 97 bcm in 2003 to almost 240 bcm in 2030 in the Deferred Investment Scenario, compared with more than 440 bcm in the Reference Scenario. Qatar sees the biggest drop in exports – from 152 bcm in 2030 in the Reference Scenario to 106 bcm in the Deferred Investment Scenario. Algerian exports also fall sharply, by 40 bcm in 2030.

MENA Oil and Gas Export Revenues

The profile of aggregate MENA oil and gas export revenues is markedly different between the two scenarios, because international oil and gas prices and production levels do not change in equal proportion over time. In the Deferred Investment Scenario, revenues are higher in the near term, because the increase in price is larger in percentage terms than the reduction in production. In short, MENA oil and gas producers as a group accrue higher revenues during the first few years. However, from around 2015, the reduction in production outweighs the increase in prices, reducing annual oil revenues. Cumulative revenues over 2005-2030 are reduced by just over $1 000 billion (in year-2004 prices) in the Deferred Investment Scenario. This compares with the reduction in investment of only $220 billion. When calculated on a net present value basis, using a discount rate of 5%, over 2005-2030, oil revenues are $230 billion lower. The fall in the net present value of oil revenues is larger if calculated using a lower discount rate or a longer period (beyond 2030).

Oil accounts for about 70% of the fall in revenues. Although gas makes up a relatively small share of total exports by value, the bigger percentage fall in gas exports means that gas accounts disproportionately for the reduction in revenues in the Deferred Investment Scenario.

MENA Energy Investment

In the Deferred Investment Scenario, the cumulative investment for oil and gas combined amounts to $830 billion over 2004-2030 (in year-2004 prices) – about $220 billion, or just over a fifth, less than in the Reference Scenario (Figure 7.12). On average, oil investment is $18 billion per year, $5 billion a year less than in the Reference Scenario. Most of the reduction will occur in the upstream sector but, as a consequence of lower oil supply, investment in the downstream sector will be lowered by 18%, falling to $4.1 billion per year compared with $4.8 billion in the Reference Scenario. The drop in gas investment, at just under 20%, is slightly less marked than for oil. Annual capital spending falls by $3 billion to an average of $13 billion over 2004-2030.

Figure 7.12: **Cumulative Oil and Gas Investment in MENA Countries in the Reference and Deferred Investment Scenarios, 2004-2030**

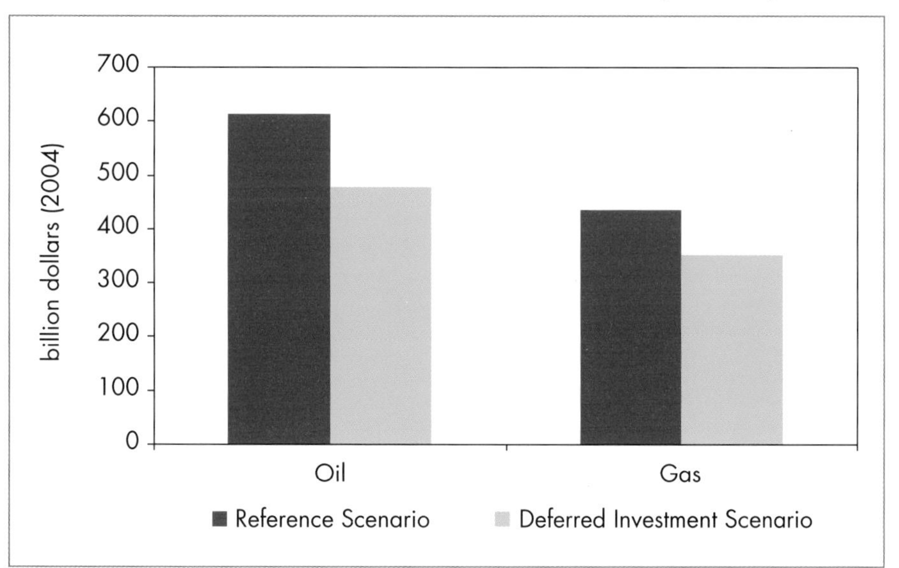

IMPLICATIONS FOR WORLD ENERGY MARKETS AND GOVERNMENT POLICY

HIGHLIGHTS

- MENA will play a major role in meeting the world's growing energy needs through 2030. In each of our three scenarios, the volume of oil and gas produced and exported by MENA countries in aggregate increases significantly.
- Despite the mutual economic benefits of increasing energy trade, growing imports from MENA countries will have major implications for energy security in consuming countries. In the short term, expanding trade will exacerbate the risks of a severe supply disruption. A particular cause for concern is the growing reliance on strategic transportation channels through which almost all the oil and gas exported by Middle Eastern countries must flow. Growing dependence on a shrinking number of large producers carries additional risks.
- Government policies in consuming countries could reduce the risk of disruptions and mitigate their consequences. Ensuring an appropriate level of emergency preparedness will remain an essential foundation of their strategies. Maintaining OECD oil stocks at 90 days of net imports would require adding 1.1 billion barrels between today and 2030 at a cost of almost $70 billion.
- Neither producing nor consuming countries would benefit economically from any constraints on upstream oil and gas investment. More vigorous efforts by producing countries to reform energy subsidies would bring important economic benefits, by promoting more efficient energy use and freeing up more oil and gas for export.
- New policies could bring about a major diversification of consuming-country energy supplies and promote more efficient energy use, lowering their vulnerability to supply disruptions and helping to tackle environmental challenges.
- In an Alternative Policy Scenario, which analyses the impact of new measures in consuming-country policies to save energy and reduce pollutant and greenhouse-gas emissions, world energy demand in 2030 falls even more relative to the Reference Scenario than in the Deferred Investment Scenario. MENA oil production is lower than in the Reference

Scenario, but still grows by more than 50%, or 16 mb/d, between 2004 and 2030.

- In the Alternative Policy Scenario, lower energy demand, resulting mainly from increased energy efficiency, depresses the price of oil by 15% on average over the projection period compared to the Reference Scenario. However, consumers have to invest some $1.1 trillion more to purchase more energy-efficient capital goods, equivalent to $15 per barrel of oil. The overall cost of oil to the consumers, around $46 per barrel on average, is therefore similar to the oil price in the Deferred Investment Scenario.

Introduction

The Reference Scenario results presented in Chapters 2 to 6 depict the energy future likely to emerge if government policies in consuming countries as of mid-2005 continue unchanged. The economic and population growth and international energy price assumptions determine the pattern of future energy demand; and the hydrocarbon-rich countries of the Middle East and North Africa are assumed to adjust their production to meet the demand for oil and gas not met from elsewhere. The results of a Deferred Investment Scenario, presented in Chapter 7, describe what might happen if MENA oil and gas production does not keep pace with demand, whether as a deliberate policy by producers or for other reasons which limit investment in productive capacity.

This chapter reflects on the broader implications of these outcomes for global energy markets. But it also adds a new dimension: what might result from deliberate policy action by the governments of oil-importing nations to change these outcomes? Such intervention could be provoked by fear of inadequate energy supply security or of excessive energy prices, or both. It could also result from determination to avoid environmental damage from too high a level of hydrocarbon consumption. A World Alternative Policy Scenario, discussed in this chapter, shows the possible outcome of decisive action by the governments of consumer countries, driven by any or all of these considerations.

Successive editions of the *World Energy Outlook* have had a consistent message: world energy demand is rising inexorably and this rise will continue unless governments take decisive action to stop this happening. The message is no different this time, whether under the Reference Scenario or the Deferred Investment Scenario. The precise outcomes do, of course, vary: starting from 10.7 billion tonnes of oil equivalent in 2003, world primary energy demand reaches 16.3 billion toe in 2030 in the Reference Scenario and 15.4 billion toe in the Deferred Investment Scenario. But, even in the latter case, demand would still be about 43% above the level of 2003. And in both cases, oil remains the leading energy source.

MENA's Role in World Energy Supply

The trends described in the Reference and Deferred Investment Scenarios are reassuring for the governments of oil- and gas-producing states. Our projections of MENA oil and gas supply support the widely-held view that this region will play an increasingly important role in meeting the world's growing energy needs through to 2030 and probably well beyond. Today, MENA's share in world oil and gas production is much lower than its share of proven reserves and the region could readily and profitably exploit its large resource endowment to meet the world's growing need for energy.

In the Reference Scenario, MENA's share of global oil supply grows substantially, on the hypothesis that MENA countries will act to provide the additional oil needed to meet rising global demand, after allowing for projected non-MENA production. MENA's market share increases from 35% in 2004 to about 44% in 2030 (Figure 8.1). In the Deferred Investment Scenario, in which oil production is constrained, the picture is markedly different: in a diminished world market, MENA's overall market share falls slightly to 33% in 2030. MENA oil exports nonetheless continue to grow in volume terms, from 22 mb/d in 2004 to 24 mb/d in 2030.

Figure 8.1: **Share of MENA in Global Oil Supply**

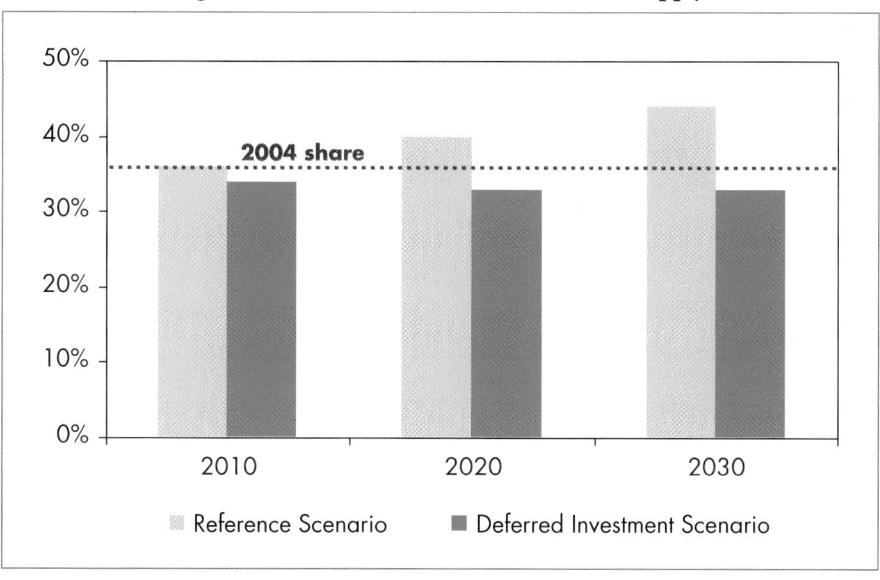

In the Reference Scenario, MENA's share of the global natural gas output grows even more quickly than that of oil, from about 14% in 2003 to 25% in 2030. This is matched by a growing share in world gas exports, from 26% to almost 50%. In the Deferred Investment Scenario, MENA's share of global gas production still grows, but less quickly, to 22%, while its share of inter-regional gas trade reaches 37%.

The degree to which oil- and gas-importing countries are dependent on MENA supplies differs markedly between the two scenarios. In the Reference Scenario, the oil-import dependence of OECD countries, taken as a whole, rises from 55% in 2004 to 68% in 2030 (Table 8.1). Much of the increase represents imports from MENA countries. In the Deferred Investment Scenario, the OECD's oil-import dependence still increases, but to only 63%. In North America, dependence hardly increases at all, as a result of higher indigenous production and the dampening effect on demand of higher road-fuel prices. Many importing countries are also large net importers of energy-intensive manufactured goods, such as petrochemicals, some of which come directly from MENA countries or from other regions that themselves depend heavily on MENA oil and gas. For example, it is estimated that the United States already imports goods whose manufacture and transportation depend on about 1 mb/d of Middle East oil, equal to about a third of US oil imports from the Middle East (Cordesman, 2004).

Table 8.1: **Dependence on Oil Imports in Net Importing Regions**

		Reference Scenario		**Deferred Investment Scenario**	
	2004	**2010**	**2030**	**2010**	**2030**
OECD	**56%**	**59%**	**68%**	**58%**	**63%**
North America	42%	42%	52%	40%	43%
Europe	58%	71%	85%	70%	84%
Pacific	93%	94%	95%	94%	94%
Developing Asia	**47%**	**58%**	**75%**	**57%**	**71%**
China	42%	57%	72%	57%	67%
India	69%	74%	88%	73%	87%
Other Asia	43%	51%	72%	51%	67%
European Union	*79%*	*85%*	*94%*	*84%*	*93%*

Many regions, including all three OECD regions, are also set to become more dependent on imports of natural gas, both in absolute terms and as a proportion of supply. MENA's share of these gas imports will also grow markedly, with most of the gas supplied in the form of liquefied natural gas. The source of imported gas is important, as gas-supply chains are less flexible than those for oil. In the

Reference Scenario, the share of MENA in gas imports rises highest in Europe, from 31% in 2003 to 58% in 2030. Volumes surge from 63 bcm in 2003 to 118 bcm in 2010 and 287 bcm in 2030. Even in the Deferred Investment Scenario, the volume of gas imported from MENA and its share in total imports in Europe rises significantly. This is also true of North America.

The availability of MENA oil and gas for export hinges not just on the expansion of installed productive capacity and short-term production policies, but also on the extent of MENA domestic demand. Demand in the MENA region, like demand elsewhere, is expected to grow, driven by economic and population growth. How quickly it will grow is uncertain, due largely to uncertainty about economic prospects – notably oil revenues in the main producing countries – and government policies on industrialisation, energy subsidies and economic reforms. The continuity of government policy assumed in the Reference Scenario is particularly uncertain in this region. MENA energy demand falls slightly in the Deferred Investment Scenario, compared to the Reference Scenario, as a result of lower revenues and economic growth and higher prices. The removal of energy subsidies – which are very large in many MENA countries, encouraging waste and discouraging investment in more efficient energy equipment – could have a much bigger impact on demand. It would also free up more oil and gas for export.

MENA's growing share in global energy use, reflected in a growing share of world GDP, carries with it certain adverse consequences. Notably, MENA's CO_2 emissions rise by 2.6% per year, doubling over 2003-2030. The region's proportionate responsibility in the search for a global solution to climate change will grow.

Policy Issues for MENA Producing Countries

The policy priority for governments of oil- and gas-producing countries everywhere in the world is to maximise the national benefit derived from their resources. But views differ as to the best route to that end, including the role of private and foreign companies in developing resources and the optimal rate of depletion. Most OECD countries believe that exploitation of hydrocarbon resources is best left to private companies, with the government setting licensing and fiscal terms. In many MENA countries, for whom hydrocarbons are a particularly valuable national resource, the government prefers to play a more direct role, involving state control at every stage and, in some cases, 100% state ownership of the companies tasked with developing the country's resources. Export revenues play a key role in those countries' economic development and, to the extent that they are used to invest in other activities, can ultimately support the diversification of their economies away from over-reliance on hydrocarbons.

Investment and Production Policies

The availability of oil and gas exports from MENA countries will depend largely on their investment and production policies. The pace at which MENA countries actually develop their hydrocarbon resources is largely a matter of policy: most MENA countries will not lack capital (see below). All countries, if they so decide, can gain access to the most advanced technologies – either by opening up their upstream sector to direct investment by international oil companies or by buying oil services. The cost of developing oil and gas fields in most MENA countries is very low by international standards, so such investments are, in principle, financially lucrative and economically viable.

Of course, MENA policy-makers must decide whether their chosen rate of expansion of production might result in an overall level of MENA or global output which might drive down prices, partially or wholly offsetting the increased revenue sought through higher production. The analysis of the preceding chapter suggests that deferring investment and constraining the expansion of productive capacity will *not* achieve higher revenues.

Most MENA countries are members of OPEC and, therefore, give consideration to the mutual interests of the organisation's membership in formulating their oil policies. Since 1982, OPEC has operated a system of short-term production targets or ceilings, aimed at adjusting supply to perceived shifts in global demand in order to prevent excessive price movements. As part of the process of determining these targets, each member country must consider trends in global oil demand and non-OPEC supply, the call on OPEC supply, capacity developments in other countries and what these trends might mean for their own future share of total OPEC supply. Related investment decisions cannot be changed as readily as the short-term output targets. The potential for each OPEC member to raise its production varies considerably, so each MENA country's share of total OPEC production will change. Iraq, which enjoys a temporary waiver from the OPEC production system, is most likely to produce at a level which gives it an increased market share in the medium to long term.

In the Deferred Investment Scenario, lower MENA oil and gas production leads to significantly higher prices over the projection period. Market dynamics could change this picture markedly. Were consuming countries to adopt new policies and measures to curb oil and gas use, as described in the Alternative Policy Scenario (see below), the fall in the call on MENA oil and gas would be considerable. Higher prices, reinforced by other policy interventions, would induce faster technological development and deployment on the supply and demand sides (Box 8.1). This could result in much lower reliance on fossil fuels, whether from MENA or elsewhere.

Box 8.1: **Price-Induced Technological Change**

Persistently higher prices, which lead to a permanent shift in consumer and producer expectations of future price levels, have a particularly strong impact on fuel mix and technological change. The development and deployment of more efficient energy technologies on the supply and demand sides accelerated after the two oil shocks of the 1970s. This contributed to a halving of the oil intensities of IEA countries in the following two decades. Higher prices of conventional fuels make alternative energy sources more commercially attractive. Developers of these technologies have an incentive to allocate more resources to their development. Nuclear energy, wind power, solar power and coal liquefaction and gasification were greatly advanced after the oil shocks.

An important reason for these advances was increased public and private research and development (R&D). IEA governments more than doubled their collective energy R&D spending in real terms during the 1970s, with spending peaking with oil prices in the early 1980s. Since then, R&D spending has more or less followed oil prices, declining in the 1980s and stabilising in the 1990s.

Energy prices also affect demand-side technology. With higher energy prices, consumers will tend to choose more energy-efficient appliances. Industrial and commercial businesses will also introduce energy-saving equipment in their factories and offices. The oil price shocks of the 1970s, together with government policies, led to significant energy savings in all energy-using sectors. The rate of improvement in energy intensity slowed dramatically after the crash in oil prices in 1986 (IEA, 2004). With the recent surge in prices, it is likely that energy savings will again accelerate as consumers and businesses give priority to energy efficiency when they invest in new equipment and buy new appliances.

Higher prices reduce the need for governmental support to help a new technology into the market, as the need to bring costs down is reduced. After a long period of only modest improvements in car fuel economy in IEA countries, the car industry is now accelerating development of more efficient technologies, such as hybrid engines. Over the last two years, General Motors has increased its annual R&D spending by $500 million and Toyota by almost $1 billion.

Empirical analysis has demonstrated a positive relationship between energy prices and technological innovation. The expectation of higher prices in the future is likely to spur faster development of advanced technologies on both the supply and demand sides in the coming years.

The rate of MENA investment in the upstream oil industry will be a key determinant of how much spare crude oil production capacity will be available in the future. For many years, spare capacity has been concentrated in OPEC countries, mainly those in MENA. Saudi Arabia has traditionally played the role of swing producer, usually holding the largest amount of spare capacity. The existence of so much spare capacity was largely the result of over-investment in installed capacity in the 1970s, when demand was checked by surging prices.

Upstream investment by MENA producers was greatly reduced in the late 1980s and 1990s, with the result that spare capacity has diminished in recent years as the call on MENA supply has grown. This has contributed to upward pressure on oil prices (Figure 8.2). There is a time-lag before investors overcome the caution induced by experience of low prices and begin to reinvest with confidence. In July 2005, spare sustainable OPEC production capacity amounted to 2.1 mb/d, of which all but 150 kb/d was in MENA countries.[1] Saudi Arabia accounted for most of MENA's spare capacity. Tighter crude oil supply, together with very high refinery-utilisation rates, has reduced supply flexibility and sharply increased upward pressure on prices. Prices are unlikely to soften without more investment in both crude oil production and refining,

Figure 8.2: **Spare Crude Oil Production Capacity Worldwide and Average IEA Crude Oil Import Price**

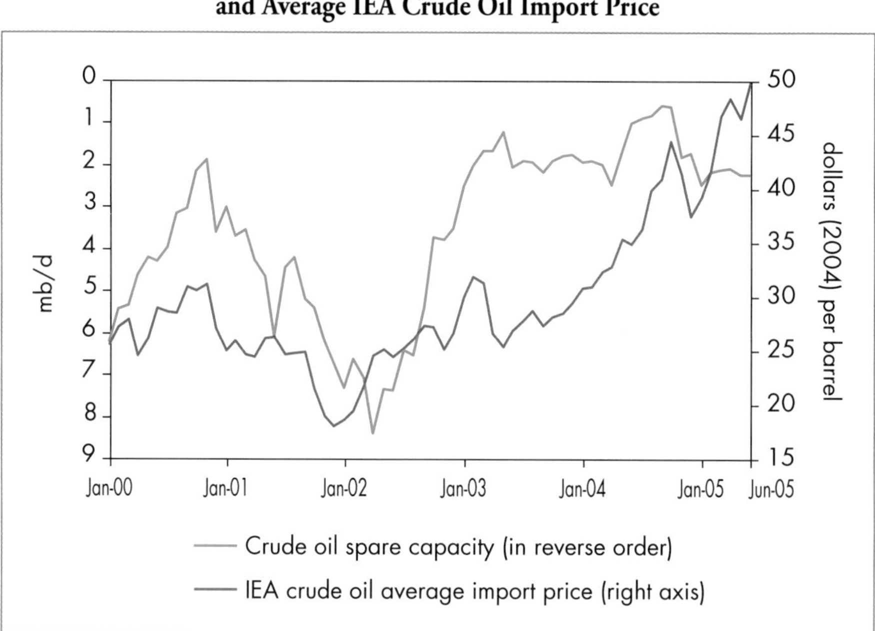

1. *IEA Oil Market Report*, August 2005.

unless demand growth slows abruptly. But sustained higher prices will give new impetus to alternative technologies, dampening growth in oil demand and, eventually, easing the upward pressure on prices.

MENA countries face difficult political choices about how much of their oil and gas revenues they can afford to set aside for social spending and how much should be held back for oil and gas sector investment. Circumstances differ, not least according to the extent of national energy resources. Even where internal finance is insufficient, capital availability should not be a constraint, given the economic attractiveness of the sector. Foreign investment can be attracted where considerations of national sovereignty do not exclude the offer of commercially attractive terms. Any state may legitimately decide that the exploitation of its resource is the exclusive right of the state or its organs; but a resource has value only when exploited and states facing capital or technical constraints may have to compromise to realise their assets. Several MENA states have already shown how this might successfully be done.

Energy Subsidies

The availability of MENA oil and gas exports is a net figure, after deducting domestic demand from marketable production. All MENA exporting countries have an interest in maximising the proportion of their production which is available for export. In almost every country, this means facing up to the issue of under-pricing of energy to the domestic market. Such subsidies are widespread and large, driving up internal energy demand, especially as the population grows.

Though an economically sound policy, reforming subsidies is fraught with difficult. Indigenous populations have become accustomed to sharing directly in the national wealth generated by hydrocarbons by paying little or nothing for their use. Even in the richest MENA countries, with increasingly diverse economies, it has been difficult for the government to persuade their people that their best interests lie in paying international market prices for their energy. Our analysis assumes that large energy subsidies will persist in many MENA states over the projection period, despite some progress in reducing their magnitude. How successful they are in coming to grips with this issue will have important implications for the availability of oil and gas for export and international prices.

Policy Issues for Consuming Countries

Energy Security

Our analysis of the outlook for MENA's role in global energy supply has important and complex implications for the energy security of consuming countries. Increased international trade in oil and gas will bring economic benefits to both exporting and importing countries. Nonetheless, the prospect of MENA claiming a bigger share of global oil and gas supply and trade naturally raises concerns about the risk of major supply disruptions because of political instability in parts of the region and past experience of disruptions to oil supplies from some countries in the region. In addition, oil and gas production will become increasingly concentrated in fewer and fewer countries. This will add to the vulnerability to a disruption and to the risk that those countries will seek to use their dominant market position to force up prices at some point in the future.

Many circumstances can drive up oil prices, including sudden supply loss (or the expectation of it), lack of spare capacity in the supply infrastructure, deliberate withholding of supplies for political reasons and co-ordinated production cutbacks. The rising dominance of MENA in global markets intensifies these risks. In practice, vulnerability depends not just on the risk and duration of a disruption and the resulting price shock, but also on the flexibility and resilience of the economy to respond to and withstand higher prices. The higher an economy's oil intensity and the less fuel-switching capability there is, the more vulnerable it will be. Experience has shown that the sudden loss of even a modest volume of oil can lead to sharp increases in prices, especially if associated with limited spare capacity or rising geopolitical tensions.

The growing inflexibility of oil demand is increasing the vulnerability of importing countries. The share of transport in total oil use continues to grow in all regions in all three scenarios described in this *Outlook*. Demand for oil-based fuels, which account for almost all transport fuel consumption, responds very little to changes in price in the short term. These fuels cannot be substituted in existing vehicles and, as their costs account for only a small share of the total costs of driving, the price of fuel affects driving behaviour and vehicle choice only to a limited extent. As a result, fuel demand tends to change very little in response to price increases, especially in the near term. For a given supply reduction, the required price adjustment needed to bring global demand down is, therefore, expected to increase, aggravating the cost to importing countries.

On the other hand, long-term energy security in importing countries would be enhanced by the timely development of MENA's low-cost resources, bringing greater mutual economic benefits in the long term – as long as the short-term supply risks can be mitigated. Slower development of those resources would lead to higher prices, hindering economic development for consumers and producers alike (see Chapter 7). Long-term security, therefore, hinges on sufficient investment occurring in a timely way to meet the world's energy needs at an acceptable cost. Adequate investment that ensures long-term energy security would also enhance short-term security, insofar as it contributes to spare capacity that can be called upon at short notice, thereby reducing the risk of a supply shortfall and price shock. The oil price collapse in 1998-1999, which at one time took the average price for crude oil imported into the IEA countries to below $10 per barrel, led to greater discipline among OPEC members in pursuit of higher prices. Now, after four years of sharply rising prices, consumer governments are again seriously concerned about the economic effects of sustained high prices. It has been estimated that a $10 rise in the oil price, if sustained for a year, cuts GDP in OECD countries by 0.4% (IEA, 2004). The adverse effect can be four times worse in very poor developing countries.

Vulnerability to Supply Shocks

Worries about the threat of disruptions to MENA energy supplies stem from the experience of the last few decades. Most significant oil-supply disruptions worldwide in recent decades have occurred in MENA. Since 1970, there have been 17 serious oil-supply disruptions involving an initial loss of 0.5 mb/d or more crude oil (Table 8.2). All but three of them were related to events in Middle Eastern or North African countries. Four major crises – the 1973 Arab-Israeli War, the 1978-1979 Iranian Revolution, the 1980-1988 Iran-Iraq War and the 1990-1991 Gulf War – led to initial shortfalls of between 3 mb/d and 5.6 mb/d. The largest disruption since the 1990s was the loss of 2.6 mb/d of production in Venezuela in late 2002 and early 2003 as a result of a strike at the national oil company. This was followed in March 2003 by the loss of almost as much supply in Iraq as a result of the invasion. These losses were compounded by the loss of 800 kb/d of Nigerian oil due to an oil industry strike. The pre-conflict level of output in Iraq has still not been fully restored.

Although the security of oil facilities in all producing countries has been stepped up in recent years in response to the growing terrorist threat, various forms of threat to MENA energy production and transport persist. For example, frequent attacks on oil facilities in Iraq by insurgents have severely affected production and exports. In previous conflicts, such as the Iran-Iraq War and the Iraqi invasion of Kuwait, oilfields and tankers were deliberately and systematically targeted. Supply interruptions can also result from deliberate political acts by the producing government.

Trade binds suppliers and customers in mutually beneficial dependence and is to be welcomed. But inter-regional oil and gas trade is vulnerable to interruption and the threat must be assessed and evaluated. A particular cause for concern in this respect is the growing reliance on MENA exports through a small number of strategic transportation channels or chokepoints – an issue highlighted in *WEO-2004*. Most of the oil and gas exported by Middle Eastern countries flows

Table 8.2: **Major Oil Supply Disruptions since 1970**

Date	Gross supply loss* (mb/d)	% of world oil demand	Reason
5/70-1/71	1.3	2.5	Libyan price controversy
4/71-8/71	0.6	1.1	Algeria-France nationalisation dispute
3/73-5/73	0.5	0.8	Lebanon civil unrest
10/73-3/74	4.3	7.1	Arab-Israeli War & Arab oil embargo
5/77	0.7	1.1	Accident at Saudi oilfield
11/78-4/79	5.6	8.5	Iranian Revolution
10/80-1/81	4.1	6.2	Outbreak of Iran-Iraq War
4/89-6/89	0.5	0.8	UK Cormorant platform accident
8/90-1/91	4.3	6.5	Iraqi invasion of Kuwait
4/99-3/00	3.3	4.5	OPEC production cutbacks
11-12/99	1.1	1.5	Iraqi oil export suspension (rejection of phase V1 extension)
12/00	1.6	2.1	Iraqi oil export suspension (price disagreement with UN)
6-7/01	2.1	2.7	Iraqi oil export suspension (rejection of UN resolution 1352)
4-5/02	1.8	2.3	Iraqi oil export suspension (rejection of UN resolution 1862)
12/02-3/03	2.6	3.4	Venezuelan strike
3/03-12/03	2.3	2.9	Iraqi conflict
8/05	1.5	1.9	Katrina Hurricane damage to US crude oil production facilities

* Initial production loss only: in some cases, this loss was quickly made up by production increases elsewhere. Disruptions involving an initial supply of 500 kb/d only are shown.
Sources: US Department of Energy and the IEA Secretariat.

Figure 8.3: **Oil Export Flows from MENA and Major Strategic Maritime Channels**

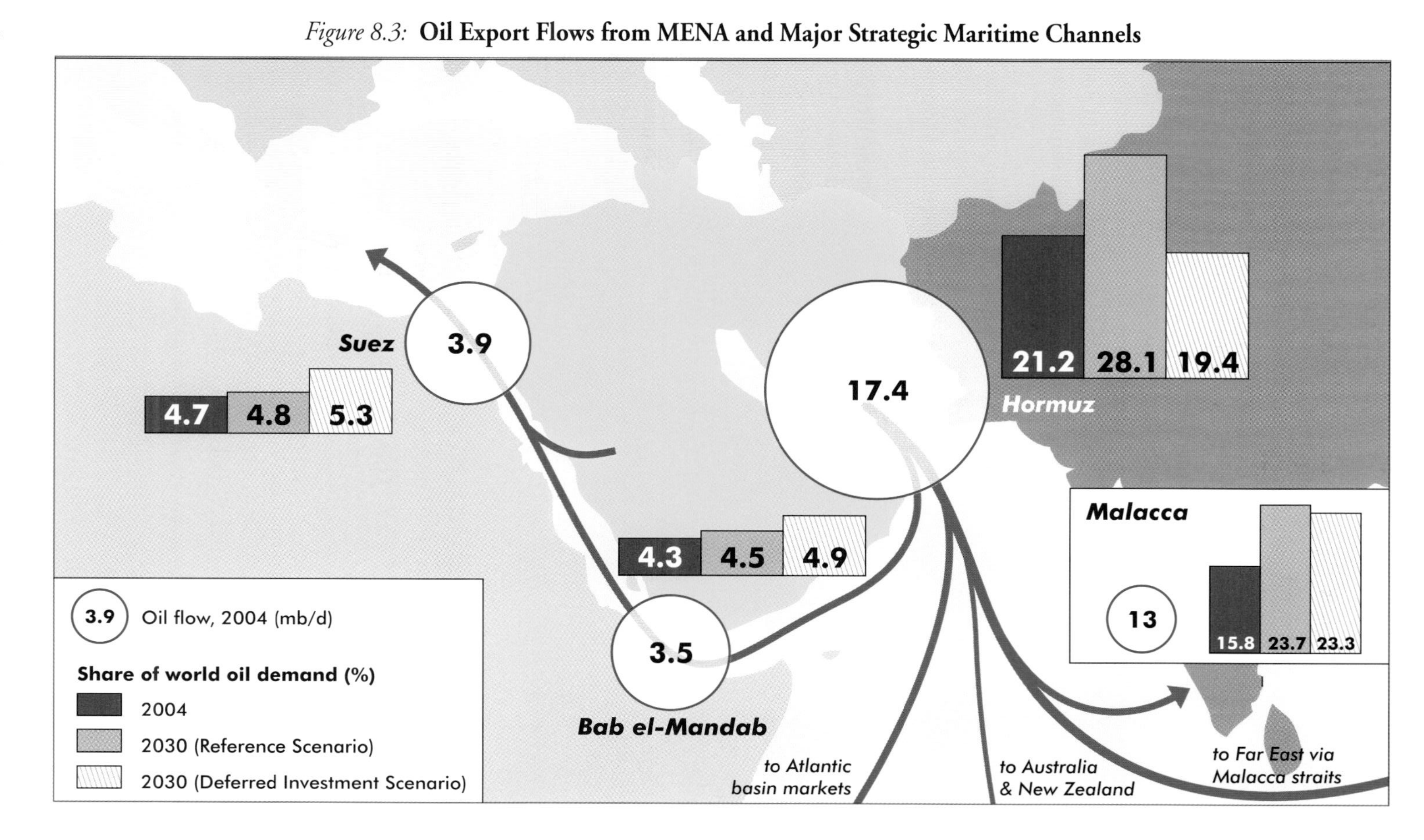

through just three channels, all of which are susceptible to sudden closure as a result of accidents, piracy, terrorist attacks or war (Figure 8.3):

- *The Straits of Hormuz*, at the mouth of the Persian Gulf. The straits comprise two 3-km-wide inbound and outbound lanes. If they were blocked, only a small share of the oil could be transported along alternative routes. Shipping through the straits was disrupted during the Iran-Iraq War. In 2004, we estimate that 17 mb/d, or 21% of the world's total oil supply, flowed along this route – the world's busiest oil-shipping route. Much of the oil – around 13 mb/d in 2004 – is subsequently shipped through the Malacca Straits between Indonesia, Malaysia and Singapore, another busy and narrow maritime route, which has been the scene of disruption of oil shipments due to accidents and piracy in the past.
- *The Bab el-Mandab passage*, which connects the Gulf of Aden with the Red Sea. In 2004, around 3.5 mb/d was shipped through this passage en route to the Suez Canal and Sumed Pipeline (for onward shipment to Europe and the United States).
- *The Suez Canal and Sumed Pipeline*, both of which connect the Red Sea to the Mediterranean. The capacity of the canal is currently about 1.4 mb/d, while the pipeline has a capacity of 2.5 mb/d. Tankers transport oil from ports on the Persian Gulf and the Red Sea. The closure of these routes or the Bab el-Mandab passage would force tankers to take the much longer route around the Cape of Good Hope in South Africa.

Table 8.3: **MENA Oil and LNG Export Flows through Major Strategic Maritime Channels**

	2004		2030			
			Reference Scenario		Deferred Investment Scenario	
	Volume	Share in global supply	Volume	Share in global supply	Volume	Share in global supply
Oil (mb/d)						
Hormuz	17.4	21%	32.5	28%	20.4	19%
Bab el- Mandab	3.5	4%	5.2	5%	5.2	5%
Suez Canal*	3.9	5%	5.6	5%	5.6	5%
LNG (bcm)						
Hormuz	28**	1%	175	4%	85	2%
Suez Canal/Bab el-Mandab	4**	0.1%	60	1.3%	38	0.9%

* Includes the Sumed Pipeline.

** IEA estimate.

On average, around 80% of total Middle East oil exports in 2004 were shipped along at least one of these three routes. The rest was shipped by pipeline to the Mediterranean. Shipments of MENA LNG through these channels amounted to about 90 million cubic metres on any one day. Much of the additional oil and LNG that will be exported in the future can be expected to be shipped along these three maritime routes, with oil shipments through the Straits of Hormuz projected to grow from 17 mb/d today to 32 mb/d in 2030 in the Reference Scenario, while LNG shipments along the same route will jump from 28 bcm to 175 bcm (Table 8.3). By 2030, as much as 28% of the world's oil supply and 4% of gas supply could flow through the Straits of Hormuz. Even in the Deferred Investment Scenario, traffic along this route increases significantly in volume terms.

Implications for Emergency Preparedness

The 1974 Agreement on an International Energy Program (IEP), which established the International Energy Agency, requires member countries to hold oil stocks equivalent to 90 days of net oil imports. It also provides for stocks to be released, demand-restraint measures to be activated and oil supplies to be shared, if necessary, in a co-ordinated way. In September 2005, IEA countries demonstrated the readiness and effectiveness of these arrangements by agreeing promptly to make available to the market the equivalent of 2 mb/d of oil from emergency stocks for an initial period of 30 days to help offset the loss of 1.5 mb/d of crude oil production and 2 mb/d of refining capacity resulting from damage from Hurricane Katrina in the United States.

To maintain the specified level of emergency coverage in a situation of rising oil-import dependence over 2003-2030 will require large increases in the volume of oil held in storage (Figure 8.4). For the OECD, total oil stocks would need to rise, in the Reference Scenario, to around 3 billion barrels in 2010 and 3.7 billion barrels in 2030 for them to be equal to 90 days of net imports – 1.1 billion barrels more than in 2004.[2]

Total spending on oil storage in OECD countries through to 2030 is projected at $69 billion, or $2.6 billion per annum (Figure 8.5). This is about 10% of upstream oil sector investment in the region over the period. It assumes OECD countries maintain their current commitment to hold oil stocks equivalent to 90 days of net imports. Over 60% of the total oil storage cost is required for purchasing the additional crude they will hold; around 25% for operation and maintenance of existing and new facilities that will be built over the projection

2. These estimates are based on net imports on a regional basis and make no allowance for differences between crude oil and products, conversion factors or any special allowances that may apply in practice. They do not, therefore, correspond exactly to the volumes that would be required under IEP rules.

Figure 8.4: **Volume of Oil Stocks Needed to Ensure 90 Days of Net Imports in OECD Regions in the Reference Scenario**

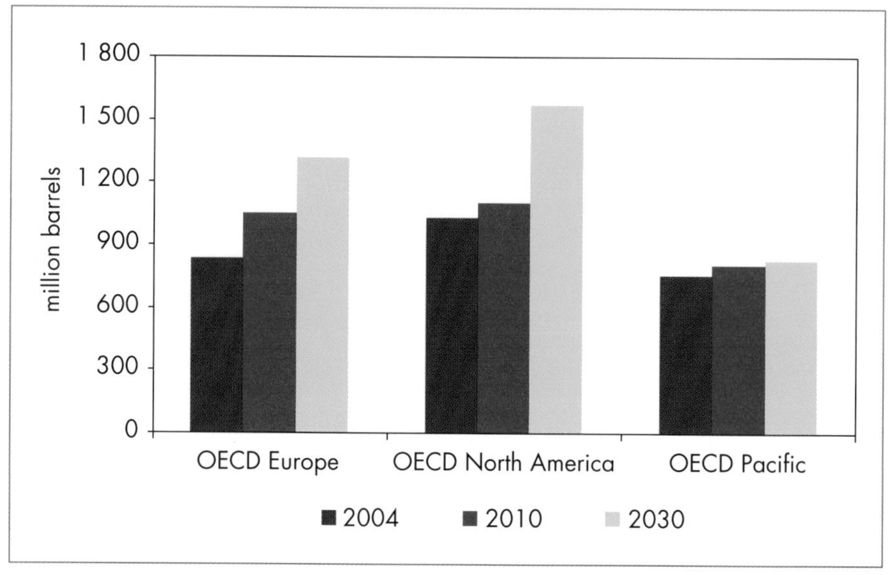

Note: Estimates are indicative and do not correspond to exact volumes that would be required under IEP rules.

Figure 8.5: **Indicative Cumulative Cost of Oil Storage, 2004-2030**

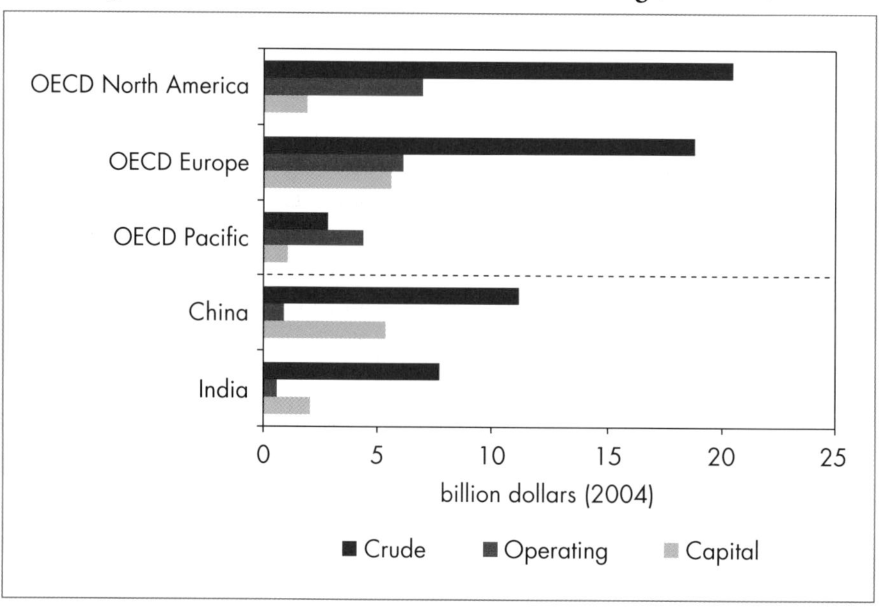

Note: The estimates for capital costs do not take account of any replacement of existing facilities that may be needed before 2030.

period; and the remainder for the construction of new capacity. Capital costs for oil storage vary significantly. They are lowest, at around $3.50 per barrel, for underground facilities such as the deep salt caverns that contain the United States' Strategic Petroleum Reserve. In comparison, capital costs for above-ground tanks are much higher, ranging from $15 to $20 per barrel. Annual operating costs are typically around $0.20 per barrel.

China and India have also started developing strategic oil storage facilities.[3] Although their longer-term plans are uncertain, it is estimated that combined spending of the two countries on strategic stocks through to 2030 could amount to $28 billion. This is based on the assumption that India builds stockpiles equivalent to 45 days of net imports and China 30 days. China is currently building four above-ground strategic oil complexes. These will have a combined volume of 102 million barrels and are expected to be in operation by 2008. Capital costs are high, at around $19 per barrel, and have been adversely affected by recent rises in the price of steel and cement. Future expansion of China's stockpiling programme is likely to involve construction of less costly underground oil facilities. India has plans to build around 35 million barrels of oil storage by 2007, covering about 19 days of net imports. They then plan to increase storage to 45 days cover at a later date. India's Ministry of Petroleum and Natural Gas is currently selecting sites for rock-cavern storage which are expected to be built at a capital cost of around $10 per barrel.

Facing the Challenge

In short, consumer governments need to face up to the full implications, including the new vulnerabilities, of concentrated reliance on MENA supply. Consuming countries must identify policies and measures aimed at reducing the risk of disruptions and higher prices, as well as mitigating their consequences. Deepening the political dialogue with producing countries, notably within the framework of the International Energy Forum, could help to address short-term security concerns, as well as to facilitate the long-term flows of capital needed to develop MENA's energy resources. Consuming-country governments also need to consider long-term policies that promote further diversification of their energy supplies as a means of both lowering their vulnerability to supply disruptions and of addressing environmental challenges, including rising greenhouse-gas emissions.

Reducing dependence on oil and gas through diversification of fuels and their geographic sources and more efficient use of energy must be central to long-term policies aimed at enhancing energy security. Any consuming country's vulnerability to an oil shock is directly linked to the oil and gas intensity of its

3. The fact that both countries are committed to building strategic stocks is sending a calming message to the markets. Other importing countries are taking note and considering this option.

economy. The challenge is made more urgent by the ever-decreasing short-term possibilities for substituting MENA oil and gas. It is not the proportionate dependence on any one fuel type which counts, but the extent of alternative sources of that fuel and the practicability of switching fuels in a crisis. In that respect, the prospects for consumers are worsening. Oil-based fuels cannot be substituted in existing vehicles. The growing concentration of oil use in transport – a result of both the Reference and Deferred Investment Scenarios – will make oil demand even less substitutable.

These challenges call for collective responses. The consequences for a net oil-importing country of a disruption in supplies, whether from the MENA region or elsewhere, depend on the extent of the price shock and the overall degree of import dependence – *not* on whether the importing country obtains its oil physically from MENA countries themselves. Oil, in its crude or refined form, is a global commodity and a shortfall in supply to one importing country affects all consuming countries, regardless of whether their supplies are directly affected or not.

The impact of a disruption in the supply of gas, on the other hand, *does* depends on the source of the gas. Gas-pipeline infrastructure is inflexible, so that a loss of supply through a particular pipeline system cannot always be made good by supplies from other sources. So consuming countries need to diversify their source of gas imports. LNG supply is, in principle, more flexible, as the loss of supply from one producer could conceivably be replaced by output from another. The growing share of LNG in world gas trade should, therefore, contribute to more flexibility in gas supply. But, in practice, there may be insufficient spare liquefaction and shipping capacity available to compensate for a large supply disruption. In addition, most LNG is at a present sold under long-term contracts, with rigid clauses covering delivery.

World Alternative Policy Scenario

Up to this point, this *Outlook* has presented systematically the projections of energy demand and supply that result from the Reference and the Deferred Investment Scenarios. These are internally consistent scenarios, deriving conclusions from stated assumptions, within a rigorous modelling framework. In both cases, no attempt is made to guess future political intentions and known government policies are assumed to continue unchanged (though the reduced investment in MENA countries in the Deferred Investment Scenario *might* be the result of deliberate policy).

In no sense, however, can either vision of the energy future be considered sustainable. G8 leaders, meeting with the leaders of several major developing states at Gleneagles in July 2005, acknowledged as much when they called for stronger action to combat rising consumption of fossil fuels and related greenhouse-gas emissions. Though both scenarios reflect the effects of policies already adopted, for example to combat climate change, governments have declared their intention to do more.

The World Alternative Policy Scenario attempts to project the energy future which might result if those intentions are given concrete effect through new policy measures. It analyses the impact of a range of policies and measures *under consideration* which are aimed at addressing environmental and energy-security concerns (see Box 1.2 in Chapter 1 and Table 8.5 for more details about the methodology and the policies analysed). In this scenario, global primary energy demand in 2030 reaches 14 658 Mtoe – 1 613 Mtoe, or almost 10%, less than in the Reference Scenario (Figure 8.6).

Figure 8.6: **Incremental World Primary Fossil Fuel Demand in the Alternative Policy Scenario and Savings Relative to the Reference Scenario, 2003-2030**

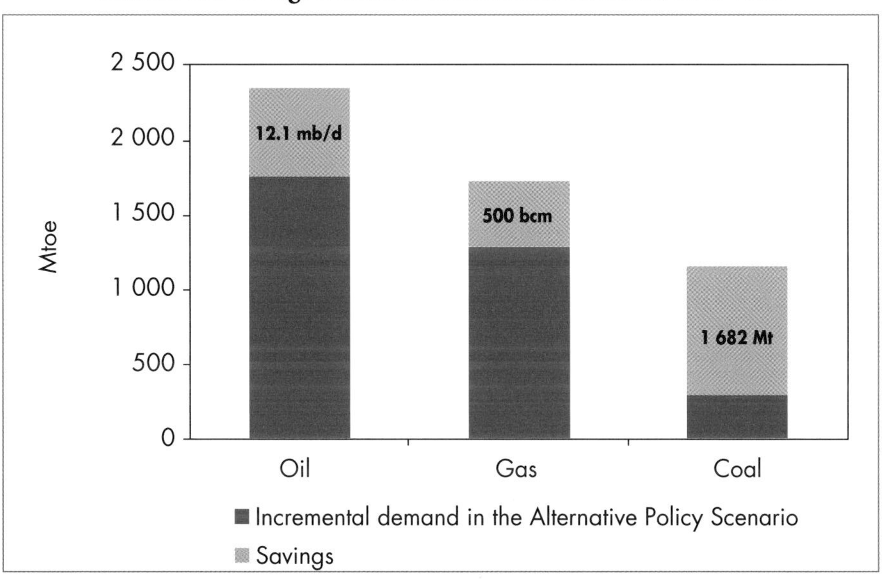

Primary energy demand grows by 1.2% per year (Table 8.4), 0.4 percentage points less than in the Reference Scenario. The reduction in demand for coal is even greater, thanks to the use of more efficient technology and switching to less carbon-intensive fuels. The effect of energy-saving and fuel diversification policies on energy demand grows throughout the projection period, as the stock of energy capital is gradually replaced and new measures are introduced. Global energy savings achieved by 2010 are very modest, at only about 244 Mtoe, or 2%.

Table 8.4: **World Primary Energy Demand in the Alternative Policy Scenario**
(Mtoe)

	2003	2020	2030	2003-2030*	Difference with the Reference Scenario in 2030
Coal	2 582	2 849	2 866	0.4%	-23%
Oil	3 785	4 727	4 967	1.0%	-10%
Gas	2 244	3 142	3 528	1.7%	-10%
Nuclear	687	821	878	0.9%	14%
Hydro	227	324	370	1.8%	0.4%
Biomass and waste	1 143	1 455	1 705	1.5%	3%
Other renewables	54	207	344	7.1%	27%
Total	**10 723**	**13 525**	**14 658**	**1.2%**	**-10%**

* Average annual rate of growth.

Demand for oil in the Alternative Policy Scenario rises to just under 5 000 Mtoe in 2030, 580 Mtoe, or 10%, lower than in the Reference Scenario. But it still accounts for 34% of world primary energy demand in 2030, the same share as in the Reference Scenario. Two-thirds of these savings come from the transport sector, as fuel efficiency improves and alternative fuel vehicles – powered by compressed natural gas or biofuels – or gasoline-powered hybrids penetrate the market. Table 8.5 details the policies taken into account in selected regions and countries. The share of oil in transport demand is significantly lower in all regions (Table 8.6).

Table 8.5: **Selected Policies Considered in the Alternative Policy Scenario in the Transport Sector**

Region	Programme/measure	Impact
OECD North America	• Increased and extended tax credits for hybrid vehicles and tax breaks for alternative fuels and vehicles • Increased and extended mandatory requirement for ethanol blending with gasoline	• Increased use of CNG, LPG, fuel-cell, hybrid powered vehicles and biofuels • Increased use of biofuels and decreased use of gasoline
OECD Europe	• Extended voluntary agreements with car manufacturers • Increased support for alternative fuels • White Paper on package of transport policies	• New car and light-truck efficiency improves • Increased use of biofuels • Slower growth in passenger and freight transport and modal shift from road and aviation to rail and bus

Table 8.5: **Selected Policies Considered in the Alternative Policy Scenario in the Transport Sector** *(continued)*

Region	Programme/measure	Impact
OECD Pacific	• Extended Top Runner programme and similar • Increased R&D and tax credits for alternative fuels vehicles • Policies to promote use of public transport systems (Japan and Korea)	• New car and light-truck fuel efficiency improves • Increased use of CNG, LPG, fuel-cell and hybrid powered vehicles and biofuels • Slower growth in passenger and freight road transport, modal shift to mass transport
China	• Tighter vehicle-fuel efficiency standards • Increased R&D and tax credits for clean vehicles • Expansion of intra- and inter-city railway networks	• Improved vehicle-fuel efficiency • Increased use of CNG, LPG, fuel-cell and hybrid powered vehicles, and biofuels • Slower growth in passenger vehicle transport and modal shift to mass transport
India	• Measures to accelerate the introduction of less polluting vehicles and fuels • Increased and extended efficiency requirement for two- and three-wheelers	• More efficient new vehicles; faster deployment of CNG, LPG, biofuels • Faster replacement of old, polluting vehicles • More efficient two- and three-wheelers fleet
Brazil	• Extended biodiesel programme • Increased R&D for alternative fuel vehicles • Incentives to adopt urban planning (Curitiba-like) in other cities	• Increased use of biofuels • Faster deployment of flex-fuel vehicles • Slower growth in passenger vehicles and modal shift to mass transport
Egypt	• Extension of measures to accelerate the introduction of CNG vehicles	• Increased use of CNG
Iran	• Stronger incentives to accelerate the uptake of CNG vehicles and of less polluting vehicles and fuels	• More efficient new vehicles; faster deployment of CNG; faster replacement of old, polluting vehicles

Table 8.6: **Share of Oil in Transport Demand in the Reference and Alternative Policy Scenarios**

	2004	2030	
		Reference Scenario	Alternative Scenario
OECD North America	97%	96%	93%
OECD Europe	98%	97%	90%
OECD Pacific	98%	98%	96%
China	92%	97%	95%
India	98%	94%	90%
Brazil	87%	84%	81%

Natural gas demand is also 10% lower in 2030 than in the Reference Scenario. Almost three-quarters of the savings by 2030 come from power generation. Demand for coal falls even more by 2030, by 858 Mtoe or 23%. Most of the reduction in coal use occurs in the power sector.

By contrast, the use of carbon-free non-hydro renewables, excluding biomass, is 27% higher in 2030 than in the Reference Scenario. Biomass, hydropower and nuclear energy also grows. Renewables partially displace fossil-fuel consumption. By 2030, global consumption of biomass is 50 Mtoe higher in the Alternative Policy Scenario than in the Reference Scenario. Most of the net increase in the use of renewables results from OECD government incentives to promote their use in the power sector and in transportation. Consumption of renewables other than biomass – mainly in power generation – increases even more, by 70 Mtoe in 2030.

Lower overall energy consumption and a larger share of less carbon-intensive fuels in the primary energy mix together yield a 16%, or 5.8 gigatonne, reduction in global carbon-dioxide emissions compared to the Reference Scenario in 2030 (Figure 8.7). The bulk of the reduction comes from lower coal use, especially in power generation in non-OECD countries. This results mainly from the reduction in electricity demand brought about by new end-use efficiency policies.

The fall in primary energy demand in the Alternative Policy Scenario leads to major changes in MENA's role in global oil and gas supply. Global oil demand in 2030 is reduced by 12.1 mb/d in 2030, leading to a 5.9 mb/d reduction in MENA production to 44.6 mb/d. Global demand for natural gas in 2030 is 500 bcm lower, resulting in a fall in MENA gas production of almost 200 bcm, to about 1 010 bcm.

Figure 8.7: **Global Energy-Related CO_2 Emissions in the Reference and Alternative Policy Scenarios**

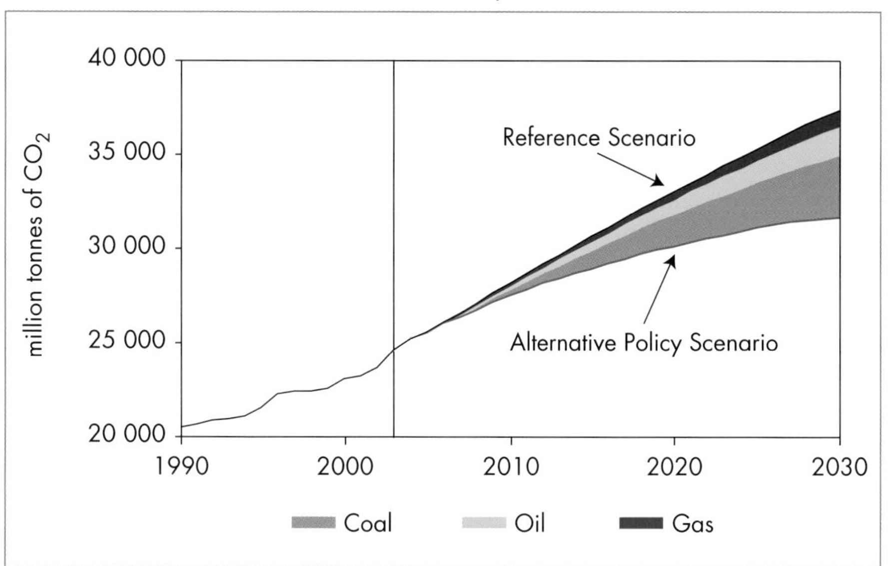

8

Comparing the Alternative Policy and Deferred Investment Scenarios

Although the assumptions underlying the Alternative Policy and Deferred Investment Scenarios are very different in nature and scope, global energy demand is significantly lower than in the Reference Scenario in both cases. However, the causes of lower demand are markedly different. In the Deferred Investment Scenario, higher international energy prices are the driver, induced by slower growth in MENA oil (and gas) production. In the Alternative Policy Scenario, government policies in consuming countries directly curb demand and, contrary to the situation in the Deferred Investment Scenario, lead to lower energy prices. Differences between the two scenarios in the size of the reduction in energy demand vary by fuel and region.

The fall in world energy demand in 2030 relative to the Reference Scenario is larger in the Alternative Policy Scenario (10%) than in the Deferred Investment Scenario (6%). The share of oil and gas in primary energy demand among consuming regions – an indicator of vulnerability to supply disruptions – falls further in most regions. In both the Deferred Investment and Alternative Policy Scenarios, oil and gas demand falls significantly in the importing regions (Figure 8.8). The exception is China, where government policies encourage switching to gas from coal. Oil and gas imports still increase in volume terms in all three scenarios.

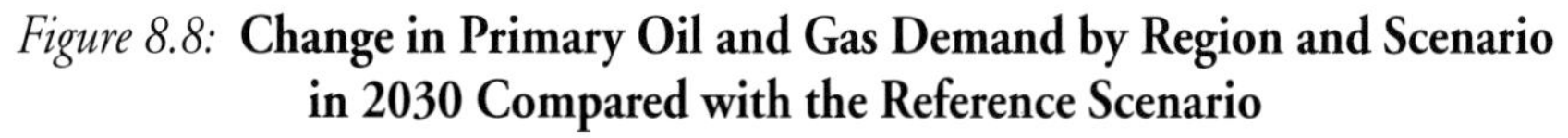

Figure 8.8: **Change in Primary Oil and Gas Demand by Region and Scenario in 2030 Compared with the Reference Scenario**

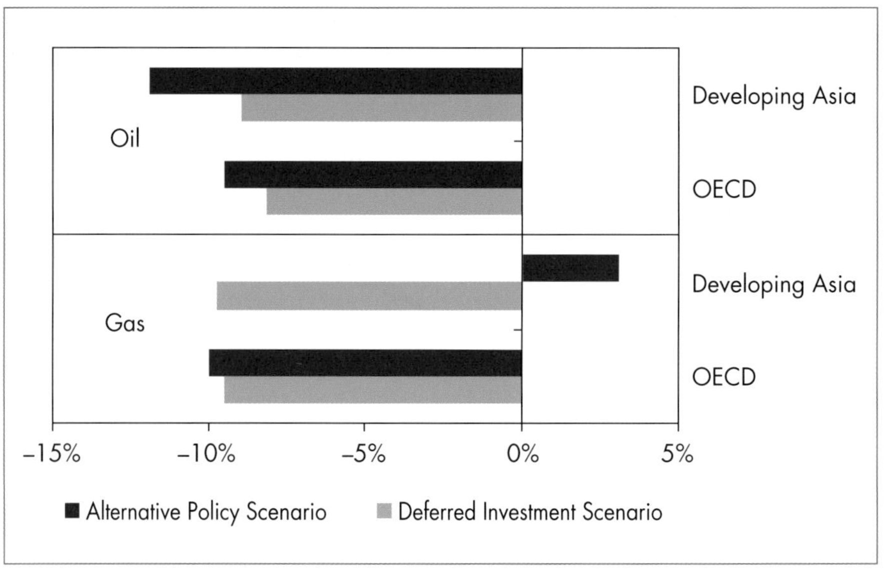

The share of MENA in total inter-regional gas trade – another indicator of short-term supply security – is lower in both the Alternative Policy and Deferred Investment Scenarios compared with the Reference Scenario. The share falls most in the Deferred Investment Scenario, to 37% in 2030 compared with 48% in the Alternative Policy Scenario and 49% in the Reference Scenario. The causes of this fall vis-à-vis the Reference Scenario are different in the two cases. In the Deferred Investment Scenario, higher oil and gas production in non-MENA regions make up part of the loss of MENA supply. In the Alternative Policy Scenario, lower gas imports from MENA result solely from policies that lower demand and, therefore, imports from all regions. As MENA is the marginal supplier to Europe, the United States and Asia, in both scenarios the fall in gas imports from MENA is larger than from other net-exporting regions.

The Deferred Investment and Alternative Policy Scenarios have different implications for effective costs to consumers. In the former scenario, the average IEA crude oil import price is projected to rise by $8 a barrel, or 20%, on average over the projection period compared with the Reference Scenario.[4] In 2030, oil prices are $13 a barrel higher in real terms. This increase in prices

4. Prices of the major benchmark crude oils, West Texas Intermediate and Brent, will be correspondingly higher. In 2004, the average IEA crude oil import price was $5.11 per barrel lower than first-month WTI and $1.89 lower than dated Brent.

would, other things being equal, feed through into higher retail prices for end-users. The oil price averages $33 per barrel in the Alternative Policy Scenario. This is $6, or 15%, *lower* than in the Reference Scenario, because lower demand depresses prices. But in this scenario, consumers will have to invest some $1.1 trillion more over the projection period to purchase more energy-efficient capital goods, brought to the market as a result of government policies. The bulk of this investment will be made in the transport sector of OECD countries, to buy more efficient cars, hybrid and biofuel-powered vehicles. This additional investment, in total, enables consumers to save some $300 billion in their oil bill. Overall, the net cost to consumers amounts to $15 per barrel saved.[5] If this sum is added to the oil price in the Alternative Policy Scenario, consumers end up with very similar oil cost in the Deferred Investment Scenario and in the Alternative Policy Scenario (Figure 8.9). However, in the Alternative Scenario importing countries would be less dependent on MENA oil and gas and global carbon-dioxide emissions would be around 16% lower (compared with only 7% lower in the Deferred Investment Scenario).

Figure 8.9: **Average Cost of Oil by Scenario, 2004-2030**

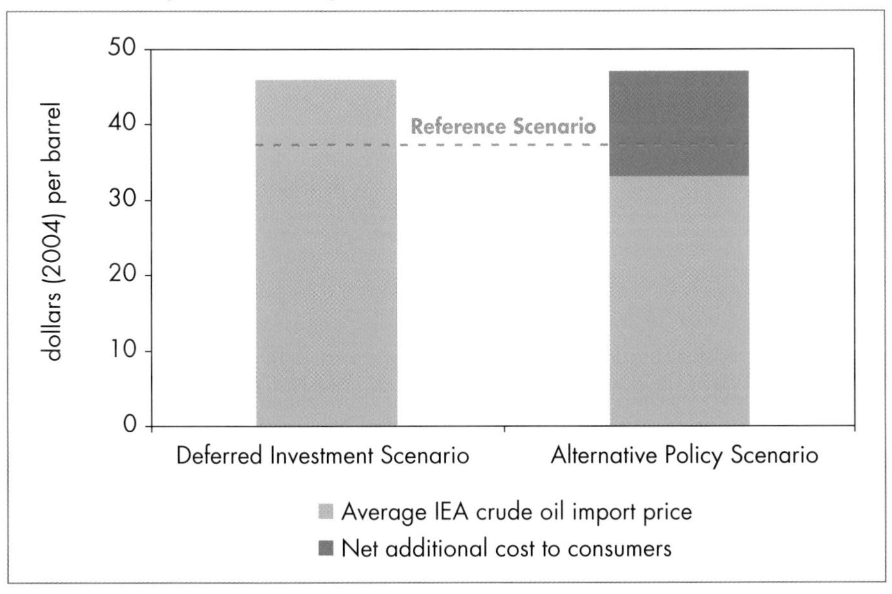

5. The net additional cost to consumers over the projection period, obtained by deducting the fuel savings from the $1.1 trillion additional investment, amounts to $800 billion. The oil saved in the Alternative Policy Scenario compared to the Reference Scenario over the same time horizon is some 52 billion barrels. The net additional cost to the consumer therefore amounts to $15 for each barrel of oil equivalent saved.

The lower call on MENA oil and gas in the Alternative Policy Scenario, together with lower international prices, results in a 21% reduction in cumulative MENA oil and gas export revenues over 2005-2030 compared with the Reference Scenario. In the Deferred Investment Scenario, revenues are 8% lower (Figure 8.10). In both scenarios, the loss of revenues is much larger than the reduction in investment.

Figure 8.10: **Cumulative MENA Oil and Gas Export Revenues, 2005-2030**

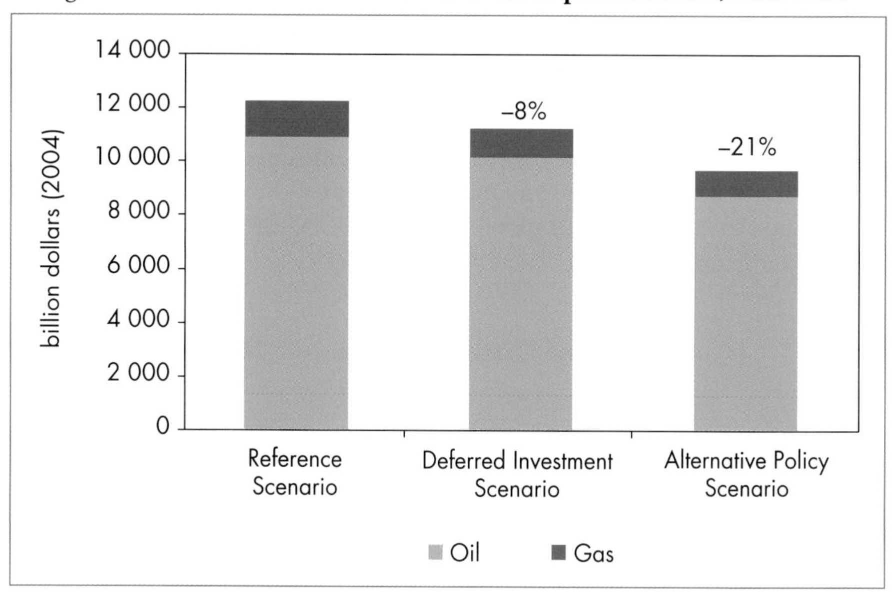

Actual trends may lie somewhere between all three scenarios. In practice, the policies of producing and consuming countries will change over time in response to each other, to market developments and to shifts in market power. *If MENA upstream investment falls short of what appears to be necessary to sustain the global economy, the more likely it is that consuming countries will adopt additional policies to curb demand growth and reliance on MENA.* This would have the effect of tempering the long-term impact on prices of lower MENA investment. It would also augment the impact on oil and gas demand. The combined impact of both lower investment and new government policies to curb energy use and promote switching to less carbon-intensive fuels would, of course, depend on the timing of the new measures and their magnitude.

Our analysis suggests that new government policies designed to reduce oil and gas imports would counterbalance the price effects of potentially lower MENA investment at very modest net cost to the importing countries. In this sense, consuming-country policies – whether motivated by energy-security or environmental concerns – and producing-country policies may be self-reinforcing. However, the more successful the importing countries' policies are, the more likely it is that the producing countries will adopt policies to sustain their production, so as to counter the effects of the importing countries' policies. The effect would be lower prices.

These interactions illustrate the case for improving market transparency, more effective mechanisms for exchanging information between oil producers and consumers, and a more profound dialogue between them. Concerns among consuming countries about security of supply are matched by concerns among producing countries about security of demand. This balance has important implications for investment flows and the development of resources in MENA countries. Together, consumer and producer governments can improve the mechanisms by which they seek to reconcile their interests and achieve mutually beneficial outcomes.

ALGERIA

HIGHLIGHTS

- Primary energy demand in Algeria is projected to grow at an annual rate of 2.8% over the *Outlook* period, reaching 70 Mtoe in 2030. Per capita demand will remain higher than the average for North Africa.
- Natural gas and oil together meet 98% of the country's energy needs. Gas, which Algeria holds in abundance, will remain the leading fuel in power generation and in non-transport final uses. Its share of total primary energy demand will remain at about two-thirds over the projection period.
- Crude oil production is projected in the Reference Scenario to rise steadily through to 2010, continuing the trend of recent years, before starting to level out and decrease. Including NGLs, oil output is projected to rise from 1.9 mb/d in 2004 to 2.2 mb/d in 2010, before dropping to 1.6 mb/d in 2030. Much of the production will be in the form of condensates and NGLs. Exports, mainly to Europe, will follow a similar pattern.
- Marketed natural gas production is projected to climb briskly to 107 bcm in 2010 and around 200 bcm in 2030. The overwhelming bulk of the incremental output will be exported, by pipeline and in the form of LNG. At 4.6 tcm, Algeria has the fifth-largest proven gas reserves in MENA, which are expected to grow rapidly as exploration efforts are stepped up.
- Electricity generation will grow by 3.4% per year between 2003 and 2030, reaching 74 TWh in 2030. The country will add 12 GW of capacity – mostly gas-fired. Investment needs, at $17 billion over 2004-2030, are large, making private participation necessary. New water-desalination plants, mainly using electricity-based reverse osmosis technology, will boost power demand across the *Outlook* period, adding 4 Mtoe to primary energy needs in 2030.
- The total cumulative investment needs in the Algerian energy sector are estimated to amount to $114 billion over the *Outlook* period. The upstream oil and gas industry will absorb most of this investment, the bulk of which is expected to come from foreign investors. A new hydrocarbon law, adopted in 2005, paves the way for increased foreign participation.
- In the Deferred Investment Scenario, crude oil production reaches 1.3 mb/d in 2030 – 240 kb/d lower than in the Reference Scenario. Natural gas production reaches 156 bcm in 2030, 42 bcm lower than in the Reference Scenario. As a result, oil and gas exports are significantly lower.

Overview of the Algerian Energy Sector

The energy sector, and notably the hydrocarbons sector, is of vital importance to the Algerian economy. The development of the sector dates back to the late 1950s, with the discovery of two giant associated oil and gas fields at Hassi-Messaoud and Hassi R'Mel. The early focus was on production of crude oil, which began in 1956. Production of natural gas started in 1961 and Algeria became the world's first liquefied natural gas (LNG) producer in 1964.

Algeria has been a member of the Organization of the Petroleum Exporting Countries (OPEC) since 1969. As the level of Algerian oil production started to be limited by OPEC quotas in the 1980s, attention was refocused onto the development of the country's large natural gas resources.

Algeria's early gas strategy was not particularly favourable to international oil companies, but by the mid-1980s, moves began to attract private investment to the upstream gas sector. Gas production and exports are expected to continue to grow in the future, with the construction of new pipelines to Europe and new LNG plants. Further expansion of the gas distribution network, and a gas-to-liquids (GTL) project, will spur growth in domestic demand.

Algeria has been successful in attracting substantial amounts of foreign investment in the hydrocarbons sector, since the upstream sector was partially

Table 9.1: **Key Energy Indicators for Algeria**

	1971	2003	1971-2003*
Total primary energy demand (Mtoe)	3.7	33.0	7.1%
Total primary energy demand per capita (toe)	0.26	1.04	4.4%
Total primary energy demand/GDP**	0.06	0.16	2.9%
Share of oil in total primary energy demand (%)	62	32	–
Net oil exports (mb/d)	0.8	1.6	2.1%
Share of oil exports in production (%)	94	87	–
Share of gas in total primary energy demand (%)	31	66	–
Net gas exports (bcm)	1.4	63.9	12.8%
Share of gas exports in production (%)	52	73	–
CO_2 emissions (Mt)	8.9	77.7	7.0%

* Average annual growth rate.

** Toe/thousand dollars of GDP in year-2004 and PPPs.

opened through the adoption of the 1986 hydrocarbon law and its 1991 amendments. A new hydrocarbon law, adopted in April 2005, further deregulates the upstream and downstream sectors and opens the way for increased foreign participation. The government is also restructuring and deregulating the electricity sector. Growth in power-generation capacity in Algeria has failed to keep pace with surging demand in recent years. Per capita generation is low compared with other MENA countries.

Political and Economic Situation

Political Developments

A former colony of France, Algeria gained independence in 1962. The National Liberation Front (FLN) dominated Algerian politics in the decades following independence, but its power has since been challenged by other political movements, notably the Islamic Salvation Front (FIS), demanding political and economic reform.

At the beginning of the 1990s, civil strife broke out. The conflict raged for most of the 1990s and claimed the lives of tens of thousands of people. Political reforms were introduced gradually. The 1989 constitution and its 1996 revision paved the way for multi-party politics. The first pluralistic presidential election was held in 1995. A policy of national reconciliation was announced in 1997, culminating in the adoption of a Civil Harmony Law in July 1999 and its endorsement in a national referendum the following September. Algeria's current president, Abdelaziz Bouteflika, was elected in 1999 and re-elected in a landslide victory in the 2004 elections.

Economic reform has progressed in parallel with political reform, in the aftermath of the 1986 oil price collapse. In 1994, triggered by an impending balance of payments crisis, Algeria embarked upon a programme of macroeconomic stabilisation and structural reform supported by the International Monetary Fund and the World Bank. The pace of reform has picked up considerably since the end of the 1990s. The reforms aim to scale back the role of the government and liberalise the economy, including the hydrocarbons sector.

Relations with the European Union (EU) have been strengthened through a political and economic association agreement which took effect on 1 September 2005 and which touches on energy sector co-operation, and through the Euro-Mediterranean Partnership, which aims to create a free trade zone by 2010. Algeria has also started negotiations on joining the World Trade Organization.

Economic Trends

Algeria is currently experiencing a strong economic upturn. Gross domestic product (GDP) has grown rapidly since 2000, largely due to rising crude oil and natural gas production and higher prices. GDP growth reached an estimated 5.3% in 2004, averaging 4.7% over 2000-2004. A surge in export revenues has fuelled an increase in government spending, boosting private sector activity and household incomes. The fiscal and external sectors have also been strengthened by the increase in hydrocarbon revenues. Despite the improved economic and political climate, serious economic and social problems remain. Unemployment in 2003 was nearly 24%, albeit down from 29.5% at the start of the decade (IMF, 2005). There is a serious housing shortage, which was exacerbated by a major earthquake in 2003. Algeria scores relatively poorly on most indicators of human development, ranking 103rd out of 177 countries worldwide on the United Nations Development Programme Human Development Index (UNDP, 2005). Poverty is a major problem: on average, 15% of the population lived on less than $2 a day during the 1990s.

The Algerian economy remains highly dependent on the hydrocarbons sector and the government's fiscal balance fluctuates in line with international oil prices. Oil was the motor for Algeria's economic development in the 1970s and, despite the increasingly important role played by natural gas, oil remains the largest single contributor to GDP. Oil and gas combined accounted for more than 36% of GDP in 2004 and 98% of export earnings. Agriculture accounts for nearly 10% of GDP and industry for about 7%. The contribution of industry to GDP growth has been falling in recent years.

In our Reference Scenario, economic growth is expected to average 3% per year over the projection period (Table 9.2). Annual growth will slow from an average of 4.4% over the rest of the current decade to 2.3% in the final decade of the *Outlook* period. This reflects an assumed drop in oil and gas prices from recent highs and a slow-down in the rate of expansion in hydrocarbon production. Oil and gas will, nonetheless, remain by far the largest sector in the economy and will support growth in non-hydrocarbon GDP. The share of oil and gas in the economy will fall to 23% by 2030.

Table 9.2: **GDP and Population Growth Assumptions for Algeria in the Reference Scenario** (average annual growth rate, in %)

	1971-2003	1990-2003	2003-2010	2010-2020	2020-2030	2003-2030
GDP	4.1	2.3	4.4	2.7	2.3	3.0
Population	2.6	1.9	1.6	1.3	0.9	1.2
Active labour force	3.7	4.0	2.8	1.8	1.5	1.9
GDP per capita	1.5	0.5	2.7	1.4	1.4	1.7

Demographic Trends

The Algerian population grew at an average annual rate of 2.6% between 1971 and 2003, reaching 32 million in 2003. The rate of population growth is projected to slow over the *Outlook* period, from 1.6% per year in 2003-2010 to 0.9% in the 2020s. The annual rate of growth will average 1.2 % between 2003 and 2030. These projections imply an average increase of 1.7% per year in per capita GDP, which will be about 60% higher in 2030 than today.

The age composition of the population creates important challenges for the government. Algeria has a very young population: almost one-third is under the age of 15 years. As a result, there will be increasing pressure for the government to take action to stimulate the rate of job creation. The rate of unemployment is much higher among the young than the rest of the working population: over 45% for the under-24s (IMF, 2005).

Energy Policy

Energy policy has been largely focused on oil and gas production, given that sector's central role in the Algerian economy. The government has intervened heavily in the sector, through ownership and regulatory controls. The national oil company, Sonatrach, was established in 1963 and Algeria became a member of OPEC in 1969. Sonatrach took over the assets of all other operators in the country when the industry was fully nationalised in 1971. Despite the gradual opening-up of the oil and gas sector, Sonatrach retains a dominant position. The government expects natural gas to play an even larger role in domestic energy supply in the future than it already does.

With the adoption in 1986 of a new hydrocarbon law, Algeria opened its upstream oil and gas industry to foreign investment through production-sharing agreements and participation contracts with Sonatrach. Foreign participation increased massively in the last decade, notably after the application of the 1991 amendments to the 1986 hydrocarbon law. The frequency of licensing rounds has increased; a seventh bidding round is currently under way and an eighth is planned for 2006. The new hydrocarbon law, adopted in 2005, is expected to lead to further involvement of international oil companies and support the development of the upstream industry (see the Oil Supply section below).

The electricity sector is also being liberalised. The monopoly of the state-owned national utility, Sonelgaz, on production, transmission, distribution and international trade in electricity, as well as on the distribution and sale of natural gas, was ended by the new electricity law of 2002. The industry is being restructured through "unbundling" of the supply chain which involves separating the functions of generation, transmission, distribution and supply. The restructuring will also open access to third-party suppliers, for example,

IPPs. The government has announced plans to split Sonelgaz into two enterprises, which will be dedicated to electricity and to gas distribution respectively. Minority private ownership of up to 30% will be possible in both. One aim is to raise funds for expanding capacity to meet rapidly rising domestic demand – particularly for electricity. Prices to end-users are to be increased to full-cost levels, a move that has been under consideration for some time. The additional income will go towards the higher cost of gas purchases from Sonatrach and towards new investment.

Energy Demand

Primary Energy Mix

Total primary energy demand in Algeria amounted to 33 Mtoe in 2003, almost ten times higher than in 1971. Demand soared during the 1970s, by nearly 15% per year, in line with the surge in GDP brought about by higher oil prices. Demand growth slowed considerably thereafter, to only 2.5% per year between 1990 and 2003. Demand is projected to expand at a similar pace over the *Outlook* period, reaching 43 Mtoe in 2010 and 70 Mtoe in 2030 – an annual rate of 2.8% (Table 9.3). At 3.7% per year, demand growth will be strongest in the current decade, tapering off to 3.1% per year in 2010-2020 and 1.9% per year in 2020-2030. Per capita demand will remain higher than the North African average, increasing by over 50% from just over 1 toe in 2003 to 1.6 toe in 2030. Primary energy use will increase at roughly the same pace as GDP, resulting in nearly constant energy intensity across the projection period – unlike the trend observed in most other MENA countries, where energy intensity decreases. This is in large part due to the expansion of the largely export-oriented hydrocarbon sector, which is particularly energy-intensive. Total CO_2 emissions in the Reference Scenario are projected to increase from 78 million tonnes in 2003 to 161 Mt in 2030.

Table 9.3: **Algerian Total Primary Energy Demand in the Reference Scenario**
(Mtoe)

	1990	2003	2010	2020	2030	2003-2030 *
Oil	9.7	10.4	13.5	17.4	21.0	2.6
Gas	13.5	21.8	28.2	39.4	47.3	2.9
Coal	0.6	0.7	0.8	0.9	1.0	1.5
Other	0.0	0.1	0.1	0.2	0.4	5.1
Total	**23.9**	**33.0**	**42.5**	**57.9**	**69.7**	**2.8**

* Average annual growth rate (%).

Algeria's energy use is dominated by natural gas and oil, which combined meet 98% of the country's energy needs. Gas is expected to remain the leading fuel, demand for which will expand at an annual average rate of 2.9% over the projection period. Its share of total primary energy demand will rise even further, from 66% in 2003 to 68% in 2030, largely at the expense of oil. Gas will remain the main fuel in power generation and in most final use sectors. The fuel requirements for water desalination will account for 35% of the incremental energy demand in the power and water sector from 2003 to 2030.

Demand from industrial, commercial and residential end-users is also expected to climb steadily. Later in the *Outlook* period, a new GTL project at Tinrhert will add to gas demand growth.

Sectoral Trends

As in other MENA countries, the largest single contribution to incremental demand will come from power generation and water desalination (Figure 9.1), driven mainly by strong end-use electricity demand. Most of the additional capacity that will be built between now and 2030 will be gas-fired. The amount of gas used will double across the period, to reach 18 Mtoe in 2030. Other energy transformation sectors – notably LNG – also see their share in primary demand increase.

Figure 9.1: **Incremental Primary Oil and Gas Demand in Algeria by Sector, 2003-2030**

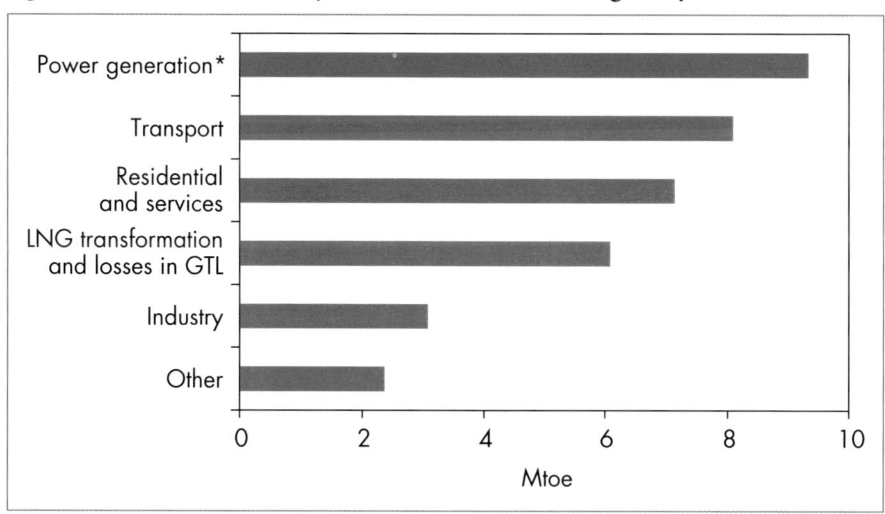

* Including water desalination.

Total final energy consumption in Algeria is projected to rise from 19 Mtoe in 2003 to 41 Mtoe in 2030, at an average annual growth rate of 2.9%. The greatest increase in volume terms will come from the residential and services sectors, where per capita energy use is currently very low and pent-up demand for energy services is very strong. The share of these sectors in final energy use will rise, from 36% in 2003 to 39% in 2030. The share of transport in final energy use will also rise slightly, reaching 36% in 2030, while the share of industry is projected to drop.

Final oil use is projected to double, to reach 19 Mtoe in 2030, though its share will fall further from just over half now to 47% in 2030, continuing the trend of the last three decades. Electricity will see the fastest increase in demand, at a projected rate of 3.5% per year, driven mainly by the residential sector. Natural gas also sees a strong increase in final demand, growing at an average annual rate of 3.1%. Gas use will more than double over the projection period, reaching 17 Mtoe in 2030.

Industry accounted for a quarter of total final consumption in 2003. About one-third of industrial energy use is in chemicals production. Natural gas is the dominant fuel source, accounting for over 60% of consumption. Natural gas will remain the leading industrial fuel of choice throughout the projection period, though its share is expected to fall slightly. This is mainly because non-chemical industries, such as the cement and textile industries, which rely less on gas and more on electricity, will grow more rapidly. Consumption of electricity in industry will increase faster than any other final energy source and will approach that of oil by 2030. New water-desalination plants, using electricity-based reverse osmosis technology, will contribute around one-third of the increase in power demand. Coking coal consumption in the iron and steel industry will increase at around the average rate for industrial energy demand and its share in total industrial energy use will remain steady across the period.

Energy demand in the *transport* sector will grow at 3% per year between 2003 and 2030, a slightly faster pace than overall total final consumption. Road transport is the main contributor to demand growth, with the rest coming mostly from oil and gas pipelines. Rising demand for mobility will stem mainly from population growth and rising per capita GDP. Most of the current vehicle stock is very old and inefficient: over 50% of all cars on the road today are more than 20 years old. Vehicle ownership has been growing rapidly in the past few years, but at 56 cars per 1 000 people it is still low compared to other MENA countries. Although car ownership levels are expected to grow significantly over the projection period, high unemployment and widespread poverty may hold this back somewhat.

The *residential and services* sectors account for over a third of total final consumption. Consumption in these sectors is projected to more than double, from 7 Mtoe in 2003 to 16 Mtoe in 2030. Gas and electricity will account for most of this growth. Electricity use will nearly triple. Around 99% of the Algerian population currently has access to electricity and, as per capita GDP increases over time, demand for electricity will increase in line with higher availability of equipment such as refrigerators and air-conditioning. Yet this increase will hinge on the success of government efforts to increase the housing stock. Widespread poverty in many parts of the country explains the current low level of per capita residential energy use. In the services sector, the bulk of the growth will be related to tourism. The sector was badly hit by the civil conflict in the 1990s, the events of 11 September 2001 and recent terrorist incidents in North Africa, but business has recently begun to recover.

Electricity Supply and Desalinated Water Production

Overview

Algeria's electricity generation reached 30 TWh in 2003. It is almost exclusively based on natural gas. There are marginal contributions from oil-fired power plants (2.3% of total generation) – used to power remote villages in the south – and hydropower (0.9%). Algeria's power sector has developed slowly in recent years. The civil conflict in the 1990s, in particular, caused substantial delays in the development of new projects (Figure 9.2). Over the past five years for which data are available, power generation increased by 4.9% per year, which is low relative to other countries in the region. Electricity generation per capita – at 928 kWh – is also among the lowest in MENA. Preliminary estimates for 2004 indicate that electricity output reached 31 TWh.

Figure 9.2: **Algeria's Power Sector Capacity Additions, 1971-2002**

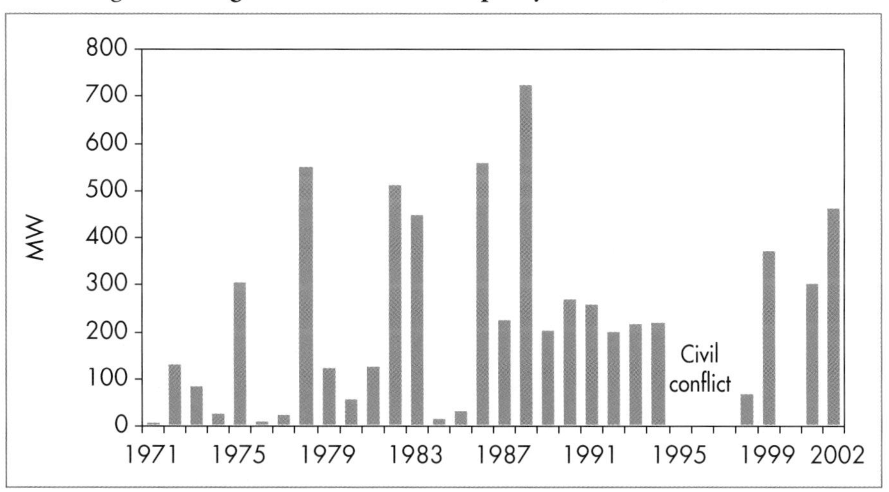

Installed generating capacity was 7 GW in 2003. Like most other MENA countries, Algeria makes extensive use of open-cycle gas turbines and steam boilers to generate electricity and has lagged behind in the use of CCGT power plants. Open-cycle gas turbines account for over half the installed capacity. Generating efficiency – at just under 30% in 2003 – is low. There is substantial scope for improving efficiency and saving gas, which could be exported to earn additional income for the country.

The electricity sector is dominated by Sonelgaz, which controls the production, transmission and distribution of electricity as well as the distribution of natural gas. Investment in the power sector has not kept pace with the increase in demand, prompting the government to draw up a national strategy to promote electricity efficiency and conservation in end-uses. Private investment is also being encouraged. Algeria faced shortages and rationing of electricity in mid-2003.

Electricity prices in Algeria, which are set by the Electricity and Gas Regulatory Commission (CREG), established in 2002, are half those of OECD Europe (Figure 9.3). Algeria is, however, one of the few MENA countries making serious efforts to bring electricity prices into line with costs. CREG made upward adjustments to electricity tariffs in June 2005, prices for large consumers increasing by 10.5%, while prices for households rose by 4.9%. Further increases are planned for December 2005. Concurrently, Sonelgaz is making efforts to reduce network losses, which are high, and to recover arrears. Losses, mainly due to theft, were reduced from about 16% of total electricity generation in 2002 to 14% in 2003. Total debts to Sonelgaz exceeded $200 million in 2003, of which some 38% was held by the Algerian government (Ministry of Energy and Mining, 2004).

Algeria's electricity network is connected with the networks of Tunisia and Morocco. These three countries formed COMELEC (Comité Maghrébin de

Figure 9.3: **Comparison of Electricity Sale Prices in Algeria and OECD Europe**

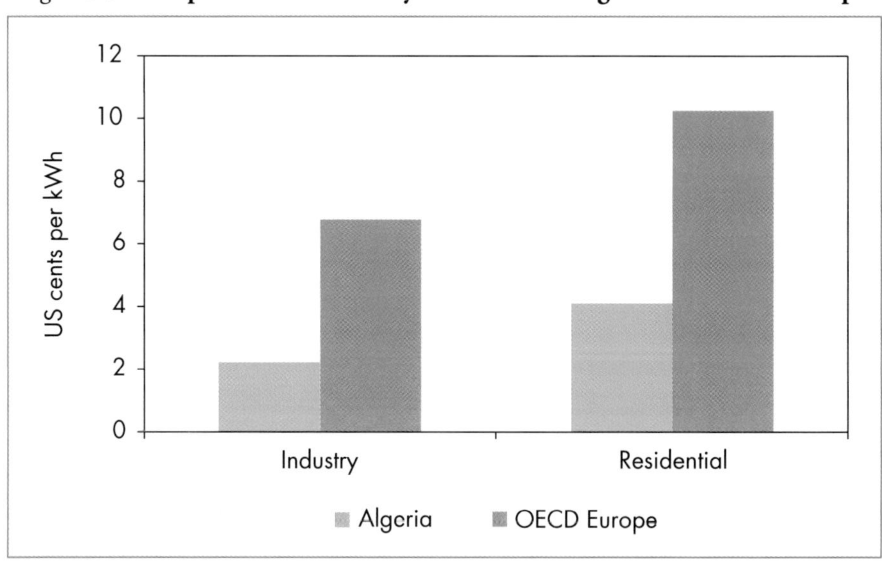

Note: Algerian sale prices are based on 2003 revenue and sales data from Sonelgaz.

l'Electricité) in 1975 to promote electricity exchanges; and they were later joined by Libya and Mauritania. The networks of Algeria, Tunisia and Morocco have been synchronously interconnected to Europe through an underwater link between Morocco and Spain since 1997 (UCTE and Eurelectric, 2005). The total amount of exchanges is very small and Algeria was a net importer of electricity in 2003. There are plans to increase Algeria's interconnections in the future, with both neighbouring countries and Europe (Italy and Spain).

Electricity Production

Electricity generation is projected to increase by 3.4% per year, rising from 30 TWh in 2003 to 74 TWh in 2030. Natural gas is expected to remain the dominant fuel in power generation. Renewable energy is expected to increase its share, but this share will remain very small, despite government initiatives to boost solar use (Box 9.1), as renewables-based electricity generation costs are much higher than those of gas. Other renewable energy sources, including wind and biomass, are also being promoted through New Energy Algeria (NEAL), a joint venture between Sonatrach and Sonelgaz set up in 2002. Total new capacity needs over 2004-2030 are expected to be about 12 GW, which is much more than the total added over the past 30 years. New capacity will be increasingly CCGT, which is expected to be the most economic way to produce electricity as the opportunity cost of gas rises.

9

Box 9.1: **Prospects for Solar Energy in Algeria**

Algeria is seeking to promote the use of solar energy in rural electrification and elsewhere. Because of its geographical location, Algeria has high sunshine levels and a very good solar energy potential (Table 9.4). Sonelgaz has an ongoing programme to install photovoltaic systems in remote villages as part of its rural electrification programme. A total of 20 villages in the southern part of the country – far from the main network – has been equipped with PV systems.[1] There are plans to develop a hybrid gas-solar thermal project. The plant, located at Hassi R'Mel, will include a 130-MW CCGT unit and a 25-MW solar component.

Algeria has set an ambitious target to produce 5% of its electricity from solar power by 2010. Achieving this target means that over 2 TWh of electricity must come from solar power plants in 2010. Building the 700 MW required to produce this amount of electricity will cost $1.3 billion. Under current policies, only the 25 MW of the Hassi R'Mel plant is likely to be operational by 2010. Solar power is projected to reach 250 MW by 2030.

1. Algeria's electrification rate approaches 99%, but a number of villages in the south remain unelectrified.

Table 9.4: **Algeria's Solar Energy Potential**

	Coast	High planes	Sahara
Area (%)	4	10	86
Average sunshine duration (hours/year)	2 650	3 000	3 500
Average annual energy received (kWh/m^2)	1 700	1 900	2 650

Source: Algerian Ministry of Energy and Mining.

Water Production

Water consumption in Algeria was some 6 244 million cubic metres in 2003. Before 1999, there was no water-management policy in Algeria, leading to a badly maintained and inefficient water system. Because of severe droughts, water has routinely been rationed in the northern part of the country.

Water losses are estimated to be 40%, 32% from technical losses and leakages and 8% from illegal connections (World Bank, 2004). The previous low fixed fee for water supply is being abandoned. Today, consumers have a choice between a high fixed fee and a water meter.

Water quality is deteriorating, particularly in the north-west of Algeria. Turning to sea-water desalination is essential to guarantee a safe drinking water supply.

Sonatrach and Sonelgaz recently launched Algeria's first independent water and power producer (IWPP) project at Arzew. The Arzew plant is a co-generation project consisting of a 320-MW gas-fired power plant and a 90 000 cubic metres per day desalination plant located near Oran. The project came on stream in September 2005.

Construction on Hamma, the first private desalination reverse osmosis (RO) water project in Algeria, began in July 2005 and the plant is expected to start operating in 2006. The project will be the largest membrane desalination plant in Africa, as well as one of the largest desalination plants in the world. Capacity will be 200 000 cubic metres per day and project costs are $270 million. The Overseas Private Investment Corporation contributed $200 million towards the project.

The Ministry of Water has recently started the construction of some 20 small-scale reverse osmosis desalination plants, each with a capacity of less than 5 000 cubic metres per day, to supply towns along the Mediterranean coast. All of the plants were intended to be operational by July 2003, but they are only now coming on stream.

Algeria also has six reverse osmosis plants that are expected to come on stream from 2005 to 2008. Cap Djinet, Mostaganem and Zeralda are each projected to supply 100 000 cubic metres per day of desalinated water. The El Tarf and Jijel plants will each supply 50 000 cubic metres per day and the Tlemcen RO plant will consist of two units, one 60 000 cubic metres per day and the other 40 000 cubic metres per day. The share of RO in total desalination capacity was 54% in 2003 and is projected to rise to 73% in 2030.

Electricity and water generation in combined water and power plants will remain exclusively gas-fired. Fuel requirements for desalination in Algeria will rise from 0.5 Mtoe in 2003 to 3.8 Mtoe in 2030, accounting for 5% of total primary energy supply. About 40% of the total electricity capacity additions will be for new combined water and power (CWP) plants.

Table 9.5: **Water and Desalination Capacity Projections for Algeria**

	2003	2010	2020	2030
Water consumption (million cubic metres)	6 224	6 605	7 230	7 780
Desalination capacity (million cubic metres)	125	542	1 307	2 004
Total fuel (oil and gas) requirements for desalination (Mtoe)	0.5	1.5	3.0	3.8

Algeria will need to invest $5 billion in new desalination plants over the projection period, amounting to nearly 45% of total investment needs for new electricity and water production plants. The government has indicated that all future desalination plants will be financed solely from local banks, significantly altering the risk profile, since any potential investor will be borrowing funds from a government lender, selling water to a government off-taker and co-developing the project with a government shareholder (Global Water Intelligence, 2005b).

Oil Supply

Policy Framework

Despite the increasing involvement of international oil companies in Algeria's oil sector, Sonatrach, the national oil company, remains the dominant company, with direct or indirect involvement in all stages of the supply chain. It explores for and develops oil (and gas) fields, in co-operation with several foreign companies, operating under association and production-sharing contracts. A new hydrocarbon law, paving the way for the restructuring of Sonatrach and the further opening-up of the industry to competition from private domestic and foreign companies, was finally approved in April 2005.

When first proposed in 2001, the law was strongly opposed by the trade union movement and many of Sonatrach's own managers. Following President Bouteflika's re-election in 2004 and the reappointment of Chakib Khelil as Minister of Energy and Mines, the law was adopted by the lower house of parliament and the senate in early 2005, with relatively few amendments.

The principal changes contained in the new law concern Sonatrach's role in implementing government policy on oil and gas. Sonatrach will no longer be responsible for licensing and supervising concessions to foreign companies. The existing production-sharing regime will be replaced by a tax and royalty system, and new arrangements for gas marketing will be introduced. Up to now, Sonatrach has effectively acted as an arm of the Algerian government, negotiating licences and contracts with foreign companies and monitoring the performance of each production-sharing contract. This role will now be performed by a new entity, to be called Alnaft (Agence Nationale pour la Valorisation des Ressources en Hydrocarbures), which will report to the energy ministry. Alnaft will also be responsible for promoting investment in exploration. Another new agency will handle more technical matters, such as regulations on health, safety and the environment, management of the process of issuing pipeline concessions, and the regulation of pipeline tariffs for third-party users.

The new tax regime will be based on four geographical zones (A to D), for which different rates will apply. The lowest rates will be applied to zone A, the least prospective, in order to encourage exploration of such areas. Operators in all zones will be subject to five different charges:

- A per square kilometre area tax.
- A royalty, on a sliding scale according to the volume produced.
- A petroleum revenue tax, based on profit, starting at 30% and rising to 70%.
- Corporation (profit) tax, currently levied at a rate of up to 30%.
- A property tax on fixed assets.

Acreage will be offered on the basis of transparent bidding rounds, with awards based on technical and financial criteria. Bidders may propose to pay a higher royalty than the minimum official rate. Foreign companies are no longer obliged to team up with Sonatrach, which will be required to compete for new acreage on the same terms. Although Sonatrach will no longer have an automatic majority stake in new licences, it will nonetheless have the option to take a 20% to 30% stake in any new discoveries. Foreign companies will also be permitted to apply to build and operate their own oil and gas pipelines. Up to now, Sonatrach has held a monopoly over all oil pipelines and export gas pipelines.

These reforms, which are expected to be implemented by the end of 2005, will enhance the appeal of upstream projects to outside investors. They will give Sonatrach more discretion over the use of its capital, much of which has up to now been committed to costly domestic projects. It is expected to choose to expand its overseas activities. The government hopes that the new legislation will stimulate investment and increase oil and gas production. The government has set a goal of increasing crude oil production capacity to 2 mb/d by 2010. Actual production will depend on the success of efforts to attract capital investment and on Algeria's OPEC production quota.

Resources and Reserves

According to the *Oil and Gas Journal* (2004), Algeria had proven oil reserves of 11.8 billion barrels at the end of 2004, of which nearly 60% are in the Hassi Messaoud field in the centre of the country (Figure 9.4). Much of the remaining reserves are located in the east, near the Libyan border. At current production rates, these reserves would last almost 17 years. The US Geological Survey (2000) estimates that undiscovered recoverable resources could add a further 9.9 billion barrels to proven reserves.

Algeria can be divided into two main geological regions. The north, "Alpine Algeria", a 200 to 300 km-wide band along the coast, has been folded and uplifted by the collision of the African and European plates. Although small hydrocarbon deposits are known to exist, the geology is very complex, as a result of the Alpine mountain formation, and the area has hardly been explored.

To the south of that region lies the immense Saharan platform. Prior to the Alpine mountain formation, this was a region of successive transgressions and regressions at the edge of the African plate, and therefore it has a thick cover (up to 6 000 metres) of sediments, dating from the Cambrian to the Triassic. This is where all the known large Algerian fields are to be found, *e.g.* Hassi Messaoud in Cambrian to Devonian sandstones, Hassi Berkine in Paleozoic and Triassic sandstones and carbonates, or the giant gas reservoirs of Hassi R'Mel in Triassic sediments. The main reservoirs correspond to large structural traps, but the region can be expected to feature a variety of different types of reservoirs at various depths in sedimentary traps as well. The region is vast and only partially explored, with a significant amount of hydrocarbons believed to be awaiting discovery in several of the basins. On the other hand, a large part of the region is remote, with a harsh climate, and lacks transport infrastructure. As a result, development is likely to be slow, moving step by step from currently exploited areas.

Sonatrach has a long tradition of technical co-operation with other countries, such as France or Russia, and is technically sophisticated. The recent allocation of many blocks to international companies will encourage the influx of efficient modern technologies.

Exploration activities have been stepped up in recent years, with increasing participation from foreign companies. There have been six upstream exploration licensing rounds since 2001, the most recent of which was concluded in April 2005 and awarded contracts to BP, BHP Billiton, Gulf Keystone and Shell.

Figure 9.4: **Major Oil and Gas Fields and Energy Infrastructure in Algeria**

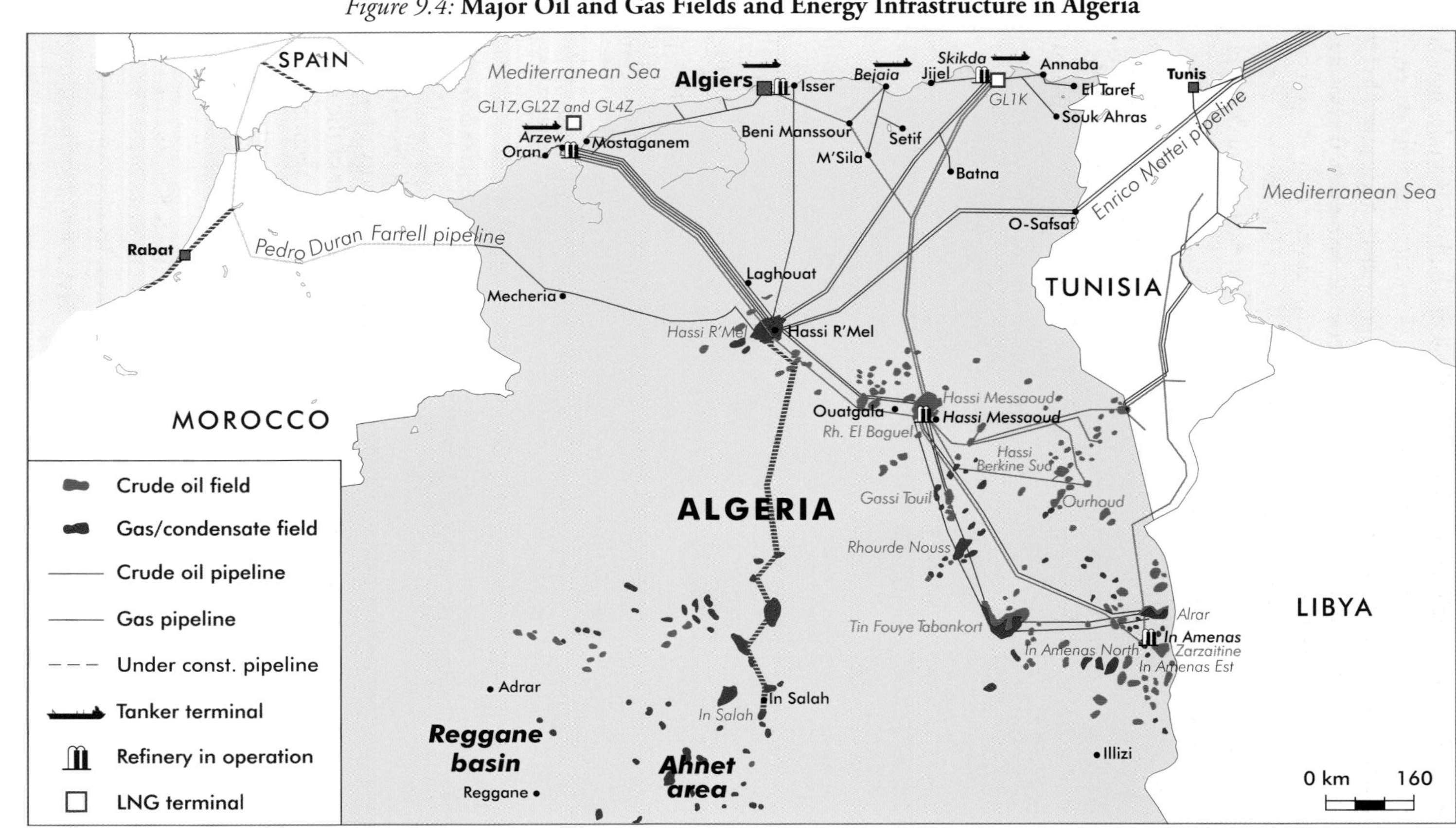

A seventh round, which will be held under the terms laid down in the new law, is planned for late 2005. A total of eight exploration contracts were signed in 2004, up from five in 2003 and seven in 2002. Today, there are some 35 foreign companies operating in the Algerian oil sector alongside Sonatrach. BP has invested the most in the Algerian upstream sector, though predominantly in gas projects. Anadarko is thought to be the largest investor in oil. The Minister for Energy and Mines, Chakib Khelil, aims to conclude an average of ten exploration deals per year.

Several major discoveries have been made in recent years. Thirteen fields – four containing oil and nine gas – were discovered in 2004, up from seven in 2003 and six in 2002 (Sonatrach, 2003, 2004a and 2005). The most significant of these discoveries are in the Berkine basin and the nearby Ourhoud field. The Ministry of Energy and Mining estimates that the field wildcat drilling success rate is averaging more than 30%. An estimated 60 exploration wells were drilled in 2004, up from 41 in 2003 and 22 in 2002.

Crude Oil Production[2]

Crude oil and NGL production is projected in the Reference Scenario to rise steadily through to the middle of the next decade, continuing the upward trajectory of recent years. Around 2015, production is expected to start to decline constrained by the limits to Algeria's reserve base. Production of crude oil (excluding NGLs) averaged 1.2 mb/d in 2004 and reached 1.3 mb/d in the first half of 2005 – well in excess of Algeria's OPEC quota of 880 kb/d. The government is currently pushing hard for an increase in its production quota to allow it to reach a targeted production (excluding NGLs) of 2 mb/d by 2010. Algeria produces disproportionately large volumes of condensates and NGLs, averaging almost 700 kb/d in 2004. We project output, including NGLs, to rise from 1.9 mb/d in 2004 to 2.2 mb/d in 2010, before dropping to 1.6 mb/d in 2030.

The biggest oilfield in Algeria, operated by Sonatrach, is Hassi Messaoud (Table 9.6). Output is currently running at around 370 kb/d and is the biggest contributor to the Sahara Blend, Algeria's main export blend. Sahara is one of the highest-quality crude oils in the world, with an API gravity of 44° to 45° and an extremely low sulphur content of about 0.05%. Most of the output of Sahara Blend is exported to Europe, where it commands a sizable premium over most other crude oils because of stringent European environmental regulations on the fuel sulphur content of gasoline and diesel. Anadarko is the largest foreign oil producer, with total output in 2004 of 63 kb/d (from Hassi Berkine South and Ourhoud). It is currently developing several other fields, with first output due in 2007.

2. All the oil-production figures cited in this section include crude oil, natural gas liquids (NGLs) and condensates.

Table 9.6: **Oilfields in Algeria**

Field	Year of discovery	Remaining proven and probable crude oil and NGL reserves at end 2004 (billion barrels)	Cumulative production to 2004 (billion barrels)	Gravity (average °API)
Alrar	1961	0.3	0.2	48.0
Gassi Touil	1961	0.1	0.6	47.4
Hassi Berkine Sud	1995	0.6	0.2	44.3
Hassi Messaoud	1956	3.9	6.5	46.1
Hassi R'Mel	1957	3.7	1.4	46.1
Ohanet	1960	0.1	0.1	45.2
Ourhoud	1994	1.9	0.1	40.4
Rhourde El Baguel	1962	0.6	0.6	42.3
Rhourde Nouss	1962	0.3	0.3	36.3
Tin Fouye-Tabankort	1961	0.1	1.0	40.3
Zarzaitine	1957	0.3	1.0	41.6
Other fields		2.4	6.2	

Sources: IHS Energy and IEA databases.

Production from currently producing fields is expected to drop over the projection period, from over 1.9 mb/d in 2004 to under 1.5 mb/d in 2010 and only 500 kb/d in 2030 (Table 9.7). At 370 kb/d, Hassi Messaoud currently produces nearly twice the output of any other field in Algeria. Although production from this field will start declining after 2010, Hassi Messaoud will remain the largest oilfield at the end of the projection period, with production of 130 kb/d projected for 2030. Production from Hassi R'Mel, mainly in the form of condensates, will rise by one-third over the current decade, reaching 240 kb/d in 2010 before declining to 110 kb/d in 2030. The Ourhoud field, after nearly doubling production over the remainder of the current decade, is expected to decline thereafter, although its projected 2030 level of 97 kb/d is still higher than the current level of production. Other fields, including Rhourde Nouss and Tin Fouye-Tabankort, will, in contrast, see production tailing off by the end of the *Outlook* period.

Much of the incremental production will be in the form of condensates and NGLs, because of the rapid development of gas and gas/condensate fields. This will substantially boost the country's exports of liquefied petroleum gas. Algeria is already the world's second-largest LPG exporter, after Saudi Arabia. New discoveries and fields that are awaiting development are expected to account for a majority share of total Algerian oil production in 2030: from making up 30% of production in 2010, these fields are projected to contribute nearly two-thirds of production at the end of the *Outlook* period (Figure 9.5).

Table 9.7: **Algeria's Crude Oil Production in the Reference Scenario** (kb/d)

	2004	2010	2020	2030
Currently producing fields	**1 930**	**1 469**	**858**	**506**
Alrar	23	23	23	23
Gassi Touil	2	0	0	0
Hassi Berkine Sud	75	70	64	58
Hassi Messaoud	370	354	212	127
Hassi R'Mel	198	240	195	107
Ohanet	2	2	2	2
Ourhoud	86	161	125	97
Rhourde El Baguel	35	60	60	31
Rhourde Nouss	46	44	14	2
Tin Fouye-Tabankort	53	10	0	0
Zarzaitine	27	25	15	7
Other fields	1 012	480	147	51
New developments	**0**	**631**	**939**	**889**
Fields awaiting development	0	347	329	133
Reserve additions and new discoveries	0	284	610	756
Total	**1 930**	**2 100**	**1 797**	**1 395**

Notes: Includes NGLs and condensates. Takes account of improved oil recovery techniques.

Figure 9.5: **Algeria's Crude Oil Production by Source in the Reference Scenario**

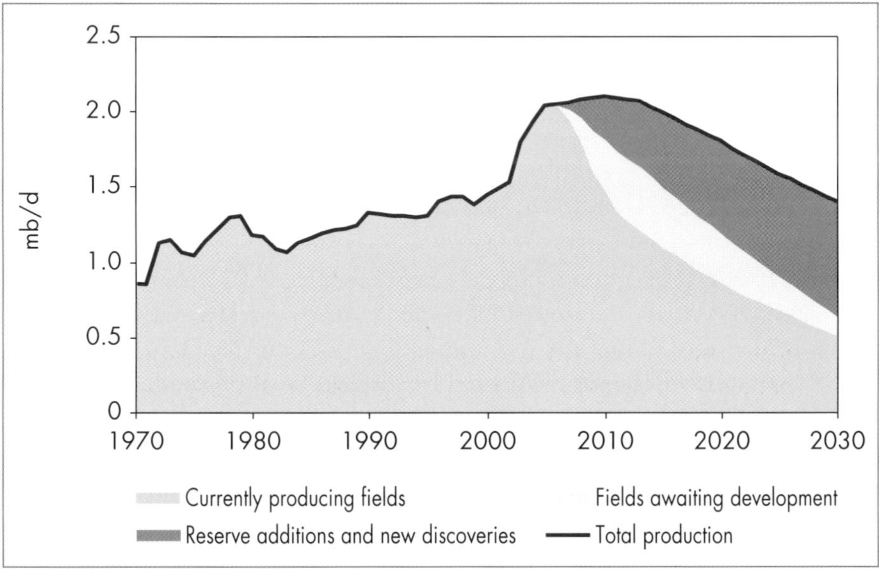

Note: Includes NGLs and condensates.

Oil Refining and GTL

Algeria currently has four refineries, with a combined capacity of 450 kb/d. The Skikda refinery is by far the largest, with a capacity of 300 kb/d (Table 9.8). All four refineries are operated by Naftec (Société Nationale de Raffinage du Pétrole), a subsidiary of Sonatrach. The refineries supply most of Algeria's refined oil product needs, though Algeria also imports small volumes of specialist products. Algeria is a net exporter of oil products.

Refining capacity is projected to expand most rapidly at the beginning of the projection period, reaching 700 kb/d in 2010 and 1 mb/d in 2030. After tenders issued in late 2003, the Skikda, Algiers and Arzew refineries are now undergoing modernisation and upgrading, which will boost distillation and secondary processing capacity. In addition, a new 13 kb/d refinery at Adrar, currently being built by the China National Petroleum Corporation (CNPC), is expected to commence operations in 2007. Sonatrach and CNPC will jointly market the output. The Algerian government is also looking to upgrade the 6.5 kb/d In Amenas refinery, which has been closed since 1986. The construction of a new export refinery of 320 kb/d to be located close to Algiers is also under consideration. We do not expect this refinery to reach full capacity production until shortly after 2010.

Table 9.8: **Algerian Oil Refineries**

		Distillation capacity (kb/d)	
	Year commissioned	**2004**	**2010**
Current capacity		**450**	**630**
Algiers	1964	60	60
Arzew	1973	60	60
Hassi Messaoud	1960	30	30
Skikda	1980	300	480
Additional capacity			**63**
New refinery at Algiers*			50
Adrar			13
Total		**450**	**693**

* The precise location has not yet been decided. Full capacity (320 kb/d) is assumed to come on line after 2010.

Refinery output will be supplemented by the output of oil products – mainly middle distillates – from a new GTL plant that it is assumed will be built during the second decade of the projection period. Sonatrach plans to build a plant on the Mediterranean, to be supplied with gas from several fields in the Tinrhert region near the Libyan border. After an initial tender process, Sasol Chevron, Shell and PetroSA have been asked to submit commercial bids for the project. GTL output from this plant is expected to reach about 35 kb/d by 2020.

Exports

The bulk of the increase in crude oil and NGL production will be exported, mainly to Europe. Oil products will account for a growing share of total Algerian oil exports. Total oil exports are projected to increase from 1.68 mb/d in 2004 to 1.72 mb/d in 2010, before decreasing to 1.08 mb/d by 2030 (Figure 9.6). By 2030, 69% of the country's production will be exported, compared with 87% now.

Figure 9.6: **Algeria's Oil Balance in the Reference Scenario***

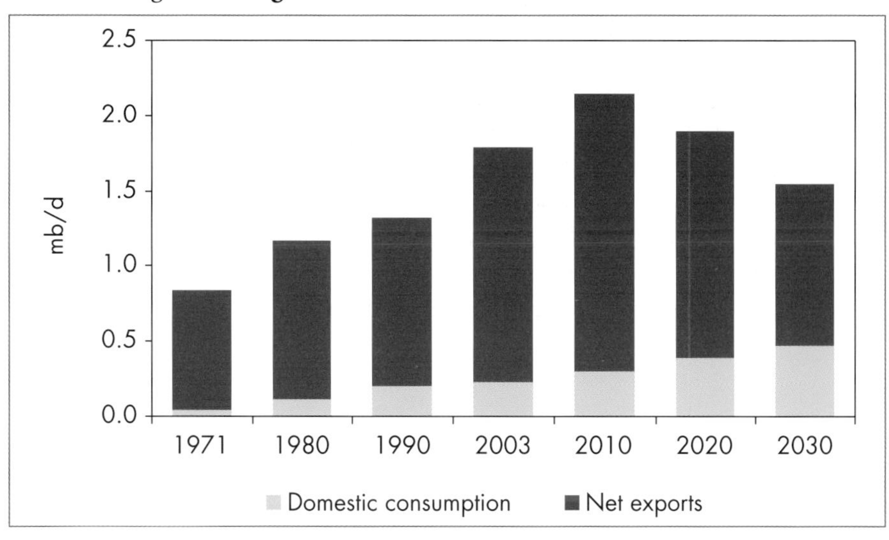

*Including output from GTL and natural gas processing plants.

Natural Gas Supply

Resources and Proven Reserves

Algeria has the eighth-largest proven reserves of natural gas in the world. Most of the reserves are associated with oil or condensates. At the end of 2004, Algerian reserves stood at 4.6 trillion cubic metres (Cedigaz, 2005). Hassi R'Mel is Algeria's largest gas field by far, with more than half of total reserves. The field, located 200 km south of Algiers, is operated by Sonatrach. Other large fields include Rhourde Nouss, Tin Fouye-Tabankort and In Salah. The potential for increasing reserves and production is thought to be high.

Sonatrach is leading gas exploration in Algeria, with most of its recent activities focused on the Reggane basin in the south of the country. The basin lies close to the In Salah region, where Sonatrach already produces gas in association with BP and Statoil. This latter venture is the first project in Algeria for which Sonatrach was not responsible for 100% of the gas production. Most other foreign companies have concentrated on exploring for and developing oil.

Gas Production

Algeria is the fifth-largest natural gas producer worldwide and the largest producer among OPEC member countries. Commercial production started in 1961, with marketed output reaching 88 bcm in 2003. Gross production is much higher, as much of the gas produced in association with oil is reinjected to enhance oil recovery. In addition, a small amount of gas is flared, though the volume has fallen sharply in recent years. Sonatrach has been working to reduce gas flaring since 1973. From around 80% in 1970, the share of flared gas in total Algerian natural gas production has been reduced to 11% in 2003 (Sonatrach, 2004b). Sonatrach aims to eliminate gas flaring altogether by 2010. In the Reference Scenario, marketed production of natural gas is projected to climb briskly to about 107 bcm in 2010 and 198 bcm in 2030. The overwhelming bulk of the incremental output will be exported, by pipeline and in the form of LNG.

Total natural gas output reached 88 bcm in 2003, up from 48 bcm in 1990. In 2003, 78% of Algeria's natural gas production came from Hassi R'Mel (Table 9.9). The rest of Algeria's gas comes from recent projects, including Tin Fouye-Tabankort, which has been in production since 1999, and In Salah, which came on stream in 2004. In Salah, a $2.5 billion joint venture involving the development of seven fields in the south of the country, produced 5.8 bcm in 2004 and is expected to reach a plateau of 9 bcm in 2005. Enel, the Italian utility, has a contract for 4 bcm, with Sonatrach taking the remaining 5 bcm to cover its domestic and export commitments. The project was originally a joint venture between Sonatrach and BP. Statoil acquired 49% of BP's stake in 2003. Output is transported by pipeline to Hassi R'Mel, a hub for all of Algeria's onshore gas production. In Salah is also the site for an industrial-scale demonstration of geological storage of CO_2.

Table 9.9: **Algeria's Natural Gas Reserves and Production**

Field/project	Year discovered	Year of first production	Remaining proven reserves (tcm)	Marketed production in 2004 (bcm)
Hassi R'Mel	1957	1961	3.0	68.6
Tin Fouye-Tabankort	1961	1999	0.2	6.8
In Salah	1958	2004	0.2	5.8
Gassi Touil/Rhourde Nouss	1961	2008*	0.2	–
In Amenas	1958	2005	0.1	–
Other			0.9	7.2
Total			**4.6**	**88.3**

* Planned.

Sources: IHS Energy and IEA databases.

Two other projects are also currently under development. The first of these is the In Amenas project, due to come on stream in late 2005. Like In Salah, it is a joint project between Sonatrach and BP/Statoil. In Amenas has reserves of around 120 bcm of gas. The project involves development of four gas fields and the construction of gas gathering and processing facilities. Sonatrach has exclusive marketing rights. Targeted production at plateau is around 9 bcm a year of gas and around 60 kb/d of liquids. The gas will be piped to Hassi R'Mel. The second is the Gassi Touil integrated gas project, which is being developed by a consortium including Sonatrach, Repsol YPF and Gas Natural. The project includes plans for development and production, as well as transport and a new natural gas liquefaction train close to Arzew. It is located on the western edge of the Berkine basin, to the north of Rhourde Nouss, whose reserves are included in the Gassi Touil development plans. Production is expected to start in 2009. Sonatrach estimates that the project will involve investment of up to $4 billion.

Production from the Hassi R'Mel, Tin Fouye-Tabankort, In Salah, In Amenas and Gassi Touil projects is expected to reach a plateau of close to 100 bcm soon after 2010. Further increases in natural gas production will come from new projects yet to be announced (Figure 9.7). These may include Ahnet, an area close to In Salah, which is being explored by a consortium including Sonatrach, Petronas and Gaz de France.

Figure 9.7: **Algeria's Natural Gas Production in the Reference Scenario**

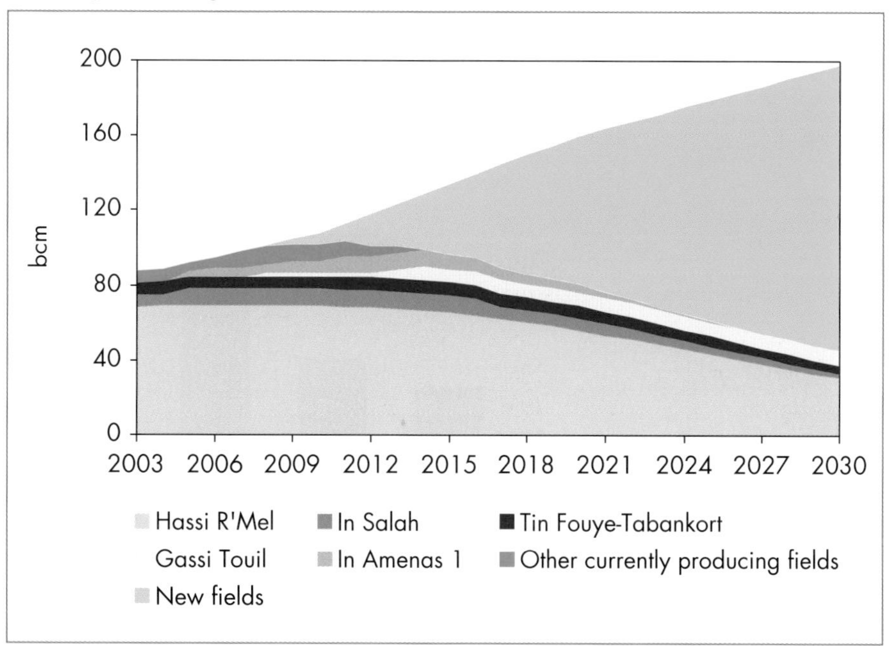

Export Plans and Prospects

Algeria is the world's third-largest exporter of natural gas. It exports LNG to Europe and the United States, and natural gas via sub-sea pipelines under the Mediterranean to Spain (via the Pedro Duran Farrell pipeline that runs across Morocco) and to Italy (via the Enrico Mattei gasline that traverses Tunisia). Algeria was the first LNG producer in the world and has two liquefaction centres, with a total of four plants: three at Arzew (GL 4Z, GL 1Z and GL 2Z) and one at Skikda (GL 1K). Sonatrach owns and operates all LNG plants and carriers, as well as export pipelines in Algeria.

Exports are due to increase sharply in the next few years as pipeline and LNG plant expansions and new pipeline and LNG projects are brought on stream. In 2003, over 40% of the 64 bcm exported was in the form of LNG. This share will rise as new liquefaction terminals are built and the Skikda plant, which was badly damaged in an accident in January 2004, is brought back into production, with higher capacity than before the accident. Europe and the United States, where demand is expected to grow rapidly, will continue to absorb all of Algeria's LNG output.

Gas exports by pipeline are also expected to increase over the projection period. The capacity of the Enrico Mattei gas pipeline (originally called Transmed) was recently increased to 27 bcm per year. There are plans to boost capacity further, to over 30 bcm in 2008 and 33.5 bcm in 2012. The capacity of the Pedro Duran Farrell gas pipeline (originally called Maghreb-Europe Gaz) was also boosted by 3 bcm to 12 bcm in 2004, helping to offset the loss of LNG

Figure 9.8: **Algeria's Natural Gas Balance in the Reference Scenario**

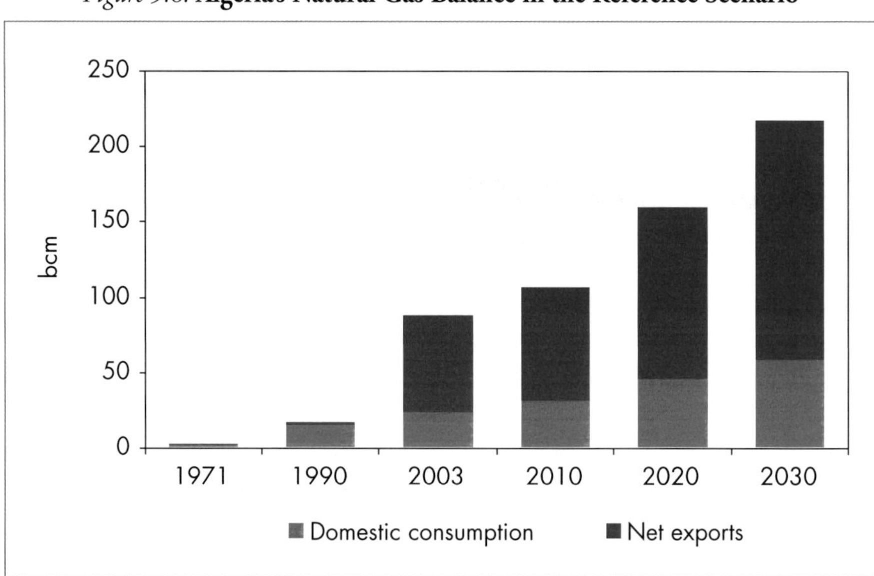

capacity following the Skikda accident. A third export pipeline, the Medgaz line running directly from Algeria to Spain across the Mediterranean, is under construction and is expected to be completed in 2009. Another pipeline, the GALSI pipeline, crossing the Mediterranean from Algeria to Italy via Sardinia, is under discussion. This pipeline, proposed by a consortium of Sonatrach, Edison Gas, Enel, Wintershall and Eos Energia would have an initial capacity of 10 bcm. In total, Algerian natural gas exports are projected to climb to 76 bcm in 2010 and 144 bcm in 2030.

Energy Investment

Algerian energy sector investments are projected to total $114 billion over 2004-2030. Investment needs are expected to be greater towards the second half of the projection period; the decade spanning from 2021 to 2030 will alone absorb around $45 billion, or 40%, of total cumulative investment. The share of investment devoted to natural gas will increase dramatically: from accounting for 36% of total investment from 2004 to 2010, it will make up 60% of investment in the third decade of the projection period (Figure 9.9).

Figure 9.9: **Algeria's Cumulative Energy Investment by Sector in the Reference Scenario**

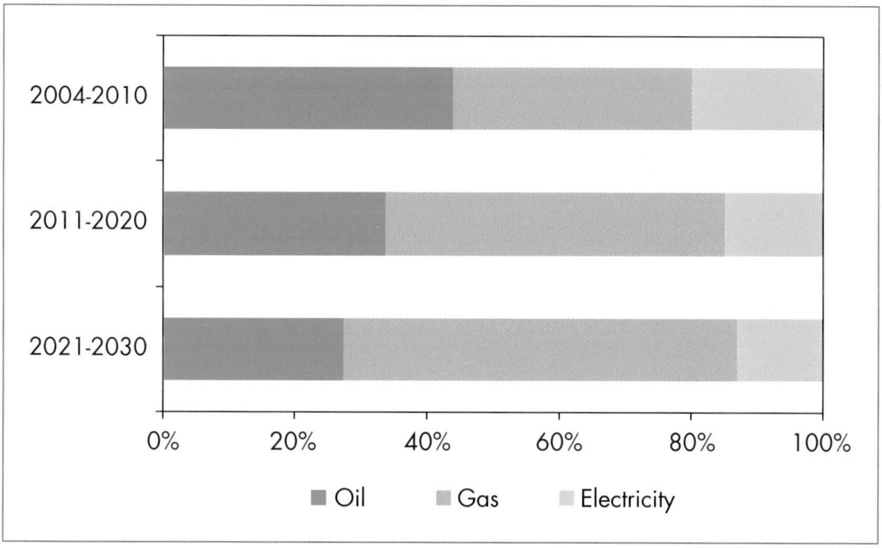

In the oil sector, total cumulative investment needs are projected to amount to $38 billion dollars over the *Outlook* period (Figure 9.10). The upstream will absorb $27 billion, or nearly three-quarters, of this investment. Annual upstream investment needs will be highest in the second decade of the projection period, when finding and development costs will be higher than current costs as a result of depletion of the lowest-cost reserves. A further $1 billion will go to GTL production and $10 billion to the refining sector – over 85% of which will go towards adding new capacity, with the remainder being invested to upgrade and convert existing refining facilities.

Figure 9.10: **Algeria's Oil Investment in the Reference Scenario**

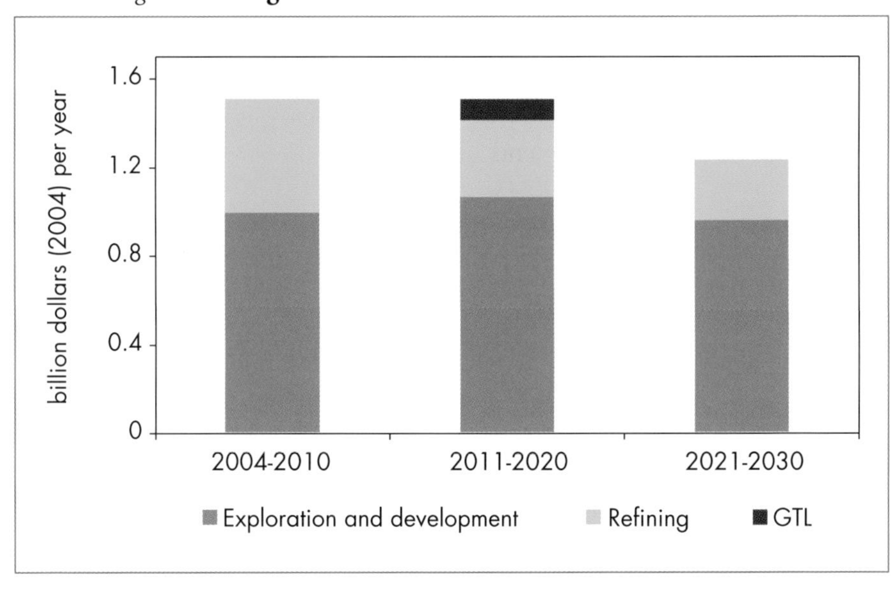

The bulk of this investment, especially in the upstream sector, is expected to come from foreign investors, reflecting recent changes to the investment and fiscal regime. Interest has recently been boosted by both high oil prices, which have pushed up profit margins, and the new hydrocarbon law. Algeria may face more intense competition for capital in the future, as Libya opens up its upstream industry to foreign investment, following the lifting of trade and investment sanctions.[3]

The natural gas sector will absorb over half of the total projected investment in Algeria over the *Outlook* period. Total cumulative gas investment will amount to $59 billion in the Reference Scenario, of which $35 billion will go to the upstream sector. Liquefaction plants will cost just over $15 billion. In total, capital needs will be highest in the last decade (Figure 9.11).

3. A recent survey of oil company professionals by Fugro Robertson ranks Algeria second worldwide behind Libya for the attractiveness of exploration, taking account of the quality and size of the resource base, exploration success rates, political risk and the number of foreign companies already active in the country.

Figure 9.11: **Algeria's Natural Gas Investment in the Reference Scenario**

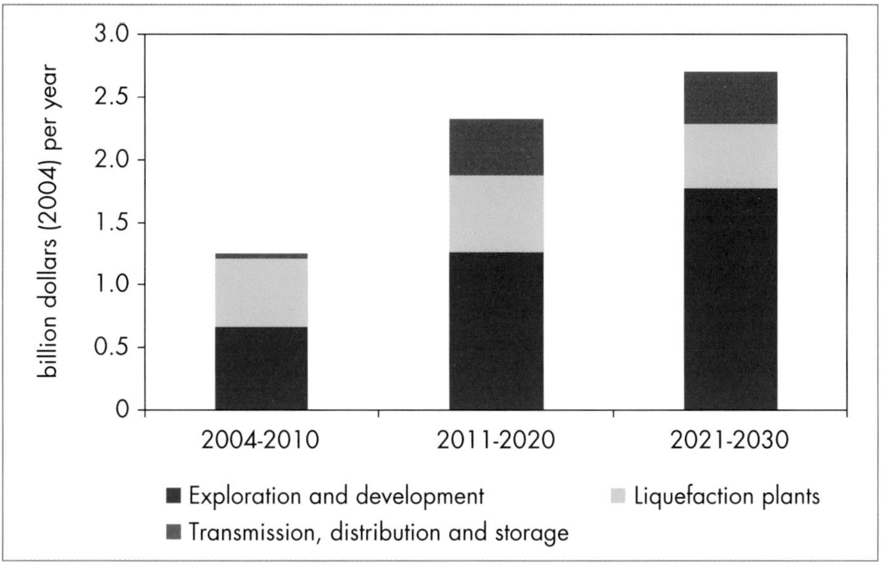

Algeria is unlikely to face major difficulties in financing the projected gas production and export capacity expansions. The country already has a relatively attractive investment climate, one that could be significantly enhanced by the new hydrocarbon law. The geographical proximity of Algeria to Europe, Algerian efforts to comply with stringent EU environmental regulations and the long-standing relationships that have been established with European buyers, are major advantages.

Algeria's power sector will need substantial investment over the period 2004-2030. Total cumulative investment is expected to amount to $17 billion. Power plants alone will cost $7 billion and networks $10 billion. Unlike the hydrocarbon sector, financing power projects will be more difficult. Sonelgaz's revenues are not expected to be sufficient to fund all new projects. As a result, the government is seeking to increase private investment in the electricity sector. A law passed in 2002 ended Sonelgaz's monopoly and a number of independent power or combined power and water projects are now being developed.

Deferred Investment Scenario[4]

Energy Demand

Algeria's primary energy demand in the Deferred Investment Scenario grows at a slightly slower pace over 2003-2030 than in the Reference Scenario,

4. See Chapter 7 for a detailed discussion of the assumptions and methodology underlying the Deferred Investment Scenario.

averaging 2.7% per year over the projection period, 0.1 of one percentage point lower than the Reference Scenario growth rate. Demand reaches 68 Mtoe in 2030, compared with 70 Mtoe in the Reference Scenario (Figure 9.12). Primary demand for both oil and gas falls, with gas accounting for most of the decline in volume terms – both because it is the leading primary fuel and because demand for electricity, which is almost entirely generated with gas, is lower. Most of the reduction in demand results from slower economic growth, the consequence of lower oil and gas production.

Figure 9.12: **Algeria's Primary Energy Demand in the Reference and Deferred Investment Scenarios**

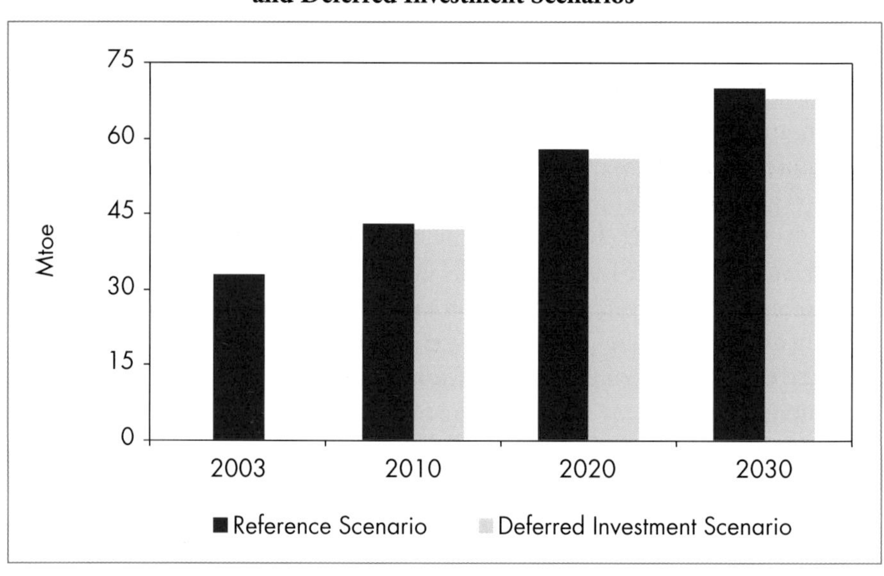

Oil and Gas Production

In the Deferred Investment Scenario, upstream oil investment in each MENA country increases much more slowly over the projection period than in the Reference Scenario. As a result, MENA oil production is lower and the international crude oil price is higher. Natural gas prices increase in line with oil prices. In Algeria, a fall in global demand for natural gas resulting from the higher prices will be the main determinant of the lower gas production levels expected in the Deferred Investment Scenario relative to the Reference Scenario, as the call on Algerian gas for pipeline and LNG exports will be reduced. Domestic demand is also lower. Consequently, total gas production is 42 bcm, or 21%, lower in 2030 in the Deferred Investment Scenario. Investment in the natural gas sector is around 25% lower than in the Reference Scenario.

Oil production is affected in two ways. Production will be lower as a result of a 10% fall in upstream oil investment compared with the Reference Scenario; and there is a fall in production of associated NGLs and condensates because of lower production of natural gas. Oil output in the Deferred Investment Scenario in Algeria falls steadily over the entire projection period, reaching 1.3 mb/d in 2030, down from 1.9 mb/d in 2004.

The expansion of oil refining capacity in the Deferred Investment Scenario is significantly slower as a result of lower domestic demand. Capacity rises from 0.5 mb/d in 2003 to 0.8 mb/d in 2030 – some 0.2 mb/d, or close to one-quarter lower than in the Reference Scenario.

Oil and Gas Exports

Oil and gas exports fall markedly in the Deferred Investment Scenario. Net oil exports fall from 1.7 mb/d in 2003 to 0.9 mb/d in 2030 – nearly 225 kb/d, or more than 20%, lower than in the Reference Scenario. Natural gas exports are even more significantly affected, because of lower gross production and slower demand growth in Europe and North America. Gas exports reach 104 bcm in 2030, 40 bcm, or 28%, less than in the Reference Scenario (Figure 9.13).

Figure 9.13: **Algeria's Natural Gas Balance in the Reference and Deferred Investment Scenarios**

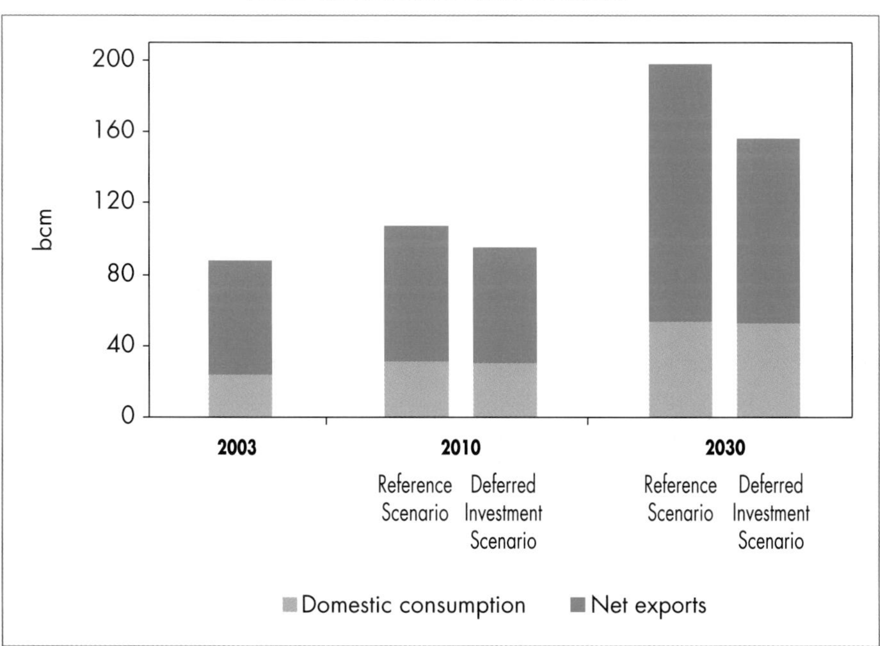

CHAPTER 10

EGYPT

HIGHLIGHTS

- Egypt is the largest economy in North Africa and is, with around 70 million inhabitants, home to almost half the region's population. Gross domestic product is expected to grow at an average 3.6% per year over the projection period, a slower pace than in the past three decades.
- In the Reference Scenario, Egypt's primary energy demand grows by 2.6% per annum, reaching 109 Mtoe by 2030. Energy-demand growth will be much lower than over the past three decades, as the economy matures and population growth slows. Natural gas will overtake oil as the dominant fuel in the second half of the *Outlook* period.
- Egyptian oil production peaked over a decade ago and has since been in decline. In our Reference Scenario, production falls from 0.7 mb/d in 2010 to 0.5 mb/d 2030 and Egypt, currently a minor oil exporter, becomes a net oil importer by around 2015.
- Egypt's strategic importance as a key transit route for oil and gas exports from the Middle East to Europe will grow over the *Outlook* period. The Suez Canal will handle 4% of global inter-regional oil trade by tanker and 9% of LNG trade by 2030, doubling its current share of global trade.
- Egypt, with 1.9 tcm of natural gas reserves, is of increasing importance in the global gas market. Production in the Reference Scenario trebles, reaching 50 bcm around 2010 and over 90 bcm by 2030. LNG exports started in 2005 and are expected to increase significantly, with total natural gas exports reaching 28 bcm by 2030. Access to foreign technology and capital will be needed to meet projected production levels.
- Electricity generation will double, from 92 TWh in 2003 to 188 TWh in 2030. Reliance on natural gas will increase, with small volume increases in renewables as well. The electricity sector will require $36 billion of investment over 2004-2030, with private financing likely to be necessary.
- Total cumulative investment in the Egyptian energy sector is projected to amount to $85 billion over the *Outlook* period. Investment in the natural gas and electricity sectors will account for the largest shares.
- In the Deferred Investment Scenario, oil production declines more rapidly, to 380 kb/d in 2030. Egypt will become a net oil importer around 2010. Natural gas production will also be much lower, reaching 78 bcm in 2030. The decrease in production will be fully reflected in lower exports.

Overview of the Egyptian Energy Sector

The first Egyptian oilfield was discovered in 1869 at Gemsa along the south-western Gulf of Suez, and was brought on stream in 1910. The first natural gas field was discovered in 1967. Egypt now ranks 29th in the world in terms of proven oil reserves and 18th in proven natural gas reserves (Cedigaz, 2005; *Oil and Gas Journal*, 2004). Most of Egypt's hydrocarbon reserves are state-owned and controlled by the Egyptian General Petroleum Company (EGPC) or the Egyptian Natural Gas Holding Company (EGAS). As oil production is now in decline, the focus of the Egyptian energy sector has shifted to the development of the abundant natural gas resources and is finding expression in major liquefied natural gas (LNG) projects and the development of a cross-continental Arab Gasline.

Egypt is currently the 21st-largest gas producer in the world and the sixth-largest in the MENA region. Egypt became a gas exporter in 2003, when the first phase of the Arab Gasline was inaugurated. Income from gas exports will contribute to future growth. Hydrocarbons currently make up 9% of Egyptian GDP (Central Bank of Egypt, 2004), a low share in comparison with most other oil- and gas-exporting MENA countries. Domestic consumption of gas is rapidly increasing as the government upgrades and extends the domestic natural gas grid. Egypt also has the potential to play an important role in the development of regional gas and electricity markets in North Africa.

Egypt's power sector is dominated by the Egyptian Electricity Holding Company (EEHC), which was set up in 2000 as part of plans to liberalise the electricity sector. These plans are progressing very slowly, even though private sector participation in power generation has been possible since 1996, notably through independent power projects.

Table 10.1: **Key Energy Indicators for Egypt**

	1971	2003	1971-2003*
Total primary energy demand (Mtoe)	7.8	53.9	6.2%
Total primary energy demand per capita (toe)	0.23	0.80	3.9%
Total primary energy demand/GDP**	0.15	0.20	0.9%
Share of oil in total primary energy demand (%)	80.9	51.0	–
Net oil exports (mb/d)	0.17	0.23	0.8%
Share of oil exports in production (%)	56.8	30.6	–
Share of gas in total primary energy demand (%)	0.9	43	–
CO_2 emissions (Mt)	20.8	122.2	5.7%

* Average annual growth rate.

** Toe/thousand dollars of GDP in year-2004 dollars and PPPs.

Political and Economic Situation

Political Developments

Egypt was under British rule from 1882, when Britain took over in a bid to gain control over the Suez Canal. It gained partial independence in 1922 and full sovereignty after the Second World War. Egypt is viewed in economic terms as a relatively open country and aims to lead other countries in the region in terms of structural reform and integrating the Arab economies into the global system (Nazif, 2005). President Hosni Mubarak has been in power since October 1981. After winning the 1987, 1993 and 1999 elections as the sole candidate, he was re-elected in the country's first multi-candidate presidential election in September 2005.

A more technocratic National Democratic Party (NDP) cabinet was installed in June 2004 and is expected to give new impetus to Egypt's commitment to developing a market-driven economy. It has already implemented changes to the income tax system and has liberalised the tariff regime. A programme for the privatisation of state-owned enterprises (SOEs) has been launched, but progress has been slow: SOEs still dominate the economy, including the upstream energy sector. The government has also undertaken financial and banking sector reforms, with a view to increasing the sector's competitiveness in order to keep pace with the needs of the evolving economy. The independence and surveillance capability of the Central Bank has been increased and joint-venture and public-sector banks consolidated and privatised. Foreign ownership and private-sector involvement in the banking sector is permitted.

Economic Trends and Developments

Egyptian economic growth has been strong in recent years and is expected to stay strong across the projection period. GDP growth in 2004 is estimated to be around 4%. Growth slowed slightly in the wake of the terrorist attacks in New York in September 2001 and terrorist incidents in Egypt, particularly in the tourism sector.

The services sector and productive sector each account for around half of Egypt's GDP. Within the services sector, the Suez Canal plays a notable role, accounting for nearly 3% of total GDP. Within the productive sector, manufacturing, agriculture, irrigation and fishing dominate: together, these contribute around 35% of GDP. The hydrocarbon sector is much less dominant than in many other MENA countries, contributing around 9% of GDP.

Oil exports and Suez Canal revenue are Egypt's main sources of foreign currency, alongside tourism. Despite efforts to free oil for export by encouraging increased domestic consumption of natural gas, oil export revenues are likely to be displaced by gas export revenues in the future, as more natural gas projects, notably in the LNG sector, come on line.

Despite policies aimed at increasing private and foreign investment, introduced over the past decade as a means of stimulating economic growth and employment opportunities, unemployment is still high, especially among the young and among women. Indeed, the rate is rising and is generating "poverty pockets" in certain parts of the country. At close to 11% of a labour force of 27.5 million people in 2004, it is one of the most serious problems confronting the economy. Egypt has a very young population. Nearly 65% of the population is under 30 years of age and more than 30% is less than 15 years of age and will soon join the ranks of those seeking work.

The Egyptian pound was floated in January 2003 and has since depreciated. The overall economic effect has been positive. There has been strong growth in activities that generate foreign exchange, such as oil exports and Suez Canal traffic. The official inflation rate in 2003/04 was 8.1%, but it could in reality be higher as the official consumer price index includes a number of subsidised commodities.

Human development indicators in relation to Egypt have been improving over the last few years and on certain key criteria, scores are higher than both the MENA and world averages (Table 10.2). However, there are still serious problems in terms of gender and income equality. Nearly 23% of the population lives below the national poverty line, measured at 50% of the average household income.[1] Moreover, there are significant differences between northern and southern provinces in terms of health, education and living standards. Combating poverty and inequality is listed as the third objective of the government's Fifth Social Plan (2002/03-2007/08), after education/literacy and employment (African Development Bank and OECD, 2004).

Table 10.2: **Human Development Indicators for Egypt, 2003**

	Life expectancy at birth (years)	**Adult* literacy** (%)	**Combined gross enrolment ratio**** (%)	**GDP per capita** ($ thousand in PPPs and year-2004 dollars)
Egypt	69.8	55.6	74	4.0
MENA***	67.0	64.1	62	5.7
World	67.1	n.a.	67	8.2

* Aged 15 and above.

** Combined gross enrolment ratio for primary, secondary and tertiary schools.

*** Includes Djibouti, the Occupied Palestinian Territories and Sudan, in addition to the countries in the WEO MENA region.

Source: UNDP (2005).

1. According to UNDP (2005), 3.1% of the population lives on less than $1 a day.

There is an extensive system of social subsidies in Egypt. Total government subsidies amounted to 26.3 billion Egyptian pounds in 2004 (roughly 2% of GDP). These cover a variety of sectors, including the energy sector (petroleum products and electricity account for the bulk of the subsidies), the housing sector (E£ 1.4 billion in 2004), medicine/health insurance (E£ 420 million in 2004) and food and staple commodities such as sugar and bread (E£ 2 billion in 2004) (Figure 10.1).

Subsidies are a burden on the budget, and encourage growth in consumption, including energy demand. The consequent rise in oil and gas consumption is pre-empting a growing part of the revenue from exports. The government has in the past year increased the price of diesel by 50%, to reach E£ 0.6/litre, but this is still below the cost. 2004 also saw the introduction of a new "super-premium" petrol, which at E£ 1.4/litre is considerably more expensive than standard gasoline, which costs E£ 1.0/litre.

Figure 10.1: **Social Subsidies in Egypt, 2004**

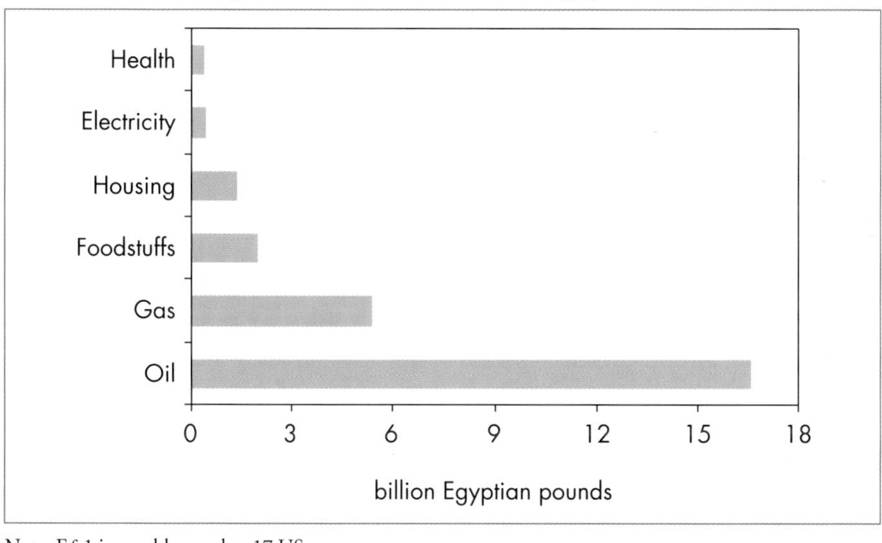

Note: E£ 1 is roughly equal to 17 US cents.
Sources: IDSC (2005); IEA analysis.

In the Reference Scenario, GDP growth is expected to average 3.6% per year over 2003-2030, slower than the 5.3% annual growth seen over the past three decades, as the economy matures and population growth slows (Table 10.3). GDP is set to grow fastest in the current decade, before slowing down progressively over the next two decades.

Table 10.3: **GDP and Population Growth Rates in Egypt in the Reference Scenario** (average annual growth rate in %)

	1971-2003	1990-2003	2003-2010	2010-2020	2020-2030	2003-2030
GDP	5.3	4.1	4.2	3.6	3.1	3.6
Population	2.2	2.0	2.0	1.6	1.2	1.6
Active labour force	2.6	3.0	2.7	2.1	1.9	2.2
GDP per capita	3.0	2.0	2.2	2.0	1.9	2.0

Between 1971 and 2003, the Egyptian population grew at an average annual rate of 2.2%, reaching some 70 million in 2003. The population is expected to continue to grow for some time yet; by 2030 it is projected to have surpassed 100 million. The labour force is expected to grow at an annual average rate of 2.2% over the projection period. Between 700 000 and 800 000 new jobs will be needed each year to keep pace.

The international energy price trends underpinning the Reference Scenario projections for Egypt are summarised in Chapter 1. We assume that energy subsidies in Egypt will *not* be reduced significantly over the projection period.

Energy Policy

The Egyptian government owns most of the country's hydrocarbon resources. Petroleum development in Egypt is based on production-sharing agreements (PSAs). Unlike many of the other countries in the MENA region, Egypt is not a member of the Organization of Petroleum Exporting Countries (OPEC), so is not bound by the organisation's decisions about oil production. A major objective of the government is to increase foreign investment in the oil and gas sector. Egypt has a relatively secure investment climate, but liberalisation efforts are progressing at a slow pace.

The Petroleum Ministry administers the Egyptian hydrocarbon sector. The ministry operates through four state-owned bodies: the Egyptian General Petroleum Company (EGPC), the Egyptian Natural Gas Holding Company (EGAS), the Egyptian Petrochemicals Holding Company (ECHEM) and Ganoub El-Wadi Petroleum Holding Company (Ganope).

The EGPC is responsible for regulating and controlling the petroleum activities of the Egyptian energy sector. This state-owned agency also negotiates licence agreements for oil exploration and production, with the target of achieving national self-sufficiency in oil and gas. The Egyptian government invites regular tenders for exploration acreage. Exploration has increased rapidly over the last few years and the largest find since 1998 was made in 2003 in the Gulf of Suez. Ganope is in charge of all upstream and downstream development in upper Egypt, which has thus far remained relatively unexplored.

Established in 2001, EGAS was formed to manage the development of upstream and downstream natural gas projects. Its importance is rising with the growing activity of the sector in Egypt, triggered by the significant upgrading of natural gas reserves in the last few years.

The government has been seeking to encourage downstream projects, notably in the petrochemical and refinery sectors, in a bid to boost economic growth, create jobs and diversify the country's economic base. A downstream liberalisation programme was initiated in late 2004.

Egypt's electricity sector is dominated by the Egyptian Electricity Holding Company (EEHC), which was created in July 2000 through the transformation of the Egyptian Electricity Authority into a joint-stock (holding) company. EEHC controls thirteen subsidiaries responsible for generation (five companies) and distribution (eight companies) and a Transmission Network Company responsible for operation and maintenance of the network. The development of renewable energy is the responsibility of the New and Renewable Energy Authority (NREA), which was established in 1986 to direct efforts to develop and introduce renewable energy, primarily solar, wind and biomass.

Egypt ratified the Kyoto Protocol in January 2005 and is therefore entitled to participate in clean development mechanism (CDM) projects.

Energy Demand

Primary Energy Mix

Energy demand in Egypt is projected to grow from 54 Mtoe in 2003 to 109 Mtoe in 2030, at an average annual growth rate of 2.6% (Figure 10.2). At 3.4% per year, demand growth will be strongest in the current decade, with total demand reaching 68 Mtoe in 2010. The growth rate will decline to 2.6% and 2.1% over the two following decades. Egypt's energy mix will continue to be dominated by oil and gas, which will account for 95% of primary energy demand in 2030.

Demand for natural gas will grow most strongly, with average annual growth of 3% over the projection period. Towards the second half of the projection period, it will overtake oil to become the dominant fuel in the primary energy mix. Gas will account for 48% of demand by 2030, in line with government policies that encourage fuel-switching for power generation and increased use of gas in petrochemicals, fertilizers and the cement industry.

Half of the 29 Mtoe increase in gas demand projected through to 2030 will be absorbed by power generation, which currently accounts for over 60% of gas demand. The final consumption sector will account for another 31% of the increase. Other transformation, including GTLs and LNG, will account for the remainder. Most of the increase in oil demand will stem from growing transportation use.

Figure 10.2: **Egypt's Total Primary Energy Demand by Fuel**

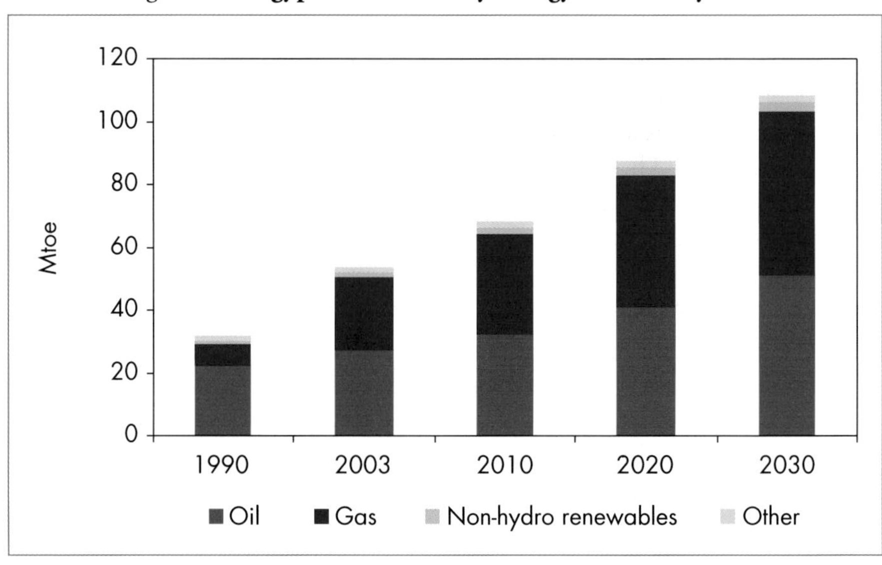

Energy demand will grow less rapidly than GDP, resulting in a fall in energy intensity from 0.20 to 0.16 toe per thousand dollars of GDP. Per capita primary energy demand, at 0.8 toe per person, is lower than the North African average. It is expected to reach 1.1 toe per person by 2030, below the projected North African average for that year.

Sectoral Trends

Total final consumption in Egypt stood at 36.7 Mtoe in 2003. Of this, 22 Mtoe, or 60%, was oil, mainly for transport and industry. These two sectors will continue to dominate final consumption of oil over the *Outlook* period, with over 85% of the increase in final consumption for oil stemming from growing demand in the transport sector. Demand for both electricity and natural gas will more than double over 2003-2030. The share of electricity in total final consumption will remain constant, but that of natural gas will increase from 18% to 20%. This will be at the expense of the share of oil, which will decrease from 60% to 58%. For natural gas, much of the increase is driven by rapidly growing demand in the industry sector, where gas-intensive industries, such as the fertilizer and petrochemical industries, will experience relatively strong growth.

Industry

Industrial energy consumption reached some 14 Mtoe in 2003, accounting for around 40% of total final energy consumption. Gas is the most used fuel, supplying 41% of industry energy needs, followed by oil at 36% and electricity at 17%. By 2030, industrial energy requirements will reach 24 Mtoe. On average, energy demand in the industrial sector will grow by 1.9% per year. However, natural gas and electricity are expected to increase at the faster pace of 2.7% per year. The shift towards these two fuels will be driven by the increasing importance of fertilizers and petrochemicals: a 20-year, $10 billion development programme for the fertilizer and petrochemical industry was announced by the Egyptian government in 2002, with the underlying aim of capitalising on the plentiful natural gas reserves of the country.

Domestic demand for fertilizers is expected to grow strongly. Egypt's per capita area of productive land is only about one-tenth of the global average. Egypt is a net importer of cereals and food, and maintaining steady agricultural output will be of the utmost importance, not least to the trade balance. More efficient technology in both petrochemicals and fertilizers will allow the ratio of industrial energy consumption to grow more slowly than the rise in industrial output. Cement will be the next most important energy-consuming industry over the *Outlook* period.

Transport

Oil demand in transport is expected to grow at 3.7% per annum to 29 Mtoe by 2030, nearly three times the current level. Vehicle ownership levels in Egypt are currently low, at approximately 34 per thousand inhabitants, compared to an average of 36 in developing countries and more than 500 on average in the OECD. During the projection period we expect the number of vehicles to triple, to more than 6 million. Diesel will remain the dominant fuel but is expected to lose some market share to gasoline due in part to a faster increase in cars than trucks. Air pollution concerns, especially in Cairo[2] and

2. Cairo accounts for almost half of the on-road vehicles in Egypt. Particulate matter, lead and sand blown into urban areas from the desert, as well as NO_x and SO_x, make air quality in Cairo a major health issue. According to the World Market Research Centre (2002), Cairo's poor air is responsible for between 10 000 and 25 000 deaths per year.

Alexandria, will support efforts to shift from oil to natural gas through the introduction of compressed natural gas vehicles. Natural gas consumption will nonetheless remain marginal in the transport sector.

Residential and Services

In 2003, the residential and services sectors accounted for around a quarter of total final consumption. Electricity consumption stood at 4.0 Mtoe and oil consumption, mainly in the form of LPG, stood at 3.9 Mtoe. These two fuels accounted for 88% of energy consumption, renewables for another 8% and natural gas for the remainder.

Energy demand in these sectors is projected to nearly double, to 17 Mtoe, by 2030. Gas is expected to grow fastest, at 7.7% per year. The number of customers connected to a natural gas distribution grid, currently almost 2 million, is set to increase considerably over the projection period, allowing natural gas to increase its share from 5% today to 18% in 2030. Natural gas will partly displace LPG. Oil demand will therefore increase at the moderate pace of 0.8% per year.

Electricity demand will increase at 2.8% per year. The increasing importance of tourism and the services sector in the Egyptian economy will spur electricity consumption. In the residential sector, per capita electricity consumption is low at 425 kWh, compared to 1 500 kWh in OECD Europe. Changing patterns of electricity use and increasing use of appliances and of air-conditioning will have a marked effect across the projection period, raising per capita consumption by nearly 45%. Some 98% of Egyptians now have access to electricity, thanks to a 34-year-long government plan directed particularly at rural areas.

Solar energy meets a small part of demand in the residential and services sectors. There are more than 200 000 solar water-heating systems in use in houses, the commercial sector, new cities and tourist villages. The potential is much larger. Solar heat is projected to reach 0.1 Mtoe in 2030.

CO_2 Emissions

In the Reference Scenario, we project an increase in Egypt's CO_2 emissions at an average annual rate of 2.6%, from 122 Mt in 2003 to 151 Mt in 2010 and 242 Mt in 2030. This is considerably slower than over the past three decades, when they grew at 5.7% per year. The transport sector alone will be responsible for 45% of the increase.

Several studies have estimated the potential for introducing clean development mechanism (CDM) projects in Egypt. Estimates of the average marginal abatement cost of CO_2 vary; for wind plants, one study sets the cost at a low €7/tonne of CO_2 (EEEA *et al.*, 2003). This compares very favourably with the current prices in the EU's new CO_2 emissions market, which in early July 2005 passed €29/tonne of CO_2. CDM projects could provide Egypt with an important source of additional finance.

Electricity Supply

Overview

Egypt's electricity generation reached 92 TWh in 2003. Although total electricity generation has been increasing very fast, averaging 8% per year between 1998 and 2003, per capita electricity generation remains quite low, at 1 358 kWh in 2003. Preliminary estimates for 2004 indicate that electricity generation reached 98 TWh.

Nearly 80% of electricity is based on natural gas. In 2003, oil-fired generation accounted for about 6% of the total and hydropower plants for 14%. The share of gas increased substantially in the late 1990s, following substantial foreign investment in Egypt's gas sector and the decline in oil production (Figure 10.3). Power plants consume over 60% of total primary gas supply.

Figure 10.3: **Egypt's Natural Gas Production and Shares of Oil and Gas in Electricity Generation**

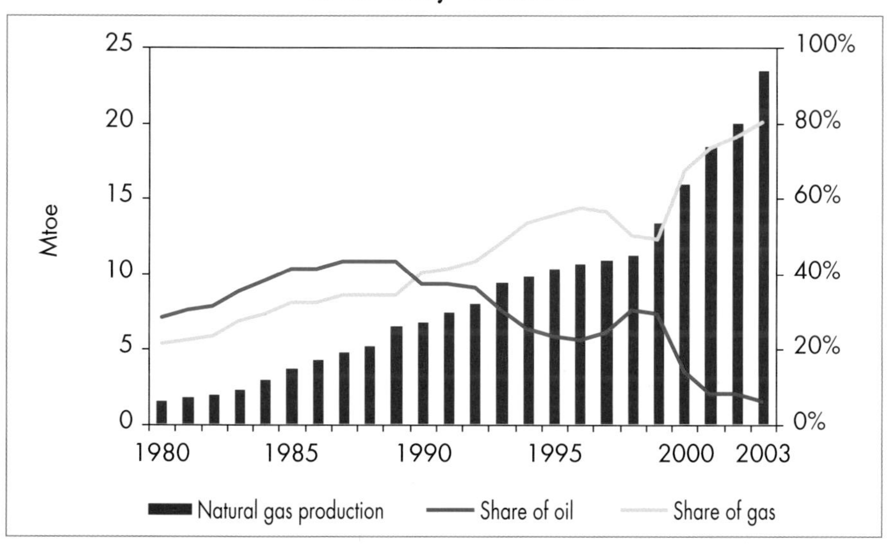

Total installed generating capacity was 18 GW in 2003. Most power plants are gas-fired boilers. There are still few CCGTs in Egypt, despite the rapid growth in gas-fired generation. There are a few hydropower stations, with an installed capacity of 2.7 GW, mostly in Aswan, where dams have been built to control flooding of the Nile River. Egypt had 93 MW of wind power capacity in 2003. This rose to 140 MW in 2004.

The Egyptian electricity grid was connected to those of Libya and Jordan in 1998. The Libya-Egypt interconnection consists of a 220-kV single circuit line that extends 250 km. The Jordanian interconnection involves mainly a 500-kV transmission line crossing the Suez Canal, a 500-kV transmission line across the Sinai Peninsula and a 400-kV submarine cable across the Gulf of Aqaba. There are plans to increase interconnections, notably with the creation of the Mediterranean Power Pool, a project that will connect the power grids of North Africa (Algeria, Egypt, Libya, Morocco and Tunisia), Spain, the Middle East (Jordan, Syria and Iraq) and Turkey. Power exchanges are still small relative to the total size of Egypt's electricity market, but exports have been rising — since 1999, by a factor of six — while imports remained more or less flat. Table 10.4 shows the power exchanges between Egypt, Jordan and Libya in 2003/2004.

Table 10.4: **Egypt's Electric Power Exchanges in 2003/2004**

	Libya	Jordan
Interconnection voltage (kV)	220	400/500/500
Exports (GWh)	213	826
Imports (GWh)	69	53

Source: Egyptian Electricity Holding Company (2003).

Since 1996, Egypt has allowed private sector participation in power generation, through build-own-operate-transfer (BOOT) projects. The law stipulates that independent power producers must sell wholesale electricity to the government-owned power company for a twenty-year period of time and transfer all assets to it at the end of that period. The country's first independent power project, a 683-MW gas-fired steam plant, was commissioned in 2002. Two more BOOT plants, at the Gulf of Suez (680 MW) and at Port Said (680 MW) started operation in 2003.

Electricity tariffs in Egypt remain subsidised. Total subsidies paid amounted to about $450 million in 2003, equivalent to around 0.2% of Egypt's GDP. The average tariff for the residential sector (across all consumption levels) was 1.84 US cents per kWh in 2003 (Table 10.5). These tariffs were increased by about 5% in 2004, the first change since 1992.

Table 10.5: **Egypt's Electricity Tariffs by Sector, 2003**

Sector	US cents per kWh
Residential	1.84
Agriculture	1.14
Commercial	2.93
Industry	0.76 – 2.50

Sources: Egyptian Electricity Holding Company (2003); IEA databases.

Electricity Production

Egypt's electricity generation is projected to increase at an annual rate of 2.7% between 2003 and 2030. Growth will be strong in the period to 2010, climbing by 4% per year, and will slow to 2.5% per year in 2010-2020 and to 2% in 2020-2030.

To meet the projected electricity demand, Egypt will need some 19 GW of new capacity by 2030. Much of it is expected to be based on CCGT technology.[3] A number of projects currently under development are CCGTs. Most new power plants are expected to be gas-fired, taking the share of natural gas in electricity generation from 80% in 2003 to 86% in 2030. The share of oil, which had already fallen to 6% in 2003, is expected to decline further, to just above 2% in 2030. Some existing oil-fired power stations are likely to be retired during the projection period, as they reach the end of their operational life.

Construction of new hydropower plants is expected to be limited so the share of hydropower in total electricity generation will decline from 14% in 2003 to 10% by 2030. Four units with a total capacity of 64 MW are under construction at Naga Hammadi, while in the longer term a pumped storage facility could be developed at Gabal Galala.

Egypt hopes to increase electricity generation from wind and solar power. There is an estimated wind potential of about 20 GW in the Red Sea area. Wind capacity is projected to rise to 1 GW by 2030.

Given its abundant solar resource, Egypt also plans to develop solar-based power generation. A project to build a combined solar thermal and natural gas facility is under way. The power plant will have a total capacity of 127 MW, of which the solar part will account for 31 MW. Egypt plans to develop similar projects in the future with the aid of international grants.

Oil Supply

Resources and Reserves

The geology of Egypt is complex with several, relatively small, hydrocarbon provinces of different origins. This gives rise to a variety of types of reservoir of various ages and lithologies. Egypt can be divided into three main hydrocarbon regions: the Gulf of Suez, the Nile Delta and the Western Desert regions.

Egypt's oil resources have been fairly well explored and development uses relatively advanced oilfield technologies.[4] This sector is largely open to

3. See Chapter 6 for a discussion on the economics of open-cycle and combined-cycle gas turbines.

4. An example is the offshore Scarab-Saffron development that has one of the longest sub-sea tiebacks in the world. With oil production in decline, activity is focused more on gas, as well as deeper, more difficult fields. There have been recent deep-water (2 000 metres water depth) oil developments in the Mediterranean and deep (5 000 metres) high-pressure gas developments.

international companies. Egypt has an estimated 3.7 billion barrels of proven oil reserves, mainly concentrated in the Gulf of Suez and Nile Delta areas. Half of Egypt's offshore resources are in the Gulf of Suez area. Although the government has indicated that the oil in place may be up to 13.2 billion barrels, the figure for proven reserves has not changed for ten years, leaving some doubt about the remaining potential. The US Geological Survey sets undiscovered oil at 3.1 billion barrels.

The focus of drilling has shifted from the Gulf of Suez to the Nile Delta in the Mediterranean, where the success rate has been increasing steadily since 1999. In July 2005, the first major oilfield (El Tamad) in the northern Nile Delta region was discovered, with estimated reserves of 12-14 million barrels of extra light crude oil and 0.6-1.4 billion cubic metres of natural gas.

Crude Oil Production[5]

In 2004, Egypt produced 708 kb/d of oil. Egypt's oil production has been in decline since it peaked at 980 kb/d in 1993, nearly one-third higher than today. This has occurred despite efforts to attract investment and to introduce measures such as enhanced oil recovery. Future oil finds are likely to be small-scale, insufficient to halt the overall decline in production. In our Reference Scenario, production falls to around 670 kb/d by 2010 and 450 kb/d in 2030. Although condensate production grows in line with increased natural gas exploration and output, it is unable to offset the decline. Egypt is expected to become a net oil importer by around 2015. Net imports will reach 500 kb/d in 2030. In our Deferred Investment Scenario, production levels are even lower (see below).

The Gulf of Suez, with nearly 100 oilfields, is currently responsible for half of Egypt's total output (Table 10.6). However, its production has been declining by about 50 kb/d per year since 1993. Other key areas include the Western Desert and the Nile Delta basin. There is also production on the Sinai Peninsula.

In the Reference Scenario, by 2010 the share of currently producing fields in total oil production will have fallen to 60%, whilst fields currently awaiting development will account for over a quarter of total output. As the projection period evolves, reserve additions and new discoveries will constitute an increasing share of total production. By 2030, these categories will account for over 90% of total output (Figure 10.4).

5. All the oil-production figures cited in this section include crude oil, natural gas liquids (NGLs) and condensates.

Table 10.6: **Oilfields in Production in Egypt, 2004**

Field	Proven reserves end-2004 (million barrels)	Production (kb/d)	% of total production
Gulf of Suez		**389**	**55**
Belayim	461	169	24
Ras Gharib	339	37	5
Zeit Bay	77	24	3
East Zeit	33	17	2
Ras Budran	30	14	2
Other Gulf of Suez	517	128	18
Western Desert	**379**	**139**	**20**
Other fields		**180**	**25**
Total		**744**	**100**

Note: Includes NGLs and condensates.
Sources: EGPC and IHS energy databases.

Figure 10.4: **Egypt's Oil Production by Source**

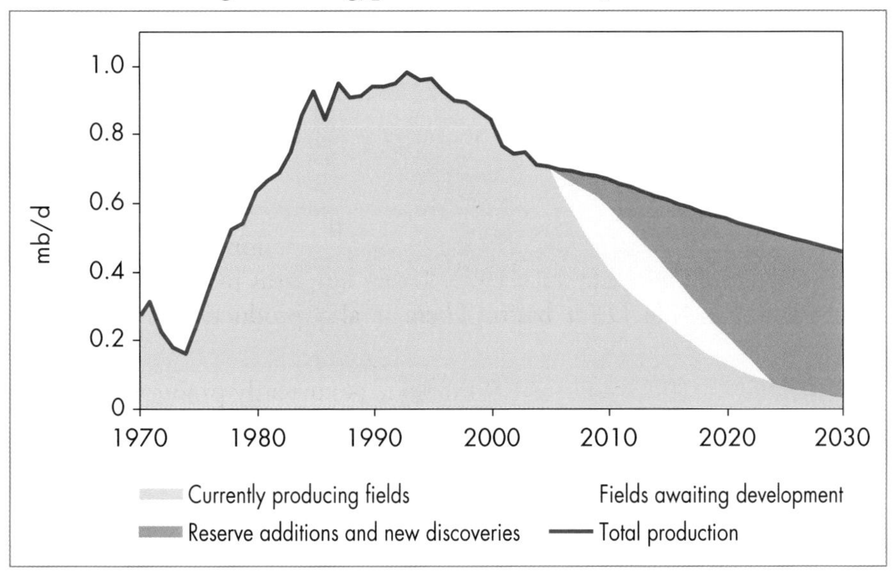

Oil Refining

At the end of 2004, Egypt had nine refineries with a total capacity of about 760 kb/d, making it Africa's largest refining centre. With the exception of the Midor-operated Sidi Krir refinery in Alexandria, each is operated by the

Figure 10.5: **Main Oil and Gas Fields and Energy Infrastructure in Egypt**

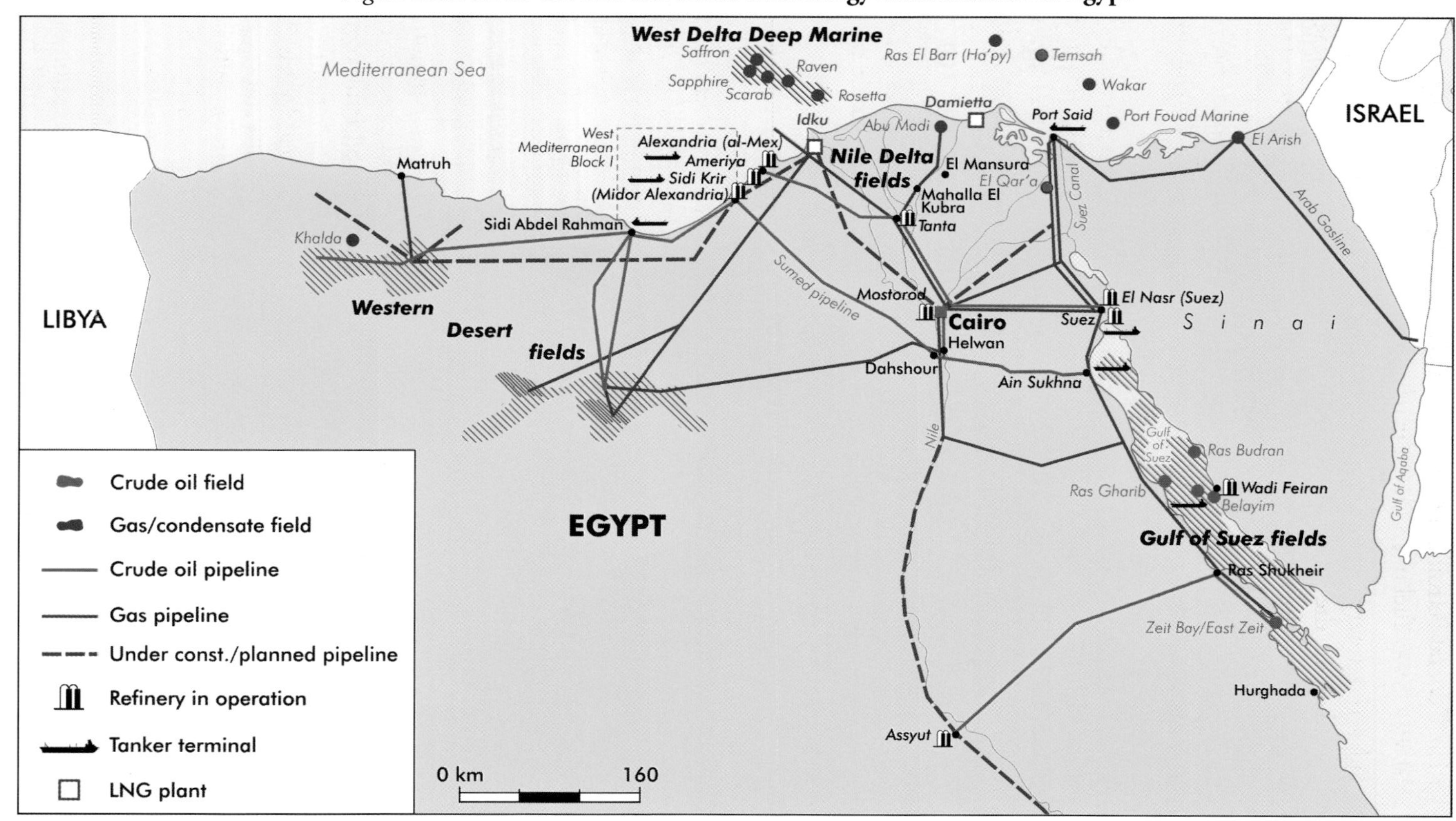

Egyptian General Petroleum Corporation (EGPC). Egypt has a large surplus of relatively heavy fuel oil, which it exports.

The Sidi Krir refinery is Egypt's newest refinery. It was opened in 2001 and was the first plant in the Middle East to comply with new EU environmental standards. Other refineries will comply with these standards as Egypt's refinery sector is upgraded. The first initial public offering in the government's downstream liberalisation programme occurred in June 2005, in the Sidi Krir petrochemical company. Plans have also been announced to sell stakes in the Sidi Krir and Alexandria refineries.

Future refinery expansions will be designed to increase production of lighter products, petrochemicals and higher-octane petrol. In addition to expanding already existing refineries, the government is currently considering building a new 350 kb/d refinery at Port Said and EGPC has plans for a new refinery close to Suez, with capacity of 130 kb/d.

In the Reference Scenario, Egypt's crude oil distillation capacity increases to over 800 kb/d by 2010 and to 1.1 mb/d in 2030. This will meet growing domestic demand and maintain exports of refined products at their current level.

Oil Trade

In 2004 Egypt exported 175 kb/d of oil.[6] The country is experiencing a squeeze on its oil exports: domestic demand has been rising whilst production has been declining. As a result, oil exports have fallen by over 50 % in the past decade. In our Reference Scenario, Egypt will become a net oil importer by around 2015. Net imports will have risen to around 510 kb/d by 2030 (Figure 10.6).

Egypt is nonetheless strategically important in international oil trade as it provides transit routes for most of the oil supplied to Europe from the Middle East. This importance is set to increase in the future as the major oil-consuming countries become even more reliant on imports. Oil passes through Egypt via both the Suez Canal and the Sumed pipeline (Suez-Mediterranean pipeline). The Sumed pipeline links the Gulf of Suez Ain Sukhna terminal with the Sidi Krir terminal on the Mediterranean coast. It is owned by the Arab Petroleum Pipeline (APP) company, in which Egypt has a 50% stake. Other partners include Saudi Arabia, Kuwait, the UAE and Qatar. In 2003, the throughput of the Sumed pipeline was around 2.5 mb/d, mainly of Saudi Arabian oil. Oil transit via the Suez Canal, which was 1.3 mb/d in 2002, is expected to triple from now to 2030 (Box 10.1).

6. Most of this oil is owned by foreign companies operating under production-sharing agreements. They sell around 0.1 mb/d to the domestic market.

Figure 10.6: **Egypt's Oil Balance**

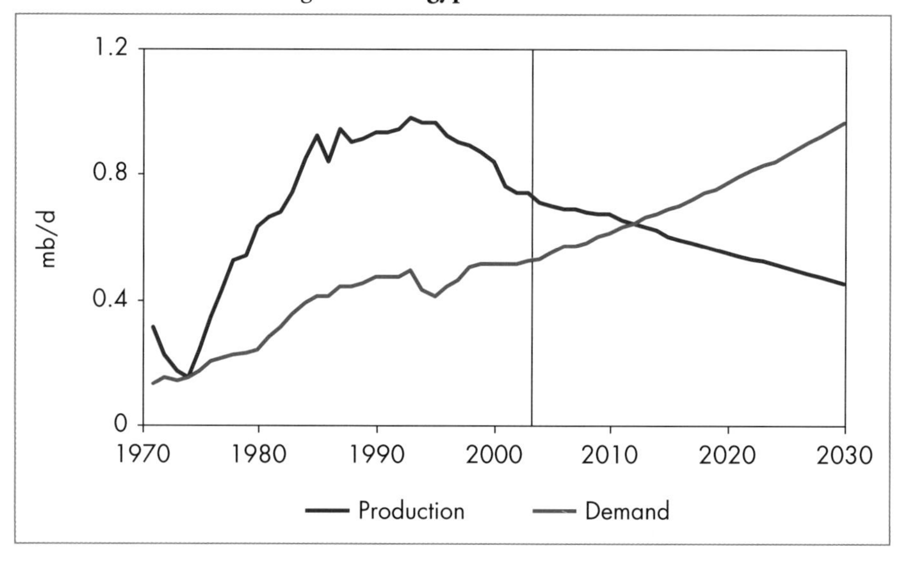

Box 10.1: **The Suez Canal**

The Suez Canal is of great practical and strategic importance to Egypt. Inaugurated in 1869, and allowing direct passage from the Red Sea to the Mediterranean Sea, the canal is an important transit route for oil and LNG from exporting countries in the Middle East to European importing countries. The canal is 192 km long; and with an average passage time of 14 hours (due to be reduced to 11 hours), it is a faster and more fuel-efficient way of transporting oil and LNG to Europe than circumventing the African Cape of Good Hope.

The canal has in the past been insufficiently wide to accommodate the largest types of oil tankers, losing out on these tankers' business to the Sumed pipeline. In order to make the Suez Canal more competitive with pipeline transport, the Suez Canal Authority (SCA) a few years ago embarked on a refurbishment programme that includes widening and deepening the canal sufficiently to accommodate larger tankers. The programme is due to be completed in 2010.

In recent years the Suez Canal has become increasingly important for LNG tankers. LNG carriers are in general smaller than crude oil tankers and therefore do not face the same difficulties in terms of size restrictions. Since 1994, the SCA has offered discounts of up to 35% on canal passage fees for LNG carriers, with greater discounts for those holding heavier cargoes and lower discounts for oil tankers.

Traffic through the canal has been growing steadily. Between 2001 and 2004, the number of vessels using the canal grew on average by 7% per year, and the total tonnage passing through increased at a rate of 11%. Since 2002, oil tonnage has increased faster than total tonnage, rising from 21% of total tonnage in 2002 to 23% in 2004. Suez Canal receipts increased by over 60% between 2001 and 2004 (Figure 10.7). In 2004, these receipts made up 22% of the Egyptian government's total revenues.

Traffic through the Suez Canal is projected to more than double over the projection period, with an even higher rate for LNG carriers, which by 2030 are expected to carry through the canal fifteen times today's volume of LNG. By 2030, 4% of global inter-regional net trade for oil tankers and nearly 9% for LNG trade will pass through the Suez Canal.

Figure 10.7: **Tonnage of Vessels Passing through the Suez Canal and Suez Canal Receipts**

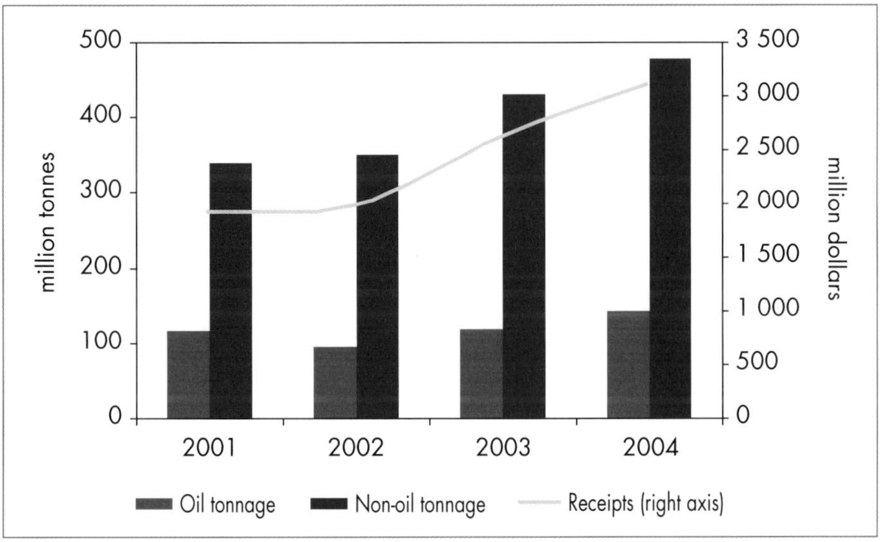

Source: Egyptian Cabinet (Monthly Economic Bulletin, various issues).

Natural Gas Supply

Resources and Reserves

Egypt has proven gas reserves of 1.87 trillion cubic metres (Cedigaz, 2005) or just over 1% of the world total. The success rate of natural gas exploration has

Figure 10.8: **Egypt's Proven Reserves of Natural Gas**

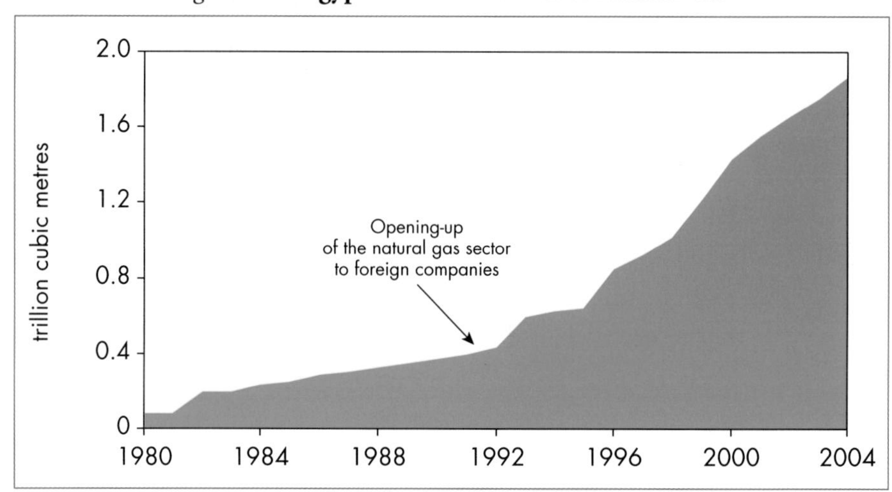

Source: Cedigaz (2005).

increased dramatically since 1991, when foreign companies were first allowed to participate in the Egyptian natural gas sector. As a result, proven reserves have more than quadrupled since 1991 (Figure 10.8).

The Egyptian government estimates probable or possible reserves to be as high as 2.8–3.4 tcm and appears determined to achieve "proven status" for these reserves. The oil minister recently announced plans to increase natural gas proven reserves by 0.85 bcm over the next five years; this will require investment in the vicinity of $10 billion. Significant gas discoveries have been made in Egypt's Western Desert. Egypt has both medium-sized and small fields. The largest include Abu Madi-El Qar'a, Port Fouad Marine area, Raven, Sapphire, Temsah and Wakar.

Gas Production

Egypt was the second-largest natural gas producer in Africa in 2003 (after Algeria), with output of 29 bcm. The West Delta Deep Marine, Port Fouad, Temsah, Ras El Barr (Ha'py) and Rosetta fields made up about 50% of this. The Nile Delta is also an important producing region. Egypt's gas production has increased rapidly over the last decade and doubled in the five years to 2003.

In the Reference Scenario, natural gas production is expected to reach 49 bcm in 2010 and 92 bcm in 2030. Production from West Delta Deep Marine, the Khalda area and Port Fouad are expected to account for nearly half of the production in 2010 and for a quarter in 2020. Development of new fields, from already discovered reserves, will account for two-thirds of production in 2030 (Figure 10.9).

Figure 10.9: **Egypt's Natural Gas Production by Source**

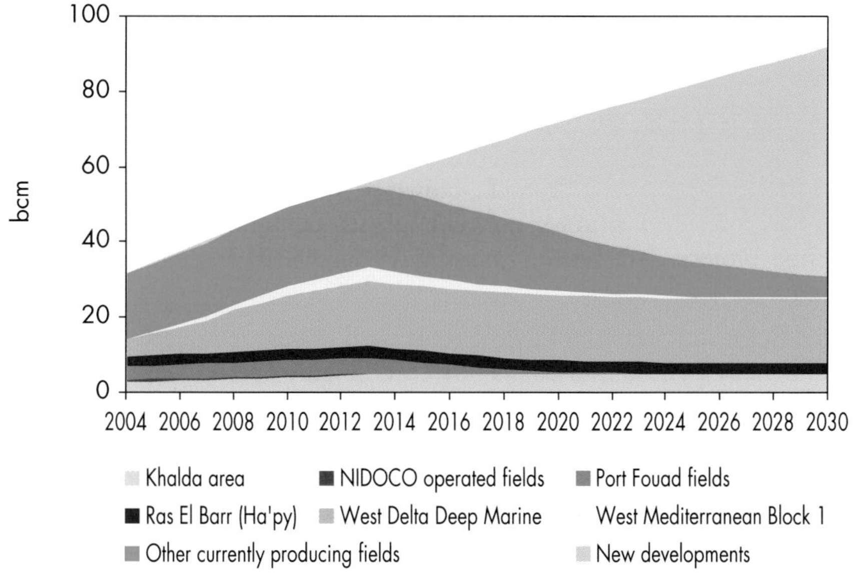

100
80
60
40
20
0
bcm
2004 2006 2008 2010 2012 2014 2016 2018 2020 2022 2024 2026 2028 2030
Khalda area
NIDOCO operated fields
Port Fouad fields
Ras El Barr (Ha'py)
West Delta Deep Marine
West Mediterranean Block 1
Other currently producing fields
New developments

10

Export Plans and Prospects

Egypt became an LNG exporter in 2005 and its LNG production capacity currently stands at 12 Mt/year. The first shipment of LNG from Egypt's Damietta (owned by the Spanish Egyptian Gas, SEGAS) and Idku (owned by the Egyptian LNG, ELNG) LNG terminals took place in January and June 2005, respectively. The LNG Damietta plant has a capacity of 4.8 Mt/y. A second train is under consideration. The Idku LNG terminal consists of two 3.6 Mt/y trains, which were built in separate phases. The second train started production in early September 2005, nine months ahead of schedule (BG Group, 2005a). Gaz de France has signed purchase contracts for the first train production and BG Gas Marketing for the second. Plans for a third train are being considered.

The first phase of the Arab Gasline, which originates in Egypt and stretches over 264 km to reach Jordan, was completed in July 2003 and marked the starting point of natural gas exports from Egypt. There are plans to extend the pipeline to reach Syria and Lebanon, and possibly Turkey. It is expected that the pipeline could eventually deliver a total of 3 bcm per year.

Israel and Egypt signed a Memorandum of Understanding on 30 June 2005 for the construction of a sub-sea gas pipeline linking the two countries. The pipeline will have an annual capacity of around 1.7 bcm and will thus be a significant addition to Egypt's export infrastructure. The agreement is for the supply of 25 bcm of natural gas to Israel Electric Corporation (IEC) over a period of 15 years, with an option to extend the deal for a further five years, and is worth an estimated $2.5 billion.

Another possible production outlet for Egyptian gas could be gas-to-liquids (GTL) projects. Shell International Gas Ltd. and EGPC have signed a joint-venture development protocol for a combined LNG and GTL plant, which would produce 75 kb/d of GTL. However, no final agreement exists. The Reference Scenario assumes no GTL plants are built, because LNG and pipeline projects will remain more cost-effective throughout the *Outlook* period.

In the Reference Scenario, Egypt's net exports of gas are projected to increase to 19 bcm by 2020 and to 28 bcm by 2030 (Figure 10.10), despite increasing domestic gas demand. The new fields to be brought on stream will be more difficult and more costly to develop. International companies might remain in Egypt, given the favourable fiscal conditions, but the opening-up of the upstream market in other MENA countries could increase competition for capital.

Figure 10.10: **Egypt's Natural Gas Balance**

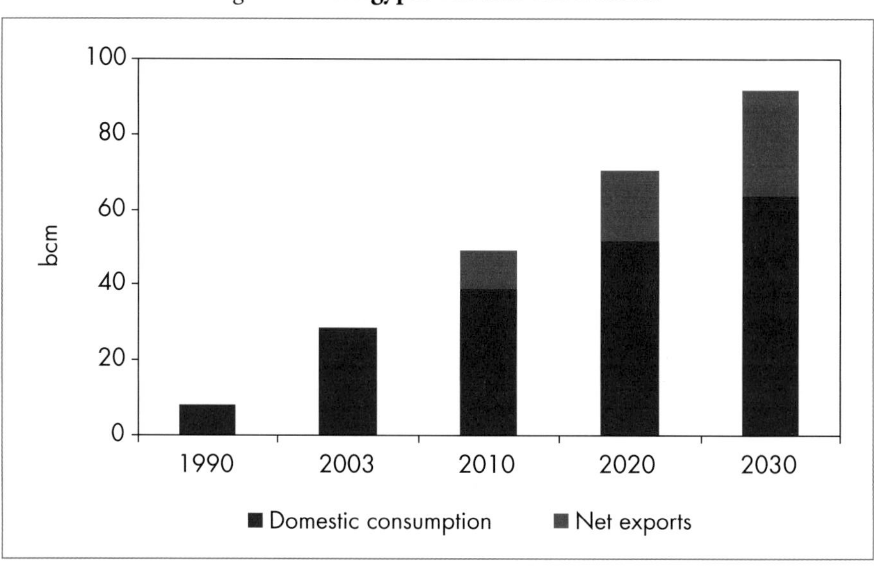

Energy Investment

To meet the Reference Scenario projections, cumulative investment of $85 billion (in year-2004 dollars) is needed in the Egyptian energy sector from 2004 to 2030. Investment needs are expected to be greater towards the second half of the projection period when the most easily accessible reserves will have been more depleted and turning reserves into production will be significantly more difficult and expensive. Investment in the natural gas sector will, alongside the electricity sector, account for the largest share of investment, growing from just under one-third in the first decade to nearly 45% in the third decade (Figure 10.11).

Figure 10.11: **Egypt's Energy Investment**

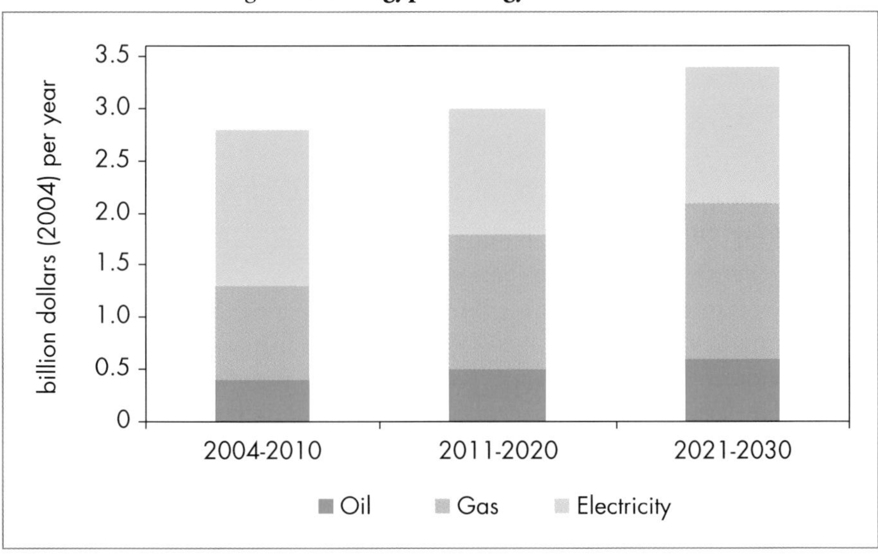

The conditions for private investment in Egypt's energy sector are fairly good and a number of foreign investors, both international majors and small and medium-sized companies, are already actively involved in the Egyptian hydrocarbon sector. Investments in the upstream oil and gas sectors are available through 20-year production-sharing agreements (PSAs) with the national energy companies. EGPC and EGHC are entitled to acquire 50% of any production venture once a production licence has been awarded; and EGPC also has priority to purchase the oil or gas produced. For successful exploration projects, cost recovery is in general available on up to 40% of production costs and 100% of operating expenses (EGPC, 2004). In the downstream sector, investors are entitled to hold up to a 100% share in companies. Egyptian investment laws ensure equality of treatment and full profit transfer. The Egyptian government also offers a number of incentives, such as a time-limited exemption from corporation tax.

Egypt came third after Libya and Algeria in a recent survey of countries preferred for exploration by international companies. This is an indication that funds and investment are likely to be available.

In the oil sector, investment needs will total $14 billon across 2004-2030, $9 billion of which will be needed in the exploration and development sector and the rest in the refining sector. In the refining sector over $3 billion will go towards adding capacity while the rest will be needed for conversion of existing facilities.

Investment needs in the natural gas sector are projected to total $35 billion. Of this, nearly 65% will go to exploration and development activities and nearly 15% to liquefaction. The rest will go to transmission, distribution and storage. Project finance for export-oriented projects can be obtained from a variety of sources. The Egyptian LNG project at Idku readily found available financing for its first train from Eygptian and international commercial banks, as well as from the European Investment Bank. The second train will be financed by a similar group of investors.

Egypt's power sector is expected to require $36 billion of investment in power generation, transmission and distribution over the period 2004-2030. Most of this investment will be in networks ($7 billion in transmission and $16 billion in distribution). How new projects will be financed is unclear at the moment. Pressured by large investment needs, Egypt has permitted private participation in its power sector. With ample domestic funds now available, EEHC is now financing its own projects, given the current large availability of domestic funds. However, the size of investment needed in the power sector suggests that the country may have to turn again to the private sector or undertake deeper reforms.

Deferred Investment Scenario[7]

Given its limited oil resources and the fact that it looks poised to become a net oil importer by 2010, Egypt is not likely deliberately to reduce investment in the hydrocarbon sector, making the country even more dependent on imports. Indeed, the government has every interest in fostering an attractive inwards investment climate, by creating the right fiscal and logistical conditions for potential investors. However, investment could shift to countries with larger

7. See Chapter 7 for a detailed discussion of the assumptions and methodology underlying the Deferred Investment Scenario.

fields and more easily exploitable resources, as well as better investment terms. This could give rise to the conditions assumed in our Deferred Investment Scenario, where investment, and the ensuing oil and gas production, are lower than in the Reference Scenario.

Energy Demand

Given the relatively low dependence of the Egyptian economy on the oil and gas sector, GDP and domestic energy demand in the Deferred Investment Scenario are only marginally lower than in the Reference Scenario. Total primary energy demand in the Deferred Investment Scenario grows at 2.5% per year across the projection period, compared with 2.6% in the Reference Scenario. Demand is projected to reach 105 Mtoe in 2030, just 4 Mtoe or 3.8% lower than in the Reference Scenario. The bulk of this decrease is accounted for by a drop in the demand for oil, with gas still becoming the leading primary fuel by the end of the projection period.

Oil and Gas Production

In the Deferred Investment Scenario, upstream oil investment in Egypt over 2004-2030 is on average 10% lower than in the Reference Scenario, resulting in a faster decline in oil production and slower growth in associated gas production. In terms of oil production, the decrease will be more noticeable towards the end of the projection period; at 620 kb/d, production will be 7% lower than in the Reference Scenario in 2010 and at 380 kb/d, it will be 74 kb/d, or 19%, lower in 2030. Egypt will have become a net importer of oil a few years earlier than in the Reference Scenario. Since Egypt's gas reserves are mainly oil-associated, cuts in oil investment and production will affect production of natural gas as well. Natural gas production is projected to reach 46 bcm in 2010 (7% less than in the Reference Scenario) and 78 bcm in 2030 (15% less).

Oil and Gas Trade

Oil and gas trade is significantly affected in the Deferred Investment Scenario. Egypt becomes a net oil importer earlier and on a more substantial scale, with imports reaching 525 kb/d by 2030, 3.2% higher than in the Reference Scenario. Natural gas exports are dramatically lower, because of both lower production and slower demand growth in Europe and North America. Gas exports reach 16 bcm in 2030, 12 bcm or 43% less than in the Reference Scenario (Figure 10.12)

Figure 10.12: **Egypt's Natural Gas Balance in the Reference and Deferred Investment Scenarios**

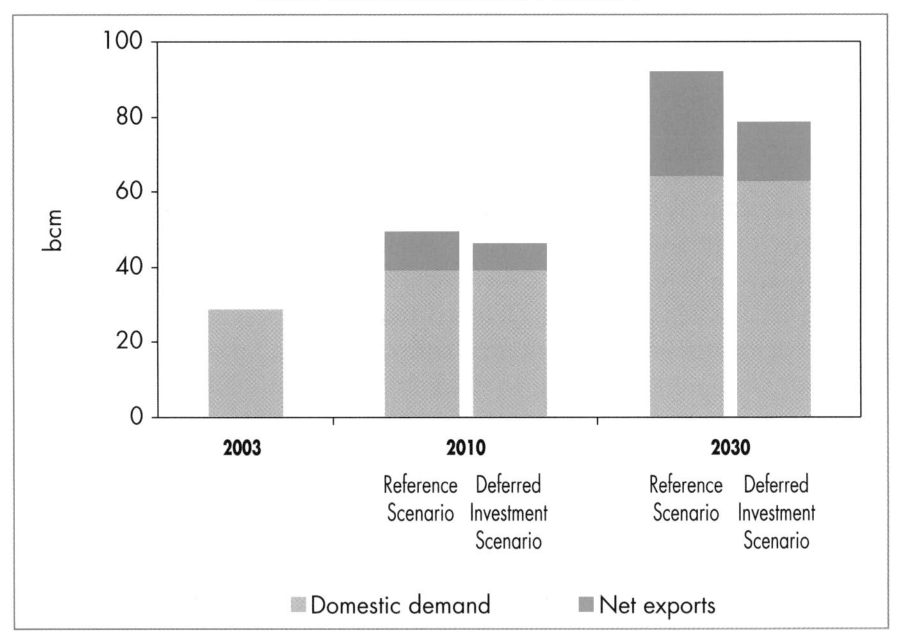

IRAN

HIGHLIGHTS

- The Islamic Republic of Iran has by far the largest population in the Middle East. The oil and gas sector currently accounts for some one-fifth of GDP, close to the Middle East average. The main challenges facing the government are reducing unemployment, combating inflation, removing energy subsidies and optimising the development of its vast oil and gas resources.
- Primary energy demand in Iran is projected to increase at an average annual rate of 2.6% in 2003-2030, down from around 5% over the past decade. This assumes the progressive removal of energy subsidies, now equivalent to a staggering 10% of GDP.
- Iran's oil reserves are the second-largest in the Middle East, after Saudi Arabia. Production is concentrated in a few giant fields with high decline rates. In the Reference Scenario, oil production is projected to grow from 4.1 mb/d in 2004 to 4.5 mb/d in 2010 and to 6.8 mb/d in 2030. Oil exports grow from 2.7 mb/d in 2004 to 4.4 mb/d in 2030. Higher output and exports hinge on access to foreign technology and capital: cumulative investment needs will amount to around $80 billion in 2004-2030.
- Iran holds the second-largest natural gas reserves in the world, almost half of which are in the super-giant South Pars field. In the Reference Scenario, gas production is expected to grow to 110 bcm in 2010 and to 240 bcm in 2030. Iran, currently a net importer of gas, will see its gas exports to Europe and Asia reach 57 bcm by 2030. These trends call for cumulative gas investments of $85 billion over 2004-2030.
- Electricity generation will increase from 153 TWh in 2003 to 359 TWh in 2030, requiring 54 GW of new generating capacity and total investment in power infrastructure of $92 billion. Mobilising this investment in a timely manner will be difficult unless the subsidies in place today are removed.
- In the Deferred Investment Scenario, crude oil production reaches 4.9 mb/d in 2030 – only 0.8 mb/d higher than current levels and 1.8 mb/d less than in the Reference Scenario. Oil demand changes slightly compared to the Reference Scenario, so oil exports remain flat at about 2.6 mb/d. Natural gas exports reach only around 30 bcm in 2030 – little more than half those in the Reference Scenario.

Overview of the Iranian Energy Sector

Iran has immense oil and gas reserves, but it is less able than other Middle Eastern countries to capitalise on them because of barriers to foreign investment and heavy subsidies. Under-pricing of energy has left a legacy of energy inefficiency, environmental degradation and excessive fuel dependence – problems that the government is now trying to address. Iran is a net gas importer, even though it holds the second-largest gas reserves in the world. Oil exports were 2.6 mb/d in 2003, down from 4.3 mb/d in 1971, and now represent a much smaller share of production (Table 11.1).

The oil and gas sector is controlled by the state-owned National Iranian Oil Company (NIOC). Foreign companies have recently been allowed access to the sector, but only through Iranian affiliates. The major oilfields in Iran are in decline and large investments are needed in enhanced recovery and to develop new fields. Exploiting the massive gas reserves will also require large inflows of foreign capital and technology.

Table 11.1: **Key Energy Indicators for Iran**

	1971	**2003**	**1971-2003***
Total primary energy demand (Mtoe)	19	136	6.3%
Total primary energy demand per capita (toe)	0.6	2.1	3.7%
Total primary energy demand /GDP**	0.11	0.28	3.0%
Share of oil in total primary energy demand (%)	84	48	–
Net oil exports (mb/d)	4.3	2.6	–1.5%
Share of oil exports in production (%)	93	65	–
Share of gas in total primary energy demand (%)	12	50	–
Net gas exports (bcm)	0	–2.3	–
CO_2 emissions (Mt)	41	349	6.9%

* Average annual growth rate.
** Toe/thousand dollars of GDP in year-2004 dollars and PPPs.

Total primary energy demand was 136 Mtoe in 2003 and per capita consumption 2.1 toe. The share of oil in primary energy demand was about 50%, down from 84% in 1971. Through heavy subsidies, the government has favoured the domestic use of gas to free up oil for export. Consequently, the share of gas in energy demand has increased considerably in recent years. In 2003, oil production was 4.0 mb/d and gas production was 78 bcm. CO_2

emissions have risen somewhat faster than primary energy demand on an average annual basis over the past three decades, reflecting a drop in the share of renewables.

Political and Economic Situation

Political Developments

The Islamic Republic of Iran was created in 1979, following a revolution that overthrew the monarchy of Mohammad Reza Shah. An eight-year war with Iraq began the following year, resulting in heavy Iranian casualties, massive infrastructural damage and a slump in oil production and exports. The revolution and war eventually led to Iran's isolation from the international community.

In the 1990s, the economy steadily recovered and political institutions were reformed along conservative lines. The elected parliament, the Majlis, rules alongside the religious leaders, such as the Grand Ayatollah Khamenei and the Council of Guardians, to ensure the compatibility of key legislation with the constitution. In August 2005, Mahmoud Ahmedinejad became president of the Republic.

In recent years, especially under the presidency of Mohammad Khatami, Iran stepped up efforts to attract foreign investment, in particular in the energy and telecommunication sectors. A foreign investment bill was introduced in 2002, paving the way for foreign ownership of up to 100% of non-oil industries. The law clarified the legal framework for investment, although uncertainties and restrictions remain.

In early 2004, the Majlis adopted a five-year plan for 2005-2009[1] calling for privatisation and economic reforms (Box 11.1). But the foreign investment climate may be soured by recent international unease over Iran's nuclear development plans. For investment inflows to increase markedly, a resolution of the nuclear issue would be required. Investment in Iran has also suffered from a 1995 US Executive Order banning US firms from involvement in the Iranian hydrocarbons sector. In addition, US law provides for sanctions on non-US companies investing more than $20 million in the sector. The Iranian government has applied to join the World Trade Organization and, in May 2005, the WTO established a working party to examine the application.

1. The Iranian fiscal year ends on 20 March of the Gregorian calendar. The Iranian year 1384 started on 21 March 2005.

Box 11.1: **The Fourth Five-Year Development Plan in Iran**

The fourth five-year plan covering the period from 2005 to 2009 sets out very ambitious economic targets (Table 11.2):

- GDP is targeted to increase by 8% per year, resulting in a decline in unemployment from the current 10.4% to 8% by 2009. Inflation is expected to be reduced to 10%, from 13% in 2004 (CBI, 2004/2005).
- Achieving these targets will require an estimated $387 billion of investment – $356 billion from domestic sources, and $31 billion from abroad. Foreign direct investment would need to increase tenfold compared with 2004. Non-oil exports are targeted to increase by 11% per year.
- The industry and mining sectors are expected to grow by 11.2% per year, increasing their share in GDP from 14% today to 16% by 2009. Priority will be given to the development of energy-intensive industries, to which the plan allocates up to $9 billion.
- The plan calls on the government to privatise all enterprises not covered by article 44 of the constitution.[2]
- Energy subsidies, which now represent 10% of GDP, will be reduced, falling to 1.7% of GDP in 2009.

Table 11.2: **Long-Term Macroeconomic Objectives of the 4th Plan in Iran**

	2004	2009	2015
GDP ($ billion in year-2004 dollars and PPPs)	480	705	n.a.
Unemployment rate (%)	10.4	8.0	7
Oil exports (million $)	22 553	25 265	31 433
Non-oil exports (million $)	7 720	12 817	32 098
Oil production (mb/d)	4.1	5.1	n.a.
Gas production (bcm)	81	91	n.a.

Economic Trends and Developments

Recent Economic Performance

GDP grew by 6.6% in 2004. Per capita income, on the basis of purchasing power parity, is around $7 300, comparable to that in Brazil. Recent economic growth has largely been stimulated by higher oil production and prices, as well

2. Under article 44 of the Iranian constitution, all large-scale industries, foreign trade, banking, insurance, energy and physical infrastructure, radio and television, postal and telephone services, and mass transportation belong to the state.

as by rising private investment flows. Nonetheless, economic growth was slightly lower than the government target under the third five-year development plan covering 2000-2004. Growth of the non-oil sector was also strong in 2004, at 7%, mainly due to higher oil revenues.

The hydrocarbons sector currently accounts for 22% of GDP, in line with the Middle East average (IMF, 2004b). Oil and gas export revenues were $33 billion in 2004, or 80% of total export earnings. Over the past decade, Iran has successfully reduced its dependence on oil production and exports, but how best to use oil revenues to promote growth and to further diversify the economy remains a major challenge.

The oil and gas sectors still provide over half of central government revenues. Fluctuations in oil revenues lead to cyclical changes in GDP. In 2000, the Iranian government established the Oil Stabilisation Fund in order to manage oil revenues better and reduce this cyclical impact (Box 11.2). Government spending is very high and is mainly directed to public-sector investment, to subsidies for essential goods like bread and gasoline and to repayment of the external debt, which amounted to $14 billion in 2004.

Box 11.2: **The Oil Stabilisation Fund in Iran**

The Oil Stabilisation Fund was established in December 2000 to cushion the government budget from oil-price fluctuations. The third five-year development plan for 2000-2004 established a ceiling, based on the expected oil price, on the oil export revenues that could be transferred to the budget. Oil revenues in excess of the budgeted amount are transferred to the fund. If the oil revenues are lower than the budget allocation, the central bank uses the fund to compensate for the shortfall. The fund may also be used to cover budget deficits arising in other ways or to finance a variety of additional public-sector projects and to import oil products.

The effectiveness of oil funds in providing protection against price fluctuations depends on the structure of the fund and how it is managed. Ideally there needs to be a clear separation between that part of the fund – meant to buffer price fluctuations – and long-term savings. The Iranian fund does not appear to have this clear separation, and sometimes the fund has been used to top up budget spending. Management of the fund is most crucial during periods of high oil prices, as profits should be saved instead of used for current government expenditures. Since its creation, withdrawals from the fund have been higher than budgeted (Table 11.3). The government has drawn systematically upon the fund to pay for public consumption.

Table 11.3: **Iran's Oil Stabilisation Fund** (million dollars)

	2000-01	2001-02	2002-03	2003-04
Oil revenue	24 280	19 339	22 966	27 355
Budgeted use of oil revenue	11 731	12 864	11 058	11 579
Actual use of oil revenue	14 726	15 279	17 800	20 949

Sources: IMF (2004b); Central Bank of Iran (2004/2005).

The economy expanded fivefold between 1960 and 1976. During this period, Iran had one of the fastest growing economies in the world. This growth was a result of domestic political stability, low inflation and rapid growth in oil production. The economy stagnated from 1977 to 1988, as a result of unrest before and during the 1979 revolution and the eight-year war with Iraq. Oil output and revenue plummeted. Economic growth recovered after 1989, averaging 5% per year in the fourteen years to 2003.

In the past decade there has been growing internal acceptance of the need for economic and social reforms and increased transparency. Actual progress with reform has been patchy and intermittent. Some major economic reforms, including exchange rate unification[3] and trade liberalisation, have been implemented. Other more contentious reforms, including removal of energy subsidies and a greater role of the private sector in the economy, are lagging behind. Although restrictions on foreign direct investment have been relaxed, the climate remains difficult because of bureaucracy, opaqueness and the relatively unattractive rate of returns that has been offered.

The key challenges facing Iran today are high unemployment, rising inflationary pressures and large energy subsidies. Unemployment declined to 10.3% in 2004/2005, from 14.1% in 2001, due to an increase in public-sector jobs and to a government-sponsored low-interest loan programme for small and medium-sized enterprises. However, the workforce is expected to double by 2030, due to population growth and the rising participation of women, which is now less than 15%. Some 700 000 new jobs will need to be created every year over the projection period just to prevent unemployment from rising. This will require an increased role for the private sector and higher-quality investment (IMF, 2004a).

A striking feature of the Iranian economy is its very high investment to GDP ratio. The ratio was very high in 1960-2002, more than 30% on average,

3. From 1997 to 2001, Iran had a multi-exchange rate system; one of these rates, the official floating exchange rate, by which most essential goods were imported, averaged 1 750 rials per US dollar; in March 2002, the system was converged into one rate.

compared to 26% in China and to 22% in Indonesia. The relatively poor efficiency of investment has nonetheless left Iran's physical infrastructure in need of upgrading and modernisation.

The current five-year plan mentions the removal of energy subsidies as one of the reforms to be introduced. Direct subsidies account for 25% of government spending and 10% of GDP. They contribute to the large share of government spending in GDP. However, the Majlis declared in 2005 its intention not to raise domestic energy prices.

Macroeconomic and Demographic Prospects and Assumptions

In the Reference Scenario, GDP is assumed to grow by an average of 4.5% per year in 2003-2010, slowing to 3.4% in 2010-2020 and 3% in 2020-2030 (Table 11.4). Rising oil and gas revenues will account for a significant proportion of this economic expansion, particularly after 2010, and will help to stimulate an expansion of the non-hydrocarbon sector. An increase in the size of the active labour force and rising labour productivity will also boost non-hydrocarbon GDP. The labour force is projected to grow at an average rate of 2.8% per year over the projection period. The population will grow from 66 million in 2003 to 90 million in 2030, an average annual growth rate of 1.2%. Currently almost 60% of the population is under 25 years old, but this share is expected to decline to 35% by 2030 (UNDESA, 2004).

Table 11.4: **GDP and Population Growth Rates in Iran in the Reference Scenario**
(average annual rate of change, in %)

	1971-2003	1990-2003	2003-2010	2010-2020	2020-2030	2003-2030
GDP	3.3	4.6	4.5	3.4	3.0	3.6
Population	2.6	1.5	1.3	1.4	0.9	1.2
Active labour force	3.1	3.1	3.5	2.9	2.3	2.8
GDP per capita	0.7	3.0	3.2	2.0	2.2	2.3

The international energy price trends assumed in the Reference Scenario are summarised in Chapter 1. In Iran, domestic prices in the Reference Scenario are assumed to increase gradually to international levels in the period to 2015, as energy subsidies are phased out.

Energy Policy

Currently, overall responsibility for the Iranian energy sector lies with the Supreme Energy Council, a supervisory body established in 2001. The Ministry of Petroleum controls the activities of all state-owned oil and gas companies from upstream to petrochemicals. The state-owned National Iranian Oil Company (NIOC) has 18 subsidiaries that cover the entire oil and gas chain.

Under the Iranian constitution, foreign companies cannot be granted rights to the country's natural resources. However, the government has developed a form of contract, the buy-back, which allows foreign contractors, operating through an Iranian affiliate, to participate in exploration and development with NIOC. The government accepts that foreign investment is needed to maintain current oil production levels in fields with high decline rates and to achieve the ambitious gas-development programme, which would allow Iran to become a major net exporter. The government also plans to diversify its economic base away from upstream oil and gas, among other things through LNG and petrochemical projects.

Iran has been a member of OPEC since 1960 and is currently the second-largest OPEC producer, after Saudi Arabia. Iran's role has evolved throughout the years as its production capacity has not increased as fast as expected over the past two decades.

The power sector is run by the state-controlled company, Tavanir. Power-plant construction is handled by the Iran Power Development Company, a wholly-owned subsidiary of Tavanir. Eventually, Tavanir may be broken up into smaller companies as part of a privatisation package. In addition to power generation, Tavanir is responsible for transmission and distribution.

The government is increasingly concerned about the high growth in domestic energy consumption and recognises that consumption must be curbed in order to free up oil for export and to mitigate environmental problems. The current five-year plan gives priority to increasing energy efficiency in buildings, to improving vehicle efficiency and to expanding the use of vehicles fuelled with compressed natural gas (CNG). The price of CNG will be set at 40% of the price of gasoline, on a calorific-value basis.

Last year, Iran expressed its interest in ratifying the Kyoto Protocol. It was suggested that its accession to the United Nations Framework Convention on Climate Change (UNFCCC) would follow Russia's ratification. The Majlis approved the ratification, but it was rejected by the Council of Guardians.

Energy Subsidies

Energy subsidies are a heavy burden on the central budget. Domestic energy prices are well below production and opportunity cost.[4] Energy subsidies were

4. The opportunity cost of a barrel of oil is defined as the border price of oil. Each marginal barrel not consumed domestically could be exported at a higher price. The difference between the domestic price and the export price represents the welfare loss of the economy, and therefore the implicit subsidy.

introduced as a social policy to support low-income households. In 2003, energy subsidies amounted to IR 138 trillion ($16.5 billion), or 10% of GDP based on the market exchange rate. This share has been stable over the past decade. The World Bank estimates that energy subsidies in Iran are the highest in the world, in both absolute and relative terms (World Bank, 2003).

Subsidies on oil products account for almost two-thirds of total subsidies (Table 11.5). Subsidy rates vary between 62% and 98%, LPG being the most subsidised fuel. The transport and residential sectors are the most heavily subsidised, accounting respectively for 31% and 25% of total subsidies. (Figure 11.1)

Table 11.5: **Energy Subsidies by Fuel in Iran, 2003**

	Unit	**Domestic price**	**FOB price**	**Subsidy rate (%)**	**Subsidies (billion rials)**
LPG	Rials/litre	29	1 336	98	4 669
Gasoline	Rials/litre	650	1 764	63	20 481
Kerosene	Rials/litre	160	1 621	90	14 185
Diesel	Rials/litre	160	1 580	90	35 598
Fuel oil	Rials/litre	80	1 213	93	16 247
Natural gas	Rials/cubic metre	80	418	81	24 511
Electricity	Rials/kWh	126	334	62	22 443
Total					**138 134**

Figure 11.1: **Energy Subsidies by Sector in Iran, 2003**

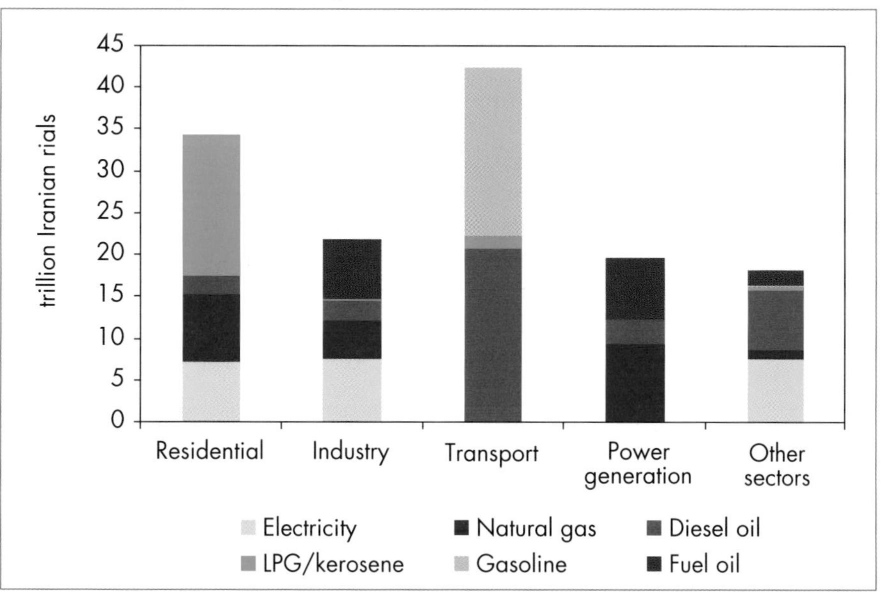

These price distortions have hampered the uptake of energy-efficient appliances, cars and industrial equipment and processes. Energy consumption per unit of output of new equipment is typically higher than in OECD countries (Figure 11.2). Subsidies have also led to high energy intensity, restricted exports, boosted fuel imports and greater pollution. Although the government recognises the benefits of removing energy subsidies, especially for oil products, political obstacles and concerns over distributional effects hinder reforms. Some 7% of the population lives on less than $2 a day. Without a comprehensive compensation package, the removal of energy subsidies would have a regressive effect, especially damaging the poorest part of the population.[5] The removal of energy subsidies would have a major impact on energy demand (see Box 11.3).

Figure 11.2: **Energy Consumption per Unit of Output of New Equipment in Iran and the OECD**

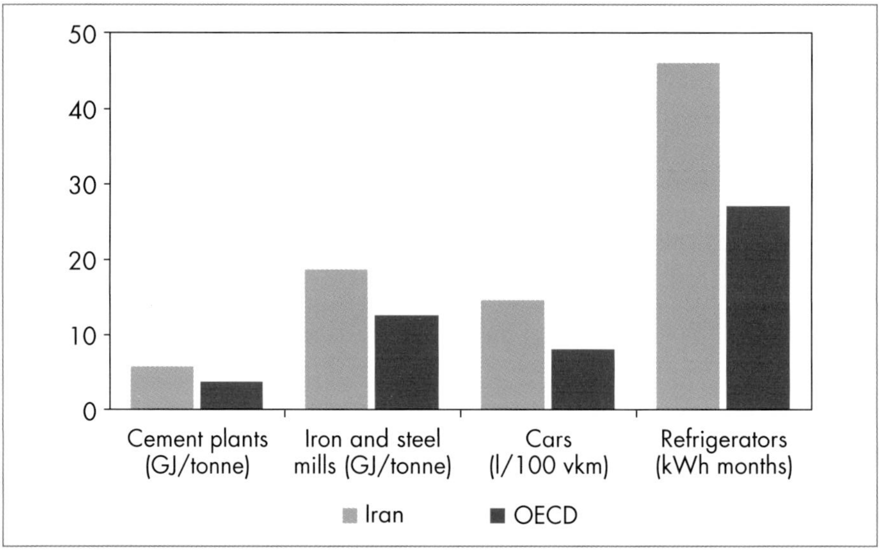

50
40
30
20
10
0
Cement plants (GJ/tonne)
Iron and steel mills (GJ/tonne)
Cars (l/100 vkm)
Refrigerators (kWh months)
Iran
OECD

Energy Demand

Primary Energy Mix

Total primary energy demand in Iran was 136 Mtoe in 2003, comparable to that of Spain. Primary energy use has climbed steadily since the Iran-Iraq conflict, averaging 5.6% annual growth in 1989-2003. Increasing incomes, boosted by crude oil export revenues, population growth and urbanisation were

5. The impact of energy subsidies removal on inflation is uncertain. See Majlis Research Centre (2004) and Von Molkte *et al.* (2004).

the main drivers of demand. Per capita energy consumption was 2.1 toe in 2003, four-fifths of the Middle East average and comparable to per capita energy demand in Venezuela. Energy intensity was 0.3 toe per thousand dollars of GDP, 30% higher than the average in OECD countries – mainly because of low efficiency.

Oil and gas accounted for 98% of primary energy demand in 2003. Renewables, mostly hydro, and coal make up the remaining 2%. The fuel mix has changed markedly over the past three decades. Oil accounted for 48% of the fuel mix in 2003, down from 84% in 1971. The government, seeking to free up crude oil for export, has strongly encouraged the substitution of natural gas for oil in all sectors, partly through subsidies. The share of natural gas in the fuel mix jumped from a mere 12% in 1971 to 50% in 2003.

Primary energy demand is projected to double to 270 Mtoe in 2030 in the Reference Scenario. Growth is expected to be most rapid from now to 2010, at 3.4% per year, and then to slow to 2.7% per year in 2010-2020 as the pace of economic expansion slows. Natural gas consumption is expected to grow strongly at 3% per year, boosting its share to 55% in 2030. The power generation and the residential sectors combined will account for two-thirds of incremental gas use. Use of gas in producing LNG and GTL – new activities that will emerge over the projection period – will account for a total of 28 Mtoe in 2030, about the same amount as for petrochemicals. The share of oil in primary demand will fall by seven percentage points, to 41%, in 2030. The contribution of renewables, coal and nuclear is expected to be marginal (Figure 11.3). More rapid removal of energy subsidies than assumed in the Reference Scenario would dampen demand growth markedly (Box 11.3).

Figure 11.3: **Iran's Primary Energy Demand**

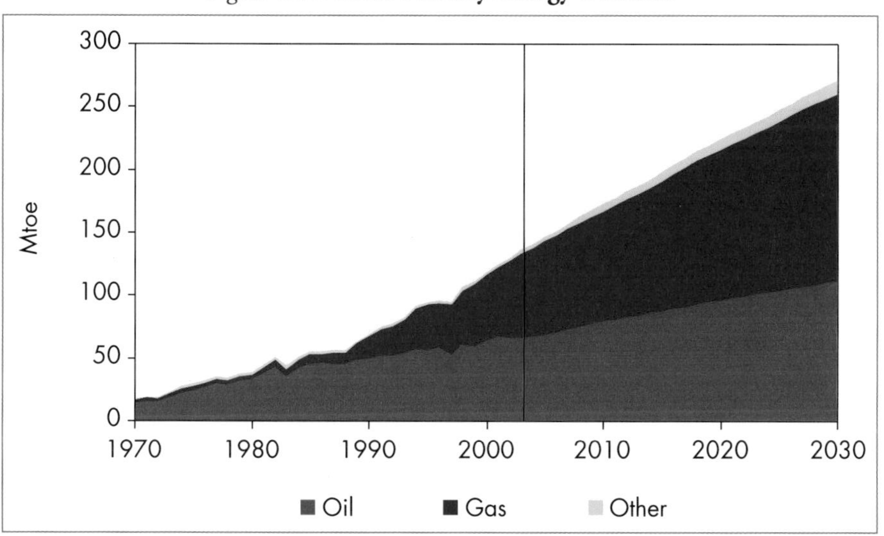

Box 11.3: **Impact on Iran's Energy Markets of Removing Subsidies in 2006**

In the Reference Scenario, subsidies are assumed to be gradually eliminated in the period to 2015. While recognising the political difficulties in implementing this proposition, we also analysed the effect on energy demand of a complete removal of all energy subsidies in 2006. Oil demand would then be 6.2% lower in 2010 and 6% lower in 2030 compared with the Reference Scenario (Figure 11.4). Reduced gasoline and diesel consumption in the transport sector accounts for the bulk of the reduction. Gas consumption would be 16% lower in 2010 and 13% in 2030. Electricity consumption – and gas inputs to power generation – would also be lower.

The oil and gas saved would be available for export. Oil exports would increase by some 150 kb/d, or 3% in 2030. Gas exports would be 20 bcm, or 30%, higher. As a result, oil and gas export revenues would be some $140 billion higher over the period 2006-2030. These revenues could be used to compensate low-income households for higher energy prices. Lower energy consumption would reduce local pollution; and CO_2 emissions would fall by as much as 65 Mt in 2030 compared with the Reference Scenario.

Figure 11.4: **Iran's Primary Energy Demand in the Reference Scenario and the Subsidy Removal Case**

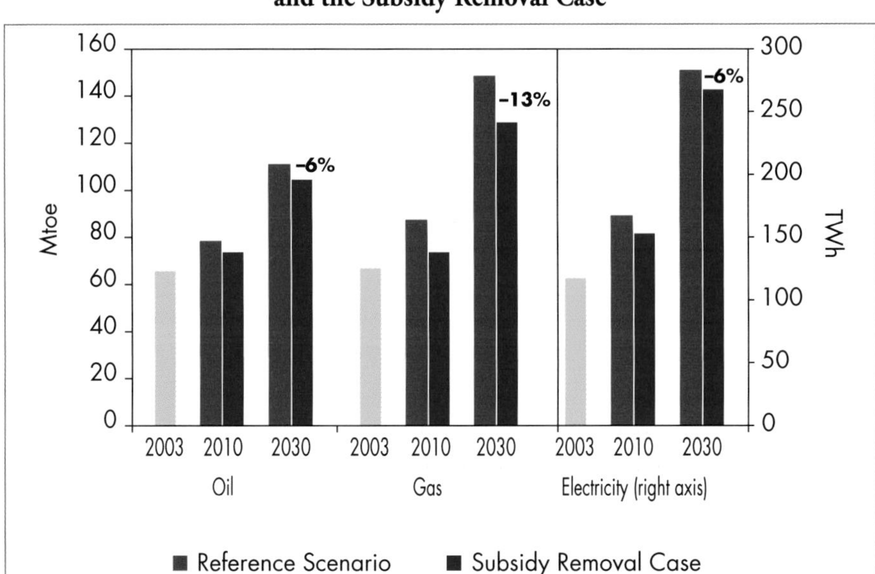

Primary energy demand will grow much less rapidly than GDP over the *Outlook* period. Primary energy intensity will fall by 0.9% per year between 2003 and 2030. In contrast, per capita demand will increase by almost 50%, in line with rising household incomes and a slow-down in population growth.

Sectoral Trends

As in most other countries, demand growth will be led by the power sector and transport. Energy inputs to power generation will grow by 2.6% per year over the *Outlook* period. Electricity demand will grow even faster, by 3.3% per year. Final energy consumption will grow by 2.4% per year over 2003-2030, down from 5.5% in 1990-2003. The share of industry will fall, despite rising demand from petrochemical production. The share of transport in primary oil use will increase from 46% today to 58% in 2030, spurred by increasing vehicle ownership. (Figure 11.5)

Figure 11.5: **Iran's Primary Oil and Gas Consumption by Sector**

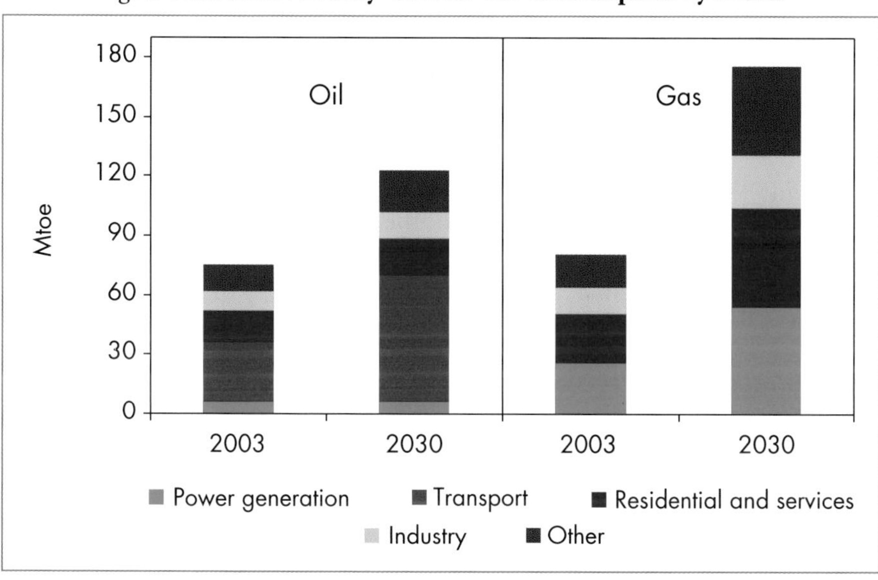

11

Industry

Energy demand in the industry sector, at 27 Mtoe, accounted for 25% of total final consumption in 2003. Non-metallic minerals, petrochemicals, iron and steel and food processing account for three-quarters of industrial production. The energy efficiency of industrial processes is well below the international average. For example, iron and steel production is 50% less energy-efficient than the world average (World Bank, 2003). Natural gas and oil account for 85% of industrial energy consumption. To take advantage of its gas resources,

Iran is expected to develop further its petrochemical industry (UNIDO, 1999) and to remain one of the world's top five petrochemical producers throughout the *Outlook* period. By 2030, industrial energy consumption is projected to reach 50 Mtoe, growing by an average of 2.4% per year. The share of electricity and gas in industrial energy use will grow.

Transport

In 2003, the transport sector accounted for 28% of final consumption and for 46% of primary oil consumption. Growth in this sector accelerated over the last decade. Vehicle ownership has been increasing rapidly, but from a very low base. In 2003, Iran had 45 vehicles per thousand inhabitants, compared to more than 500 per thousand inhabitants on average in the OECD. Car sales in 2004 were 500 000 units, double the number in 2000. But the efficiency of the vehicle fleet is very low by international standards. About two-thirds of vehicles in Iran are more than 15 years old (Ministry of Road and Transport, 2005). In the Reference Scenario, the number of vehicles on the road nearly triples over the projection period. As new vehicles are expected to be more efficient, fuel use will rise less rapidly, by 2.8% per year on average.

Pollution from vehicles in large cities is a serious problem. Four thousand people die from air pollution-related diseases each year in Teheran – one of the most polluted cities in the world. To reduce demand for refined products and lower air pollution, the government is seeking to introduce compressed natural gas vehicles. In the Reference Scenario, the introduction of CNG vehicles is expected to continue over the projection period, but their share in the vehicle stock will remain marginal, accounting for less than 1% of road-fuel demand in 2030.

Residential and Services Sectors

The residential and services sectors together account for 44% of final energy consumption – well above the average in OECD countries. Demand has risen strongly in recent years, partly due to subsidised prices. Natural gas accounts for half of the total energy demand of 48 Mtoe in the two sectors, equivalent to gas demand for power generation. The natural gas distribution network has been significantly expanded over the past decade and it covered 80% of the urban population in 2003 (Institute for International Energy Studies, 2004). Oil is the second-most important fuel, at 16 Mtoe, but its share has been falling steadily. Electricity consumption increased very rapidly over the period 1990-2003, by 6.4% per year, spurred by increasing ownership of appliances and rising household wealth.

In the Reference Scenario, demand in these sectors is projected to reach 87 Mtoe in 2030. Natural gas use will grow by 2.7% per year and its share of total energy demand in these sectors will rise by six percentage points, mainly at the expense of oil. The gas distribution network will be further extended and will cover 90% of the urban population by 2030. Electricity is also expected to grow briskly, at 3.1% per year, spurred by growth in the ownership of appliances and the increasing importance of the services sector in the economy.

Electricity Supply

Overview

Iran's electricity production was 153 TWh in 2003 and preliminary estimates indicate that it reached 165 TWh in 2004. Output grew by around 8% annually over the past decade but per capita electricity generation is still one of the lowest in MENA, at 2 299 kWh in 2003. This level is one-third that of Saudi Arabia and about the same as in Lebanon. Over three-quarters of electricity generation is based on natural gas. Iran has made steady progress away from oil since the 1970s, when more than half of the country's electricity was produced in oil-fired power plants. By 2003, the share of oil had fallen to 16%.

Iran relies on hydropower to meet about 7% of its electricity needs, producing 11 TWh in 2003. Economically exploitable resources are estimated at 50 TWh a year and the theoretical potential at 176 TWh a year (World Energy Council, 2004). Wind power started contributing to Iran's electricity generation in 2001. Installed capacity was 16 MW in 2003, producing 30 GWh. The best sites are in the north (Guilan Province), north-east (Khorasan Province) and the south. Iran started manufacturing 600 kW wind turbines in 2004.

Iran has power interconnections with Azerbaijan, Turkey, Armenia, Turkmenistan and Iraq. Power exchanges are, however, small, at less than 1% of the country's total electricity generation. In 2003, Iran exported 919 GWh, while it imported 1 489 GWh.

Iran's installed capacity was 33 GW in 2003. An increase of about 3 GW was expected in 2004. Most of the power is generated in steam boilers, although the share of combined-cycle gas-turbine (CCGT) plants has been increasing (Figure 11.6). While power plants use mostly natural gas, fuel oil is still used in steam boilers. This is because of the pronounced seasonality of natural gas demand. In the autumn and winter, gas is used mainly for heating in households and consequently there is less gas available for power generation. Nearly half of the fuel used in power plants at that time is oil. Steam power plants consumed about 6 Mtoe of fuel oil in 2003. Gradual replacement with natural gas would free up more oil for export and would also reduce the environmental impact of burning oil.

11

Figure 11.6: **Iran's Electricity Generation by Technology, 2003**

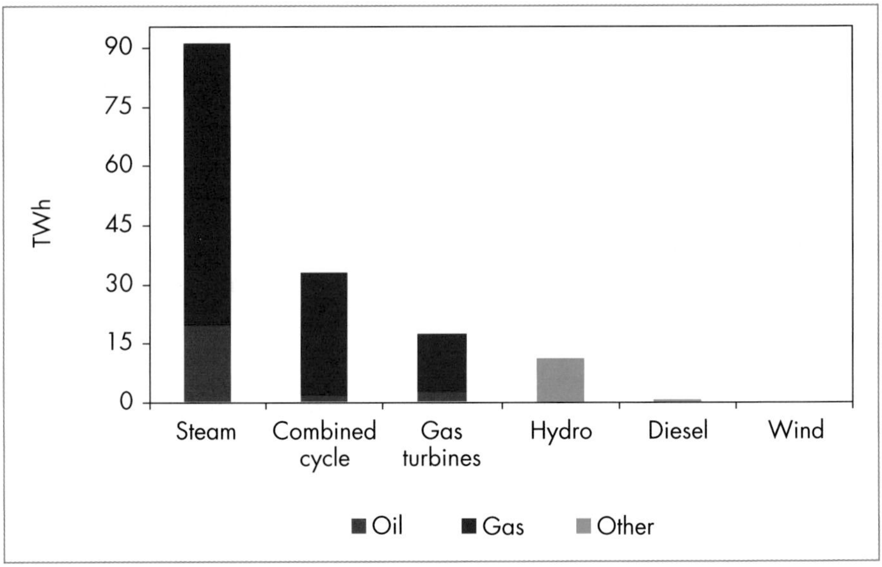

Sources: Tavanir Company (2005); IEA databases.

The Iranian power sector makes extensive use of open-cycle gas turbines, which accounted for 22% of installed capacity and produced 12% of total electricity in 2003. This share is higher than in most other countries. Gas turbines are typically used for peak load in OECD countries and account for smaller shares of capacity and electricity generation.[6] They have the advantage of low initial costs but they also have low generating efficiency and high running costs (see the section on *Power Plant Economics* in Chapter 6). The main reasons for the large number of gas turbines in Iran are their low initial costs and their short construction period. They were built at a time when population and electricity demand were growing rapidly (Meibodi, 1998). The efficiency of gas turbines was 28% in 2003, compared with 38% for boilers and 44% for CCGT plants. The inefficient use of natural gas accentuates the problem of shortages during winter. Some gas turbines have been or are scheduled to be converted to CCGT plants.

State companies are responsible for electricity generation, transmission and distribution in Iran. Tavanir, under the control of the Ministry of Energy, is in overall charge of the entire sector, controlling 16 regional electricity companies, 42 distribution companies, 27 generating companies and five other companies dealing with electricity development, power plant project

6. For example, in the United States, gas turbines accounted for 15% of installed capacity in 2003 but for only 1% of total generation.

management, plant maintenance, new energy sources and energy efficiency. Tavanir produces 98% of the country's electricity. Various industries produce the remaining 2%. Tavanir is responsible for developing plans for the expansion of the power sector. The current planning period goes to 2011 and includes:

- Increased use of natural gas in electricity generation and further development of combined-cycle power plants.
- Domestic production of such plants.
- Improvements in generating efficiency and increased productivity.
- Completion or construction of hydropower projects, including the country's first pumped storage plant.
- Creation of a competitive environment for electricity generation.
- Increased contribution of the private sector in electricity generation through build-operate-transfer and build-own-operate projects.
- Private investment in transmission.
- Expansion of interconnection; one with Russia is planned for 2006, which would allow the exchange of 500 to 800 MW.

Electricity prices are heavily subsidised in Iran, with consumers paying only a fraction of actual costs (Table 11.6). The agricultural and residential sectors pay the lowest tariffs, equivalent to 0.2 US cents and 1.1 US cents per kWh in 2003. Subsidised electricity consumption cost the government about $2.6 billion in 2003, while payment by consumers amounted to $2.5 billion. With heavy subsidies, consumers have little incentive to save electricity and the burden on the government budget is enormous. Subsidised electricity limits investment, since the power company makes no profit to reinvest, and yet the need for investment in new power plants grows. Moreover, the wasteful use of energy caused by subsidies worsens environmental degradation.

Table 11.6: **Electricity Tariffs and Costs in Iran, 2003** (US cents/kWh)

	Tariff	Cost
Residential	1.1	5.0
Public	1.8	3.9
Agricultural	0.2	3.9
Industrial	2.0	3.5
Commercial	4.8	4.8
All users	1.56	4.0

Source: Data obtained directly from Tavanir, converted using an exchange rate of 1 dollar= 8 194 rials.

Electricity network losses are high in Iran, at 17% in 2003 (about 5% in transmission and 12% in distribution) – twice those in OECD countries. Technical deficiencies and electricity theft from distribution networks explain the high losses. Lack of funds for maintenance has often been blamed for the poor performance of the electricity grid. The estimated increase in the cost of electricity because of these high losses is about 10%. The share of losses has increased in the last decade.

Peak load has been rising faster than overall demand, increasing the need for investment in new capacity. Demand for cooling in the summer has contributed substantially to this increase. Tavanir estimates that a one-degree rise in temperature in the summer raises capacity needs by 170 MW, while each degree fall in temperature in the winter calls for an additional 40 MW.[7] In general, daily demand peaks in the evening, while the annual peak load occurs in July or August.

Electricity Production

Iran's electricity generation is projected to increase at 3.2% per year over the period to 2030. Most new capacity is expected to be gas-fired, principally CCGT. They require comparatively low initial investment of about $500/kW, a major advantage where funds for investment are limited. CCGTs also have high generating efficiency, which can help moderate domestic gas consumption. The share of natural gas in electricity generation is projected to rise from 77% now to 80% in 2030.

The availability of natural gas for power generation is not expected to increase substantially in the medium term. Iran will, therefore, continue to rely on oil for part of its generation. In the longer term, as natural gas production increases, more and more gas will be available for power generation, replacing oil. The share of oil is projected to fall to 7% in 2030.

Iran does not currently produce any electricity from nuclear power but considers that there should be a nuclear component in the electricity-generation mix. Its nuclear power programme was launched some thirty years ago. At that time, Iran considered building up to 23 reactors (UK House of Commons, 2004). In 1975, construction of two nuclear units, based on a German design, started at Bushehr (located in south-western Iran, on the Persian Gulf). Construction stopped following the Islamic revolution in 1979. Part of the construction site was destroyed during the Iran-Iraq War. After the war, Iran sought to resume construction and, in 1995, reached agreement with Russia to finish the reactors. Currently, one is under construction. Its

7. The increase in electricity demand for heating in the winter is not so large because heating is mostly based on natural gas.

completion was originally scheduled for the end of 2003 or the beginning of 2004, but it has been delayed. Iran now expects the plant to be operational in 2006. In February 2005, Iran signed an agreement with Russia under which Russia will provide the fuel needed to run the Bushehr plant and will accept the spent fuel back in Russia.

Despite international objections to the country's nuclear programme, Iran plans to install a second unit at Bushehr, though there is no current activity to complete it. In addition, the Iranian Nuclear Energy Organisation is carrying out research to determine a site for a 5 GW plant. The possibility of adding 20 GW of nuclear capacity has also been raised. Our Reference Scenario assumes that only one unit at Bushehr is completed.

Hydropower is projected to rise to 30 TWh by 2030. Several projects are now at varying stages of construction or planning. These new projects are expected to bring Iran's hydropower capacity to 11 GW by 2030. Iran plans to develop other renewable sources as well. A wind atlas is being developed as a step towards exploiting its wind-power resources, which could amount to about 10 GW.[8] The cost of electricity production from wind farms is estimated at 5.2 US cents per kWh (Government of Islamic Republic of Iran and UNDP, 2005), higher than that of a CCGT plant, which implies that the development of wind power will require some form of financial support, especially if the private sector is to be involved. Wind capacity is projected to reach 2 GW in 2030. A 55-MW geothermal power plant is also being built and there are plans to develop a 17-MW solar thermal power plant. There were about 177 kW of photovoltaics in 2002 (World Energy Council, 2004). Solar generating capacity is projected to reach about 260 MW in 2030.

Oil Supply

Overview

Crude oil production (including natural gas liquids) in Iran averaged 4.1 mb/d in 2004. The oil industry has been the engine of Iranian economic development, with production commencing early in the twentieth century. Over the past three decades, production fluctuated with political events. Production peaked in 1974 at 6 mb/d. The war severely affected Iran's oil production, particularly in the oil-rich Khuzestan Province. Economic growth faltered and investment in the oil sector slumped, with production capacity falling to 3 mb/d in 1980. Investment and production started to rise in the mid-1990s.

8. Earlier estimates were on the order of 6.5 GW.

The current Iranian plan provides for an increase in oil production to 4.5 mb/d by the end of 2005, 4.8 mb/d by 2010 and 5.8 mb/d by 2015. These targets can be achieved only with much higher investment in exploration, production and enhanced oil recovery than at present. Given the current depletion rate of existing producing fields, the financial situation of NIOC and the state of technology in the sector, Iran is unlikely to attain these productions targets without significant inflows of foreign capital and technology. In the Reference Scenario, oil production is projected to grow to 4.5 mb/d in 2010 and to 6.8 mb/d in 2030. Oil exports are expected to be 2.8 mb/d in 2010 and 4.4 mb/d in 2030. To achieve even these levels and to raise refining capacity to meet growing domestic demand, cumulative investment of around $80 billion will be needed in 2004-2030. In the Deferred Investment Scenario, oil production will increase only marginally to 4.9 mb/d in 2030 and exports will be almost stable at around 2.6 mb/d.

Policy Framework

Article 44 of the Iranian constitution prohibits ownership of hydrocarbon resources by foreign companies. However, "buy-back" contracts, introduced in the late 1990s, allow foreign companies, operating through an Iranian affiliate, to enter into exploration and development contracts with NIOC. The first major projects under such contracts involved Italy's Eni and France's Total. Under a buy-back arrangement, the contractor funds the required investments for the project and NIOC repays the costs, including operating expenditures and an allowance for capital and accrued bank charges, in the form of oil or gas actually produced from the project. The contractor transfers operation of the field to NIOC when the contract term – typically five to seven years – is completed.

The effectiveness of these contract arrangements is uncertain. They do not appeal to most of the major oil companies, who find that they are required to carry too much of the project risk, while not benefiting from any profit from higher international oil prices. Moreover, they do not allow the companies to book reserves as their own. Nonetheless, Iran has been able to attract $20 billion in investment under buy-back deals. The Iranian government has attempted to make the arrangement more appealing to foreign companies by increasing the reward elements and the length of the contracts.

In 1997, the Ministry of Petroleum reorganised the oil sector, creating a clearer distinction between policy and commercial responsibilities. There are now four main state-owned companies in the oil and gas sector: NIOC, which has responsibility for hydrocarbon exploration and production, the National Iranian Gas Company, the National Petrochemical Company and the National Iranian Oil Refining and Distribution Company. NIOC subsidiaries include the National Iranian South Oil Company, which handles exploration and production in oil-rich Khuzestan, and the Pars Oil and Gas Company, which handles development of the South Pars gas field.

Resources and Reserves

The *Oil and Gas Journal* puts Iran's reserves at 125.8 billion barrels,[9] the second-largest in the Middle East, after Saudi Arabia. The US Geological Survey puts undiscovered recoverable resources at 67 billion barrels. Two-thirds of the country is comprised of sedimentary basins with hydrocarbon potential (Figure 11.7). More than 60% of proven reserves are concentrated in six super-giant fields : Agha Jari, Ahwaz, Bibi Hakimeh, Karanji, Gachsaran and Marun.[10]

The major petroleum producing province is the Zagros/Arabian basin, but the Caspian Sea region and the Central basin also have vast potential. The Zagros and Arabian basins lie in the west of the country along the Persian Gulf and the Iraqi border. The sedimentary column is very thick, up to 12 000 metres, with reservoirs ranging from Permian to Oligocene-Miocene. Many of the reservoirs are in carbonates, often heavily faulted and fractured, as a result of the tectonic stresses linked to plate collision.

The Luristan-Khuzestan basin, located in the Southern Province north-west of Bushehr, contains around 85% of total oil reserves. Offshore fields in the Persian Gulf account for the remainder. The super-giant South Pars gas field, an extension of the Qatari North Field, is thought to have natural gas reserves of 13 tcm and condensates reserves of 3 to 4 billion barrels. Iran also has some 100 million barrels of proven reserves in the Caspian, where oil in place is thought to be around 33 billion barrels. The Caspian region is attracting a lot of interest, but projects in the Iranian portion are lagging behind those in the northern Caspian. Development of this region will require modern, deep-water offshore technologies.

Recent discoveries include the super-giant Azadegan field, a geologically complex field in the Khuzestan region and Iran's largest discovery in 30 years. Reportedly it contains reserves of 26 billion barrels. Another major discovery has been the Darkhovin onshore field, near Abadan, thought to contain up to 5 billion barrels. In 2001, NIOC announced the discovery of the Dasht-e-Abadan field, in shallow waters near Abadan. It could contain reserves comparable in size to Azadegan. Despite these recent discoveries, all of which were made by NIOC, some geologically interesting regions have yet to be explored.

9. The Iranian Oil Ministry estimates reserves at 130.8 billion barrels.
10. Super-giant fields are defined as containing more than 5 billion barrels of proven reserves.

Figure 11.7: **Main Oil and Gas Fields and Energy Infrastructure in Iran**

Crude oil field
Crude oil pipeline
Gas/condensate field
Gas pipeline
Refinery in operation
Tanker terminal or loading platform

0 km 160

Crude Oil Production[11]

Iran has 68 oilfields currently in production, with onshore fields accounting for 80% of production. Offshore, however, production is rising thanks to field development contracts awarded to foreign companies. Over the past 15 years enhanced oil recovery facilities have been installed at many ageing Iranian oilfields suffering from declining reservoir pressure and water encroachment. All the fields currently in production were discovered before the 1970s. The average natural decline rate of these fields is relatively high compared to fields in the rest of MENA. Onshore fields are estimated to be declining on average at 8% per year and offshore fields at 10%.[12] The natural decline in production is thought to be about 270 kb/d per year. In addition to high decline rates, the average oil recovery rate in Iran oilfields is low, at 27%. This is mainly due to poor maintenance and outdated technology. For example, Iran has drilled only about 100 horizontal wells. The sulphur content of Iran's crude oil is generally regarded as medium, while gravities are mainly in the 28°-35° API range. Heavier oil is found in the offshore field, Soroosh/Nowruz (20° API), and lighter oil in the onshore field, Naft-e-Safid (44.5° API).

In 2003, the six super-giant oilfields accounted for some 60% of oil production in Iran. The three largest onshore producing oilfields are Ahwaz, with 765 kb/d, followed by Gachsaran, with 560 kb/d, and Marun, with 458 kb/d. These three fields account for two-thirds of total onshore production. In the absence of the application of advanced techniques, production from these fields will drop sharply. For this reason, NIOC is using a combination of infill drilling, well workovers and enhanced oil recovery techniques at all three fields. Gas injection is used on a large scale at the Marun field, and it is likely to be used at Ahwaz. The Doroud and Aboozar fields combined account for more than 70% of offshore production. However, Soroosh and Nowruz, awarded to Shell, are expected to produce some 190 kb/d by the end of 2005, outpacing all other offshore fields.[13]

In the Reference Scenario, crude and NGL production is expected to reach 4.5 mb/d in 2010 and 6.8 mb/d in 2030. Production from currently producing fields is expected to increase by 7% to 2010. In 2010-2030, gas and water reinjection in a number of fields will increase recovery rates and the

11. All the oil-production figures cited in this section include crude oil, natural gas liquids (NGLs) and condensates.

12. *Energy Business Review*, Vol. 13, No. 3, January-March 2003, and industry sources.

13. Iran has no oil or natural gas operations in the Caspian region, although Petrobras has discussed with NIOC the possibility of production-sharing agreements to produce in deep water offshore. Iran continues to insist on the validity of regional treaties signed in 1921 and 1940 with the former Soviet Union, which set the shares of the Caspian resources allotted to each country.

Table 11.7: **Major Oilfields Currently in Production in Iran**

Field	Crude oil* remaining proven and probable reserves at end-2004 (billion barrels)	Sustainable capacity (kb/d)	Gravity (average °API)	Date of discovery
Onshore				
Ahwaz	16.8	1 040	32	1958
Marun	14.6	570	34	1963
Agha Jari	8.5	190	34	1937
Parsi, Karanji	6.5	200	34	1963
Gachsaran	6.4	600	32	1928
Rag-e-Safid	3.5	100	29	1963
Bibi Hakimeh	3.4	150	30	1961
Kupal	1.8	55	32	1965
Pazanan	1.2	80	35	1961
Binak	0.7	50	30	1959
Lab-e-Safid	0.4	40	35	1968
Naft-e-Safid	0.3	25	45	1938
Ramshir	0.3	30	28	1962
Chashmeh Khush	0.1	37	29	1967
Offshore				
Soroosh, Nowruz	3.2	190	20	1962
Doroud	2.4	150	34	1961
Aboozar	0.9	200	26	1969
Foroozan	0.5	29	35	1966

* NGLs and condensates are not included.
Sources: IEA database; Arab Petroleum Research Centre (2005); IHS Energy databases; *Energy Business Review*.

drilling of new wells will reduce the decline in production (Table 11.8). These fields are expected to play a key role in Iran's oil production outlook.

Production from Ahwaz, the largest currently producing field, is expected to continue to decline, reaching some 400 kb/d by 2030. Reinjection of associated gas, through installation of gas compression facilities under a buy-back contract, will stop production declining as fast as in recent years. We expect production from Gachsaran to increase to over 700 kb/d by 2010, mainly through new development wells and workovers. Gachsaran is expected

Table 11.8: **Iran's Crude Oil Production in the Reference Scenario** (kb/d)

	2004	2010	2020	2030
Currently producing fields	**4 149**	**4 450**	**4 507**	**3 934**
Ahwaz	761	607	547	408
Gachsaran	588	721	685	578
Marun	412	412	373	282
Doroud	190	176	82	25
Soroosh-Nowruz	190	158	85	64
Karanji	187	187	150	84
Agha Jari	157	405	405	405
Bibi Hakimeh	139	139	123	95
Aboozar	136	92	21	4
Rag-e-Safid	98	86	67	52
South Pars	87	147	272	422
Other fields	1 204	1 320	1 697	1 515
New developments	**0**	**0**	**1 042**	**2 816**
Fields awaiting development	0	0	833	2 253
Reserve additions and new discoveries	0	0	208	563
Total	**4 149**	**4 450**	**5 549**	**6 750**

Notes: Projections assume improved oil recovery techniques are employed. Includes NGLs and condensates.
Source: IEA analysis.

to peak around 2010 and decline to less than 600 kb/d by 2030. Production from Iran's most mature field, Marun, is expected to decline steadily through to 2030. Water cut in the field is high and, despite ongoing tertiary recovery initiatives, the decline is expected to continue. Reinjection programmes at the declining Agha Jari field, with gas supplied from Phases 6-8 of the South Pars development, are expected to increase its production to some 400 kb/d.

Most of the additional production will come from development of new oilfields and from the installation of enhanced recovery facilities at mature fields. Current gas reinjection requirements stand at some 35 bcm per year.[14] They could reach 60 bcm by 2030 in order to maintain pressure at ageing fields. Through enhanced oil recovery and the introduction of new technology, currently producing fields could still account for almost 60% of oil

14. Gas consumption for reinjection is not included in our energy balances and projections.

production in 2030.[15] In the second half of the *Outlook* period, fields already discovered but not yet developed will represent an increasing proportion of production, accounting for almost half of total production in 2030 (Figure 11.8). The Azadegan field will account for the bulk of this increase. NGL production is expected to increase markedly, in line with South Pars development.

In the short term, the projected growth rate in oil production in the Reference Scenario is considerably slower than Iran's official target in the current five-year plan. Delays in assigning concessions to foreign companies and an unfavourable investment climate are expected to hinder the speed with which improved oil recovery technologies can be introduced in producing fields. Further ahead, foreign capital and technology become even more important. In Iran, more than in any other MENA country, gas development will be linked to oil development. The timely development of gas fields, for reinjection purposes, will be essential to allow oil production to increase to the levels

Figure 11.8: **Iran's Crude Oil Production by Source in the Reference Scenario**

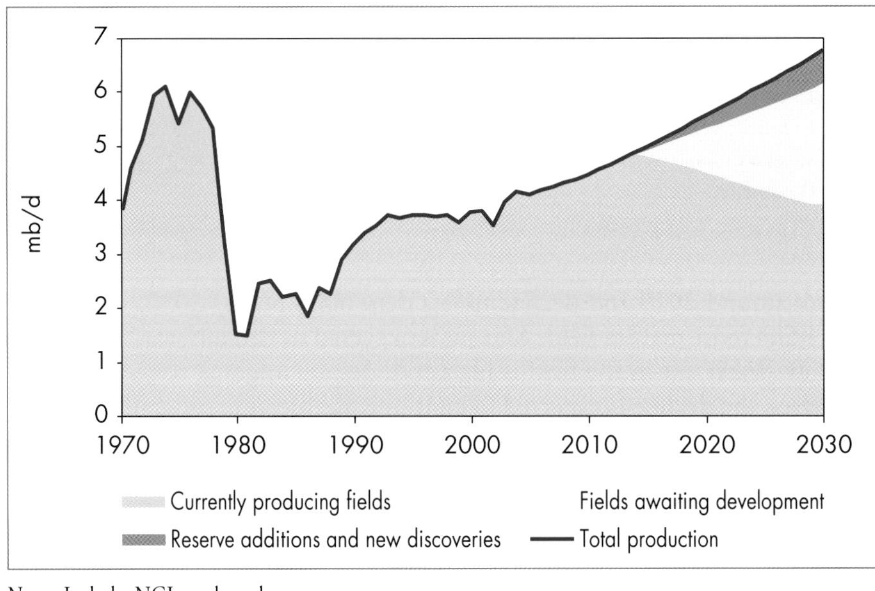

Notes: Includes NGLs and condensates.

15. There are significant opportunities in Iran to increase production with suitable investment and technology. An example is a recent study by the Norwegian Institute SINTEF on the West Zagros field, a small mature reservoir near the Iraqi border. The study indicates that by drilling 35 new wells, roughly doubling the number of wells in the field, and applying modern reservoir management techniques, production could be increased from the current 24 kb/d to about 130 kb/d and recovery increased from 5-10% to 10-30%.

projected here. In addition, a large share of future oil production will be in the form of NGLs. Thus, attracting foreign capital, access to modern technology and effective project management will be critical to expanding oil production in Iran.

Oil Refining

In 2004, crude oil distillation capacity totalled 1.5 mb/d. Refinery output equalled 1.4 mb/d, indicating a very high utilisation rate to meet soaring domestic demand. The refining industry in Iran is generally old and inefficient and lacks conversion capacity. Only 13% of refinery output is gasoline, half the average for a typical European refinery. Of the country's nine refineries, most are in need of modernisation, which will require foreign investment and modern technology.

Iran's refineries were badly damaged by the Iran-Iraq War. In 1980, pre-war capacity totalled 1.3 mb/d. In less than two years, capacity was halved, with the destruction of the Abadan refinery. Reconstruction began only at the beginning of the 1990s because of financing difficulties.

The current refining capacity and configuration is not adequate to meet the rapid growth of gasoline demand, fuelled by increased car ownership and very low prices. In consequence, in 2003 Iran imported 95 kb/d of gasoline, worth some $1.1 billion.[16] Gasoline output was 60% lower than domestic demand, while residual oil production was much higher than demand (Figure 11.9).

Figure 11.9: **Refined Product Surpluses in Iranian Refining, 2003** (%)

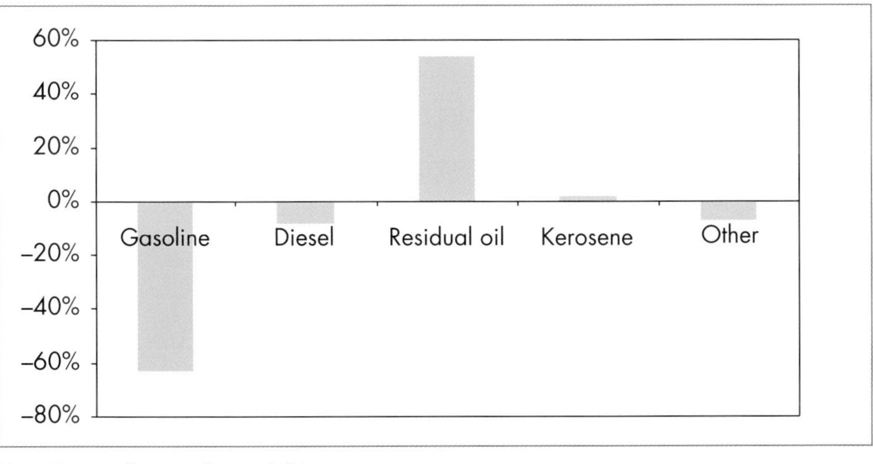

Note: Negative figures indicate a deficit.

16. According to preliminary estimates, gasoline imports in 2004 were 160 kb/d, costing $4.5 billion.

Adding a grassroots refinery capable of producing the amount of gasoline that was imported in 2003 would cost some $4.3 billion. The combined value of the gasoline import bill and gasoline subsidies in the same year corresponds to 85% of this investment. For this reason, the National Iranian Oil Refining and Distribution Company (NIORDC) has announced a number of new projects, including adding new distillation units, revamping existing refineries, increasing conversion capacity and adding condensate splitters and CNG facilities. These projects would bring the composition of output more into line with the structure of domestic demand and would allow refineries to process more crude from producers around the Caspian. If all of the projects were realised, almost 900 kb/d of additional crude distillation would come on stream by 2010.

Table 11.9: **Crude Oil Distillation Capacity in Iran**

	Start-up year	**Capacity 2004** (kb/d)	**Light product share** (%)	**Planned expansions** (kb/d)	**Expected completion date**
Abadan	1970	400	44.5	50	2005
Arak	1993	170	58.3	80	2010
Bandar Abbas	1998	232	59.0	608	2010
Isfahan	1979	284	56.9	108	2010
Kermanshah	1971	25	51.9	–	–
Lavan	1977	30	60.6	–	–
Shiraz	1974	40	59.2	–	–
Tabriz	1978	100	61.9	50	2010
Tehran	1968	225	60.9	–	–
Total		**1 506**	**57.0**	**896**	

Since February 1999, private domestic and foreign companies have been allowed to invest in new projects in Iran. However, foreign companies cannot hold a majority in a refinery asset and domestic companies do not have sufficient financial resources to undertake projects. So far, only NIORPDC has been involved in the Iranian refinery sector. NIORPDC buys the crude from NIOC, processes it and distributes the refined products. Because of the high capital cost of building new refineries, NIORPDC has announced plans only to expand and modernise existing refineries.

In our Reference Scenario, refining capacity is expected to reach 2.6 mb/d by 2030. We expect only a small part of the announced expansions to take place in the near term, most of the investment in that period being directed towards upgrading existing refineries to match the demand slate for domestic products. The refinery sector will require cumulative investment of $16 billion over the period 2004-2030, three-quarters for new capacity. The remainder will be needed for conversion units. Given the current legislative and policy framework in the energy sector, it will be a challenge for Iran to ensure the financing of the multi-billion dollar grassroots projects that will soon be needed to meet rising domestic demand.

Oil Exports

Iran currently exports around 2.7 mb/d of oil, a little less than half of which goes directly to Europe. Other major customers include Japan, China, South Korea and Chinese Taipei. Iranian crude oil has not been exported to the United States since mid-1997. Crude oil swaps are in place between Iran and Turkmenistan, Kazakhstan and, more recently, Iraq.

Iran has seven export terminals with a total capacity of 7 mb/d for oil and condensates. The largest terminal is on Kharg Island. Refined products are exported via the Abadan and Bandar Mahshahr terminals. Iranian heavy (31° API, 1.7% sulphur) makes up half of Iran's exports and Iranian light (34.6° API, 1.4% sulphur) accounts for some 40%. Lavan blend (34°-35° API, 1.8-2 % sulphur) and Foroozan blend/Sirri (29-31° API) make up the remainder.

In the Reference Scenario, Iran is projected to export some 2.8 mb/d of crude oil and products by 2010 and 4.4 mb/d by 2030. By 2030, Iranian exports will account for 12% of Middle East exports and 7% of global oil trade by 2030.

Figure 11.10: **Iran's Oil Balance in the Reference Scenario**

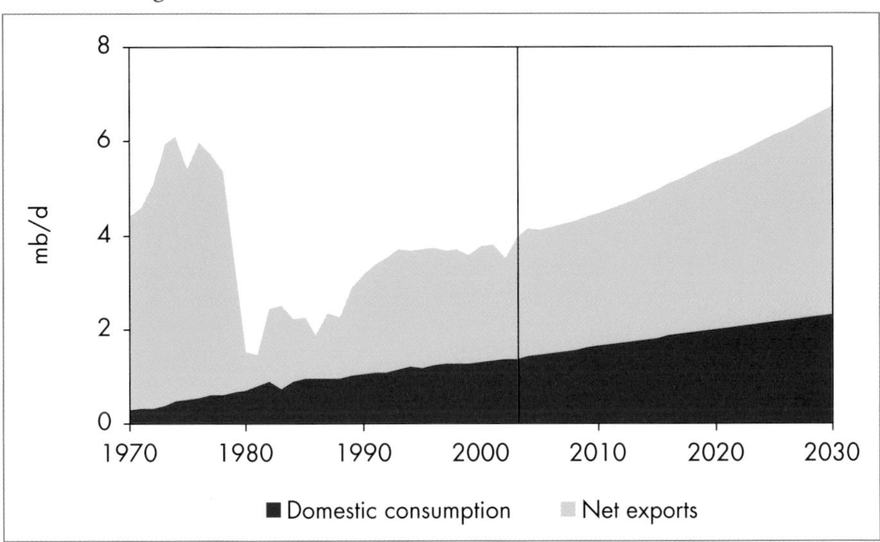

Natural Gas Supply

Resources and Reserves

Iran holds the world's second-largest gas reserves, after Russia. Cedigaz estimates that Iran's proven reserves on 1 January 2005 were 28 tcm – 16% of global gas reserves. Undiscovered resources are estimated by the US Geological Survey at just under 9 tcm, a potentially low estimate reflecting the limited exploration effort to date (USGS, 2000). Two-thirds of proven reserves, or 17 tcm, are in non-associated gas fields.

The country's largest field, the offshore South Pars gas field, is thought to contain 13 tcm of natural gas and up to 4 billion barrels of condensate (Table 11.10). South Pars is a geological extension of Qatar's giant North Field. North Pars is the second-largest field in Iran, with 1.4 tcm of reserves.

Iran's total gas reserves have increased by 12% since 2000, mainly because of the re-evaluation of South Pars and the discovery of several smaller fields. These include two major discoveries in 2000: Tabnak, which contains 850 bcm of natural gas and 540 million barrels of condensate, and Hama, which contains 133 bcm of natural gas and 58 million barrels of condensate.

Table 11.10: **Proven Natural Gas Reserves and Production in Main Fields in Iran**

	Gas reserves at 1 January 2005 (bcm)	**Marketed gas production in 2004** (bcm)
South Pars	13 026	21.3
North Pars	1 416	–
Tabnak	850	–
Kangan	613	22.9
Khangiran	475	11.8
Aghar	240	7.3
Nar	233	12.3
Assaluyeh	224	–
South Gashdo	208	–
Sarkhoun	179	–
Other	10 736	5.7
Total	**28 200**	**81.3**

Source: IEA database.

Gas Production and Distribution

In 2003, Iran's marketed gas production was 78 bcm, more than that of any other Middle Eastern country. Gross natural gas production was much higher, at 124 bcm. Reinjection accounted for some 35 bcm, flaring for some 5 bcm and shrinkage for about 6 bcm. Non-associated gas accounts for 75% of Iran's production. Production increased strongly in 2003, reflecting the first full year of production from South Pars, which produced 20 bcm, all of which was marketed. Most of the remaining marketed gas production was derived from five onshore non-associated gas fields: Kangan, Nar, Aghar, Khangiran and Sarkhoun. Other major gas fields, such as Dalan, are largely dedicated to reinjection. Iran first began producing natural gas over 30 years ago, but an organised development plan was only put in place in the late 1990s. Between 1990 and 2003, gas production increased at 10% per year on average.

The Iranian Ministry of Petroleum has set a production target of 292 bcm by 2010, with gas coming mainly from South Pars. The initial phases of Iran's multi-billion dollar South Pars development are aimed at boosting gas supplies to the growing domestic market, in order to free up oil for export, and to provide for reinjection into ageing oilfields, which currently represents the most lucrative way of using natural gas in Iran. When compared to the return in the domestic and LNG markets, the value of one cubic metre of gas is much higher if it is used to increase oil recovery and thus oil exports (Figure 11.11).

Figure 11.11: **Economic Value of Gas According to Use in Iran**

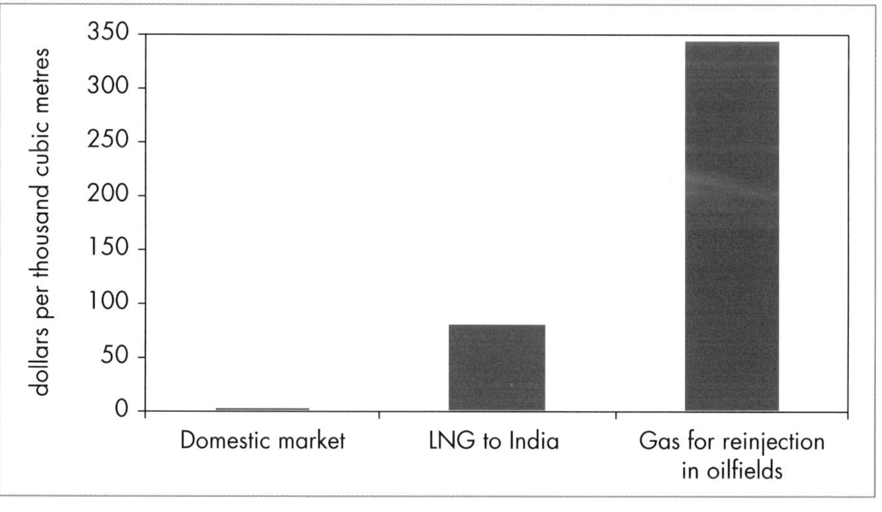

Note: The implicit gas price in the case of reinjection has been calculated quantifying the incremental oil recovery achievable in fields currently equipped with gas reinjection facilities. The oil price is assumed to be $28 per barrel, equivalent to the price threshold in the LNG price formula to India.

The progressive introduction of advanced reservoir techniques will diminish the scope for increasing oil recovery rates by this means. The economic advantage of gas reinjection compared to export will therefore decrease in the future.

The Pars Oil and Gas Company, a subsidiary of NIOC, has jurisdiction over all South Pars-related projects. It has entered into buy-back contracts or production-sharing agreements with foreign companies for most phases. In June 2005, contracts for 18 development phases had been awarded, out of the 30 phases planned by the ministry (Table 11.11).

In the Reference Scenario, marketed production is projected to reach 110 bcm in 2010 and 240 bcm in 2030, three times the current level (Figure 11.12). The first phases of South Pars development will account for most of the increase in production through to 2010, lifting the share of South Pars in marketed production from about a quarter today to more than 40%. Natural gas production from Kangan, Khangiran and Nar fields combined will still represent about half of total marketed production in 2010, but will fall to only 10% by 2030.

New field developments will be needed in the second half of the projection period. By 2030, 60% of total production will come from new developments. New South Pars phases are likely to represent the bulk of this incremental production. No additional discoveries are needed for Iran to reach the production levels projected here.

Figure 11.12: **Iran's Natural Gas Production by Source in the Reference Scenario**

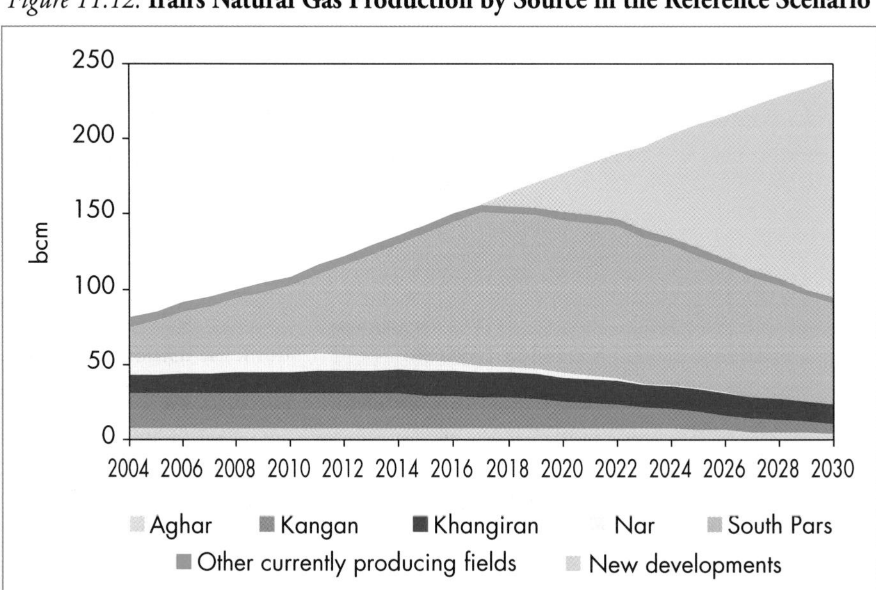

Table 11.11: **South Pars Development Plan**

Phase	Contractors	Status	Estimated output
1	Petropars, Petronas	Production started in early 2004	9 bcm of natural gas per year, 40 kb/d condensate for domestic market
2, 3	Total (40%), Petronas (30%), Gazprom (30%)	Started production in 2002. All facilities handed over to NIOC	29 bcm of natural gas, 80 kb/d of condensate
4, 5	Agip (60%), Petropars (30%), NaftIran (20%)	Production started in 2004. The two phases, with a capacity of 20 bcm, were fully operational by April 2005	20 bcm of natural gas per year, mainly reinjection and 80 kb/d of condensate
6, 7, 8	Petropars, Statoil (40%)	After Shell withdrawal, it is scheduled to come on stream in 2006	31 bcm of natural gas, mainly for reinjection, 120 kb/d of condensate
9, 10	Petropars, GS E&C, OIEC and IOEC	Project finance, loan worth $1.75 billion signed in 2003	20 bcm of natural gas per year, mainly reinjection and 80 kb/d of condensate
11	Total, NIOC	Agreement signed. Expected on stream 2010-2011	LNG exports, 2 trains of 5 Mt per year
12	Petropars	Development to be determined	LNG exports and reinjection
13	Shell, Repsol, NIOC	Under discussion with NIOC, but not signed	LNG exports, 2 trains of 7 Mt per year
14	Sasol	Design phase	GTL
15, 16	Tender open	Tender open, after Kvaerner withdrawal in 2005	50 mcm gas a day for domestic use, 80 kb/d condensate for export
17, 18	Petropars, OIEC, IOEC	Awarded in 2005. Expected implementation time is 52 months	18 bcm to the domestic grid, 25 bcf feedstock for petrochemicals, 74 kb/d of condensates and 1 Mt of LPG for export

Two main uncertainties surround these projections. As with oil, the timely execution of projects, particularly in the South Pars field, will be essential. A clear and stable market framework will be needed to attract foreign capital and technology for upstream development. The quantity of gas needed for reinjection (which is not included in marketed production) is also uncertain. If the government underestimates the volume of gas needed for enhanced oil recovery, more South Pars phases will be devoted to reinjection, reducing the volume that can be marketed.

Export Plans and Prospects

In 2003, Iran was a net *importer* of natural gas, to the tune of 2.3 bcm. Iran exported 3.5 bcm to Turkey but imported 5.7 bcm from Turkmenistan, to supply natural gas to the northern provinces for winter peak demand. Deliveries to Turkey started in 2002, following several years of delays. The 23-year contract between NIGC and Botas, the Petroleum Pipeline Corporation in Turkey, provides for 10 bcm of gas to be delivered by 2007. But Turkey has faced problems in paying for the gas, since gas demand has been much lower than expected.

Iran could export gas to Western Europe via Turkey. In 2002, Greece and Iran signed an agreement to provide for extending the natural gas pipeline from Iran to Turkey into northern Greece. Gas could also be transported to Europe via Bulgaria and, possibly, Romania. Discussions have also taken place between Iran, Italy and Austria. In 2004, Armenia and Iran began construction of a gas pipeline that will deliver gas to Armenia for 20 years from 2007. Initial flows

Figure 11.13: **Iran's Natural Gas Balance in the Reference Scenario**

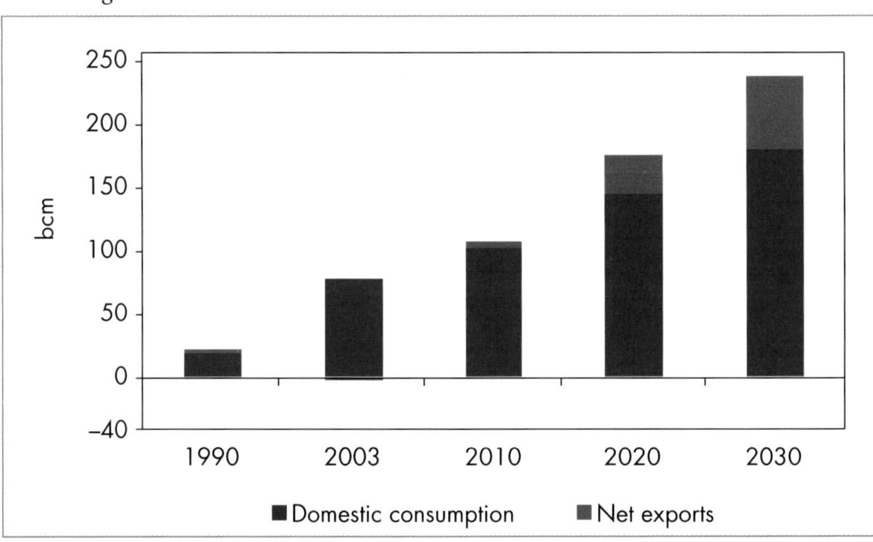

will be 5.5 bcm a year. Iran has also discussed possible pipeline developments with Pakistan and India. In addition, several phases of South Pars are designed for LNG exports.

In the Reference Scenario, Iran is set to become a net exporter by 2010, with net exports of 5 bcm. These are projected to reach 31 bcm in 2020 and nearly 57 bcm in 2030, mostly in the form of LNG. Throughout the *Outlook* period, priority will be given to supply for the domestic market, which will take three-quarters of the increase in production from now to 2030. Europe and Asia are expected to represent the main export markets for Iranian gas.

Energy Investment

Oil and Natural Gas

Cumulative investment needs in the Iranian oil industry are estimated at \$77 billion (in year-2004 dollars) over 2004-2030. The upstream sector will require about 75% of oil investment over the projection period, or \$59 billion. Investment needs will be much larger in the third decade of the projection period, when crude oil production capacity is expected to grow fastest. The unit cost for greenfield development is estimated to be around \$2 per barrel onshore and \$4 offshore.

Annual investment requirements in upstream oil will amount to around \$2 billion per year on average over the *Outlook* period, but they will vary over the projection period (Figure 11.14). Annual investment in the third decade will be as much as three times higher than in the current decade, exceeding

11

Figure 11.14: **Iran's Oil and Natural Gas Investment in the Reference Scenario**

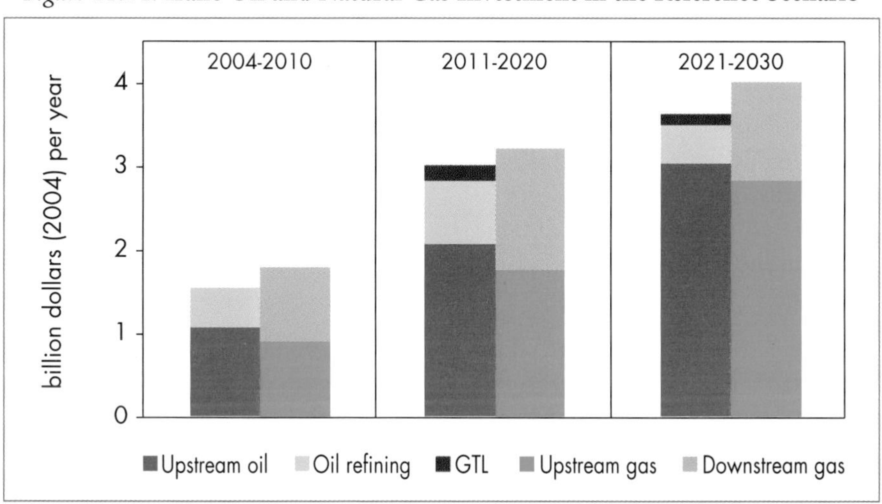

$3 billion per year. New field developments and the widespread utilisation of advanced production techniques will be needed to enable production levels to reach 6.8 mb/d in 2030. We also assume that one GTL project will come on stream between 2015 and 2020, requiring additional investment of some $3 billion.

The refinery sector will call for investment of $16 billion over the projection period, one-fifth of total oil investment requirements. Refinery investments are urgently needed, to increase crude distillation capacity and for additional treatment and conversion units. The timeliness of this investment will be essential for Iran to cease importing gasoline and to cope with local environmental pollution.

The projected trends in natural gas production and exports call for levels of investment similar to those for oil – $85 billion over the period 2004-2030. The upstream sector will absorb around 60% of this investment. Annual expenditures on upstream gas are expected to treble, reaching $2.8 billion in the third decade. The expansion of pipelines and the construction of LNG terminals will require some $32 billion over the period 2004-2030. Most of this investment will be required in the second and third decades.

The current five-year plan allocates some $47 billion to the oil and gas sectors in the period to 2009. A quarter of this investment is expected to be financed through domestic resources and the rest through buy-back contracts and external financing schemes. Buy-back contracts are expected to contribute some $25 billion, more than the total amount of capital invested under these contracts since they were first introduced. The investment needs identified in the plan are much higher than we project in our Reference Scenario, reflecting the higher oil and gas production profiles of the five-year plan.

Electricity

Iran's power sector will need to invest $92 billion in generation, transmission and distribution over 2004-2030. Power generation will take about $43 billion (including $4 billion in refurbishment). Transmission will need $15 billion and distribution networks $34 billion. Total investment will be equivalent to 1.3% of Iran's cumulative GDP in 2004-2030. This share is expected to be higher, at 2.2%, in the period to 2010. Mobilising this investment will be an enormous challenge.

Iran has turned to the private sector to develop several new power generation projects, on a build-operate-transfer basis. But it needs to manage these projects well, to keep costs down. Reform of the distribution sector will be needed to reduce network losses. Cost-reflective tariff policies will be even more important as investment needs rise over the projection period.

Deferred Investment Scenario[17]

Foreign capital and technology will be essential for Iran to reach the oil, gas and electricity production levels projected in the Reference Scenario. A deferral of the investment in Iranian upstream oil may occur for a variety of reasons. The contractual arrangements offered to foreign companies might not be considered sufficiently appealing, the investment climate might be soured in other ways and NIOC might face difficulties in financing new projects if national debt increases faster than expected in the Reference Scenario. Currently upstream oil investment represents as much as 7% of total Iranian investment. In the Deferred Investment Scenario, which assumes that the ratio of upstream oil investment to GDP remains stable at the average level of the last 10 years throughout the projection period, cumulative upstream spending falls by 24% over 2004-2030, compared with the Reference Scenario. Spending would be reduced from $59 billion to $44 billion.

Energy Demand

In the Deferred Investment Scenario, total primary energy demand in Iran grows at an average annual rate of 2.4% per year over the projection period – only 0.2 percentage points lower than in the Reference Scenario. Demand in 2030 is only some 10 Mtoe lower than in the Reference Scenario. This is mainly due to the fact that the rate of GDP growth is very similar in the two scenarios. Although oil and gas export revenues are slightly lower in the Deferred Investment Scenario, they make a relatively small contribution to GDP. Primary demand for gas falls most, because gas consumption in the energy transformation sector, mainly LNG and GTL, is lower than in the Reference Scenario because of lower global gas demand.

Oil Production and Exports

The lower level of investment in upstream oil means that the increase in oil production up to 2030 is very modest. Oil production reaches 4.2 mb/d in 2010 and 4.9 mb/d in 2030. Output is 1.8 mb/d, or 27%, lower in 2030 compared with the Reference Scenario. Lower investment causes output from fields already in production to fall by 1.1 mb/d compared with the Reference Scenario. Fall in output from fields awaiting development accounts for most of the remainder of the difference. The delayed introduction of improved recovery techniques and of up-to-date technology for secondary and tertiary production explains the difference between the two scenarios.

17. See Chapter 7 for a detailed discussion of the assumptions and methodology underlying the Deferred Investment Scenario.

Oil exports fall even more than production in percentage terms, but slightly less in volume terms – because lower inland consumption helps to compensate for the fall in production. Net oil exports are 2.7 mb/d in 2030, a reduction of 40% compared to the Reference Scenario.

Gas Production and Exports

Iran's natural gas production in the Deferred Investment Scenario is lower than in the Reference Scenario, mainly because of the fall in global gas demand. In 2030, natural gas production reaches 205 bcm – 35 bcm, or 14%, less than in the Reference Scenario. As domestic gas demand is only 4% lower in 2030, exports are reduced by almost half to some 30 bcm in 2030, compared to 57 bcm in the Reference Scenario (Figure 11.15). Exports to Europe are reduced by a quarter and to Asia, by half. Gas production in associated fields is reduced because of lower oil production, falling by a third in 2030 compared with the Reference Scenario. Production from non-associated fields, notably in South Pars, also drops sharply. The phases of South Pars devoted to LNG exports would be most susceptible to delay.

Figure 11.15: **Iran's Natural Gas Balance in the Reference and Deferred Investment Scenarios**

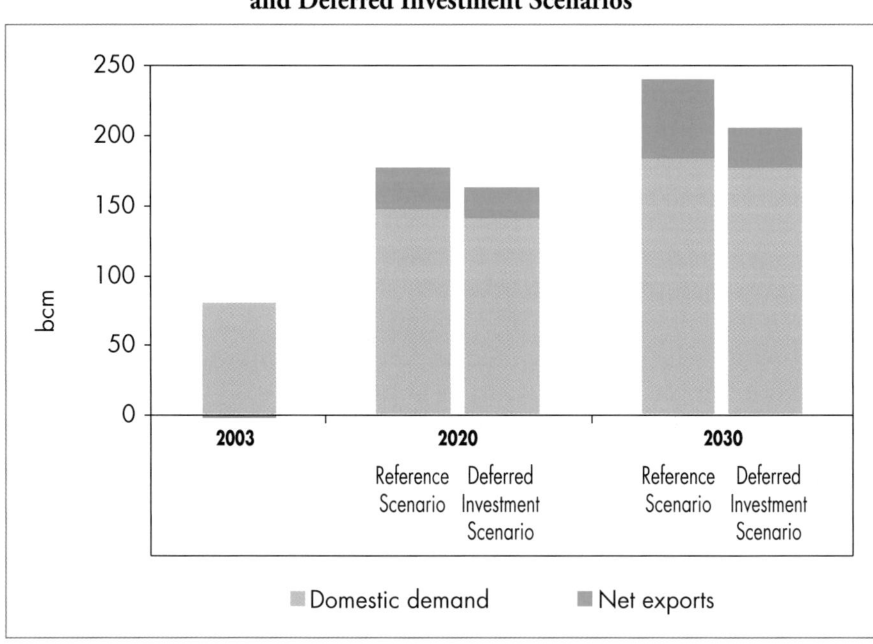

IRAQ

HIGHLIGHTS

- The successful rebuilding of Iraq's economy hinges on restoring political stability and security, in order to advance the development of the country's vast oil and gas resources. Oil-export revenues, which make up most of the country's GDP, will help finance non-oil infrastructure projects, social programmes and other forms of ongoing government spending.
- Iraqi energy consumption is projected to grow strongly over the *Outlook* period, driven by economic recovery and rapid population growth. In the Reference Scenario, primary energy demand is expected to more than double between 2003 and 2030, growing at an average annual rate of 3.3%. Oil will remain the dominant fuel, but the share of gas will grow.
- Iraq's proven oil reserves, at 115 billion barrels, are the fourth-highest in the world. Development costs are low, but how quickly these reserves can be tapped depends on domestic security, as well as on the legal and commercial framework that will emerge. Large-scale foreign investment will not occur until insurgent attacks on people and facilities subside.
- In our Reference Scenario, oil production grows from about 2 mb/d in 2004 to 3.2 mb/d in 2010 and 7.9 mb/d in 2030, on the assumption that security is gradually restored and reserves are developed progressively. Oil exports grow from 1.4 mb/d in 2004 to 2.5 mb/d in 2010 and 6.9 mb/d in 2030. These trends, which are below official targets, call for cumulative investment of $59 billion (in year-2004 dollars) over 2004-2030, more than 85% for upstream projects. In the Deferred Investment Scenario, production grows much more slowly, reaching only 4.1 mb/d in 2030.
- Marketed production of natural gas will grow even faster than that of crude oil, as more gas will come from non-associated fields. We project gas production to rise from 2 bcm in 2003 to 4 bcm in 2010 and then surge to 32 bcm in 2030. Exports could grow rapidly later in the *Outlook* period.
- Power generation is expected to pick up gradually, as reconstruction of Iraq's battered industry progresses. In the near term, generation will continue to be mostly oil-based, but gas use will grow as the gas network is restored and gas production expands. Electricity output will rise from 28 TWh in 2003 to 40 TWh in 2010 and 82 TWh in 2030. Cumulative investment needs for the electricity sector to 2030 will exceed $26 billion.

Overview of the Iraqi Energy Sector

Oil has shaped Iraq's past and holds the key to its future. The country is believed to have the third-largest oil reserves and the fifth-largest natural gas reserves in the MENA region, though the reliability of the reserves data is uncertain. How quickly those reserves will be developed is also extremely uncertain, due to political and social upheavals in the country. The outcome will affect not just the pace and nature of the country's economic and social development, but also the evolution of the international energy market and the health of the global economy.

Three decades of war, international sanctions, economic mismanagement and corruption curbed oil production, devastated the economy and drastically reduced living standards. The US-led invasion of Iraq in 2003 raised expectations that investment in the oil industry would be mobilised quickly, paving the way for rapid increases in output and export revenues. But continuing violence and insecurity are severely hampering reconstruction efforts and deterring inward investment. The outlook for the development of the country's vast oil reserves and, therefore, its economic development depends critically on how quickly political and social stability is restored.

Table 12.1: **Key Energy Indicators for Iraq**

	1971	2002	2003	1971-2003*
Total primary energy demand (Mtoe)	4.5	30.0	25.8	5.6%
Total primary energy demand per capita (toe)	0.47	1.24	1.04	2.5%
Total primary energy demand/GDP**	0.05	0.88	0.96	9.4%
Share of oil in total primary energy demand (%)	82	93	95	–
Net oil exports (mb/d)	1.62	1.52	0.88	–1.9%
Share of oil exports in production (%)	95	75	66	–
CO_2 emissions (Mt)	12.3	77.4	68.8	5.5%

*Average annual growth rate.
** Toe/thousand dollars of GDP in year-2004 dollars and PPPs.
Note: Data for 2003 are distorted by the invasion.

Political and Economic Situation

Political Developments

Since its founding as an independent state in 1932, Iraq has been plagued by political turbulence and conflict. Saddam Hussein seized power in 1979, eleven years after the Baath Party had overthrown the previous regime in a military coup. An eight-year war with Iran, which started in 1980, resulted in a colossal loss of human lives, a slump in national income – partly due to lower oil production and exports – and a deterioration in public infrastructure and welfare services. By the end of the war, the country had accumulated large external debts. Iraq invaded Kuwait in 1990, but was expelled six months later by coalition forces after a short war, which inflicted further damage to Iraq's infrastructure. A subsequent uprising against the regime in the south was brutally crushed, while a similar rebellion in the largely Kurdish north resulted in a degree of autonomy from Baghdad underpinned by military protection from coalition forces. UN economic sanctions imposed after the Kuwait war aggravated the dire economic situation. The country's highly repressive centralised economic structure also stifled economic growth and development.

Following the ousting of Saddam Hussein by a US-led military intervention in April 2003, Iraq was governed initially by the Office of Reconstruction and Humanitarian Assistance. This soon gave way to the Coalition Provisional Authority (CPA), headed by an American official. In March 2004, the CPA reached agreement with the Iraqi Governing Council, a consultative body made up of Iraqi leaders, on the Transitional Administrative Law, which sets out a timetable and a set of administrative arrangements for democratic and constitutional reform. A first major step in implementing this law was taken when power was transferred to an Iraqi interim government three months later. Elections to a transitional national assembly were held in January 2005, and a transitional government was finally approved by the president and the assembly in May. Many Sunnis, which make up around a fifth of the population, were either unwilling to vote or were unable to because of local violence.

The transitional assembly approved a draft constitution in August 2005. Approval of the constitution by a national referendum, originally scheduled for 15 October 2005, would lead to new elections to a permanent assembly and the formation of a permanent government. The constitution will pave the way for a new hydrocarbons law, establishing the legal framework for the future development of the industry. UN economic sanctions were lifted in May 2003.

Efforts to rebuild Iraq's judicial system, a key element in restoring political and economic stability, are continuing. Judicial independence – suppressed during Saddam's rule – has been formally established and attention is now focused on legal education, supported by donor projects and initiatives, and the security of jails. Corruption remains a major problem, reflecting institutional deficiencies. Transparency International ranks Iraq as the 17h most corrupt country in the world out of 145 nations covered (Transparency International, 2004).

Insurgent attacks on coalition troops and the civilian population continue in 2005. Casualties have been high and security remains the predominant concern for Iraqis. The violence has been centred on the middle of the country, although every major population centre – including towns and cities in the Kurdish north of the country and occasionally Basra in the south – has been hit. Oil installations, notably pipelines, have been badly affected. The bulk of the insurgents are disaffected Sunni members of the old Iraqi army, motivated largely by the suppression of their posts, by opposition to the role of the United States, or by a desire to resist Shiite dominance.

No timetable has been established yet for the withdrawal of coalition forces, because of difficulties in restoring security. The capacity of Iraqi security forces to deal with the insurgency is improving, but they are still far from ready to take over completely from coalition forces. Ending the insurgency and restoring law and order remains vital to the country's reconstruction efforts and is a major source of uncertainty surrounding the prospects for the rehabilitation and development of the country's oil and gas industry. Achieving that goal hinges largely on resolving political differences and negotiating a power-sharing deal acceptable to all ethnic groups.

Our analysis assumes that political and economic reform and reconstruction will move forward, if slowly, and that the Iraqi government will be able to contain the insurgency, if not completely suppress it, initially with continued coalition support. Nonetheless, the risk remains high that political instability and insurgency will continue for several years. Such an outcome would severely hamper reconstruction efforts and economic development and retard the exploitation of the country's hydrocarbon resources.

Economic Trends and Developments

Decades of war, sanctions and repression have crippled the economy. In real terms, the country's gross domestic product in 2003 was a mere 14% of what it had been in 1979. By the end of the 1970s, Iraq was among the most advanced countries in the MENA region; today, it is one of the least developed (Table 12.2). Most Iraqis today have only limited access to essential public services, including electricity, water supply and sanitation, and refuse collection. The lack of such services has contributed to public unrest and the general lack of security in many parts of the country. Nonetheless, anecdotal evidence suggests that there has been improvement in living standards since 2002.

Economic reconstruction and recovery are progressing despite the insurgency. Preliminary data show that GDP grew by around 35% in 2004, after slumping by 30% in 2003. The CPA's and the interim government's initial economic priorities were to restore essential public services and to stabilise national

Table 12.2: **Human Development Indicators, 2002**

	Life expectancy at birth (years)	**Adult literacy** (% ages 15 and above)	**Gross school enrolment** (%)	**GDP per capita** (PPP US$ nominal)
Iraq	60.5	39.7	57	1 317
Arab states	66.3	63.3	60	5 069
World	66.9	n.a.	64	7 804

Sources: UNDP (2004); IEA databases.

finances. The CPA introduced a new standardised national currency, the new Iraqi dinar, and a new set of banknotes in January 2004 and granted independence to the Iraqi Central Bank. New laws have been adopted liberalising trade and banking, establishing rules for public finance, audit and procurement and setting up inter-ministerial commissions on reconstruction, privatisation, oil and economic reform. The government's fiscal and monetary policies have been broadly anti-inflationary, although a surge in public spending is thought to have driven inflation back up to around 30% in 2004 (World Bank, 2005). Unemployment remains high at around 30%, but is thought to be falling as infrastructure projects get underway and private commercial activity expands. A nationwide food-rationing system and subsidies to essential goods and services, including electricity and oil products, remain in place.

The total cost of reconstruction has been estimated to amount to $55 billion for the period 2003-2007 (Tarnoff, 2004) and possibly as much as $100 billion in total. Most of the funding for economic reconstruction is initially coming from the United States, aided by other donor countries and development banks. But oil-export earnings are expected to become the major source of finance in the longer term. Two emergency US congressional bills provided for grants to the Iraq Relief and Reconstruction Fund of $2.5 billion for the financial year 2003 and a further $18.4 billion for 2004, although the bulk of these allocations have not yet been disbursed because of delays in implementing projects. The fund was initially managed by the CPA and is now controlled by the interim Iraqi government.

Other countries and international development organisations had pledged a total of $14.4 billion, of which at least $9 billion were loans, by mid-October 2004. Japan, the World Bank, the International Monetary Fund and Saudi Arabia are the biggest single contributors. Some aid is directed to trust funds run by the World Bank and the United Nations. Iraq's own resources, including the Development Fund for Iraq (DFI), totalled $28.4 billion by the end of September 2004. DFI deposits come largely from oil export earnings.

Most of these funds have already been spent on a wide variety of projects, including repairs to oil and electricity-supply facilities, though few international companies have been prepared to risk sending personnel to the country. Work that is taking place is slow and expensive, due to the additional costs of security.

Iraq took on a large amount of debt during Saddam Hussein's rule. Commercial debt is thought to amount to around $125 billion (Tarnoff, 2004), not including unpaid reparation claims stemming from Iraq's invasion of Kuwait in 1990. However, much of Iraq's debt is expected to be written off and the outstanding reparations renegotiated. The Paris Club, a group of official creditors that co-ordinates efforts to reschedule debts of nations facing payment difficulties, agreed in December 2004 to write off $31 billion in Iraqi debt. Half of this amount is dependent on the completion of an International Monetary Fund reform programme.

The success of efforts to rebuild Iraq's economy depends on its ability to export crude oil, especially in the short to medium term. Export revenues will help to finance non-oil infrastructure projects, social programmes and other forms of government spending. In the Reference Scenario, Iraqi GDP is assumed to grow by an average of 10.8% per year in 2003-2010, slowing to 5.7% in 2010-2020 and 5.5% in 2020-2030 (Table 12.3). Most of this economic expansion results directly from rising oil production and indirectly from rising oil revenues, particularly after 2010, as oil prices rise again (see Chapter 1).

The growth of the oil and gas sector is expected to stimulate expansion of the non-hydrocarbon sector. Strong growth in the active labour force and rising labour productivity also boost non-hydrocarbon GDP. The labour force is assumed to grow at an average rate of 3.1% per year over the projection period, underpinned by rapid growth in the overall population. Total population is assumed to grow rapidly in line with the latest UN projections, from 24.8 million in 2003 to 45 million in 2030 – an average annual growth rate of 2.2%.

Table 12.3: **GDP and Population Growth Rates in Iraq in the Reference Scenario**

(average annual rate of change in %)

	1971-2003	1990-2003	2003-2010	2010-2020	2020-2030	2003-2030
GDP	–3.5	–5.2	10.8	5.7	5.5	6.9
Population	3.0	2.5	2.7	2.3	1.8	2.2
Active labour force	3.0	3.3	3.5	3.2	2.8	3.1
GDP per capita	–6.3	–7.5	7.9	3.4	3.7	4.6

Energy Policy

Though the unstable and unpredictable political situation in Iraq makes it extremely uncertain how government energy policies will evolve in the longer term, it is very likely that future governments will seek to develop as rapidly as possible the country's vast hydrocarbon resources as a way of generating the funds needed to rebuild infrastructure and to pay for social welfare programmes. That is the stated near-term goal of the current Iraqi government. In the longer term, Iraq's relationship with OPEC and the country's role in collective efforts to manage oil supply and prices may affect depletion and investment policies (see below).

The extent to and pace at which the government liberalises the energy sector is a critical issue. The current government has indicated that it would favour equity participation by foreign companies in developing hydrocarbon resources and rebuilding the electricity infrastructure, but no law has yet been adopted that would allow such a development. The manner in which foreign companies can participate in energy projects remains to be decided. Energy-price reform and market restructuring to encourage competition are expected to proceed. The government has prepared a proposal to liberalise the market for oil products, under which retail prices would remain controlled during a transition period of up to two years, after which they would be increased gradually to international levels. Energy and some food items are the only goods whose prices are yet to be fully deregulated. Prices are typically set well below cost, leading to waste, smuggling and inefficiencies.

Energy Demand

Iraq depends almost entirely on oil for its primary energy needs, though natural gas produced mainly in association with oil has played a growing role since the 1970s. Total primary demand reached an all-time high of 30 Mtoe in 2002, before dropping back to 25.8 Mtoe in 2003 – the year of the invasion. Around 93% of the energy consumed in 2003 was oil, with gas accounting for almost all the rest. Hydropower is the only other source of energy, but provides less than 0.2% of the country's primary needs and 2% of its electricity. The rest of the country's electricity is generated exclusively from oil.

Historical data on final energy use are patchy (Box 12.1). According to the latest data available for 2003, oil accounts for 80% of total final consumption, natural gas for 7% and electricity for around 13%. Transport accounts for 53% of final oil use, industry for 20% and residential and commercial uses for most of the rest. Gas is used mainly in industry, mostly in petrochemical and fertilizer plants. No data are available for the sectoral breakdown of electricity use, though anecdotal evidence suggests that the main consumer is the residential sector.

Box 12.1: **Iraqi Energy Data Deficiencies**

There are few reliable economic statistics to hand for Iraq. War, sanctions and political interference greatly impeded the capacity of Iraq's statistical agencies to collect and publish information on a regular basis (World Bank, 2004). The energy sector was no exception. Most national statistics collected since 1991 excluded the Kurdish northern region. And there are no records of trade in energy or other goods carried out in violation of UN sanctions. The old regime published no detailed energy statistics and any records that were kept were largely destroyed or lost during the invasion in 2003.

IEA energy statistics for Iraq are derived from various sources, including OPEC and the Organization of Arab Petroleum Exporting Countries (OAPEC). Final consumption data for electricity are not available at a disaggregated level. Since 2003, detailed electricity production data have been available. There is an urgent need for the Iraqi government to build modern systems for data collection and dissemination so as to provide policy-makers and investors with reliable information on which to base their decisions.

Iraqi energy consumption is projected to grow strongly over the *Outlook* period, driven by economic recovery and a rapidly growing population. In the Reference Scenario, primary energy demand is expected to more than double between 2003 and 2030, growing at an average annual rate of 3.3% (Table 12.4). Growth is expected to be most rapid in the rest of the current decade, at 4.6% per year, and to slow to under 3% per year in the 2020s as the pace of economic expansion slackens. CO_2 emissions are projected to grow by 2.4% per year over 2003-2030.

Table 12.4: **Iraq's Total Primary Energy Demand in the Reference Scenario**
(Mtoe)

	1990	2003	2010	2020	2030	2003-2030 *
Oil	17.2	24.4	32.5	40.2	48.8	2.6
Gas	1.6	1.3	2.6	6.7	12.5	8.8
Hydro	0.2	0.0	0.1	0.5	0.8	11.9
Total	**19.1**	**25.8**	**35.3**	**47.4**	**62.0**	**3.3**

* Average annual growth rate.

Oil will remain the dominant energy source in Iraq over the projection period. Primary oil use is projected to rise from 24 Mtoe in 2003 to 49 Mtoe in 2030. As in other countries, transport will account for the lion's share of the growth in oil demand. The use of natural gas is expected to grow more rapidly in percentage terms. Output of both non-associated and associated gas, much of which is currently flared, is expected to grow strongly, providing a cheap and readily available supply of energy to meet rising demand from power stations and industry. Petrochemical and fertilizer plants are expected to absorb much of the increase in gas supply. By 2030, oil's share of primary demand will have fallen by 16 percentage points, to 79%, while the share of gas will have increased by 15 percentage points to 20%. Hydropower is expected to remain a marginal source of primary energy (Figure 12.1).

Figure 12.1: **Iraq's Primary Energy Demand by Fuel in the Reference Scenario**

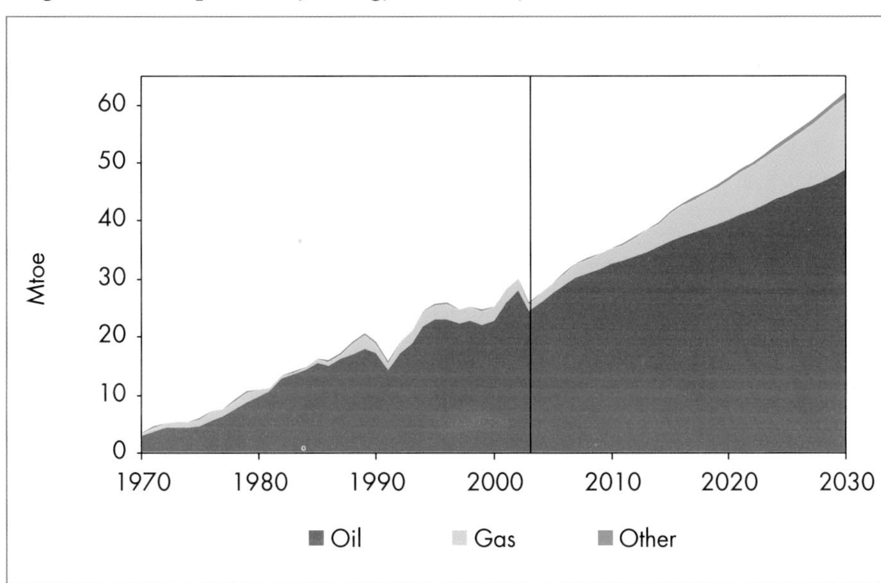

12

Final energy demand grows more rapidly, by 3.9% per year over 2003-2030, though the fuel mix is expected to change less significantly. The share of oil is projected to fall from 80% at present to 75% by 2030. Consumption of gas is projected to grow more rapidly than any other final energy source – by 5.5% per year on average over the *Outlook* period – and its share will increase from 7% to 11%.

The share of the power sector in total primary demand is projected to drop slightly, from 28% to 23% over the projection period. Relatively brisk growth in final demand for electricity will be more than offset by large improvements in the thermal efficiency of power plants as new, more efficient stations are built and the most inefficient plants are shut down.

Energy intensity will fall substantially, as demand will grow much more slowly than GDP (Figure 12.2). Primary energy intensity – energy use per unit of GDP – will fall by about two-thirds between 2003 and 2030, but will still be about 50% higher than the average for MENA and two-and-a-half times higher than that for developing countries. By 2030, intensity will be back to the level of the late 1980s. Per capita energy use will nonetheless increase steadily in line with rising household incomes and economic activity.

Figure 12.2: **Energy Intensity Indicators in Iraq in the Reference Scenario**

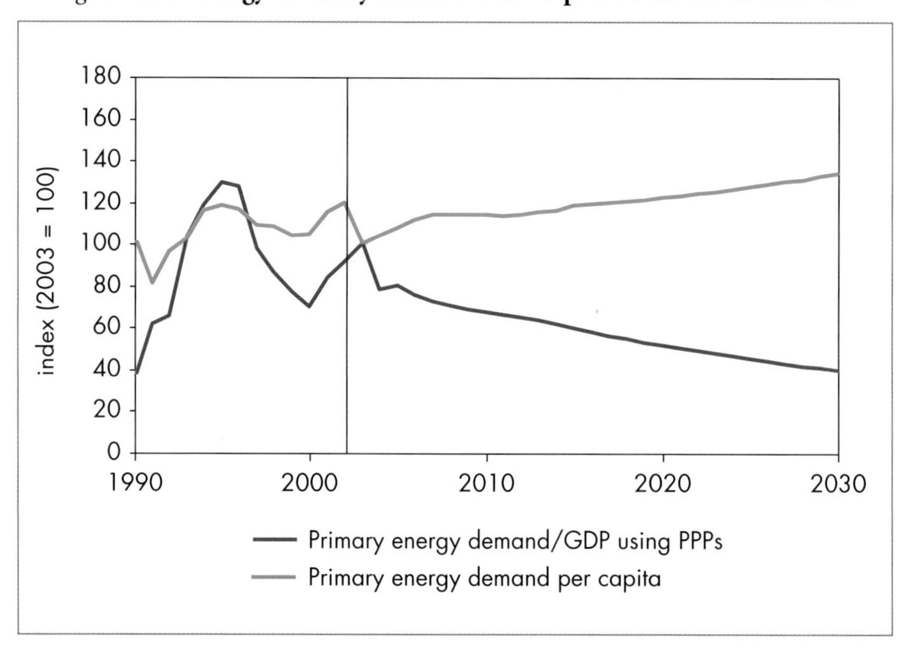

Electricity Supply

Overview

Iraq's electricity generation is estimated at 28 TWh in 2003, after having been flat at around 30-32 TWh in the late 1990s.[1] Electricity is almost entirely based on oil, with less than 2% met by hydropower. Installed power-generation

1. The exact level of electricity generation is not known. See also Box 12.1 on the accuracy of Iraqi data.

capacity is estimated between 8 and 10 GW, based on data from UNDP, World Bank and Iraq's Ministry of Electricity. Table 12.5 shows Iraq's generation capacity by type. Most electricity is generated in oil-fired boilers. The available (de-rated) capacity of power plants is much lower than the nameplate (rated) capacity. In addition to central stations, Iraq had in the past a number of power plants on industrial sites, but their exact capacity and condition are not known.

Table 12.5: **Iraq's Installed Power-Generation Capacity** (MW)

	Nameplate capacity (MW)	**De-rated capacity** (MW)
Thermal	5 415	3 275
Msaiyiab	*1 200*	*900*
Nassirya	*840*	*600*
Beji	*1 320*	*560*
Najibya	*200*	*200*
Hartha	*800*	*360*
Dura	*640*	*380*
Baghdad South	*355*	*240*
Dibs	*60*	*35*
Gas turbines	2 181	1 383
Hydropower	2 518	n.a.
Diesel units	387	n.a.
Total	**10 501**	**n.a.**

n.a.: not applicable
Sources: World Bank and IEA databases.

Wars, sanctions and sabotage have had a devastating effect on Iraq's power sector. Most of Iraq's power plants were built in the 1970s and 1980s (Figure 12.3). Consequently, a third of installed capacity is over 20 years old. During the 1990-1991 Gulf war, Iraq's electricity infrastructure was heavily bombed, and power-generation capacity – around 9 GW at the time – was drastically reduced. Some power plants were rehabilitated later in the 1990s but available capacity remains substantially lower. This resulted in severe power shortages throughout the country, though supplies to Baghdad were less affected. While residents of the capital would enjoy largely uninterrupted electricity supply, people in other parts of the country would have much less, sometimes for just a few hours a day (USAID, 2004). The lack of capacity – resulting from sanctions and a lack of investment – explains the very low average per capita electricity consumption in Iraq. It averaged only about 1 400 TWh in 2002, compared with 3 016 TWh in the Middle East as a whole.

Looting and vandalism after the 2003 war further aggravated Iraq's power supply problems. The electrical grid was particularly affected as components of it were torn down. The various organisations now working in Iraq are progressively restoring power capacity and the grid, despite continuing insurgent attacks. Besides improving the existing infrastructure, they have carried out a number of projects aimed at emergency support, including the provision of diesel generators and mobile substations. Available capacity had reached about 5 GW in September 2005.

Figure 12.3: **Power-Generation Capacity Additions in Iraq** (MW)

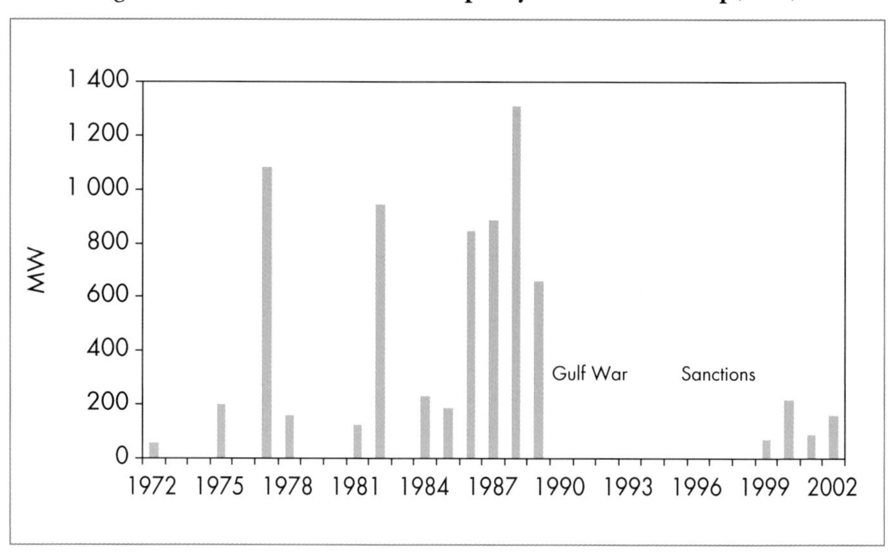

A small part of current demand is met by imports. Turkey started supplying electricity to northern Iraq in 2003. It currently makes available some 200 MW of capacity, which is expected to increase by another 1 000 MW by 2006 (PEi, 2005). Syria and Iran also supply electricity to Iraq, and Kuwait and Jordan (by on-selling power from Egypt through Jordan) could also do so.

Responsibility for electricity production, transmission and distribution lies with the Ministry of Electricity, which was established in 2003. Before that, the electricity sector was controlled by the Commission of Electricity. The ministry sets the tariffs for electricity, which are currently very low (Table 12.6).[2]

2. See Chapter 6 for an overview of electricity tariffs in MENA countries.

Table 12.6: **Current Electricity Tariffs in Iraq** (US cents/kWh)

Residential	0.07 – 2.00
Commercial	0.13 – 1.66
Industry	0.13 – 0.57
Government	0.13 – 0.33
Agriculture	0.33

Source: Ministry of Electricity website (http://www.iraqelectric.org/aboutministry.htm).

Projects to rehabilitate and expand electricity generation, transmission and distribution notably include the following:

- 27 power generation projects that include the construction of gas-fired power stations in various locations around the country.
- 50 rehabilitation projects to refurbish, rebuild and maintain existing power plants.
- 67 projects to improve the transmission grid, including the refurbishment of existing transmission lines and substations and the construction of new 400 kV and 132 kV lines and substations.
- 23 projects to revamp the distribution network, including rehabilitation of the network, cables and substations, the expansion of the network and the construction of control centres.
- 31 projects to build communication and control centres, including the reconstruction of the primary control centre in Baghdad.

Electricity Production

Iraq's electricity generation is expected to pick up again gradually, as more generation and transmission capacity becomes available. It is projected to increase at an annual average growth rate of 4% between 2003 and 2030, to deliver 82 TWh in 2030. Total new capacity requirements over 2003-2030 will be 12 GW.

In the near term, electricity generation will continue to be mostly oil-based, but as Iraq's gas network is restored and gas production capacity expands, more gas will be available for power generation. The share of gas in the electricity generation mix is projected to rise to 53% by 2030. Most new power plants are expected to be combined-cycle gas turbines. Some existing steam boilers that now use oil are likely to be switched to gas.

In the Reference Scenario, hydropower capacity is assumed to be restored progressively, with output reaching 2 TWh in 2010. In the 1990s, Iraq was planning to build new hydropower plants, but their development was suspended or delayed. The country has a total technical potential of 90 TWh, the highest in MENA (WEC, 2004). More hydropower plants could be built in the long term, as the economy recovers and demand for electricity rises. Hydropower capacity is projected to rise to 5 GW by 2030, contributing 11% of Iraq's electricity generation.

Oil Supply

Overview

Crude oil production has fluctuated enormously in recent decades. Production reached 3.9 mb/d in December 1979, but fell sharply with the start of the Iran-Iraq war to a low of 900 kb/d in 1981. Output recovered rapidly later in the decade, reaching 2.8 mb/d in 1989, but fell again after the 1991 Gulf War. Trade sanctions held production down to well under 1 mb/d during 1992-1996, but it recovered to around 2.6 mb/d by 2000 with the introduction of the UN Oil for Food Programme. Production fell back to 2 mb/d in 2002 because of disputes over the programme. The coalition forces that invaded Iraq in 2003 took care to inflict no more material damage than was necessary to ensure a quick military success, with minimal coalition casualties, so as to assist the post-war reconstruction. But fighting and subsequent looting nonetheless seriously damaged oil/gas separation plants, oil-pumping stations, gas compressors, production facilities, water supply and injection facilities and pipelines.[3] Sabotage and attacks on personnel have since caused new and repeated damage to the country's oil facilities and disrupted operations. As a result, oil production (including NGLs) dropped to an average of 1.3 mb/d in 2003, recovering to 2 mb/d in 2004. Insurgents continue to target oil installations, including the main export pipeline system that runs from Kirkuk to Ceyhan in Turkey, which has been out of action for much of 2005. As a result, crude oil production fell back from a post-war high of 2.4 mb/d in March 2004 to only 1.87 mb/d in July 2005.

The Iraqi authorities aim to increase production to 2.5 mb/d by late 2005. But achieving this will depend largely on securing oil facilities against insurgent attacks, so permitting the completion of repairs to pipelines and other facilities and of rehabilitation work on oil wells, separation plants and water-treatment facilities. Production in 2005 is likely to average about 1.9 mb/d. In the longer term, further significant increases in production capacity will require tens of billions of dollars of investment in new wells at existing fields and developing new fields. Iraq's proven oil reserves are estimated at 115 billion barrels – the fourth-largest in the world. Many fields are thought to be relatively easy and cheap to

3. Zainy, M-A, "Iraq's Oil Industry: Short-Term Restoration Plan", in *Middle East Economic Survey*, 13 October 2003.

develop. But how quickly these reserves can be tapped hinges on domestic security, as well as on the legal and commercial framework that will emerge.

Our Reference Scenario assumes security is gradually restored and reserves are opened up in a progressive manner, allowing oil production to grow to 3.2 mb/d in 2010 and 7.9 mb/d in 2030. Total oil exports (crude and refined products) grow from 1.4 mb/d in 2004 to 2.5 mb/d in 2010 and 6.9 mb/d in 2030. Production and exports are assumed to grow much more slowly in the Deferred Investment Scenario (see the final section of this chapter).

Policy Framework

Restructuring the oil industry is a major challenge facing the government and a source of critical uncertainty. Oil policy is in the hands of the Supreme Council for Oil and Gas Policy, established in July 2004 and headed by the prime minister. It is responsible for planning, investments, financing, marketing of crude oil and products, pricing of domestic product sales and contracts with foreign companies. The Energy Council, set up in May 2005, is responsible for carrying out the functions of the Supreme Council. However, the Transitional Administrative Law specifically precludes the government from making deals affecting the long-term development of the country, with the exception of steps towards debt reduction. This means that any major change in oil policy, including any large-scale opening-up of the industry to private Iraqi and foreign investment, will not be politically feasible until a ratification of the new constitution and installation of a permanent government. A hydrocarbon law will then need to be adopted.

The extent to which foreign companies are able and willing to participate in the future development of the oil industry will have a major impact on the pace of production-capacity increases. At present, the upstream oil industry is in the hands of the state-owned South Oil Company (SOC) and North Oil Company (NOC), both remnants of the Iraqi National Oil Company (INOC), which was dismantled by Saddam Hussein in 1987. The government is expected to approve a plan to re-establish INOC, under the supervision of the Ministry of Oil. INOC would take over the SOC, NOC and the State Oil Marketing Organisation, which continues to handle the supply of crude oil to refineries and oil exports. Privatising the companies separately or as a reunited entity, an option initially favoured by the United States, is unlikely to gain political or popular support in the foreseeable future. But foreign company involvement in assisting in the rehabilitation of existing producing fields – beyond the repairs already being undertaken by US contractors – as well as in developing new fields is expected to be more publicly acceptable.

Iraq needs to attract international oil companies as a source of both capital and technical and project management expertise. However, future Iraqi governments may judge that the Iraqi industry is capable of pursuing projects without foreign assistance and that, with the large rents available from new field development, the national Iraqi companies should be able to borrow enough money from the banks, using incremental oil as collateral. If international companies are given the opportunity to invest in Iraq, their involvement could take the form of buy-back or production-sharing contracts. The legal status of oilfield development agreements concluded by the Saddam regime with a number of foreign companies, including production-sharing contracts with Russia's Lukoil and the China National Petroleum Corporation remains unclear. Some could remain in place with renegotiated terms. A committee was formed in June 2005 to review these contracts.

Political developments and the restoration of security will determine how quickly a new oil policy is agreed, the industry restructured and a new legal and commercial framework put in place The international oil companies have indicated that they will not send personnel to Iraq as long as security remains a major risk, though some smaller independents have expressed some willingness to do so. Yet the international companies are cultivating relationships with the Ministry of Oil and providing some technical assistance, often free of charge, in order to position themselves for any future opening-up of the industry.

The government may impose some temporary limits on future oil developments in order to respect OPEC production ceilings or targets. This will not be an issue for several years and will most likely not arise until Iraq's share of total OPEC production approaches the level prevailing at the end of the 1970s. Our analysis assumes that, in the event that Iraq decides to remain a member of OPEC, the country will be able to negotiate progressive increases in its quota, allowing it to increase its share of overall OPEC production.

Resources and Reserves

Iraq's proven reserves at the end of 2004 are generally accepted to be 115 billion barrels (BP, 2005; *Oil and Gas Journal*, 2004; OPEC, 2004; and *World Oil*, 2005). However, the reliability of current reserve estimates is thought to be much less than for most other MENA countries. Some 60% of reserves are in fields that have yet to be developed. The country's production potential is widely accepted to be much higher, as large parts of the country have yet to be properly explored. The introduction of modern production techniques could also significantly raise recovery rates. IHS Energy puts proven plus probable reserves at 98.7 billion barrels, while the US Geological Survey estimates ultimately recoverable resources at 160 billion barrels (USGS, 2000). The French Institut du Pétrole, in a 2004 study, puts undiscovered resources at 60 to 200 billion barrels. The Ministry of Oil estimates ultimately recoverable resources at 214 billion barrels. Iraq's reserves vary widely in quality.

Figure 12.4: **Iraq's Oil and Gas Resources and Infrastructure**

Of the 73 discovered oilfields in Iraq, six are classified giant or super-giant, containing at least 5 billion barrels of proven reserves. Five of these fields – Rumaila, Kirkuk, East Baghdad, West Qurnah and Zubair – are in production, while Majnoon is awaiting development. A further 23 fields each hold more than 500 million barrels (Arab Petroleum Research Centre, 2004). Most of these fields and the bulk of the other proven reserves are in southern Iraq. Two-thirds of discovered fields have been appraised only down to the Tertiary or Cretaceous levels, which contain 99% of remaining proven reserves. Northern reservoirs are generally at less than 1 000 metres, while southern fields generally lie between 2 000 and 4 000 metres deep.[4] However, drilling to deeper strata in the late 1990s confirmed the potential for light crude and condensates. The chance of finding deeper reserves in the Western Desert region, which has hardly been explored, is thought to be particularly strong. To date only about 10% of the country has been actively explored (Arab Oil and Gas Research Centre, 2005).

Crude Oil Production[5]

There are around 20 fields currently in production in Iraq (Table 12.7). The largest by far are Rumaila, which has remaining proven and probable reserves of 15 billion barrels and production in 2004 of more than 1 mb/d, Kirkuk (11 billion barrels of reserves and production of 0.25 mb/d) and East Baghdad (18 billion barrels). Kirkuk, located in northern Iraq, is the country's oldest field and one of the first to be developed in the Middle East, having started producing in 1934. It normally produces a light, sour crude oil, with an API of around 36° and 2% sulphur. API gravity has fallen, and the sulphur content and water cut have risen since late 2002, due to technical problems caused by poor management of the field and a lack of equipment, resulting partly from sanctions. The Rumaila field, whose southern tip lies across the border with Kuwait, has 660 wells and normally produces three streams of crude oil: Basra light (normally 34°API), Basra medium (30° API) and Basra heavy (22-24° API). Because of production problems since the 2003 conflict, crude oil streams have been blended into a single export grade, with an average quality of around 30° API and 2% sulphur. The water cut at Rumaila has risen to about 25% on average, and the natural decline rate is reported to have accelerated to about 10%. The Kirkuk and Rumaila fields together produced 72% of total Iraqi oil production in 2004, up from about two-thirds in 2000.

4. G.Haider, "Economics of Oil Fields' Development Ventures in Iraq", in *Middle East Economic Survey* (23 February 2004).

5. Unless otherwise stated, all the oil-production figures cited in this section include crude oil, natural gas liquids and condensates.

Table 12.7: **Oilfields in Iraq**

Field	**Year of first production**	**Remaining proven and probable oil reserves at end-2004** (billion barrels)	**Cumulative production to 2004** (billion barrels)	**Gravity** (average °API)
Abu Ghirab	1976	0.6	0.4	24.0
Ain Zalah	1951	0.1	0.1	31.0
Ajeel	1999	0.7	<0.1	37.5
Bai Hassan	1960	1.5	0.7	34.0
Buzurgan	1976	2.4	0.1	24.0
East Baghdad	1989	18.0	<0.1	23.0
Jambur	1959	2.8	0.2	38.0
Khabbaz	1994	2.0	<0.1	37.8
Kirkuk	1934	11.3	13.7	36.0
Luhais	1977	1.0	0.1	33.0
Naft Khaneh	1927	0.3	0.1	42.5
Nahr Umr	1975	3.5	<0.1	42.0
Qaiyarah	1936	0.4	<0.1	16.0
Rumaila	1954	15.5	8.5	33.0
Subba	1990	2.2	<0.1	28.8
Sufaya	1984	0.2	<0.1	25.0
West Qurnah	1999	9.4	<0.1	27.1
Zubair	1952	6.1	1.6	35.0
Other fields		7.0	<0.1	
Total		**84.9**	**25.8**	

Sources: IHS Energy and IEA databases; Verma *et al* (2004).

Efforts to restore oil-production capacity in the aftermath of the 2003 war were severely hindered by looting and, more recently, by insurgent attacks on oil installations. These problems have been made worse by the damage to reservoirs that has resulted from field management practices aimed at maximising short-term flow rates since the late 1990s. Both the North and South Oil Companies adopted reservoir-management techniques normally considered unacceptable in the oil industry, including excessive water flooding (injection of water in excess of what is necessary to maintain pressure, in order to boost short-term oil-flow rates) and, in the case of Kirkuk, injection of surplus refined heavy fuel oil into reservoirs. Some wells have been damaged beyond repair. Lack of maintenance, partly due to a lack of spare parts, has also contributed to production difficulties.

The rehabilitation programme agreed in the immediate aftermath of the war involved repairing and modernising the above-ground infrastructure, with the aim of bringing sustainable capacity back to the pre-war level of around 2.8 mb/d. A Restoration Plan was drawn up in July 2003 by the US Army Corps of Engineers, its main US contractor, Kellogg Brown and Root (KBR), and Iraqi Oil Ministry officials. KBR was hired to carry out the bulk of the work in the south and Parsons, another American firm, in the north. Key actions included fixing the Qarmat Ali water injection facility, which has enabled higher output at Rumaila and other southern fields. Intermediate production targets were met, but the final target of 2.8 mb/d by April 2004 was not, due to worsening security problems, inadequate power supplies, labour unrest and administrative inefficiencies. In addition, insurgent attacks on the Kirkuk-Ceyhan pipeline have led some Kirkuk wells to be shut in, or oil to be reinjected for lack of an outlet, after extraction of much-needed associated natural gas and gas liquids. Many of the projects that had initially been identified have still not been completed. Nonetheless, sustainable production capacity by mid-2005 would have been in the region of 2.5 mb/d, had it not been for the closure of the Kirkuk-Ceyhan line.

The next stage of the rehabilitation programme involves work-overs of old wells, drilling new production wells and building new infrastructure to support an increase in output, mainly from the southern fields already in production, particularly Rumaila. In particular, new gas/oil separation plants and water treatment facilities are needed. KBR is building a separation plant and LPG train near Basrah at a cost of $130 million. In total, about $3 billion has so far been set aside for oil rehabilitation projects, but much of this has gone to repairing damage caused by insurgent attacks and to security arrangements.

The Oil Ministry has commissioned a number of foreign oil companies to carry out technical or operational reservoir studies. Shell is conducting a study of Kirkuk and is drawing up a master plan for natural gas. BP and ECL, a UK engineering firm, are studying the Rumaila field. These studies are not expected to be completed before 2006. In no case will staff be sent to work in Iraq because of security risks. Other firms are offering free consultancy or training in order to be in position when major development projects are put out for tender. A few engineering, procurement and construction contracts have been awarded, mostly to small independent or local firms. The ability of these firms to complete the work on time and within budget is in doubt.

The ministry initially expected the rehabilitation programme, when completed, to boost production from existing fields to around 3 mb/d by the end of 2005, but delays in implementing projects – in part due to security problems – will most likely mean that this level of production is not reached until later in the current decade. A lack of funding has limited the number of well work-overs and new wells drilled. As a result, production capacity has barely increased in the last year. Actual production of crude oil is expected to fall slightly in 2005, to 1.9 mb/d compared with just over 2 mb/d in 2004, though this is largely due to disruptions to pipelines in the north. The current target for the end of 2005 is 2.5 mb/d. Further increases in production from existing fields – notably Rumaila, West Qurnah and Nahr Umr – are expected in the medium term, with additional drilling. But from the next decade, most of the growth in Iraq's oil-production capacity will have to come mainly from new field developments, which will be more expensive and will take longer to implement. Of nine new fields that have been earmarked for early development, Majnoon is the largest, with proven reserves of over 12 billion barrels (Table 12.8). The South Oil Company is targeting initial capacity at that field, located close to the Iranian border, of 600 kb/d. Ultimate capacity could be as high as 2 mb/d.

Table 12.8: **Major Undeveloped Oilfields in Iraq Earmarked for Development**

Field	**Year discovered**	**Proven oil reserves** (billion barrels)	**Oil in place** (billion barrels)	**Official targeted production** (kb/d)	**Planned number of production wells**
Majnoon	1977	12.10	38.12	600	300
Halfaya	1975	4.61	16.11	250	120
Ratawi	1950	3.13	13.63	200	113
Nassiriyah	1975	2.62	8.03	300	158
Tuba	1959	1.53	4.64	180	86
Gharaf	1975	1.13	3.92	100	68
Rafidain	1976	0.69	3.05	100	74
Amara	1980	0.49	1.69	80	44
Al Ahdab	1979	n.a.	4.54	90	n.a.
Total		**26.3**	**93.73**	**1 900**	**960**

Sources: IHS Energy databases; Verma *et al* (2004).

The potential cost of developing new fields is very low by international standards, although precise estimates are not available. A recent study estimates development costs at around $750 per barrel per day of capacity in Kirkuk and

6. The results of the study, carried out by Tariq Shafiq, a former vice president of INOC, are summarised in "Iraq Oil Development Policy Options: In Search Of Balance", *Middle East Economic Survey*, 15 December 2003.

the surrounding areas, $1 570 per b/d in the south around Zubair and Rumaila, and $3 130 per b/d for the smaller fields in the north-west.[6] The average for the country as a whole is put at $1 040 per b/d. Other studies suggest development costs may be somewhat higher, at over $3 000 per b/d. But even this higher figure is far lower than almost anywhere else in the world. Our projections assume an average cost of just over $4 000 per b/d (in year-2004 dollars), or about $1 per barrel of oil produced assuming ten years of production.

How quickly production can be increased depends largely on political developments. There is no doubt that reserves are large enough and the development costs low enough to support a rapid increase in capacity. But raising production will call both for security to be restored and for an appropriate commercial and legal framework to be put in place. The oil industry has endured more than 200 insurgent bombings in the last two years, resulting in total export-revenue losses and repair costs of up to $10 billion.[7]

In our Reference Scenario, Iraqi crude oil production (including NGLs) is projected to grow from 2 mb/d in 2004 to 3.2 mb/d in 2010, 5.4 mb/d in 2020 and 7.9 mb/d in 2030. These projections are derived from a bottom-up assessment of new oilfield developments and rehabilitation projects, combined with a top-down analysis of longer-term development prospects (see Chapter 4, Box 4.4). The latter is based on an assessment of development costs, as well as judgments about the country's future capacity and willingness to boost upstream investment, taking account of Iraq's growing importance on the world oil market. The projected rate of production increase is considerably lower than that targeted by Iraqi officials.

Production from currently producing fields is expected to grow steadily over the projection period, reaching almost 6 mb/d in 2030. The introduction of modern techniques and practices, such as horizontal drilling and advanced completions, will boost recovery rates and the drilling of new wells will raise production rates. Improved and enhanced oil recovery techniques are also expected to increase production towards the end of the projection period. The Rumaila field will remain the single most important contributor to Iraqi production until the last decade of the projection period. Several smaller producing fields – notably West Qurnah, Zubair and East Baghdad – will see their shares rise significantly (Table 12.9). The rest of the increase in Iraqi production is expected to come from several discovered fields that have yet to be developed (Figure 12.5). The most important among them are likely to be

7. Hussain, A., "Future Challenges and Prospects of the Iraqi Oil Industry", in *Middle East Economic Survey*, 4 July 2005.

Figure 12.5: **Iraq's Oil Production by Source in the Reference Scenario**

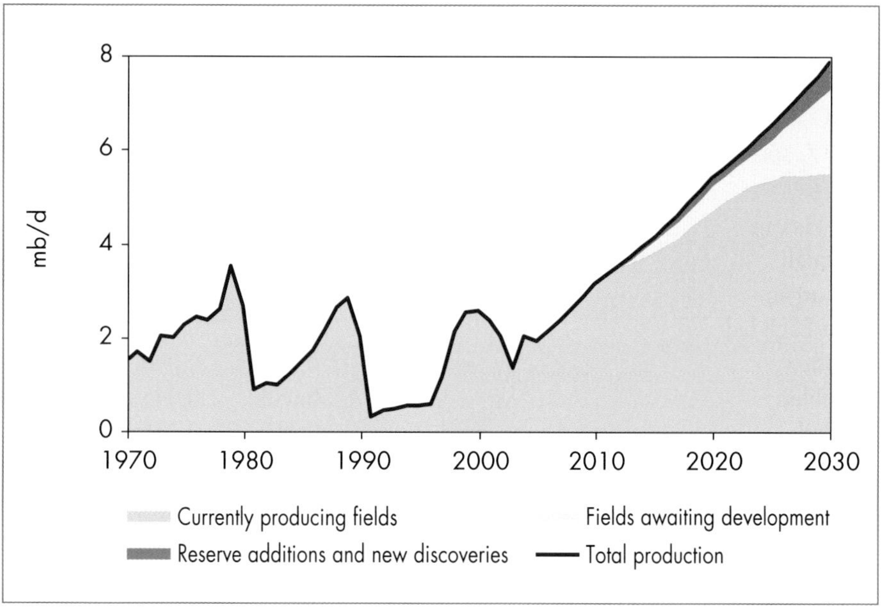

mb/d

8
6
4
2
0

1970 1980 1990 2000 2010 2020 2030

Currently producing fields
Fields awaiting development
Reserve additions and new discoveries
Total production

Majnoon, Halfaya and Nassiriyah. Although fields yet to be discovered may well contribute to production before 2030, the bulk of new production is expected to come either from existing producing fields or discovered fields that are awaiting development.

There are considerable risks to our Reference Scenario projections. In the short term, persistent insecurity would hinder the speed with which existing production facilities can be restored. Further ahead, new field developments will call for the introduction of modern technology, training of the local workforce and large investments. How quickly the necessary conditions can be created is very hard to predict. Even if the political and investment climate is conducive to attracting large amounts of capital, there will be practical constraints on how fast projects can be launched and brought on stream. The faster the pace of development targeted by the government, the more the country will have to rely on foreign involvement, both for technical services and capital. The Iraqi government will also need to take into account the impact on world markets of a rapid expansion of capacity, whether or not it is bound by any collective production ceiling. The impact of a lower level of investment on production prospects is analysed in the Deferred Investment Scenario (see the last section of this chapter).

Table 12.9: **Iraq's Oil Production in the Reference Scenario** (mb/d)

	2004	2010	2020	2030
Currently producing fields	**2.010**	**3.150**	**4.729**	**5.508**
Abu Ghirab	0.003	0.040	0.050	0.021
Ain Zalah	0.001	0.008	0.007	0.003
Ajeel	0.025	0.084	0.084	0.025
Bai Hassan	0.080	0.175	0.175	0.064
Butmah	0.000	0.005	0.004	0.002
Buzurgan	0.095	0.120	0.188	0.133
East Baghdad	0.000	0.040	0.100	0.357
Jambur	0.059	0.090	0.090	0.087
Khabbaz	0.015	0.040	0.045	0.045
Kirkuk	0.250	0.315	0.276	0.226
Luhais	0.025	0.050	0.100	0.066
Naft Khaneh	0.000	0.030	0.032	0.003
Nahr Umr	0.001	0.050	0.202	0.149
Qaiyarah	0.030	0.034	0.024	0.011
Rumaila	1.200	1.127	1.084	0.804
Subba	0.000	0.080	0.100	0.100
Sufaya	0.001	0.015	0.015	0.015
West Qurnah	0.125	0.259	0.660	1.050
Zubair	0.100	0.100	0.300	0.402
Other fields	0.000	0.488	1.193	1.943
New developments	**0.000**	**0.000**	**0.671**	**2.342**
Fields awaiting development	0.000	0.000	0.523	1.827
Reserve additions and new discoveries	0.000	0.000	0.148	0.515
Total	**2.010**	**3.150**	**5.400**	**7.850**

Note: Includes NGLs and condensates.
Source: IEA analysis.

Oil Refining

At the start of 2005, available crude oil refining capacity in Iraq stood at 587 kb/d (Table 12.10). The country has eight refineries, all but one of which (a tiny mothballed plant in Kirkuk) are operated by administrative units within the Ministry of Oil. Three refineries at Baiji, Basrah and Daurah account for almost all the capacity. A ninth refinery at Samawa, which has a capacity of 20 kb/d, has recently been rebuilt and has restarted. None of the refineries suffered any damage during the 2003 war, but all of them have experienced operating problems due to either looting, sabotage, shortages of crude oil or electricity supply problems. As a result, throughput averaged only about

Table 12.10: **Iraq's Refining Capacity at 1 January 2005** (kb/d)

Location	Crude distillation	Vacuum distillation	Catalytic reforming	Catalytic hydro-cracking	Lubricants
Kirkuk	2.0	–	–	–	–
Baiji	310.0	105.0	46.0	38.0	5.0
Basrah	150.0	20.0	16.0	–	100.0
Daurah	100.0	20.0	15.0	–	120.0
Haditha	7.0	–	–	–	–
Khanaqin	12.0	–	–	–	–
Mufthia	4.5	–	–	–	–
Qaiyarah /Mosul	2.0	–	–	–	–
Total	**587.5**	**145.0**	**77.0**	**38.0**	**225.0**

Source: *Oil and Gas Journal,* 20 December 2004.

400 kb/d in 2003. The configuration of the refineries is poorly adapted to the needs of the domestic market, due to lack of investment over many years. Output of light products – gasoline, diesel and liquefied petroleum gas – is running well below domestic demand, so Iraq has to import the deficit. Gasoline imports averaged around 40 to 50 kb/d in 2004, almost half total consumption. Because gasoline is sold at heavily subsidised prices, these imports cost the treasury around $2 billion that year and may cost as much as $1.5 billion in the first half of 2005.[8] In contrast, substantial volumes of heavy fuel oil are exported, as output far exceeds local demand. The quality of locally produced gasoline remains a serious problem.

To meet the growing shortfall in light products, the Iraqi government is soliciting interest from foreign contractors to build a 150-kb/d refinery on a build-operate-transfer basis. Output from the new plant, to be sited at Nahrain, just south of Baghdad, would permit the phased renovation of the Baiji and Daurah plants, to lift their capacities back to nameplate levels. The government has already signed contracts with foreign companies for work costing $100 million to increase the capacity of the Daurah refinery and improve the quality of the gasoline it produces. Some new conversion units at Baiji and Basrah to increase production of lighter products are planned. The State Company for Oil Projects issued a tender in June 2005 for the design, procurement and construction of the Nahrain refinery. On the assumption that this refinery is indeed built by the end of the current decade, refining capacity

8. "Iraq Must Focus on Domestic Priorities, Not Exports, Says Issam al-Chalabi", *Middle East Economic Survey*, 7 February 2005.

is projected to rise to about 800 kb/d in 2010. The government envisages building a second refinery of 200-300 kb/d, but this is unlikely to be built before 2010. We project a further increase in total distillation capacity to about 1.0 mb/d in 2020 and 1.2 mb/d in 2030.

Oil Exports

Oil exports recovered from an average of 0.9 mb/d in 2003 to an estimated 1.4 mb/d in 2004, close to the average for 2002, and averaged about 1.4 mb/d in the first half of 2005. Exports would have been higher but for attacks on the Kirkuk-Ceyhan pipeline and for power and weather-related problems at the country's export terminals on the Arab Gulf. Exports are still well below the peak of 3.3 mb/d reached in 1979. Virtually all the oil exported at present is crude oil, with heavy fuel oil accounting for the rest. Exports, again mainly in the form of crude oil, are projected to rise steadily over the projection period in line with higher production, on the assumption that the existing export infrastructure is restored and that new facilities are built. In our Reference Scenario, total oil exports climb to 2.5 mb/d in 2010 and 6.9 mb/d in 2030 (Figure 12.6).

Boosting exports via the Gulf will require an expansion of crude oil storage and loading facilities. Two terminals at Khor al Amaya and Basrah are currently operational at reduced capacity of about 1.8 mb/d. A third terminal at Fao was destroyed during the Iran-Iraq war. Repairs to Khor al Amaya, when completed, could add about 200 kb/d of capacity.

Figure 12.6: **Iraq's Oil Balance in the Reference Scenario**

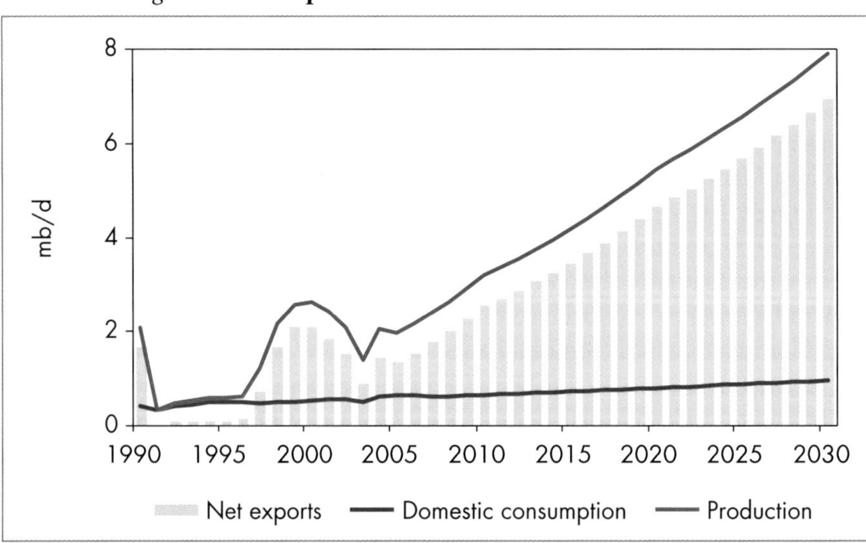

Pipelines are likely to make a significant contribution to higher exports over the projection period. Iraq has an extensive export pipeline infrastructure, but it has suffered extensive damage as a result of the 1991 war, a lack of maintenance and disuse due to sanctions imposed after that war, looting after the 2003 war and recent insurgent attacks. Most systems require extensive rehabilitation work, including modernisation using the latest telecommunications and remote control technologies. Local political factors as well as economic considerations are likely to play a major role in determining which pipelines are eventually repaired and reopened and what new lines are built. In the short term, expanding pipeline capacity will hinge on restoring security.

Repairing and upgrading the 1.65 mb/d[9] Kirkuk-Ceyhan system remains a priority. The pipeline was shut between December 2004 and June 2005, following severe damage inflicted by insurgents. In the longer term, capacity could be boosted by the construction of a new line to run parallel to the two existing lines. Congestion at the Turkish port of Ceyhan, however, could constrain future shipments through the system. Other lines that have been shut for decades could also be reactivated, depending on geopolitical developments. The 1.65 mb/d IPSA system that runs through Saudi Arabia to Yanbu on the Red Sea, which was expropriated by the Saudi government in 2001, is thought to be in good condition on the Saudi side of the border,[10] but requires major repairs on the Iraqi side. Another three-line system built in the 1940s and 1950s, which runs across Syria, with branches to the Mediterranean ports of Banias in Syria and Tripoli in the Lebanon, could also be used. The line was reactivated in 2000 to smuggle oil to Syria in defiance of UN sanctions. It is currently capable of handling about 300 kb/d, but capacity could be raised to its original level of 800 kb/d and even above 1 mb/d with the addition of pumping stations. A new line is also being considered. Iraq also plans to swap crude oil with Iranian oil products under an intergovernmental deal signed in July 2005.

Natural Gas Supply

Resources and Reserves

Natural gas resources in Iraq are very large, both in absolute terms and in relation to current production. According to Cedigaz, Iraq had 3.1 trillion cubic metres of proven reserves at the beginning of 2005, equal to around 2% of the world total. BP's figure is 3.2 tcm. The US Geological Survey estimates that undiscovered resources amount to a further 2.9 tcm. More than two-thirds of the reserves that have been identified are associated gas in oilfields. The

9. Design capacity. Effective capacity, when operational, is currently only about 1 mb/d.
10. Part of the Saudi system is thought to have been converted to transport natural gas.

largest reserves of associated gas are in the Kirkuk, Jambur and Bai Hassan fields in the north, and in the Rumaila and Zubair fields in the south. USGS estimates that roughly half of undiscovered gas is associated. Ten non-associated gas fields have been discovered, five of which are in the north-east.

The potential for producing gas is thought to be much higher than these reserve estimates suggest, as there has been very little exploration for and appraisal of non-associated gas resources in the country. In the Western Desert region, which has been largely unexplored, geological studies have revealed several highly prospective areas for natural gas. Two gas fields, each containing around 60 bcm, have already been discovered.

Gas Production and Distribution

Gas production is set to grow rapidly, in line with rising output of crude oil, with which most gas supply is associated. How much of this gas is marketed will depend on the development of infrastructure to gather, process and distribute gas to domestic and export markets, as well as meeting needs for reinjection to maintain reservoir pressure at ageing oilfields. We project marketed production to rise from 1.6 bcm in 2003 to 3.8 bcm in 2010 and 32 bcm in 2030. Most of the additional gas will go to meet domestic needs, mainly in power generation and for the production of petrochemicals and fertilizers. If investment in gas facilities lags the growth in oil production, surplus gas will have to be reinjected or flared.

Gas production was running at about 3 bcm per year from the late 1990s through to the early part of the current decade, falling sharply in 2002 and 2003. Around a quarter of all gas produced, two-thirds of which was produced in association with oil, was flared before the 2003 war. The only non-associated gas field so far developed is the Anfal field in northern Iraq, which was producing around 200 mcm per year early in the current decade.

There are two separate gas networks in Iraq. The Northern Area Gas Project centred on the Kirkuk field was commissioned in 1983, but suffered considerable damage during the Iran-Iraq war and conflict with the Kurds following the 1991 war. Gas from Kirkuk and Anfal is processed at the North Gas plant. The system also includes 250 km of pipelines to distribute LPG to households and industrial customers in Baghdad and other cities. When repaired, the system will have the capacity to process up to 6 bcm/year of dry gas.

The Southern Area Gas Project was brought on stream in 1985, substantially increasing the country's marketed gas supply. It comprises nine gathering stations for up to 16 bcm/year of gas from the Rumaila and Zubair fields and three processing plants – one at each field and a third at Basrah, connected by a transmission line – together with LPG storage and loading facilities. The system largely broke down immediately after the 2003 war, so that most gas was flared. The Rumaila processing plant started up again at the end of 2003, with capacity of about 5 bcm/year. Dry gas from the system is piped through a 2.5-bcm/year transmission line that runs from the West Qurnah field to Baghdad.

Once fully repaired, Iraq's gas-processing capacity will amount to about 22 bcm/year, which is more than sufficient to meet domestic requirements in the near term. But new gas-processing facilities will be needed to handle increased gas flows from Zubair and West Qurnah in the south. In addition, new gas gathering and distribution pipelines will be needed to connect producing wells to processing plants and centres of demand. The Ministry of Oil also plans to develop the non-associated Mansouriya gas field close to Kirkuk and to raise output at Anfal to provide fuel for power generation.

Export Plans and Prospects

Although the Iraqi government will undoubtedly continue to give priority to exports of oil, we expect exports of gas to recommence in the near future and grow steadily over the projection period. International sanctions prevented Iraq from exporting any gas before 2003. In our Reference Scenario, gas exports – entirely to Kuwait – reach 0.6 bcm in 2010. Exports to and via Turkey are expected to push total exports up to 17 bcm in 2030 (Figure 12.7). There is one export pipeline system to Kuwait already in place, but it requires some rehabilitation. Agreement has already been reached between the Iraqi and

Figure 12.7: **Iraq's Natural Gas Balance in the Reference Scenario**

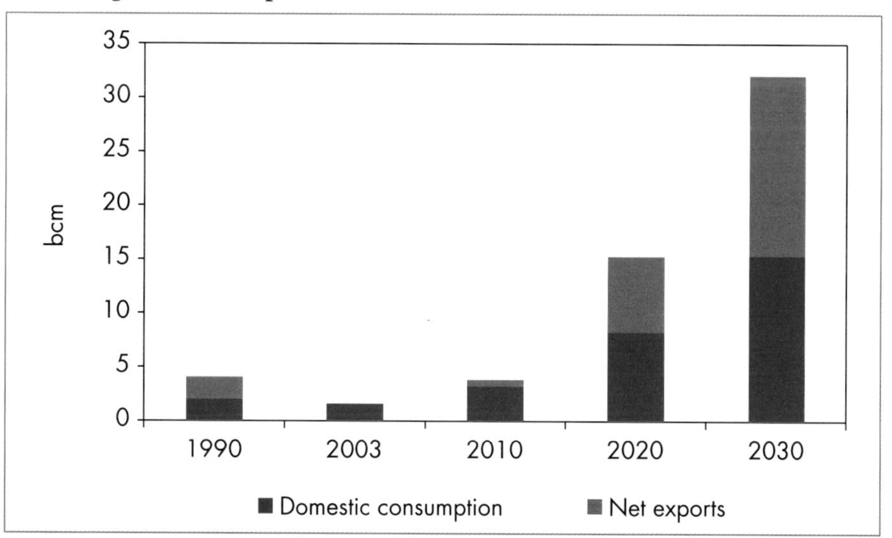

12

Kuwaiti governments on reviving exports that were halted by the 1991 war and exports are likely to restart in the next few years. The first phase of work on restoring the line, which will have an eventual capacity of around 11 mcm per day, is expected to be completed by the end of 2005 at a cost of $27 million. The second phase, which will take up to three years to complete, will be much more costly, at close to $800 million. Gas will come initially from the Rumaila field.

The prospects for starting gas exports to Turkey are more uncertain than exports to Kuwait, given current oversupply in the Turkish market and the need to repair and upgrade the Northern Gas System. Security in the north of the country is also a major obstacle at present. Gas could come from the non-associated Anfal, Chem Chemal, Jaria Pika, Khashm al-Ahmar and Mansouriyah gas fields. We expect exports to resume after 2010 and rise steadily through to 2030. Iraqi gas could be transported to European markets via Turkey. Iraq could also develop export pipeline projects to Syria, Jordan and Lebanon. In September 2004, Iraq joined the Arab Gasline consortium, which is developing a regional gas pipeline network for delivering Egyptian gas to Jordan, Syria, Lebanon and possibly Turkey.

Energy Investment

Overview

In the Reference Scenario, cumulative investment needs are projected to amount to about $96 billion over 2004-2030, or $3.6 billion per year in real 2004 dollars (Figure 12.8). The bulk of this capital will be needed in the oil

Figure 12.8: **Iraq's Energy Investment in the Reference Scenario**

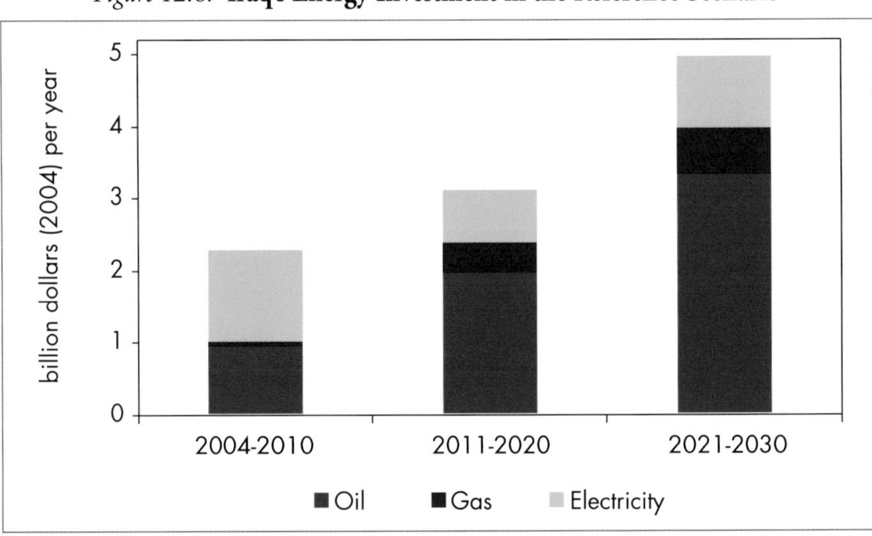

sector. The electricity sector accounts for most of the rest. Annual electricity investment needs will be particularly large over the rest of the current decade, because of the dilapidated state of the industry.

Oil and Gas Sector

Cumulative investment needs in the Iraqi oil industry are estimated at almost \$59 billion over 2004-2030. More than three-quarters, or \$51 billion, will be needed for upstream projects (Figure 12.9). Investment needs will be much larger in the second half of the projection period, when crude oil production capacity is expected to grow the most. New field development costs will also be higher than the costs of boosting output at existing fields.

Figure 12.9: **Iraq's Oil Investment in the Reference Scenario**

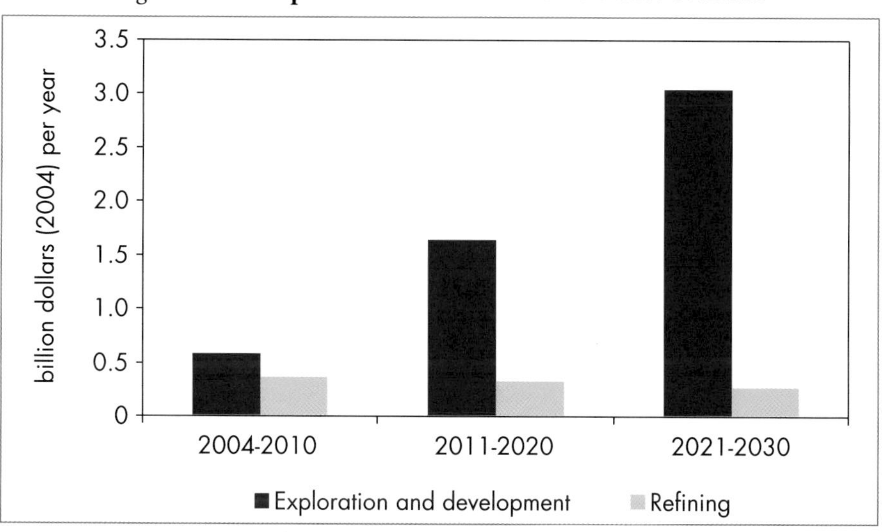

About 40% of the oil investment that will be required in Iraq over 2004-2010 will be in the refining sector, mostly for the construction of a new refinery that is expected to be completed by the end of the decade. That project and other refinery upgrades are expected to cost in the region of \$2.6 billion. Upstream investment needs amount to \$4 billion over the same period.

Much of this investment will have to come from foreign investors, as the government is highly unlikely to allocate such large sums out of its central budget. The government has indicated that it intends to open up new field developments, and possibly redevelopment of existing fields too, to international oil companies. The Ministry of Oil is in discussion with several companies with a view to commencing negotiations once the country's constitution is ratified and parliament has approved a new hydrocarbon law. The Iraqi government may also be able to borrow from multilateral lending institutions and regional banks.

The projected trends in gas production and exports imply a need for investment of around $11 billion over 2004-2030. Nearly 60% of this amount will be needed in the 2020s. As with oil investment, much of this capital is likely to come from foreign investors.

Electricity Sector

Investment needs in Iraq's power sector over the period 2004-2030 will amount to around $26 billion. Some $8 billion will go to new power generation, $4 billion to new transmission projects and $8 billion to new distribution (Figure 12.10). The cost of rebuilding damaged existing infrastructure – included in total investment – is estimated to be $6.6 billion, of which $4.2 billion is for power generation and $2.4 billion for networks. This estimate assumes that the refurbishment cost for power generation will be roughly half that of the cost of a new power plant of the same technology. The cost of rebuilding networks is assumed to be about $250 per kW. The reconstruction cost estimates given in this report are similar to those published in a joint UN/World Bank study in 2003. That study put the reconstruction cost at $6.4 billion.[11] Our estimate is slightly higher because we assume that the cost of refurbishing transmission and distribution networks will be higher.

Because the cost of rehabilitating power stations is likely to be high, scrapping old boilers and replacing them with new CCGT plants may be cost-effective. This, however, will require rapid development of the gas network.

Figure 12.10: **Iraq's Cumulative Electricity Investment, 2004-2030**

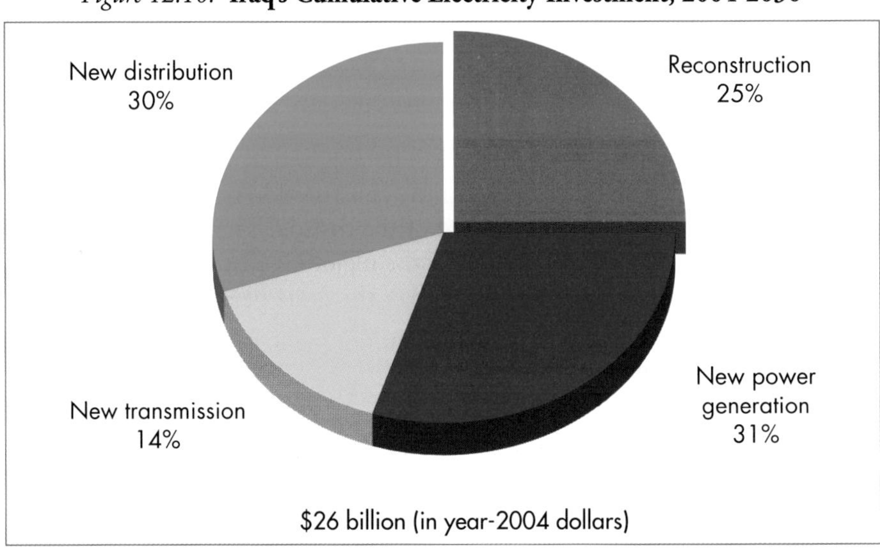

11. See UN/World Bank (2003). Investment has been converted to 2004 dollars.

At present, Iraq relies on international aid for the rehabilitation of its power sector. According to USAID, which is heavily involved in Iraqi reconstruction efforts, restoring and improving Iraq's electricity supply is the biggest and most costly challenge (USAID, 2003). Increases in the level of available capacity have been below expectations. At the same time, demand for power has been increasing because sales of electric appliances are on the rise.

The future structure of the electricity sector in Iraq remains uncertain. If prices are kept below cost, new projects will have to be financed out of public funds, as the power company's cash flows will be insufficient and private capital will not be forthcoming Oil-export earnings could provide a source of finance for electricity projects, but competition for public funds from other sectors of the economy will be intense.

Deferred Investment Scenario[12]

Energy Demand

In the Deferred Investment Scenario, total primary energy demand in Iraq grows at an average rate of 3% per year over the projection period – 0.3 percentage points lower than in the Reference Scenario. Demand reaches 57 Mtoe in 2030, compared with 62 Mtoe in the Reference Scenario (Figure 12.11). Primary demand for gas falls most, because slower growth in electricity demand reduces the need for new gas-fired power stations.

Figure 12.11: **Iraq's Total Primary Energy Demand in the Reference and Deferred Investment Scenarios**

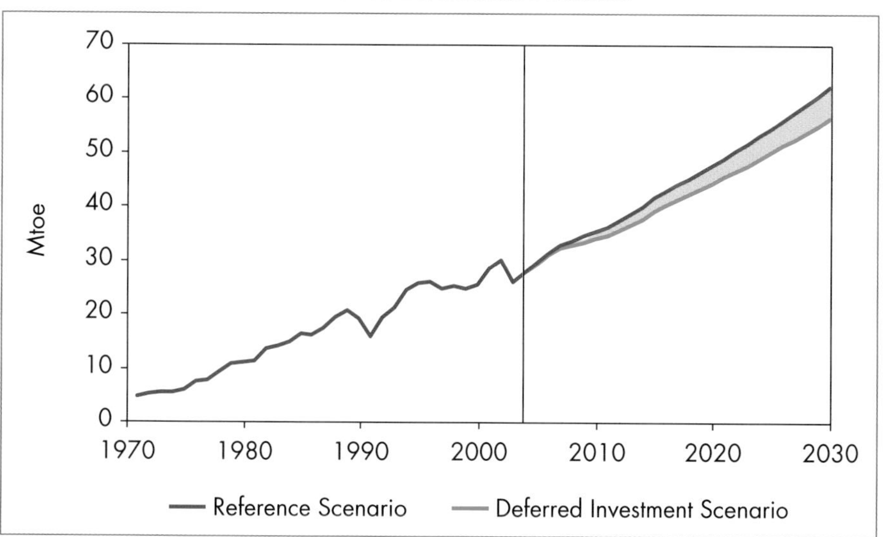

12. See Chapter 7 for a detailed discussion of the assumptions and methodology underlying the Deferred Investment Scenario.

Differences in GDP growth rates explain most of the divergence in energy demand trends in the two scenarios. Higher end-use prices in Iraq also contribute marginally to slower demand growth.

Oil and Gas Production

The assumption we have adopted, that upstream investment over the projection period is constant in relation to GDP at the level of the last ten years, results in an average level of investment some 41% lower than in the Reference Scenario. Unsurprisingly, this leads to much slower growth in production of both crude oil and natural gas, much of which is associated with oil. As a result, crude oil production is 3.8 mb/d, or about 48%, lower in 2030 compared with the Reference Scenario (Figure 12.12).

Lower investment has a slightly larger impact on production from fields already in production. Output from these fields falls furthest in relative and absolute terms compared to the Reference Scenario, rising to about 3 mb/d at the beginning of the 2020s and then tapers off to around 2.7 mb/d by 2030. Output from additional reserves in existing fields falls slightly less. Production from fields awaiting development is 0.8 mb/d lower in 2030 than in the Reference Scenario. The share of these fields in total Iraqi oil production in 2030 is 25% in the Deferred Investment Scenario compared with 23% in the Reference Scenario.

Figure 12.12: **Iraq's Oil Production by Source in the Deferred Investment Scenario**

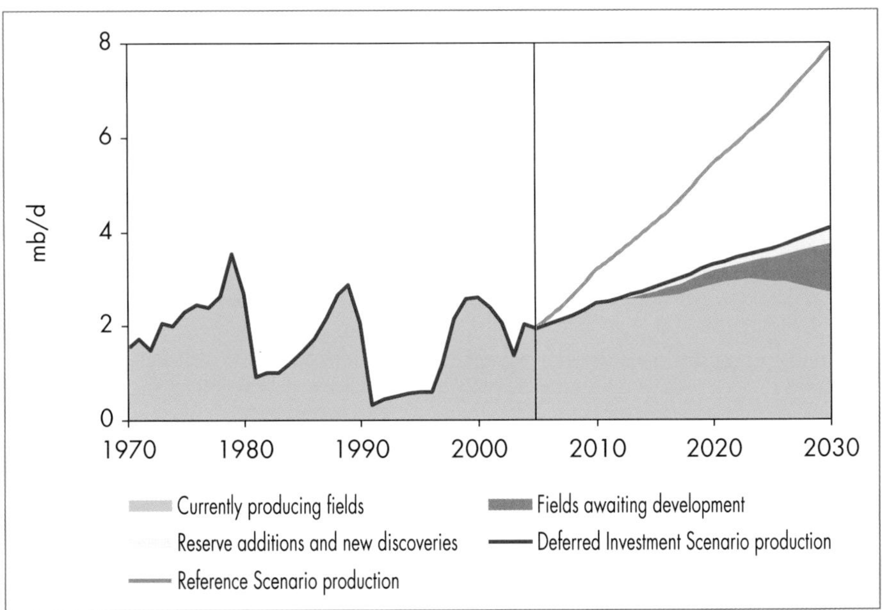

The slower growth in domestic oil demand reduces the need to expand refining capacity in the Deferred Investment Scenario. Capacity grows by 65% over 2003-2030, reaching 1 mb/d, compared with a doubling to 1.2 mb/d in the Reference Scenario.

Lower oil output results in lower natural gas production from associated oil/gas fields. Gas production from non-associated fields also increases much less in the Deferred Investment Scenario, as domestic and export demand is lower. Total gas production is 15 bcm, or 47%, lower in 2030 than in the Reference Scenario.

Oil and Gas Exports

Oil exports fall even more than production in percentage terms, but slightly less in volume terms – because lower domestic consumption helps to compensate for the fall in production. At 3.2 mb/d in 2030, net oil exports are 3.7 mb/d, or more than half, lower in 2030 in the Deferred Investment Scenario (Figure 12.13). Natural gas exports are also markedly lower, because of slower demand growth in the net-importing regions, notably Europe. Exports reach 3.3 bcm in 2030, 13.4 bcm less than in the Reference Scenario.

Figure 12.13: **Iraq's Oil Balance in the Reference and Deferred Investment Scenarios**

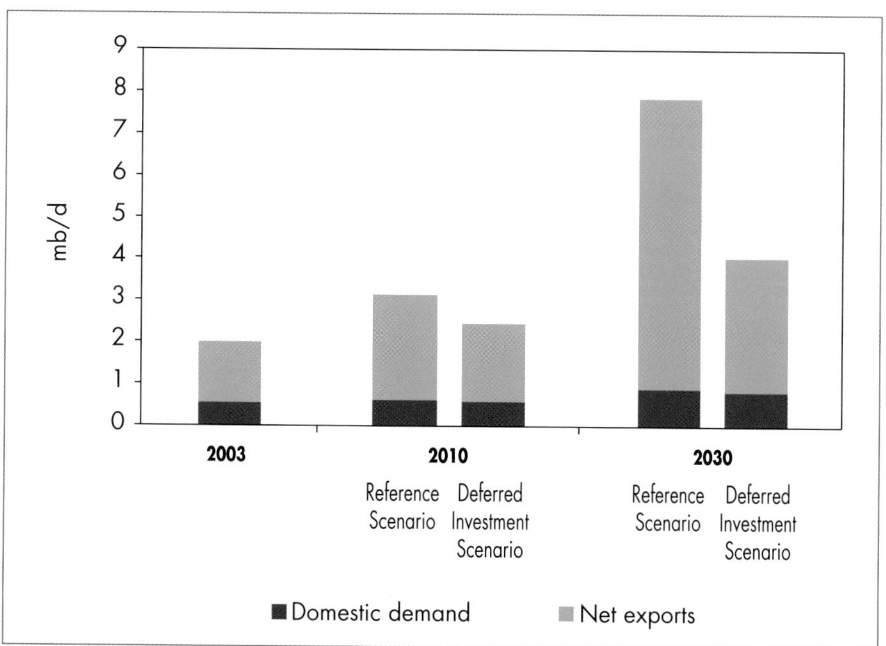

KUWAIT

HIGHLIGHTS

- Kuwait accounts for just one-hundredth of one per cent of the world's land area but a staggering 8% of the world's proven oil reserves. On the back of this wealth it has prospered and become a key supplier of oil and refined products to world markets.
- Per capita energy consumption in Kuwait is extremely high, twice the OECD average. This is due to the harsh climate, reliance on energy-intensive sea-water desalination and heavily subsidised energy prices. In our Reference Scenario, primary energy demand more than doubles between 2003 and 2030, growing on average by 2.8% per year.
- Power generation will expand rapidly to meet growing demand for electricity, which is already amongst the highest in the world on a per capita basis. In the near term, generation will be mostly oil-based, but gas use will grow once more supplies become available. Electricity output will rise from 40 TWh in 2003 to 82 TWh in 2030.
- In the Reference Scenario, Kuwait's oil and NGL production grows from 2.5 mb/d today to 2.9 mb/d in 2010 and 4.9 mb/d in 2030. Oil exports reach 2.5 mb/d in 2010 and 4.4 mb/d in 2030. International oil companies are expected to play a key role in increasing production so long as the long drawn-out process of opening up Kuwait's oil sector succeeds.
- Kuwait's refining capacity will expand to 1.3 mb/d in 2010, in line with announced plans. Much of this new capacity will produce feedstock for the country's power stations. Capacity is then projected to reach 1.4 mb/d by 2030 and the product slate to progressively shift in favour of higher-value products for export.
- These trends call for cumulative investment in Kuwait's energy sector of $86 billion (in year-2004 dollars) over 2004-2030, or $3.2 billion per year. The oil and gas sector will absorb around four-fifths of the total and the power sector the remainder.
- In the Deferred Investment Scenario, oil production grows more slowly, reaching only 3.3 mb/d in 2030. Oil exports and refining capacity are also reduced. Such an outcome could arise from delays in inviting major oil companies back into the country or the abandonment of the plan altogether.

Overview of Kuwait's Energy Sector

Kuwait is one of the smallest countries by size and has an extremely arid climate and few sources of freshwater. Yet it is an extremely prosperous nation thanks to its vast petroleum reserves. Not surprisingly, oil has played a key role in all aspects of Kuwait's development. The first oil discovery was made in 1938 and the first exports commenced in 1946, following delays posed by the Second World War. Kuwait was a founding member of OPEC in 1960 and is also a member of the Organization of Arab Petroleum Exporting Countries.

Kuwait now ranks fifth in the world in terms of proven oil reserves, twelfth for oil production and ninth for oil exports. The emirate is currently engaged in a long drawn-out process to open up its oil sector to foreign companies so as to increase its oil production significantly. Foreign partners are being sought to bring in expertise and technology, not due to any shortage of capital.

Kuwait's natural gas reserves are very small compared to many of its Gulf neighbours and are found almost entirely in association with oil. As such, their exploitation is inextricably linked to oil production levels and, by extension, OPEC quotas. Kuwait is struggling to meet domestic natural gas demand from its modest indigenous production and is actively pursuing import options.

Per capita energy consumption in Kuwait is among the highest in the Middle East and around twice the OECD average. Energy demand has grown strongly since the 1970s due to the country's rapid development after the first oil shock and the generous energy price subsidies which discourage efficient energy use. The power sector's share of energy consumption has grown particularly strongly, owing to the harsh climate which makes the use of air-conditioning essential and to reliance on energy-intensive sea-water desalination. Oil is the

Table 13.1: **Key Energy Indicators for Kuwait**

	1971	2003	1971-2003*
Total primary energy demand (Mtoe)	6	23	4.3%
Total primary energy demand per capita (toe)	7.5	9.5	0.8%
Total primary energy demand/GDP**	0.18	0.58	3.7%
Share of oil in total primary energy demand (%)	28	66	–
Net oil exports (mb/d)	3.2	2.0	-1.4%
Share of oil exports in production (%)	99	88	–
Share of gas in total primary energy demand (%)	71	34	–
CO_2 emissions (Mt)	23.2	58.3	2.9%

* Average annual growth rate.
** Toe/thousand dollars of GDP in year-2004 dollars and PPPs.

major domestic primary energy source. If imported supplies become available, gas should start to play a larger role in the demand mix, thus freeing up more oil for export.

Political, Social and Economic Trends

Political and Social Developments

Kuwait is a constitutional monarchy and has been ruled by the Al-Sabah family since 1756. The current head of state, or Emir, is Sheikh Jaber Al Ahmad Al Jaber Al Sabah. The government consists of the prime minister and a council of ministers that are collectively responsible for the general policy of the state. Unlike most countries in the Gulf region, Kuwait has an elected parliament. Known as the National Assembly, it is made up of 50 members who are elected every four years and has significant freedom and power. Laws put forward by the government must be approved by both the Emir and the National Assembly before coming into force. After years of debate, women were granted full political rights in May 2005. They can now stand and vote in both parliamentary and local elections.

To resolve a dispute over sovereignty of a desert region between Saudi Arabia and Kuwait, in 1971 the two countries established the disputed area as a Partitioned Neutral Zone. Kuwait has responsibility for administering the northern part of the zone and Saudi Arabia the southern part. Revenues from oil production in the zone are shared equally.

Kuwait has been actively pursuing economic diversification and structural reforms aimed at developing a more market-based economy, increasing the role of the private sector and reducing the economy's vulnerability to fluctuations in world oil prices. In many cases these efforts have been slow or unsuccessful. Privatisation of state-owned enterprises is obstructed by the need to safeguard the jobs of Kuwaiti citizens.

Life in Kuwait has been shrouded by events in neighbouring Iraq since the Iraqi invasion in August 1990. In 2003, Kuwait was a key coalition partner in Operation Iraqi Freedom, making available considerable land area and donating substantial funds. Kuwaiti companies have successfully competed for reconstruction-related business activity in Iraq.

Kuwait's Public Authority for Civil Information (PACI) estimates the total number of inhabitants at the end of 2004 as 2.75 million. During that year, the population grew by a rapid 8%, following a 5% increase in 2003. This strong growth is due to a sharp increase in the number of expatriate workers in response to the general pick-up in economic activity in both the private and the public sectors. PACI data indicate that expatriate workers currently represent around 66% of Kuwait's population. In the Reference Scenario, Kuwait's population is assumed to grow in line with the latest UN projections to 4 million in 2030 – an average annual growth rate of 1.9%. The Kuwait government provides its citizens with extensive public services at heavily subsidised prices and does not levy income taxes. Benefits include retirement income, marriage bonuses, housing loans, virtually guaranteed employment, free medical care, and free education at all levels. Expatriate workers receive some of these services.

Kuwait's labour force grew by 11% in 2004 to 1.6 million. Non-Kuwaitis represent 82% of the total, reflecting the reluctance of Kuwaiti nationals to take menial jobs. The wages on offer for such positions are attractive to a vast and eager labour pool from other Arab countries as well as South and East Asia. Over 90% of Kuwaiti nationals who are in employment work directly or indirectly for the government. Reducing dependence on government employment is one of the aims of the plan to privatise state companies.

Economic Trends and Developments

Kuwait is a wealthy country, with average GDP per capita in 2004 of around $17 200, the third-highest in the region after Qatar, and the UAE. Kuwait's economy experienced robust growth in the early part of 2005, following strong growth of nearly 7% in 2004 in response to higher oil prices and increased oil production which bolstered export revenues.

In 2004, oil accounted for around 50% of Kuwait's GDP, over 80% of government revenue and around 95% of export earnings. In August 2005, the National Bank of Kuwait expected oil revenues for the fiscal year 2005/06 to total between $44 billion and $47 billion. Although oil revenues have surged in recent years, in inflation-adjusted terms, on a per capita basis, they are far below the peaks reached in the late 1970s and early 1980s.

Kuwait operates a Reserve Fund for Future Generations. This was initially created as a means of providing a financial cushion should oil revenues decline. Its balance is currently over $80 billion – the bulk of which is held in investments in the United States, Western Europe, Japan and South-East Asia. Each year 10% of government revenues are allocated to the fund. During the Iraqi occupation in 1990, the Kuwaiti government-in-exile depended upon its overseas investments as its primary source of capital and to help pay for the subsequent reconstruction.

Regime change in Iraq has reduced the perceived territorial threat and created a sense of optimism in Kuwait's domestic economy. This is demonstrated by record price and liquidity levels on the stock exchange and continuing strength in the construction market. Increased government expenditure has also provided a stimulus to domestic economic activity. In April 2005, the International Monetary Fund commended Kuwait's economic performance, particularly its healthy rate of GDP growth, the increase in its per capita income and its stable rate of inflation and unemployment. In the Reference Scenario, Kuwait's GDP is assumed to grow by an average rate of 3.1% per year during 2003-2030 (Table 13.2).

Table 13.2: **GDP and Population Growth Rates in Kuwait in the Reference Scenario** (average annual rate of change)

	1971-2003	1990-2003	2003-2010	2010-2020	2020-2030	2003-2030
GDP	0.6	3.2	3.6	3.1	2.6	3.1
Population	3.5	0.9	2.7	1.8	1.4	1.9
Active labour force	4.5	1.5	3.1	2.3	1.7	2.3
GDP per capita	–2.8	2.3	0.9	1.2	1.2	1.1

Sources: Kuwait Public Authority for Civil Information; UNDESA (2004).

Energy Policy

By virtue of Kuwait's constitution, the state now owns and controls all oil resources. In the past, private companies held rights to oil resources under various concessions. These holdings were declared void through a series of agreements and legislation made during the nationalisation of the Kuwaiti oil industry in 1975. Foreigners have been prohibited since then from investing in the oil sector outside the Neutral Zone, though several majors, including BP, Shell and Chevron, continue to work in the country under limited service contracts, providing technical assistance.

Following nationalisation, the Supreme Petroleum Council (SPC) was formed to set the general policies of the oil sector within the framework of national economic and social development objectives. The Kuwait Ministry of Energy exercises policy-making powers in conjunction with the SPC and supervises all public institutions involved in the oil sector. The ministry was created by merging two previous ministries – the Ministry of Oil and the Ministry of Electricity and Water – after the National Assembly elections of July 2003.

At an operational level, the major player in the Kuwaiti oil sector is the Kuwait Petroleum Corporation (KPC). It was established in 1980 as a state-owned entity and is involved in oil and gas exploration, production, refining, transportation and marketing both within Kuwait and internationally. KPC's subsidiary, the Kuwait Oil Company (KOC), handles exploration and production activities within Kuwait and in its half-share of the Partitioned Neutral Zone.

Since 1993, Kuwait's legislative bodies have been considering allowing foreign companies to participate in the country's upstream oil sector (see Project Kuwait below). Assistance is being sought on the grounds that some of the country's oilfields, particularly those in the north and west, no longer have sufficient pressure for the oil to be easily recovered, both because of their maturity and the damage inflicted on them during the Gulf War, and that KPC does not have the necessary technical experience to enable production to continue at the most efficient rate and for recovery to be maximised.[1] Such a move is expected to help to minimise operating costs and capital investment, provide training opportunities for Kuwaiti nationals and develop strategic and economic ties. The proposal has been very controversial, resulting in delays which have stretched out for more than a decade. Some influential lawmakers are stridently opposed to the plan on the grounds that it violates the constitution's decree on the ownership of natural resources.

In recent years, the Kuwaiti government has stepped up efforts to increase the supply of natural gas to the domestic market. This is a pressing issue as there is a looming gas shortage. It would also free up additional oil that is currently used for generating electricity for export. Dedicated efforts to discover non-associated gas deposits have been largely unsuccessful. Kuwait has also engaged in talks and signed several preliminary deals with other Gulf states for natural gas imports.

Project Kuwait

Project Kuwait is a controversial plan to allow the first significant participation by international oil companies in Kuwait's upstream sector since the emirate nationalised its oil wealth in 1975. The proposal is to span twenty years and has the objective of increasing oil production capacity to 3.0 mb/d by 2010, 3.5 mb/d by 2015 and 4 mb/d by 2020. The capital cost of the project is estimated at $3 billion and the operating costs at a further $5 billion.[2] The Kuwaiti government will retain full ownership of oil reserves, control over oil production levels and strategic management of the venture.

The first stage of the proposal would increase output from the northern fields from around 530 kb/d currently to 1 mb/d by 2010. The fields involved are Abdali, Ratqa, Raudhatain, and Sabriyah. The Burgan field – Kuwait's "crown jewel" – is to remain strictly off limits to foreign interests. It has been reported that cumulative oil production over the full term of the contract could exceed

1. Hashim M. El-Rifaai, Executive Assistant Managing Director, Kuwait Oil Company, Euro Gulf Energy Forum, Kuwait, April 2005.
2. Information provided by Mr. Mohammed Al Shatti, Kuwait Petroleum Corporation, August 2005.

5 billion barrels, more than 40% greater than if the fields are developed solely by KOC. If KOC undertook the project alone, it would be exposed to many challenges with which it has limited experience, such as water and gas injection, water handling and corrosion management. KOC's recent handling of similar challenges has resulted in unexpected delays in execution and unsatisfactory performance.

For the proposal to move ahead, Kuwait's parliament must approve an enabling law, a first draft of which it rejected in 2002. A revised version currently under consideration confines foreign involvement to the northern oilfields – giving parliament an opportunity for further review before extension elsewhere in the country. It also requires the winning consortium to include a Kuwaiti state-owned entity. The possibility that the draft law will not be passed should not be understated. The outcome will have direct implications for the timing, efficiency and cost of increasing Kuwaiti oil output.

In anticipation of approval by parliament, KPC has pre-qualified several foreign oil companies to bid for the project and provided them with technical data. The companies have been grouped into three consortia that have submitted basic development plans and will eventually compete in a commercial bidding process (Table 13.3). The winning consortium will operate under an operating service agreement with reimbursement in the form of cash, not in oil. They will be paid a per-barrel fee with allowance for capital recovery and an incentive for increasing reserves.

Table 13.3: **Pre-Qualified Project Kuwait Consortia**

Consortium One	Consortium Two	Consortium Three
Chevron Texaco 50%	BP 65%	ExxonMobil 37.5%
Total 20%	Occidental 25%	Shell 32.5%
PetroCanada 10%	Indian Oil Corp. 10%	ConocoPhillips 20%
Sibneft 10%		Maersk 10%
Sinopec 10%		

13

Energy Demand

Kuwait's primary energy demand stood at 23 Mtoe in 2003. Demand levels have only recently returned to their level prior to the Gulf War of 1990 – a direct result of the significant and sustained reduction in population that followed the invasion. Kuwait's primary energy needs are met entirely by oil and natural gas, all of which is indigenously produced.

In the Reference Scenario, primary energy demand in Kuwait is set to grow by 2.8% per year to 49 Mtoe in 2030 – more than twice current levels (Figure 13.1). This is a considerably slower rate than the rate of 4.3% per annum seen over the past three decades. The rate of growth is projected to decline through the projection period as the Kuwait economy matures and the rate of population growth slows.

Figure 13.1: **Kuwait's Total Primary Energy Demand**

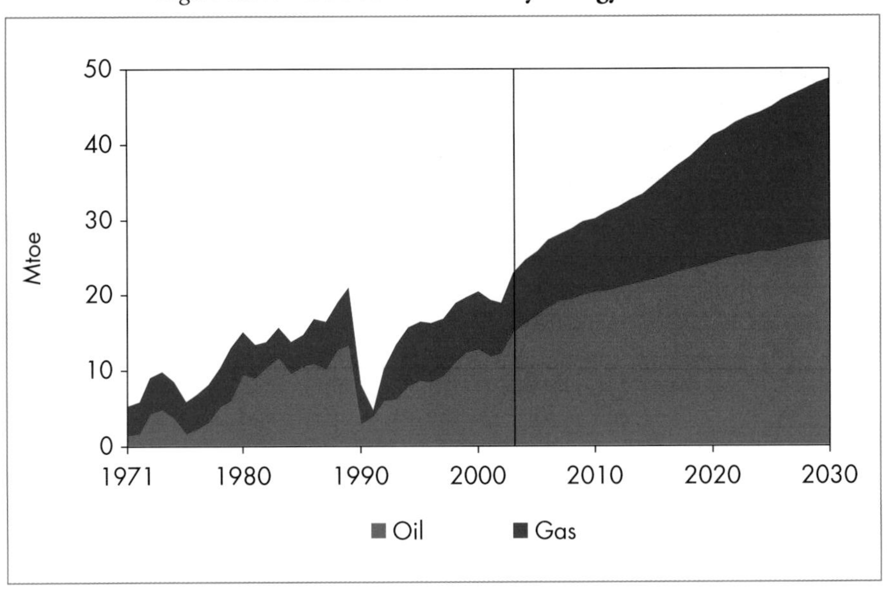

The historical increase in energy intensity in Kuwait has been quite striking and now stands at 0.58 Mtoe per thousand dollars of GDP – around three times the OECD average. It is expected to increase marginally through to around 2015 and then decline to slightly less than the current level by the end of the projection period (Figure 13.2).

Kuwait's per capita annual energy consumption is also very high at 9.5 toe, around twice the OECD average. This elevated level is due to the harsh climate, which makes air-conditioning and reliance on energy-intensive water desalination indispensable. But demand is also high because energy is used inefficiently, largely because it is heavily subsidised. Per capita energy consumption is projected to further increase to over 12 toe per annum by 2030.

Figure 13.2: **Energy Intensity Indicators in Kuwait**

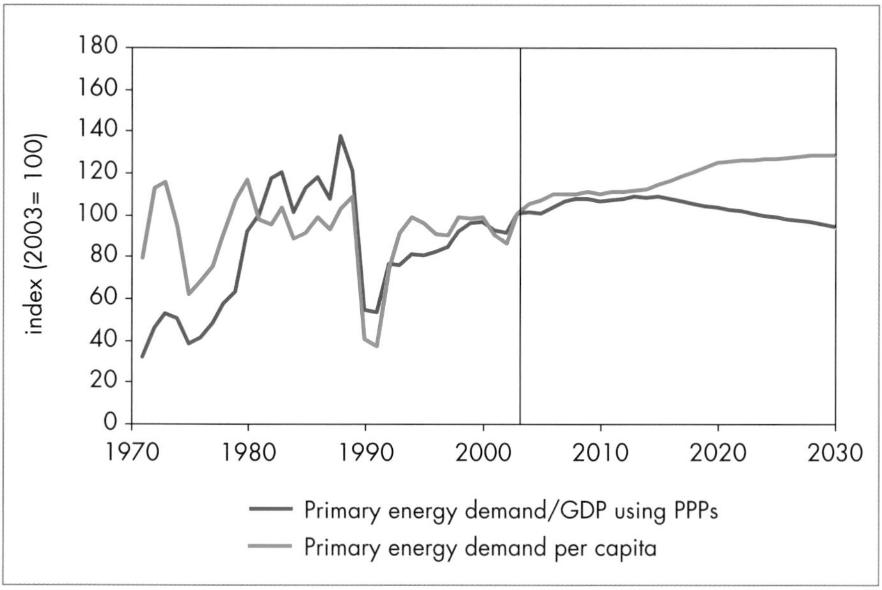

In 2003, oil met almost two-thirds of Kuwait's primary energy demand and natural gas met the remainder. Natural gas is poised progressively to increase its share. By 2030, demand for gas will have increased by 3.8% per annum to 21 Mtoe. This assumes that the country is able to overcome its gas shortage through imports. Nevertheless oil will remain the dominant fuel, with demand reaching 27 Mtoe by 2030, a growth of 2.2% per annum.

Combined power-generation and water-desalination plants represent the major source of energy demand in Kuwait. Demand from this sector will increase from 9 Mtoe in 2003 to 18 Mtoe in 2030, a growth of 2.7% per annum. Electricity demand in Kuwait has grown strongly since the 1970s as air-conditioning has become more widespread. Another factor in the growth in the power sector's share of energy consumption is the country's limited freshwater resources, which necessitates heavy reliance on sea-water desalination. Natural gas is projected steadily to increase its share in the fuel mix for power generation and water desalination, from 20% today to 43% in 2030, as indigenous gas production rises and imports become available. The fuel requirements for water desalination will account for 27% of the increase in energy demand in the power and water sector from 2003 to 2030.

The industrial sector in Kuwait currently accounts for around 28% of total final energy consumption. The main users include a number of large export-oriented petrochemical producers and a range of smaller firms manufacturing commodities such as urea, fertilizer, bricks and cement. Industrial energy demand is projected to increase by 3.7% annually to 2030, reaching 7 Mtoe. Kuwait is actively pursuing expansion of its petrochemical sector which is still modest owing to the scarcity of appropriate feedstocks and the uncertain overall business economics. Two new projects are scheduled for completion in 2008, both joint ventures between KPC's Petrochemical Industries Company and Dow Chemical. The first is known as Olefins II and will double the capacity of an existing olefins complex. The second project will produce aromatics.

Demand in Kuwait's transport sector is projected to increase by 2.8% per year, from 3 Mtoe in 2003 to 7 Mtoe in 2030. Transport will remain solely fuelled by oil. Transport's share of total energy demand has grown strongly, driven by low fuel costs, high incomes and rapid population growth. Kuwait currently has one of the highest levels of vehicle ownership in the MENA region, with more than one in three people owning a car.

A significant proportion of the energy used in Kuwait goes to keeping its oil production and refining capacity operating. Oil and gas for use in oil extraction and refining together accounted for around one-third of total primary energy demand in 2003. In line with rising oil production and refinery output, use of oil and gas in this sector is expected to increase at a rate of 2.8% per year, reaching over 14 Mtoe in 2030.

Household energy use in Kuwait is made up of electricity and the refined oil products LPG and kerosene. Demand in this sector will increase by 2.7% annually through to 2030, reaching 4.7 Mtoe. Households are expected to continue to switch away from oil products to electricity because of its cost advantages and convenience.

Prospects for solar energy in Kuwait are good as it has a seven-month summer and few cloudy days. Nevertheless, renewable energy is expected to play a very minor role in household supply as heavily subsidised electricity prices deny end-users any savings through installation of technologies such as solar water heaters.

Electricity Supply and Desalinated Water Production

Overview

Kuwait's power plants produced 40 TWh of electricity in 2003 and, according to preliminary data, 41 TWh in 2004. Some 80% of generation is based on oil, while the remainder comes from natural gas. Most electricity is produced in steam boilers. Installed capacity was 10 GW in 2003, with steam boilers representing over 90% and gas turbines and diesel engines the rest. Most power plants are integrated with water desalination.

Figure 13.3 shows the development of Kuwait's power sector. The graph clearly shows the impact the Iraqi invasion had on demand for power and power plant construction. The development of the Sabriyah project started in 1988 but was suspended following the invasion. The project got back on track in 1993 and its first units came on line in 1997. Demand for electricity dropped in 1990-1991 and came back to pre-war levels in 1994. The principal reason for this drop was the decrease in the country's population.

A key feature of Kuwait's power sector is rapidly growing demand for electricity, despite the fact that electricity consumption per capita is one of the highest in the world. Average annual growth over the period 1998-2003 was 1.9%, bringing per capita electricity generation to around 16 600 kWh, the highest in the MENA region and double the OECD average. Demand for power is twice as high in the summer as in the winter because of air-conditioning. This puts stress on the system, requiring large amounts of reserve capacity. Blackouts and brownouts occur frequently when demand surges.

The power sector is controlled by the Ministry of Energy. There are no immediate plans to encourage private-sector involvement in the power sector. Electricity prices in Kuwait are heavily subsidised, which encourages waste and inefficient use.

Figure 13.3: **Power Sector Development in Kuwait**

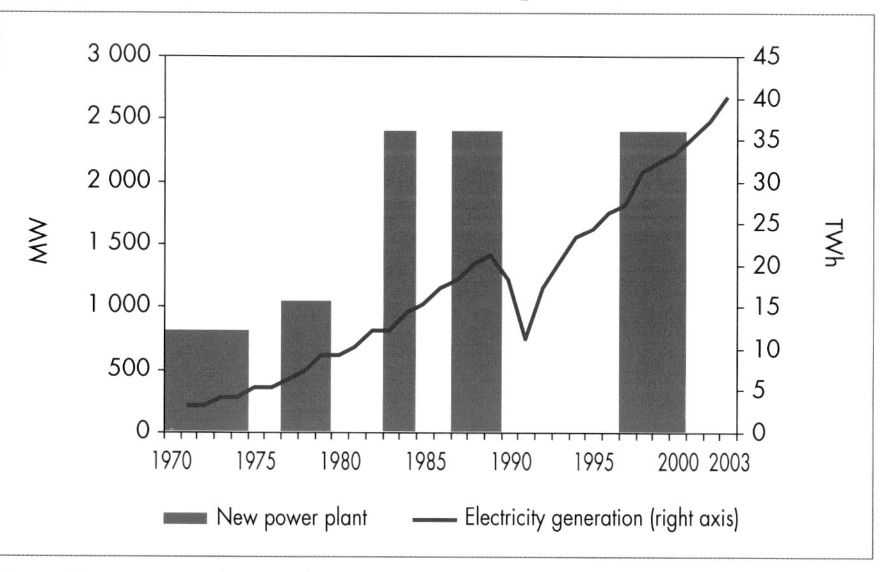

Sources: Kuwait Ministry of Energy; Platt's (2003); IEA database.

Electricity Production

Kuwait's electricity generation is projected to increase at an average rate of 2.7% per year over the period 2003-2030, doubling by 2030. Growth in the period to 2010 will be much higher, averaging 4.3% per year between 2003 and 2010. Electricity generation is expected progressively to increase its reliance on natural gas. Kuwait's government hopes to reduce oil consumption in the power sector so as to increase exports.

The share of natural gas in electricity generation is projected to increase from 20% in 2003 to 41% in 2030. The increase in natural gas consumption is expected to take place as more pipelines are built to bring gas from neighbouring countries and is expected to be more pronounced after 2010. Some of the existing power plants that now burn oil are expected to switch to burning gas once they have access to it.

To meet the projected demand for electricity, Kuwait will need to build 13 GW of new power plants. Most new capacity, about 8 GW, is expected to be gas-fired and most of this will be CCGT power plants. Although Kuwait uses no renewable energy now, it could readily produce electricity from solar energy, since its annual level of solar radiation is one of the highest in the world, at 2.2 MWh per m^2. The Reference Scenario projects about 50 MW of installed solar photovoltaic capacity by 2030.

Kuwait will be part of the Gulf Cooperation Council (GCC) interconnection grid. The first phase of this project, which could be completed by 2008, will link the power grids of Kuwait, Saudi Arabia, Bahrain and Qatar. On the Kuwait side, the project involves building a double-circuit 400 kV, 50 Hz line from Al Zour in Kuwait to Ghunan in Saudi Arabia with an intermediate connection at Al Fadhili in Saudi Arabia. This will give Kuwait the possibility to exchange up to 1 200 MW of power.

Water Production

Kuwait has no permanent surface water flows and relies almost entirely on desalination plants for its drinking water supply. Per capita water consumption rose by almost 250% between 1980 and 2000, partly in response to a decline in the relative price of water and an increase in leakage through the distribution network (World Bank, 2005). The World Bank estimates that water subsidies in Kuwait in 2000 amounted to $830 million, or 6% of oil export revenues. The emirate has unsuccessfully sought to diversify its reliance on desalination by piping water from both Iran and Iraq.

In 2003, Kuwait produced 433 million cubic metres (mcm) of desalinated water. To meet rising demand and replace ageing facilities, around 1 000 mcm of new water desalination capacity will need to be added by 2030 (Table 13.4). Kuwait has already announced plans to add around 474 mcm of new capacity by 2008 (Global Water Intelligence, 2004). The main driver of growth in desalinated water demand will be the residential sector where consumption is projected to rise from 335 mcm in 2003 to 632 mcm in 2030. Currently, almost 90% of desalinated water is produced through multi-stage flash facilities but the share of reverse osmosis is projected to rise to around 20% by 2030. Fuel requirements for desalination will rise from 3.4 Mtoe in 2003 to 4.9 Mtoe in 2030, or 10% of total primary energy supply. Over two-thirds of new power plants will be fitted with desalination units. To meet these projections, Kuwait will need to invest $2.9 billion in new desalination plants.

Table 13.4: **Water and Desalination Capacity in Kuwait**

	2003	2010	2020	2030
Water consumption (mcm)	679	768	884	1 007
Desalination capacity (mcm)	582	934	1 006	1 088
Energy needs for desalination (Mtoe)	3.4	3.7	4.2	4.9

Oil Supply

Resources and Reserves

Kuwait is part of the Arabian basin, with a geological history similar to that of Saudi Arabia. The country is relatively well explored and claims proven crude oil reserves of 101.5 billion barrels, around 8% of the world's total (*Oil and Gas Journal*, 2004). IHS Energy's estimate of Kuwait's proven crude oil reserves is significantly lower at 52.3 billion barrels. Currently producing reservoirs are mostly Cretaceous, with remaining exploration potential in the deeper Jurassic sediments onshore and both Jurassic and Cretaceous offshore.

Kuwait's oil wells include both shallow low-pressure reservoirs and much deeper high-pressure reservoirs. The shallow wells, of which there is a majority, account for approximately 50% of total production. The remainder of production comes from high-pressure wells which individually produce up to 10 000 barrels per day. Many of Kuwait's operating oilfields were discovered and brought into production more than 40 years ago and are now considered mature.

The bulk of Kuwait's proven and probable oil reserves are in the 31-billion barrel Greater Burgan area which has been in production since 1946 and is the second-largest oilfield in the world after Ghawar in Saudi Arabia. Kuwait's other major fields include Raudhatain, Sabriyah and Minagish (Table 13.5). The Neutral Zone, which is shared with Saudi Arabia, holds an estimated 5 billion barrels of reserves, half of which belong to Kuwait.

In spite of having already produced about 30 billion barrels of oil, the Greater Burgan field is still under primary recovery, due to its strong underlying aquifer. This is also the case in many of the other large fields in Kuwait. Although some fields have now been fitted with sea-water injection plants and gas-lift plants to maintain production rates, the country has significant reserve growth potential through further implementation of secondary recovery techniques. Overall extraction is relatively easy and production costs are amongst the lowest in the world; the Kuwait budget assumes a crude oil production cost of $2.85 per barrel. Average production costs from offshore operations are thought to be around $4.50 per barrel.

Table 13.5: **Oilfields in Kuwait**

Field	**Year of first production**	**Remaining proven and probable oil reserves at end-2004** (billion barrels)	**Cumulative production to 2004** (billion barrels)	**Gravity** (average °API)
Bahrah	1956	0.3	0.0	26.1
Greater Burgan	1946	31.0	29.1	30.8
Minagish	1961	3.4	0.6	34.4
Ratqa	1977	0.9	0.1	27.5
Raudhatain	1960	6.2	2.6	34.4
Sabriyah	1961	4.4	1.0	36.3
Umm Gudair	1962	4.1	0.6	26.9
Other fields	–	2.1	0.4	–
Total Kuwait	**–**	**52.3**	**34.5**	**–**
Khafji	1961	1.7	1.8	28.2
Umm Gudair South	1968	0.2	0.2	24.3
Wafra	1953	0.5	1.2	21.2
Other Neutral Zone fields	–	0.3	0.6	–
Kuwait share of Neutral Zone	**–**	**2.6**	**3.7**	**–**
Total Kuwait including share of Neutral Zone	**–**	**54.9**	**38.2**	**–**

Sources: IHS Energy and IEA databases.

Crude Oil Production[3]

Kuwait is the world's twelfth-largest oil producer and, in August 2005, output, at 2.5 mb/d, was at near full capacity. Production peaked at 3.4 mb/d in 1972 and then declined to a low of 0.9 mb/d in 1982. Production all but ceased during the Iraqi invasion in August 1990 but rebounded to pre-war levels by the second half of 1992, albeit after investment of more than $5 billion to repair damage done to the oil infrastructure.

Kuwait has four major crude oil producing areas: north Kuwait, west Kuwait, south-east Kuwait, and the Neutral Zone (Figure 13.4). The first three are wholly owned and operated by KOC, while production in the Neutral Zone is shared equally by Kuwait and Saudi Arabia. North Kuwait consists of two major fields, Raudhatain and Sabriyah, and currently produces around 600 kb/d. West Kuwait contains several minor fields that make up the two major fields, Minagish and Umm Gudair, and produces around 400 kb/d. South-east Kuwait contains the multi-reservoir Greater Burgan field that is producing around 1.4 mb/d. Kuwait's share of production from the Neutral Zone is currently around 300 kb/d.

In our Reference Scenario, Kuwait's total crude oil production is projected to reach 2.9 mb/d in 2010 and 4.9 mb/d in 2030 (Figure 13.5). These projections are derived from a bottom-up assessment of new oilfield developments and rehabilitation projects in the coming years, combined with a top-down analysis of longer-term development prospects. By 2030, Kuwait will have moved up to become the world's sixth-largest oil producer. The Reference Scenario projection for 2020 is broadly in line with Project Kuwait targets but does not rely on that proposal moving ahead in its current form.

Greater Burgan is expected to remain the single most important contributor to production throughout the projection period, but the share of smaller producing fields – such as Raudhatain and Umm Gudair – will rise. Excluding additional output through advanced technologies, Greater Burgan's output is expected to increase steadily to around 1.6 mb/d by 2015 and then decline slightly to around 1.5 mb/d by 2030. KOC does not want to risk reducing ultimate recovery in the Burgan field by increasing its production too quickly, so will rely on newer fields to meet its targets. The Raudhatain field will remain the second-largest source of production. It is expected that production from currently producing fields will account for more than half of output in 2030, or 2.7 mb/d (Table 13.6). The remainder will be production made possible through reserve additions, enhanced recovery and use of other advanced techniques at currently producing fields, or production from fields that have not yet been discovered.

3. Unless otherwise stated, all the oil-production figures cited in this section include crude oil, natural gas liquids and condensates.

Figure 13.4: **Kuwait's Oil and Gas Resources and Infrastructure**

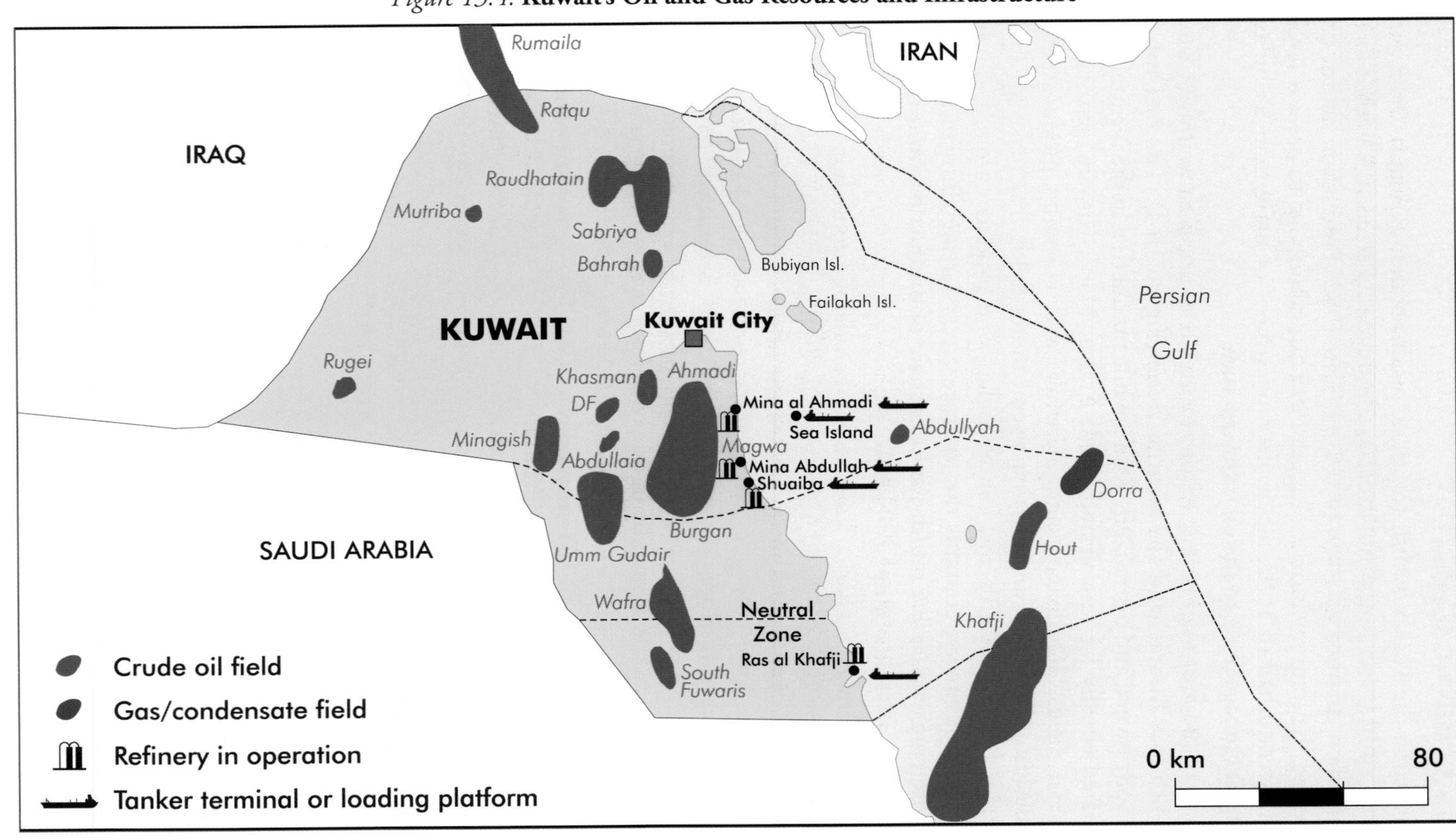

Figure 13.5: **Kuwait's Crude Oil Production by Source in the Reference Scenario**

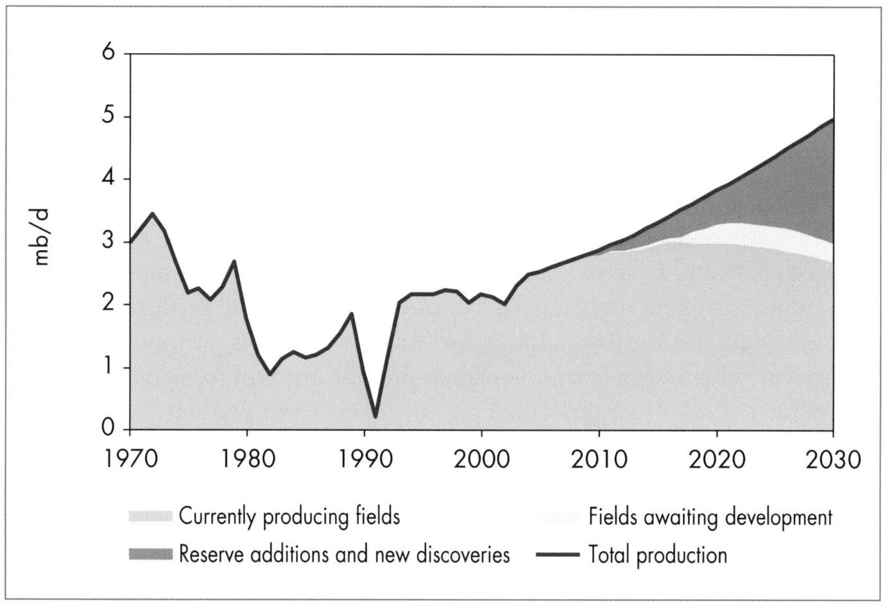

Table 13.6: **Kuwait's Crude Oil Production in the Reference Scenario** (mb/d)

	2004	2010	2020	2030
Currently producing fields	**2.47**	**2.80**	**2.99**	**2.69**
Bahrah	<0.01	0.03	0.03	0.03
Greater Burgan	1.35	1.52	1.64	1.53
Minagish	0.13	0.20	0.20	0.20
Ratqa	0.05	0.06	0.07	0.06
Raudhatain	0.29	0.38	0.50	0.40
Sabriyah	0.03	0.03	0.10	0.11
Umm Gudair	0.12	0.19	0.24	0.24
Other fields*	0.50	0.38	0.21	0.13
New developments	**-**	**0.05**	**0.83**	**2.26**
Fields awaiting development	-	0.01	0.32	0.29
Reserve additions and new discoveries	-	0.04	0.52	1.96
Total	**2.47**	**2.85**	**3.82**	**4.95**

* Includes Kuwait's share of the Neutral Zone.
Note: Includes NGLs and condensates.
Source: IEA analysis.

Oil Refining

Kuwait has oil refining capacity of just under 900 kb/d. Each of its three refineries – Mina Al Ahmadi (443 kb/d), Mina Abdullah (257 kb/d) and Shuaiba (190 kb/d) – is operated by the Kuwait National Petroleum Company (KNPC). The country is the second-largest exporter of refined oil products in the region after Saudi Arabia, with the majority of its exports going to Asia.

KNPC has announced plans to invest $9.5 billion in its domestic refining sector by the end of this decade. This investment programme will improve refinery safety and reliability, respond to increased demand for fuel complying with more stringent product specification and, in the medium term, to increased demand for low-sulphur fuel oil to be used as an input for power generation. The longer-term investment programme will focus on converting low-value fuel oil into high-valued products for export markets.

The centre-piece of KNPC's downstream investment strategy is to construct a new refinery at Al-Zour with an initial capacity of 600 kb/d. Construction is scheduled to run from 2007 to 2010 and cost $6 billion. This new refinery will initially produce 225 kb/d of fuel oil for Kuwait's power plants. Its product slate will shift in favour of higher-valued lighter products for export once much-needed natural gas imports become available. Kuwait also plans to scrap its oldest refinery (Shuaiba) and upgrade its other two existing refineries.

In addition to its domestic assets, KPC (through its subsidiary Kuwait Petroleum International) owns downstream marketing and refining operations in Western Europe and Thailand. These form a convenient outlet for Kuwaiti oil and refined product exports. During 2003, KPC signed an agreement with Shell to explore joint downstream oil sector ventures worldwide. More recently, in March 2005, Kuwait signed an agreement with BP to pursue joint investment opportunities in China and elsewhere in Asia. Areas of co-operation include oil product supply, refining, marketing and distribution. This represents a strategic move to cement Kuwait's position as a key supplier of oil and refined products to the world's most rapidly growing markets.

In the Reference Scenario, we project Kuwait's refining capacity will reach 1.3 mb/d in 2010, in line with KNPC's announced plans. It is then projected to rise to 1.4 mb/d by 2030 through minor expansions at existing facilities although scope exists to surpass this level given the more significant increase expected in crude oil production. To meet our Reference Scenario projections, total refinery investment of $11.7 billion will be needed through 2030.

Oil Exports

Kuwait's net oil exports totalled 2.2 mb/d in 2004, of which 1.4 mb/d were crude oil (OPEC, 2005). In global terms, Kuwait ranks seventh in crude oil exports and sixth in refined product exports. Around three-quarters of Kuwait's total oil exports currently go to the Asia-Pacific region, with Japan alone receiving over 20%. Kuwait also exports smaller volumes to Western Europe and the United States. The majority of Kuwait's oil is marketed as a single export blend (Kuwait Export), which is considered as typical medium-heavy Middle Eastern crude, with an average API gravity of 31° and a sulphur content of 2.5%. Kuwait also exports a small volume of light crude from its Marat field and from the Neutral Zone.

In the Reference Scenario, Kuwait's oil exports are projected to increase to 2.5 mb/d by 2010 and almost 4.4 mb/d by 2030 (Figure 13.6). The Asia-Pacific region, particularly China and South Korea, will remain Kuwait's key growth area for oil exports. The share of refined products in Kuwait's total oil exports is projected progressively to decline.

Figure 13.6: **Kuwait's Oil Balance**

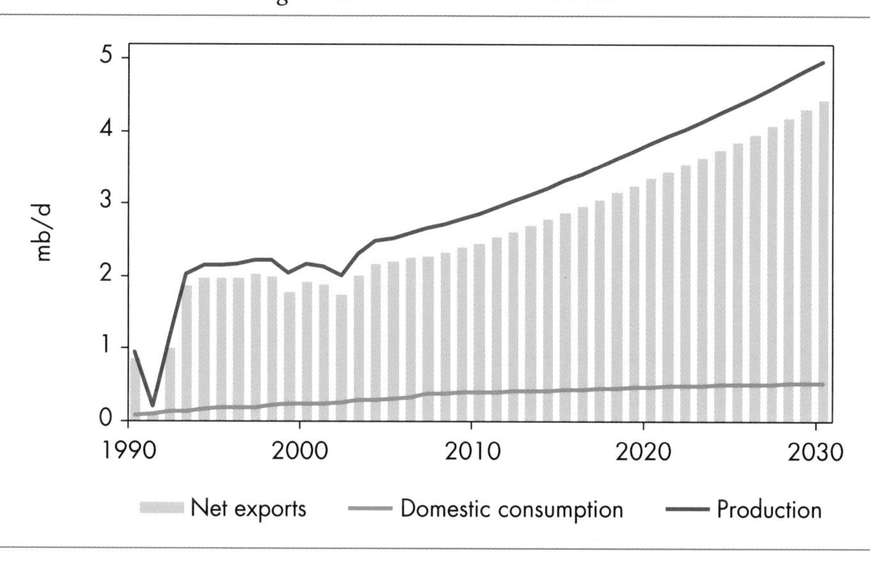

Kuwait has well-developed infrastructure for exporting crude oil, refined products and LPG. Oil flows by natural gravity through underground pipelines from where it is extracted to one of the three refineries or directly to the main port at Mina Al Ahmadi or smaller ports at Shuaiba and Mina Abdullah. Each refinery is also connected by pipeline to export terminals. In step with plans to increase oil production, Kuwait is also expanding the capacity of this supporting infrastructure. Existing gathering centres are being enhanced and new ones are being developed. Export terminal capacity is being increased, including a planned $900 million expansion for the largest port at

Mina Al Ahmadi. This will raise Kuwait's crude oil export capability to 3 mb/d. The Kuwait Oil Tanker Company – the largest tanker company in an OPEC country – has launched a major project to modernise and enlarge its fleet through the addition of a further fifteen carriers of different sizes.

Natural Gas Supply

Gas Reserves

Kuwait has 1.56 tcm of proven natural gas reserves, slightly less than 1% of the world total (Cedigaz, 2005). This places the country just outside the top-20 holders globally and its reserves are small in comparison to many of its MENA neighbours.

Kuwait's gas reserves are almost entirely found in association with crude oil. Their exploitation is accordingly linked to oil production levels and, by extension, OPEC quotas. This has contributed to past difficulties in sourcing sufficient gas to meet local demand. In recent years, KOC has stepped up efforts to discover non-associated gas, but these have been largely unsuccessful. By contrast, significant potential remains to increase reserves of associated gas. The focus of current drilling activity is on areas that are thought to contain gas-rich deposits beneath the Raudhatain oilfield in the north. Kuwait is also engaged in talks with Iran to resolve a long-running dispute over maritime borders, which would facilitate the development of the 200-bcm Dorra gas field which straddles Kuwaiti, Iranian and Saudi waters.

Gas Production and Imports

Kuwaiti gas production fluctuates in line with oil output. Gas production in 2003 totalled 9.6 bcm, more than half of which came from Greater Burgan. Kuwait has made significant efforts to increase the share of its natural gas production that is marketed. Gas flaring has been reduced substantially and is targeted to be reduced to less than 1% by 2009. Natural gas gathering systems have been installed at each of Kuwait's oilfields. These are connected to a pipeline network which delivers the gas to KNPC's facilities, where it is processed into feedstock for power generation and for petrochemicals and fertilizer production.

In the Reference Scenario, Kuwaiti natural gas production is projected to increase broadly in line with oil output – despite an expected increase in gas reinjection, particularly in the northern oilfields. Gas production is projected to reach 11.2 bcm in 2010 and 20.5 bcm in 2030. Gas imports will be necessary to fill the growing gap between demand and indigenous production. By 2010, Kuwait is projected to be importing 0.6 bcm of natural gas. Imports will rise steadily to 5.7 bcm by 2030 (Figure 13.7).

Figure 13.7: **Kuwait's Natural Gas Balance**

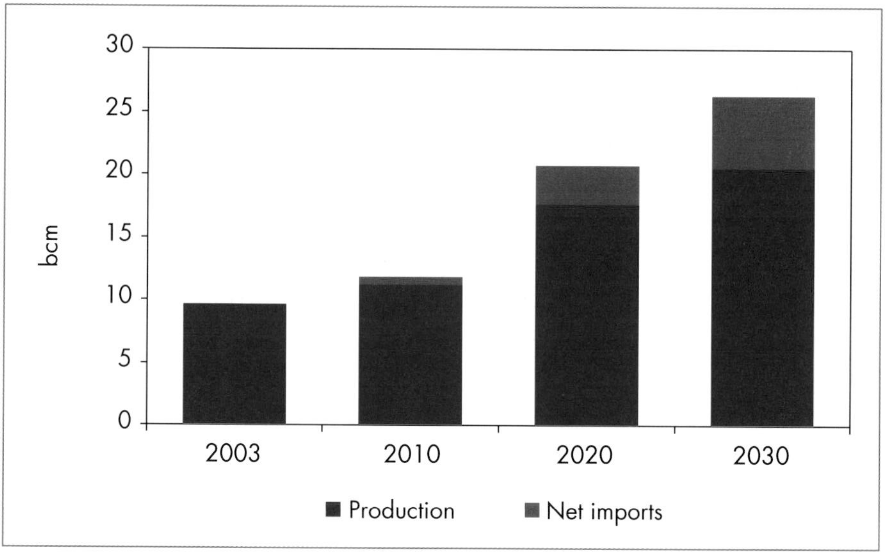

Import Plans

Kuwait has been seeking natural gas imports from its Gulf neighbours since the early 1980s as a means of easing its gas supply shortage and of freeing up for export some of the oil that it is now burnt in power plants. A number of import options are currently under consideration, although it is highly unlikely they will all be completed (Table 13.7).

Table 13.7: **Kuwait's Natural Gas Import Proposals**

Supplier	**Volume** (bcm per year)	**Announced start date**	**Status**
Iraq	0.3 – 1.9	During 2006	Awaiting rehabilitation of pipeline.
Iran	2.9	Late 2007	Final terms yet to be agreed. Pipeline yet to be constructed.
Qatar	11	-	Awaiting approval for pipeline to pass through Saudi Arabian waters.

The most promising option is the resumption of gas imports from Iraq's southern Rumaila oilfield. These were abruptly halted on the eve of the invasion of Kuwait in August 1990. Since then, the pipeline has been idle. Iraq's Southern Oil Company is scheduled to recommence supplies of 0.3 bcm per year during 2006, subject to successful rehabilitation of pipeline facilities on both sides of the Kuwait-Iraq border. It is planned that this will increase to 1.9 bcm by 2010, but we assume that they will reach only 0.6 bcm. In early 2005, Kuwait signed a preliminary agreement, not a binding contract, for a deal worth $7 billion to import 2.9 bcm per year of natural gas from Iran over a 25-year period commencing in late 2007. If the deal is finalised, the next step will be to construct the 260-km pipeline needed to transport the gas from Iran's South Pars field into Kuwait.

Kuwait has a Memorandum of Understanding for the import of an initial 11 bcm per year of gas from Qatar's offshore North Field. This project has experienced lengthy delays owing to difficulties in obtaining Saudi Arabia's approval for the necessary pipeline through Saudi territorial waters. Due to these delays, Kuwait has also looked into the option of importing the gas in the form of LNG.

Energy Investment Needs and Financing

The Reference Scenario projections call for cumulative energy sector investment of $86 billion (in year-2004 dollars) over 2004-2030, or $3.2 billion per year. Kuwait's oil and gas sector will require 78% of the total and the power sector the rest.

Investment in Kuwait's oil and gas sector would amount to $67 billion. This comprises spending on exploration and development projects, construction of new oil refineries and upgrading existing ones. Capital spending will rise in line with increasing oil and gas production and refining capacity. Upstream oil will require 80% of the total oil sector investment with the remainder in refining. These projections assume an average cost of new onshore oil production capacity in Kuwait of slightly more than $5 000 per barrel per day of capacity and an average oilfield natural decline rate of 7%. Sufficient capital is readily available internally in Kuwait to cover the required oil and gas sector investment. Nonetheless, it is expected a substantial share of the spending will be made by international oil companies in response to Kuwait's endeavours to facilitate foreign involvement.

The expansion of the power sector over the period 2004-2030 will require $19 billion of investment. In the absence of market reforms, Kuwait's power sector is expected to continue to rely on government funds. There are no plans to reform subsidies, so electricity will continue to be used inefficiently, increasing investment needs and discouraging private-sector participation, even if this were contemplated. Considerable potential exists for reducing the growth rate in electricity demand and reducing the financial burden arising from the government subsidy on electricity prices through implementing conservation strategies.

Deferred Investment Scenario[4]

Energy Demand

Kuwait's primary energy demand in the Deferred Investment Scenario grows at an average annual rate of 2.6% per year to 46 Mtoe in 2030, 5% less than in the Reference Scenario. The reduction in energy use is a result of slower economic growth due to reduced oil export revenues. Primary demand for gas falls most, as slower growth in electricity demand reduces the need for new power stations, the major driver of gas demand in Kuwait.

Oil and Gas Production and Trade

The Deferred Investment Scenario assumes that the ratio of upstream investment to GDP through to 2030 will be constant at the level of the last ten years. This translates into average investment over the *Outlook* period that is some 24% lower than in the Reference Scenario. Further delays to Project Kuwait, or its abandonment, could lead to such an outcome. Alternatively it might arise if Kuwait decides to divert revenues to support the growth of its non-oil sector and to provide social services to its nationals.

In the Deferred Investment Scenario, crude oil production reaches 3.3 mb/d by 2030, a 33% reduction on the Reference Scenario. Natural gas production is reduced by 28% to 15 bcm in 2030. Expansion of Kuwait's oil refining capacity also slows slightly in response to lower global demand for refined products, reaching 1.3 mb/d in 2030 compared to 1.4 mb/d in the Reference Scenario.

In terms of the absolute reduction in oil production, lower investment has a larger impact on currently producing fields. This peaks at around 2.7 mb/d around the middle of the projection period and then declines to just below 2 mb/d by 2030. But, in relative terms, the reduction in production is greater from new developments. Their share of total Kuwait oil production in 2030 is 40% in the Deferred Investment Scenario compared with 45% in the Reference Scenario.

Oil exports fall more than production in percentage terms, but slightly less in volume terms. Net oil exports are 2.8 mb/d in 2030 compared to 4.4 mb/d in the Reference Scenario (Figure 13.8). Natural gas imports are markedly higher, as reduced indigenous production has to be supplemented through trade. Gas imports reach 10 bcm in 2030, 4.3 bcm more than in the Reference Scenario.

4. See Chapter 7 for a detailed discussion of the assumptions and methodology underlying the Deferred Investment Scenario.

Figure 13.8: **Kuwait's Oil Balance in the Reference and Deferred Investment Scenarios**

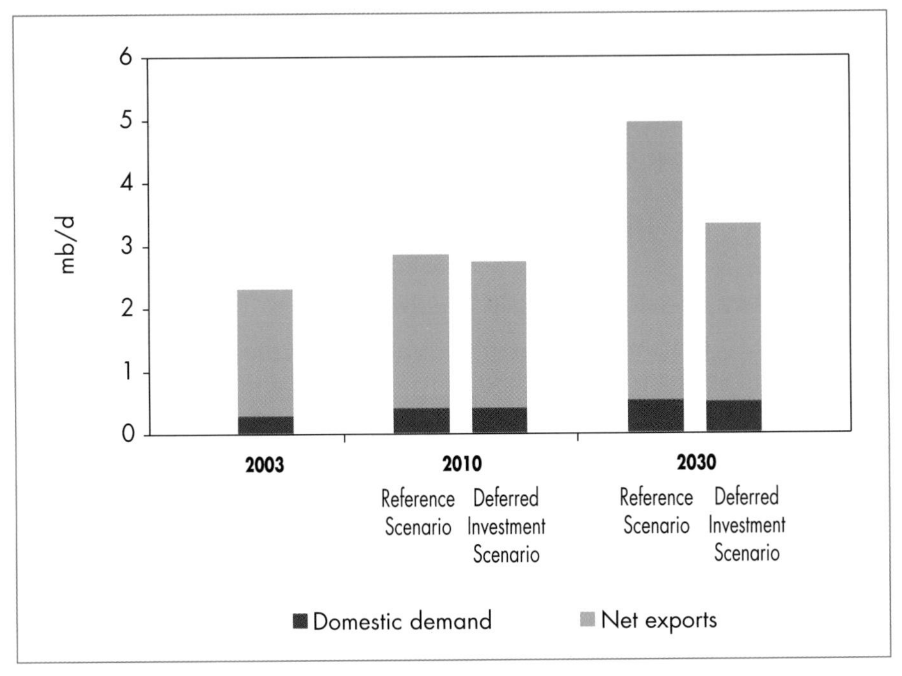

LIBYA

HIGHLIGHTS

- After 20 years of underinvestment and isolation, the Libyan energy sector is experiencing a revival. International sanctions were lifted in 2003-2004 and foreign firms are now competing for access to Libya's extensive oil and gas reserves.
- Energy consumption in Libya is projected to grow strongly over the *Outlook* period, driven largely by rapid economic growth. In the Reference Scenario, primary energy demand will increase by 28 Mtoe between 2003 and 2030, growing at an average annual rate of 3.6%. Oil will remain the dominant fuel, but the share of gas will increase, to 41% of total energy demand by 2030.
- Libya's proven oil reserves, at 39 billion barrels, account for 40% of total oil reserves in Africa. Compared to reserves in other countries, most of the fields in Libya are relatively cheap to develop. Successful development, however, will depend on Libya's capacity to attract foreign investors.
- In the Reference Scenario, oil production will rise from 1.6 mb/d in 2004 to 1.9 mb/d in 2010 and to 3.1 mb/d in 2030. Oil exports grow from 1.4 mb/d in 2004 to 1.5 mb/d in 2010 and to 2.5 mb/d in 2030. In 2004-2030, cumulative investment of $41 billion will be needed in the oil sector, 89% for upstream projects. In the Deferred Investment Scenario, production grows much more slowly, reaching only 2 mb/d in 2030.
- The rate of increase in production of natural gas will be faster than that of crude oil production. Marketed gas production will rise from 6 bcm in 2003 to 12 bcm in 2010 and then surge to 57 bcm in 2030. Cumulative investment requirements in the gas sector will be $21 billion in 2004-2030. In the Deferred Investment Scenario, gas investment will total $14 billion for a production level of 33 bcm in 2030.
- In the near term, power generation will continue to be mostly oil-based, but gas use will overtake oil as gas production expands. Electricity output will rise from 19 TWh in 2003 to 27 TWh in 2010 and to 60 TWh in 2030. Cumulative electricity investment needs to 2030 will exceed $18 billion.
- Desalinated water will be increasingly required over the *Outlook* period, accounting for 11% of total water demand in 2030. The energy requirements for the desalination process will represent 6% of total energy demand by then.

Overview of the Libyan Energy Sector

Libya has some of the world's largest and least-explored oil and gas reserves. It is located close to European energy markets and is already a major oil exporter. Oil production peaked at about 2.7 mb/d in 1971, but has stagnated, at an average of 1.5 mb/d, over the past three decades. Nevertheless, the hydrocarbons sector accounts for 95% of total exports, 30% of GDP and 75% of total fiscal revenues. The infrastructure in Libya is deficient after 35 years of a centrally-planned economy and 20 years of international sanctions. Inadequate investment is responsible for the poor state of roads, hospitals, schools, telecommunications, ports and water infrastructure. Unemployment approaches 30%.

In 2003, Muammar al-Qaddafi embarked on a path of economic reform, including the privatisation of hundreds of state-owned companies. The reform process is designed to strengthen global economic ties and attract more foreign direct investment, both of which have gained momentum since the lifting of international sanctions at the end of 2003 and in 2004. Libya plans to restore oil production to the level of the early 1970s and to upgrade its ageing refineries.[1] Expansion of natural gas production is also a priority. Libya has extensive natural gas resources and is seeking to increase exports, particularly to Europe (OME, 2004).The government also plans to use more gas domestically to free up oil for export.

Table 14.1: **Key Energy Indicators for Libya**

	1971	2003	1971-2003*
Total primary energy demand (Mtoe)	1.7	18	7.7%
Total primary energy demand per capita (toe)	0.8	3.2	4.5%
Total primary energy demand/GDP**	0.04	0.6	8.5%
Share of oil in total primary energy demand (%)	40	73	–
Net oil exports (mb/d)	2.7	1.2	–2.4%
Share of oil exports in production (%)	99.5	83.7	–
Share of gas in total primary energy demand (%)	54.2	25.7	–
Net gas exports (bcm)	0.5	0.8	1.3%
Share of gas exports in production (%)	31	12	–
CO_2 emissions (Mt)	3.7	43.1	7.9%

* Average annual growth rate.
** Toe/thousand dollars of GDP in year-2004 dollars and PPPs.

1. Restoring refinery production to 1970 levels will require a doubling of the current output of 1.5 million barrels a day by the end of the decade, according to the chairman of the National Oil Company.

Political and Economic Situation

Political Developments

Muammar al-Qaddafi has been the leader of Libya since 1969. Because of high oil revenues and a relatively small population, Libya has one of the highest rates of GDP per capita in Africa. Import restrictions and the inefficient allocation of resources, however, have led to periodic shortages of basic goods and food.

Over the past decade, industry has moved beyond processing mostly agricultural products to the production of petrochemicals, steel and aluminium. These sectors now account for about 20% of GDP. Freshwater scarcity and poor soil quality severely limit agricultural output and today Libya imports about 75% of its food requirements.

The UN Security Council lifted UN sanctions on Libya in September 2003. Diplomatic ties with the United States were restored in April 2004 and US sanctions were lifted in September. The first shipment of Libyan crude arrived in the US in May 2004. The United States and Libya do not yet have full diplomatic relations. The elimination of sanctions has boosted economic output, but Libya still faces many challenges, including growing popular discontent over political restrictions and an ageing and inefficient infrastructure. Economic reform is high on the government's agenda. The role of Prime Minister Shukri Ghanem has been bolstered by the appointment of reformists to key cabinet roles.

Economic Trends and Developments

GDP rose by 5% in 2003 and by 1% in 2004, when it was equivalent to $33 billion dollars. Government revenues increased by more than $4 billion dollars to $13.5 billion over the same year. Oil and gas receipts have consistently accounted for about a third of GDP and more than 80% of government revenues.[2] Despite efforts to diversify the economy, it is still sharply exposed to oil price fluctuations. Non-oil economic activity has almost doubled in the past ten years, but its percentage of GDP has fallen from around 76% in 1980 to 65% today. Libya is seeking to move away from reliance on oil and gas receipts.

The pace of GDP growth in Libya is projected to accelerate to 3.6% per year from 2003 to 2010, compared with a growth of 2% per year on average over the period 1990 to 2003. Rising domestic consumption and investment flows – mainly foreign – will drive economic growth over the projection period. The economy is expected to be attractive to international companies keen to

2. *Middle East Economic Digest*, "Standing at the Crossroads", 13-14 August 2004.

contribute to the rehabilitation of Libya's ageing infrastructure. Domestic confidence should be sustained well into the next decade, with both government and private consumption remaining strong. As a result, economic growth in 2010-2020 is projected to average 3.3% per year. The pace of economic growth will slow to 2.7% over 2020-2030, due to higher imports of industrial and agricultural goods.

Energy Policy

The government aims to attract foreign investment in all sectors of the economy, in particular the oil sector. It is keen to push ahead with the allocation of exploration and production concessions in order to boost oil production capacity. Foreign firms have been competing intensely for commercial opportunities in the oil sector. The government has no intention to privatise the sector, but it will be progressively opened to foreign investment.

The electricity sector is managed by the General Electricity Company of Libya (Gecol). At present, Gecol is the sole agency responsible for generation, transmission and distribution. To meet the expected surge in electricity demand, Gecol plans to double installed generating capacity over the coming decade with an investment programme of about $10 billion, to be distributed. There are also plans to expand the distribution and transmission systems. With government support, Gecol is self-financing the improvements, with some additional concessionary lending support from local banks. The cost of the programme exceeds the scope of the state budget. As a result, initial steps are being taken to open up the electricity market in order to pave the way for foreign investors. Longer-term plans include exporting electricity through the Mediterranean Ring, which will create a regional power market among North African and Southern European states.

Libya intends to abolish subsidies on electricity, fuel and basic food, which now cost the government about $5 billion per year (AFP, 2004). In order to diminish the impact on low-income households, the government plans to double the national minimum wage from 150 Libyan dinars ($116) a month to 300 Libyan dinars, and to lower taxes.

Energy Demand

Primary Energy Mix

In 2003, total primary energy demand in Libya was 18 Mtoe. Per capita consumption, at about 3.2 toe, is well above the North African average of 0.8 toe. Libya does not have a large population, extensive agricultural potential or a well-established industrial base like other North African countries such as Algeria, Egypt, Morocco and Tunisia. Libya does, however, have abundant energy resources. Given the country's small population, 5.6 million in 2004, and its large oil and gas reserves, Libya's energy situation resembles that of the small oil-exporting Gulf countries more than that of its North African neighbours. In 2003, oil and gas accounted for over 99% of energy demand. This share is not expected to change over the *Outlook* period. Total energy demand will reach 46 Mtoe in 2030, growing by nearly 3.6% per year from 2003 to 2030.

The fuel mix is expected to change, however, with the share of gas increasing (Figure 14.1). Gas now accounts for 26% of total demand, but by 2030 its share will be 41%. The share of gas will rise as a result of government initiatives to free up oil for export and of recent upward revisions to estimates of proven gas reserves. Most of the increase in gas demand will occur in the power and water sector, where 51% of total generation will be gas-fired by 2030. The fuel requirements for water desalination will account for 23% of the increase in energy demand in the power and water sector from 2003 to 2030.

Figure 14.1: **Libya's Primary Energy Mix**

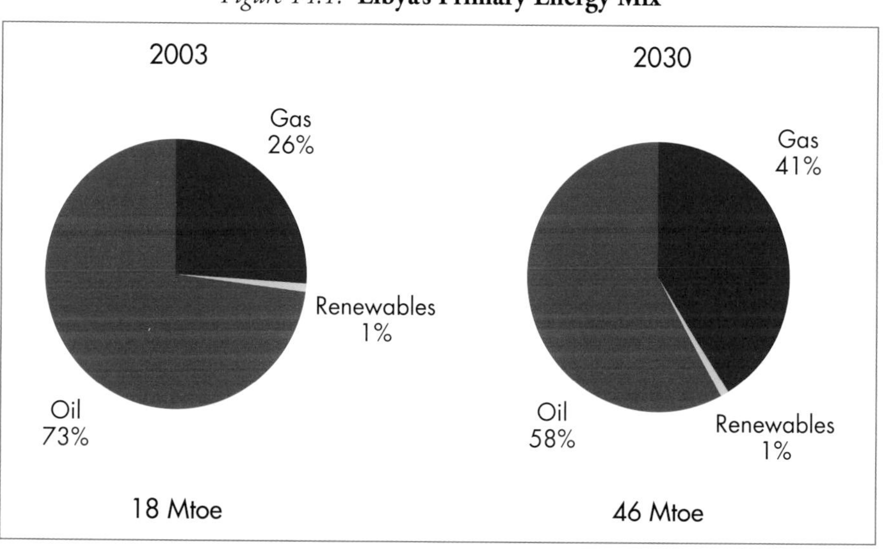

Sectoral Trends

Total final consumption of energy is expected to increase from some 10 Mtoe in 2003 to 28 Mtoe in 2030, growth of 3.9% per year over the projection period. Energy demand in the industry and transport sectors represents nearly 80% of total final consumption, but demand in the services sector will grow fastest over the *Outlook* period, at 5.9% per year on average. Energy demand in the industry sector will grow by 3.9% per year and in the transport sector by 3.7% per year.

Industry

The industrial sector in Libya accounts for 40% of total energy consumption. Demand was 4 Mtoe in 2003 and it is projected to increase to 11.2 Mtoe by 2030, reflecting the government's push to diversify the economy away from oil by promoting downstream industries. Despite the large investments needed to build petrochemical plants and Libya's slow progress in expanding this industry since the 1970s, energy demand in the petrochemical industry will grow at 4% per year; but other industries, such as construction and public infrastructure, will also play an important role, with the expected take-off of tourism and an urgent need for public infrastructure.[3] Demand in industries other than petrochemicals will reach 4.4 Mtoe in 2030 – nearly 40% of total industry demand.

Transport

Economic growth will spur energy demand in the transport sector. Vehicle ownership was about 170 cars per 1 000 inhabitants in 2003 compared to over 500 for OECD countries. Oil consumption in the transport sector is projected to increase by 3.7% per year on average to 2030. No other fuels are expected to be used for transport. Oil for transport will represent over 65% of total Libyan final oil consumption in 2030 (figure 14.2).

Figure 14.2: **Final Energy Consumption in Libya**

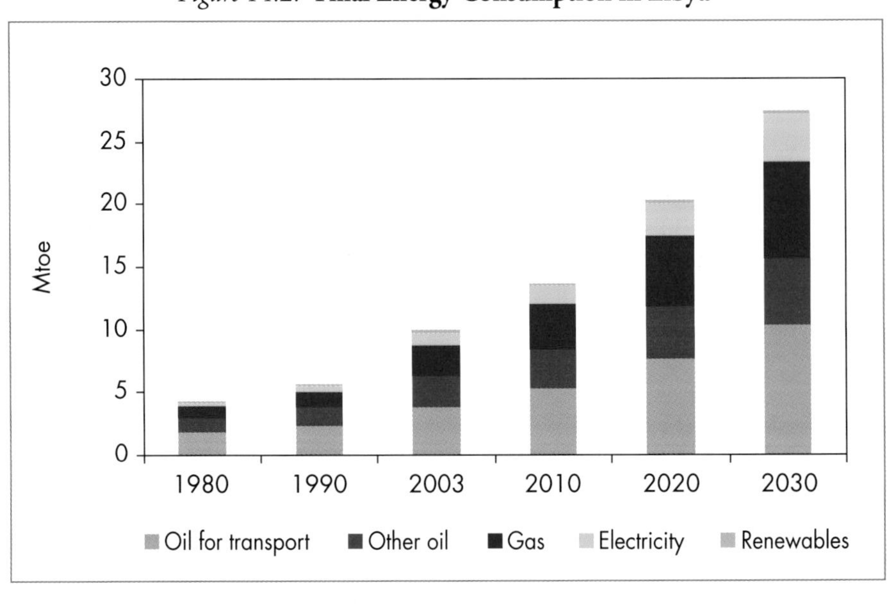

3. "More than Words Needed", *Middle East Economic Digest* (29 Oct.- 4 Nov. 2004), UK.

Residential and Services Sectors

The residential sector accounts for only 17% of total final consumption, 1.7 Mtoe in 2003. Consumption is projected to rise by 4.2% per year and reach 5.1 Mtoe in 2030. The increase in residential energy consumption will result from rising per capita incomes. Per capita energy demand in this sector is expected to be on a par with other countries in the MENA region towards the middle of the *Outlook* period.

While the services sector is and will remain relatively small in absolute terms, consumption will triple from 0.1 Mtoe in 2003 to 0.3 Mtoe in 2030. The increase in demand in this sector will, in large part, be attributable to growth in tourism.

Agriculture

Agricultural development in Libya has been inversely related to the development of its oil industry. In 1958, agriculture supplied over 26% of GDP, and Libya exported food. As agricultural output has remained relatively constant, increasing oil revenues have resulted in a decline in agriculture's overall share in national income. In 2003, this sector accounted for only about 3.5% of GDP and Libya imported over 70% of its food needs. Achieving self-sufficiency in food production is unlikely over the next 25 years, but agricultural development is high on the government's agenda. The Great Man-made River Project (GMRP), using fossil water in the south to irrigate fields in the north, is one example of government efforts to push development in the sector. Energy demand in the agriculture sector is expected to double, reaching 0.2 Mtoe in 2030.

CO_2 Emissions

As a result of increasing energy demand, CO_2 emissions will more than double over the projection period from 43 Mt in 2003 to 104 Mt in 2030. The 3.3% annual average growth in emissions over the *Outlook* period will however be slower than the 3.6% growth in demand owing to the switch to gas in power generation.

Electricity Supply and Desalinated Water Production

Overview

Libya's power plants produced 19 TWh of electricity in 2003. Estimated production in 2004 was 20 TWh. Libya's electricity generation per capita is the highest in North Africa, although, at 3 411 kWh in 2003, it is still quite low compared with OECD countries. Electricity demand has been increasing rapidly, following a sluggish period in the early 1990s.[4]

Most power generation is based on oil (80%), while the remainder comes from natural gas. The current share of oil is one of the highest in the MENA region. Libya is planning to increase the share of gas-fired generation, but progress has been slow. The share of gas has been hovering around 20%-22% since the mid-1990s.

Installed capacity was 6 GW in 2003, a mix of gas turbines, steam boilers and diesel engines, accounting for 47%, 44% and 9% of installed capacity respectively. About two-thirds of power generating capacity is over 20 years old because capacity increases in the 1990s were lower than in the 1970s and the 1980s (Figure 14.3). All new capacity added in the 1990s was in the form of gas turbines. These plants are quick to build and have low initial costs. The technology mix has kept electricity generation efficiency rather low. The average conversion efficiency of oil-fired plants was 25% in 2003, while that of gas was 32%, relatively low compared to the 42% and 43% efficiencies of OECD oil-fired and gas-fired plants.

Figure 14.3: **Power-Generation Capacity Additions in Libya**

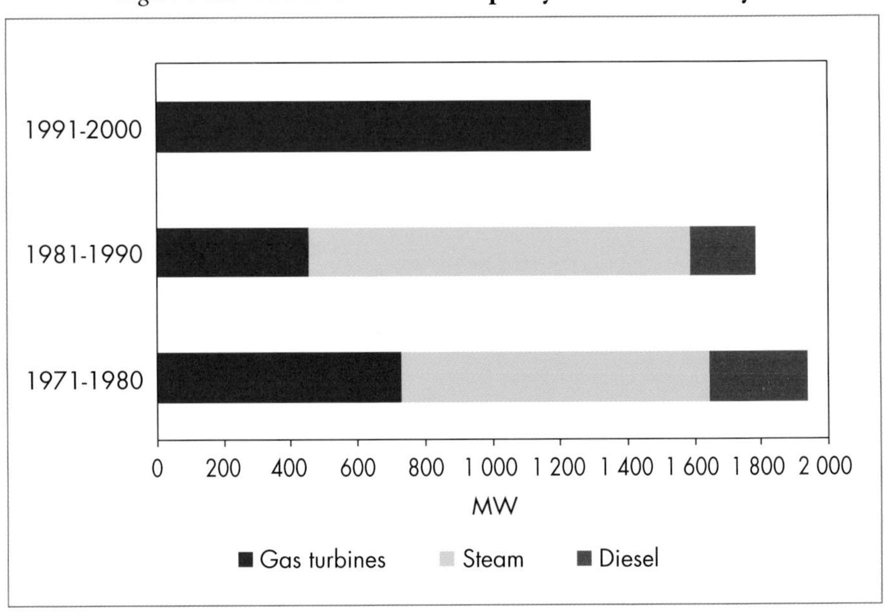

4. The average annual growth in the period 1998-2003 was 7%.

Libya's electricity network has been linked to Egypt's since 1998 and to Tunisia's since 2002. At present, Libya is a net importer of electricity from Egypt. The government would like to increase interconnection capacity with both Egypt and Tunisia, but how these projects will be financed is yet to be determined.

Electricity generation, transmission and distribution in Libya are the responsibility of the government-owned General Electricity Company of Libya (Gecol). Gecol owns over 80% of Libya's installed capacity. The remaining capacity is owned by various industries.

Libya's power sector is in urgent need of investment. Recent capacity increases have not kept up with the increase in demand, and insufficient maintenance of its power plants and the transmission network keeps the availability of power plants low. This constrained supply situation resulted in a massive blackout in summer 2003.

Gecol's revenues are very low because of low tariffs and unpaid bills. Consequently, the company relies on soft loans from organisations such as the

Table 14.2: **Libya's Existing and Planned Electricity Interconnections**

		Type	kV	MW	Length (km)	Year in operation
Existing						
From:	To:					
Tobruk	Saloum (Egypt)	AC-double	220	600	165	1998
Abukamash	Medenine (Tunisia)	AC-double	220	400	210	2002
Rowis	Tataouine (Tunisia)	AC-double	220	200	160	2004
Planned	To Egypt:					
Gdabia	Tobruk	AC-single	400	1000	375	2010-2015
Gwarsha	Tobruk	AC-single	400	1000	350	2010-2015
Tobruk	Aldbaah	AC-single	400	1000	-	2010-2015
	To Tunisia:					
Rowis	Buchemma	AC-single	400	1000	350	2010

Note: AC-single: alternating current single circuit ; AC-double: alternating current double circuit.
Sources: Eurelectric and UCTE (2005); IEA database.

Arab Fund and the Islamic Bank and on the state budget to finance new investments.[5] Libya has no concrete plans at the moment to restructure its power sector.

Electricity Production

Electricity generation in Libya is projected to increase by 4.4% per year between 2003 and 2030. The period 2003-2010 will see the highest growth rate, at 5.1% per year. Growth is expected to slow down over time, falling to 4.7% per year in the period 2010-2020 and to 3.5% per year in 2020-2030.

Libya is expected to shift gradually to greater use of natural gas in power generation. With most new capacity likely to be gas-fired, the share of gas in electricity generation is projected to rise from 20% in 2003 to 56% in 2030. The share of oil is projected to drop, but in absolute terms, oil-fired generation will increase from 15 TWh in 2003 to 26 TWh in 2030.

Libya will need 15 GW of new power-generation capacity in the period to 2030. The country is expected increasingly to construct CCGT power plants. Some of the recent open-cycle gas-turbine power plants have been designed to be converted eventually to combined-cycle plants.

Water Production

Water consumption in Libya was nearly 4.9 billion cubic metres in 2003, despite the fact that its surface water resources are very limited. About 95% of Libya is covered by desert. Groundwater is extracted at a rate estimated to be about 8 times that of the renewal of the groundwater resource, and Libya depends heavily on fossil groundwater. The coastal aquifers are the only ones that are being recharged by rainfall, but uncontrolled groundwater development from these aquifers exceeds the annual replenishment.

The government has chosen to address water scarcity through implementation of the GMRP. The project, which commenced about 25 years ago, is designed eventually to transport 2 300 mcm per year of water from the Nubian aquifers in the south to the urban population centres thousands of kilometres to the north. Delays, technical problems, corroding infrastructure and financial difficulties have plagued the five-phase scheme. Only phases 1 and 2 have been completed. The transported water was originally intended to be used primarily for irrigation, but urban water demand is growing rapidly and most of the fossil water is being used in the residential sector.

The Libyan government is also planning to expand desalination capacity to supplement water from the GMRP. Thermal desalination plants were first introduced in the late 1960s. By 2003, total installed desalination capacity was

5. World Report International Surveys (2004), *Libya*, World Report.

272 million cubic metres. Actual production is estimated to be some one-third of this, due to the fact that most plants are not in good operating condition. The desalination sector suffered from inappropriate plant designs and materials, and lack of spare parts, during the sanction years. Desalinated water accounted for only some 2% of total water supply in 2003.

Water consumption is projected to increase to 5 713 million cubic metres in 2030. Even if the GMRP were to fulfil expectations, the transported fossil water would meet only 40% of projected demand. Consequently, Gecol is implementing a $1 billion programme aimed at the installation of 310 mcm of desalination capacity by 2010.[6] Because financing difficulties are expected, the projections here are for capacity to increase by 193 mcm by 2010 and by an additional 307 mcm by 2030. About one-third of the total electricity capacity additions will be for new combined water and power (CWP) plants with desalination units. At the end of the projection period, desalinated water is expected to account for 11% of total water demand. Most of the increase in capacity will be in reverse osmosis plants. The share of RO was 26% in 2003 and is projected to rise to 41% in 2030.

Fuel requirements for desalination will rise from 0.6 Mtoe in 2003 to 2.8 Mtoe in 2030, accounting for 6% of total primary energy demand.

Table 14.3: **Water and Desalination Capacity Projections for Libya**

	2003	**2010**	**2020**	**2030**
Water consumption (million cubic metres)	4 867	5 051	5 383	5 713
Desalination capacity (million cubic metres)	272	465	532	772
Oil and gas requirements for desalination (Mtoe)	0.6	1.4	1.9	2.8

Libya will need to invest $2 billion (in year-2004 dollars) in new desalination plants over the projection period, amounting to some 20% of total investment needs for new electricity and water production plants.

Oil Supply

Overview

Libya has proven oil reserves of 39 billion barrels, 40% of Africa's oil reserves and 3% of the world total. Oil exploration in Libya began in 1953 and since

6. "Special Report Libya", *Middle East Economic Digest*, 22 August 2003.

the early 1960s the petroleum industry has increasingly dominated the economy. In 1977, Libya was the seventh-largest oil producer in the world, but its position declined somewhat in the early 1980s in line with a cut in OPEC production quotas. By 2004, Libya was the world's fifteenth-largest producer of crude oil. Now that the sanctions have been lifted and market reform is under way, Libya's prospects as a major oil producer are set to improve.

Libya aims to produce 2 mb/d by 2010 and 3 mb/d by 2015, thus restoring its production to the level of the 1970s. A large proportion of Libyan oil and gas acreage is unexplored. It is estimated that only a quarter of the country's territory has ever been licensed and much of this has not been explored using modern techniques. However, we expect production to rise slightly less quickly than targeted, due to delays in implementing economic reforms and unattractive contract terms. Libyan production is projected to reach 1.9 mb/d in 2010 and 3.1 mb/d in 2030. We project exports to reach 2.5 mb/d in 2030. Most exports will go to Western Europe. In 2030, Libya will rank sixth among MENA oil producers and exporters. Libya is one of the few countries with the capability to double its oil production over the next 25 years.

Policy Framework

Libya has been an OPEC member since 1962. During the late 1970s, production rose slightly, only to fall again in the 1980s when OPEC reduced its members' production quotas. In March 1983, Libya had an OPEC quota of 1.1 mb/d. In November 1984, this figure was revised downward to 990 kb/d. Libya's current OPEC quota is 1.5 mb/d.

The National Oil Company of Libya (NOC) is responsible for upstream and downstream oil activities, under the Petroleum Law of 1955, which was amended in 1961, 1965 and 1971. In July 1970, NOC's operations were expanded by legislation that nationalised the foreign-owned Esso, Shell, and ENI marketing subsidiaries, and a small local company, Petro Libya. In 1981, Exxon withdrew from Libya, pulling out its subsidiary operations. Mobil followed suit in 1982, when it withdrew from its operations in the Ras al Unuf system. These withdrawals gave NOC an even greater share of the overall oil industry. The oil sector was further nationalised in 1986 when the US government forced American companies to dispose of their operations in Libya. NOC is the principal instrument of government policy in the oil sector and controls about two-thirds of Libya's total oil production (Arab Oil and Gas Directory, 2004).

The Libyan oil sector was previously known for its complex administrative procedures, which delayed clearance for most projects. The government is gradually removing many of these.

The risk to investments in Libya has lessened and this should have a positive impact on the next round of negotiations. A new contract structure has been drawn for the country's fourth round of exploration and production-sharing agreements, known as EPSA-IV (Box 14.1).

Box 14.1: **Libya's Exploration and Production Sharing Agreements-Round 4 (EPSA-IV)**

Unlike the previous generation of EPSAs, EPSA-IV is an attempt to introduce a more investor-friendly model and incorporates some clauses included in a new petroleum law. EPSA-IV is a complex formula that builds on the previous models under which some aspects were left open for negotiation. The EPSA-IV formula is designed to help significantly encourage an enhanced recovery rate.

Under EPSA-IV, contracts will be based on a 5-6 years exploration phase, followed by a 25-year development period. NOC claims that the new contract model will enable agreements to be negotiated far more quickly than previously. Exploration work will be funded entirely by the private-sector operator or consortium. The foreign oil company bears 100% of costs for an initial minimum period of 5 years, while NOC retains exclusive ownership. Management is assigned to a committee comprised of two NOC representatives and one from the outside investor; voting is unanimous. If commercially viable discoveries are made, the operating company and NOC will set up a joint venture with equity divided 50:50 while 60-70% of the production will go to NOC. Other features of EPSA-IV include: open competitive bidding and transparency; joint development and marketing of non-associated natural gas discoveries; standardised terms for exploration and production (Khodadad, 2005). Under EPSA-IV, contracts are awarded on the basis of the production share acceptable to the competing oil companies. Despite improvements in contract terms, EPSA-IV remains relatively tough compared to contract terms in some other MENA countries.

Libya invited bids for 44 blocks in June 2005, with the winners to be announced in October. Fifteen exploration licences were awarded to foreign companies in January, of which 11 went to US companies. The issuing of new licences in Libya represents the first significant opening to foreign oil companies in decades by a member of OPEC.

Libya's proven oil reserves are estimated at 39 billion barrels. At the level of output (1.61 mb/d) in 2004, Libya has a reserve/production ratio of over 66 years. The country has six large sedimentary basins: Sirte, Murzuk,

Figure 14.4: **Main Oil and Gas Fields and Energy Infrastructure in Libya**

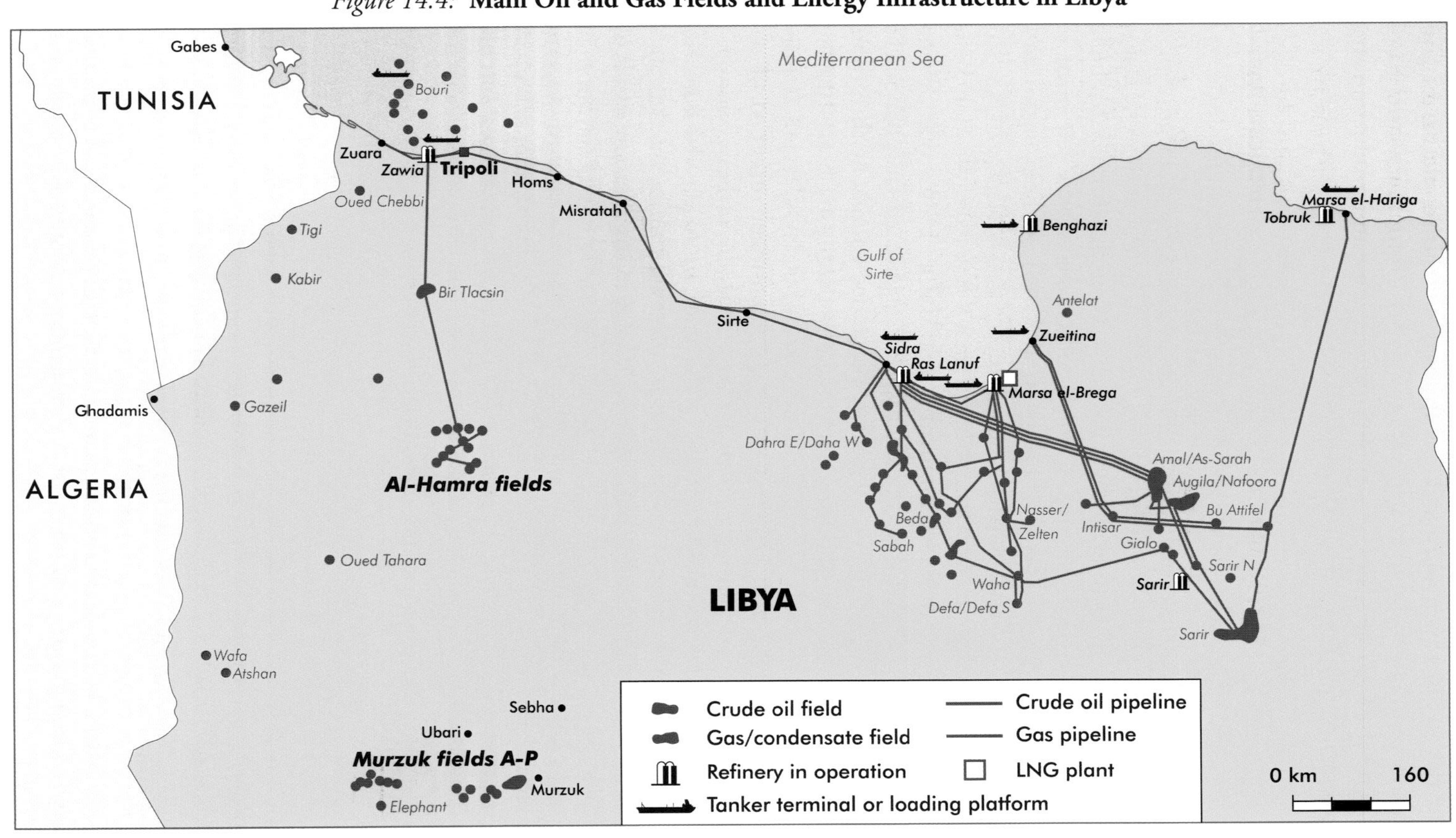

Ghadames, Cyrenaica, Kufra and the offshore. These could contain total oil-in-place of as much as 220 billion barrels, representing an undiscovered potential of 107 billion barrels.

- The leading oil regions in Libya are to be found in the Sirte basin which has been more extensively explored than the others and accounts for 87% of proven reserves. The Sirte basin is closest to the coast and is the main producing area, accounting for close to 80% of current production. It is relatively well explored, although some of the sub-basins further from the coast still have exploration potential. It is home to several giant fields, including the Sarir field, the largest in Libya with 8 billion barrels of proven reserves. It produces mostly from Cretaceous and Tertiary carbonates and sandstones.
- The Ghadames basin comes second in terms of drill-rig activity. It is an extension of the Berkine basin in Algeria. Already partly developed, it has significant remaining exploration potential in Silurian, Devonian and Triassic reservoirs. The Ghadames basin accounts for 3% of Libya's proven reserves.
- The Murzuk basin in the south-west has seen some discoveries but exploration there is just starting. The Kufra basin in the south-east has similar geology, but is essentially unexplored. Distance to markets, logistics and the desert environment are likely to make hydrocarbon production in those areas significantly more expensive than in other parts of the country. The Murzuk basin accounts for 3% of proven reserves.
- Finally, a number of smaller basins, extending offshore along the western and eastern coasts of the country, have hardly been explored but have good potential for development with modern offshore techniques. The major offshore reserves are located in the Al-Bouri oilfields next to the Tunisian border and account for 5% of proven recoverable reserves.

Trade restrictions have to some extent limited access to advanced technologies. The recent opening to international companies, however, is likely to make modern technologies fully available for exploration and, eventually, for production. Development of the more remote areas in the south will call for an influx of technology and logistical support to make them cost-effective.

Active exploration started in Libya in 1953 after oil was discovered in neighbouring Algeria. The first well was drilled in 1956 in western Fezzan, and oil was struck in 1957. Oil flowed by pipeline from Esso's concession at Zaltan to its export facilities at Marsa al-Brega in 1961. Major discoveries were made in the Sirte basin, one of the world's largest oilfields, south-east of the Gulf of Sidra. Oil was also found in the Ghadames sedimentation basin (400 km south-west of Tripoli) in 1974 and in offshore fields 30 km north-west of Tripoli in 1977.

Since 1977, efforts to tap new deposits have concentrated on Libya's offshore fields. The large Al-Bouri field was brought on stream by the NOC and AGIP (Azienda Generale Italiana Petroli), a subsidiary of the Italian company ENI, in late 1987. Other offshore exploration ventures were launched following the settlement of maritime boundary disputes with Tunisia in 1982 and Malta in 1983.

Since the mid-1980s, and up to 2001, two or three finds were regularly announced each year. Few oil discoveries were reported in Libya in 2002 and 2003, partly because of the limited amount of exploration activity. Exploration in Libya began to pick up after the award of 17 new blocks to three foreign consortia in 2003, followed by a major agreement with Shell in March 2004. The giant Elephant oilfield was discovered and brought on stream in March 2004 and now produces 50 000 b/d.

Crude Oil Production[7]

Like Algerian oil, Libyan crude oil, while having rather high wax content, is lighter and easier to handle than most crudes. It has a low sulphur content, which involves less processing to meet the internal combustion engine environmental and operability requirements in consuming regions. For this reason, and because Libya is one-third closer to European markets than countries in the eastern Mediterranean, Libyan crudes have had a receptive market in Europe. Libya's geography allows for easy pipeline access to its ports.

Libya's crude oil production was 2.69 mb/d in 1971, but only 1.61 mb/d in 2004. Spare capacity exceeded 250 kb/d in 2003. Oilfields and recovery rates were badly affected by the international sanctions, which restricted the use of advanced technologies, and by underinvestment resulting in ageing production facilities and poor maintenance.

Most new field developments are being undertaken by foreign companies. NOC estimates that the flow rates of some large reservoirs could be doubled through the application of new technologies and the refurbishment and upgrading of production facilities. From 2003 to 2004, production increased by 8.5%, reaching 1.61 mb/d.

Libya plans to increase its production capacity to 2 mb/d by 2010 and to 3 mb/d by 2015. Considering the investment and upgrading needed in the oil sector, and given the failure to aggressively implement economic reform and to finalise some contract negotiations, we do not expect the oil sector to develop as quickly as planned. Accordingly, Libyan oil production is projected to reach 1.9 mb/d by 2010. The projections are also more conservative in the longer

7. All the oil-production figures cited in this section include crude oil, natural gas liquids (NGLs) and condensates.

Table 14.4: **Libya's Crude Oil Production in the Reference Scenario** (kb/d)

	2004	2010	2020	2030
Currently producing fields	**1 614**	**1 850**	**1 706**	**1 505**
Amal (012-B/E/N/R)	39	53	53	53
Augila-Nafoora (102-D/051)	42	45	30	3
Beda (047-B)	16	16	15	14
Al-Bouri (NC041-B)	65	64	1	0
Bu-Attifel (100-A)	113	113	107	70
Dahra East-West (032-F/B/Y)	11	11	11	10
Defa (059-B/071-Q)	106	57	9	1
Gialo (059-E/4M/5R/6K)	97	88	60	33
Intisar (103-A)	4	4	4	3
Intisar (103-D)	21	14	7	3
Messla (065-HH/080-DD)	99	10	0	0
Nasser (006-C/4I/4K/4G)	35	29	15	6
Sarir (065-C)	233	233	134	52
Waha (059-A)	20	59	58	54
Other fields	712	1 055	1 201	1 201
New developments	**0**	**0**	**794**	**1 545**
Fields awaiting development	0	0	381	1 066
Reserve additions and new discoveries	0	0	413	479
Total	**1 614**	**1 850**	**2 500**	**3 050**

Note: Includes NGLs and condensates.
Sources: IHS Energy database; IEA analysis.

term, with production of 3 mb/d towards the end of the second decade. By 2030, production capacity in Libya is expected to be just under 3.1 mb/d, placing Libya sixth among MENA oil producers.

Libya's oilfields are small and numerous and only three fields, Sarir, Bu-Attifel and Defa, have production exceeding 100 kb/d. In 2004, the two largest producing fields were Sarir (14%) and Bu-Attifel (7%). In 2010, Sarir will still remain the largest operating field in Libya, with production of 233 kb/d, but its share in total production will have fallen to 12%. Production in the Gialo and Bu-Attifel fields will be higher than in the Defa field. By 2030, the combined production of Sarir and Bu-Attifel will be less than 4% of total Libyan oil production.

Most of the existing fields have not yet reached their full production potential and investment and maintenance efforts will be concentrated on them well into the next decade. Production in existing fields is expected to start declining in the next five to ten years. Nearly all of the 14 major producing fields in 2004 will still be producing in 2030, but their share in total production will decline from 56% in 2004 to only 10% in 2030. Most new fields are expected to start production after 2010, and their combined production is projected to be higher than production in existing fields by 2015 (Figure 14.5).

Figure 14.5: **Libya's Crude Oil Production by Source in the Reference Scenario**

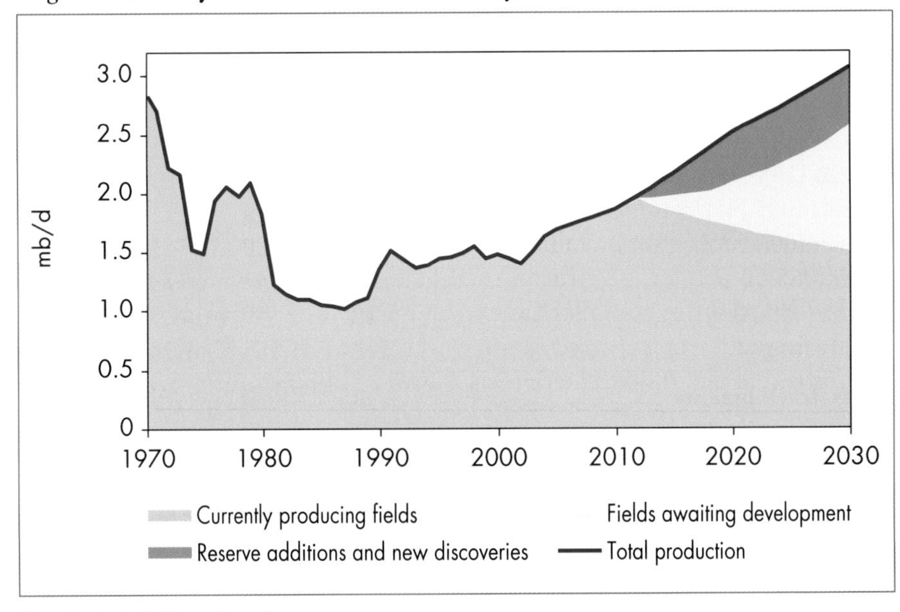

Note: Includes NGLs and condensates.

Oil Refining

Libya has five oil refineries, with a combined capacity of 380 kb/d (Table 14.5). All of the refineries are operated by NOC and all are in urgent need of upgrading. Conversion capacity is low and the refineries are unable to produce environment-friendly products such as unleaded and reformulated gasoline and low-sulphur products. Libya will need to upgrade the quality of its products to comply with new EU specifications. Because Libya was unable to import replacement parts for its refineries during the sanction years, current processing capacity at all five plants is well below design capacity.

The government has plans to upgrade its largest and oldest refineries. In the short term, the Ras-Lanuf facility will be upgraded at an expected cost of $2 billion to $3 billion. NOC also plans to expand the Zawiya refinery at an

Table 14.5: **Current and Planned Crude Oil Distillation Capacity in Libya**
(kb/d)

	2004	**2010**
Ras-Lanuf	220	230
Zawiya	120	140
Tobruk	20	20
Al-Brega	10	10
Sarir	10	10
Total	**380**	**410**

Source: IEA database.

estimated cost of $250 million. The government has, for the moment, abandoned its plans to construct a 20 kb/d plant at Sabha to serve the domestic market. None of the upgrades have commenced. The time-frame is uncertain but the upgrades are expected to take place before 2010. There are also plans to construct a new 200 kb/d refinery at Misurata to target the European market some time after 2010.

Libyan refining capacity increased dramatically in 1985, when the export refinery at Ras-Lanuf came on stream with a 220 000-b/d capacity. From 1985 until the early 1990s, Libya produced more than double its refined product needs (Figure 14.6). However, because of refinery specificity, there has always been a mismatch between output and demand for some products. Since the beginning of the 1990s, refining capacity has increased very slowly.

Domestic demand for oil products was 13.2 Mtoe in 2003 and total production of oil products was 17.4 Mtoe, of which 16 Mtoe came from refineries and the rest from gas-processing plants. In aggregate, Libya produces 38% more products than it consumes. Domestic demand in Libya is set to increase by 2.7% per year over the *Outlook* period, reaching 27 Mtoe in 2030. Refinery output is expected to reach 30 Mtoe in 2030, with an average annual growth of 2.4% from 2003 to 2030.

With growing domestic and export demand, refining capacity will need to be both upgraded and expanded. Capacity is projected to reach 410 kb/d by 2010. The construction of a new export refinery at Misurata — along with further upgrading of the existing capacity — will lead to an increase in refining capacity of 200 kb/d, bringing total capacity to 560 kb/d by 2020. Increasing domestic demand for refined products and the growing European market will push capacity to just under 670 kb/d by 2030, nearly 76% higher than the current level.

Figure 14.6: **Oil Product Output and Oil Demand in Libya in the Reference Scenario**

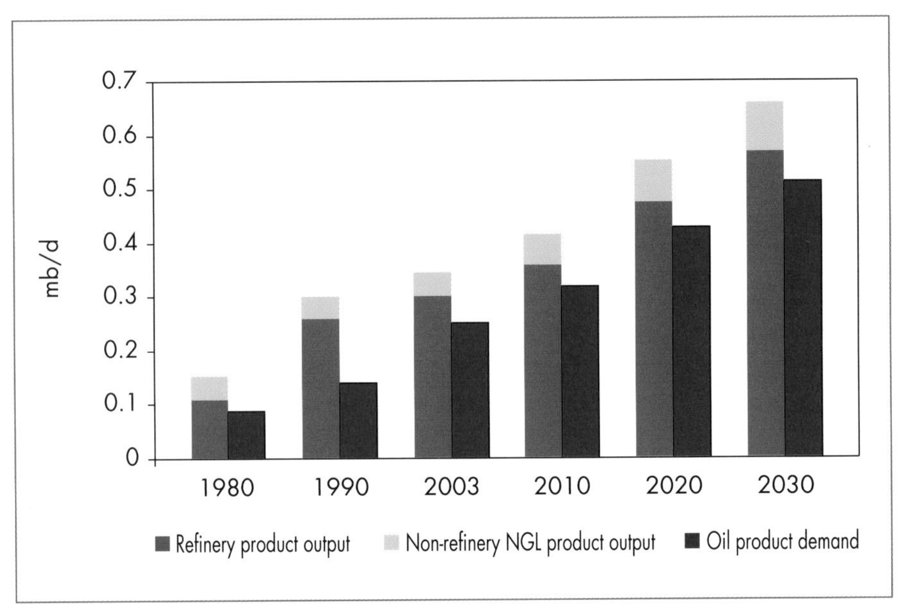

Exports

In 2004, Libya exported 1.35 mb/d of oil, of which 7% was refined products. Western Europe is the primary market for Libyan oil exports. Italy, Germany and Spain accounted for over 70% of all Libyan crude oil exports in 2003. European refineries like Libya's light sweet crude because of its suitability for production of light products. NOC's largest single market is its traditional trade partner, Italy. NOC sells crude directly to refiners to limit the amount traded on spot markets. In 2030, total oil exports will have nearly doubled to 2.54 mb/d. The share of exports in total oil-product output remains constant at about a quarter over the projection period.

Libyan oilfields are served by a complicated network of oil pipelines leading to the six principal export terminals at Marsa al Burayqah, As Sidra, Ras al Unuf, Marsa al Hariqah, Zuwarah and Az Zuwaytinah. Pipelines to these terminals serve more than one company, thus mixing different oil blends that are then standardised for export. The share that an individual company receives is determined by the amount and quality of the oil that enters the common pipeline.

Libya started to develop a refining and distribution network in Europe in the late 1980s, acquiring assets in Italy, Germany and Switzerland. Libya has captured 5% of Italy's retail market through NOC's subsidiary, Tamoil. There are also distribution and marketing channels for Libyan petroleum in London and Frankfurt. Libya has also invested in the downstream sector in Pakistan and in the development of service stations and pipelines in Egypt.

Natural Gas Supply

Resources and Reserves

As of 2004, Libya had proven natural gas reserves of 1 453 bcm (Cedigaz, 2005). In December 2000, important reserves were discovered near the town of Al-Brega. In the last two years, reserves have been revised upwards by 11% and some estimates put total gas reserves in the range of 2 000 bcm. Currently, 30% of gas is reinjected, 15% flared, 5% used in gas production and processing and 50% marketed. In 2003, associated gas, found mainly in six Sirte basin onshore fields accounted for three-quarters of gas production.

Production and Distribution

In 2003, natural gas production in Libya was 6.4 bcm, about four times the level of 1971. Production of natural gas received a major boost in 1971, when a law was passed requiring oil companies to stop flaring gas and to store and liquefy the gas condensate from their wells. However, over the past three decades, natural gas production has lagged far behind oil production owing to the high costs of transport and liquefaction.

The Western Libya Gas Project of Italy's ENI is the largest gas project to date. While some gas from this project will be dedicated to the domestic market, the keystone of the project is the Green Stream pipeline from Mellitah in Libya to Gela in Sicily and then on to the mainland in Italy. It was officially inaugurated in October 2004 but encountered some technical difficulties in early 2005 that delayed full operations. The 550-kilometre pipeline will transport 8 bcm of gas a year.

Gas is expected to play a major role in the Libyan energy scene over the *Outlook* period and the NOC, which has previously concentrated its attention on oil structures, is now devoting more attention to natural gas. By 2010, total annual gas production is expected to nearly double to 12 bcm. But most of the increase is expected to occur in the last 10 years of the projection period. By 2030, gas production will be 57 bcm per year.

There are 12 major gas fields in operation in Libya accounting for 98% of production, but most of them are mature and are projected to decline substantially in the next ten years. By 2010, these fields will account for only 57% of total production and by 2030 they will represent less than 7%. Intisar (103-D) and Bu-Attifel account for over 61% of total gas production today, but will account for less than 30% of production in 2010 and only 3.2% in

2030 (Figure 14.7). Production from Intisar (103-D) and Bu-Attifel fields will decrease from 4.3 bcm in 2004 to 3.5 bcm in 2010 and to 1.8 bcm in 2030.

In the short to medium term, the decline in production in mature fields will be offset by the Western Libya Gas Project (WLGP) which is expected to account for 43% of total gas production by 2010. While WLGP production will increase in volumetric terms to 8.8 bcm in 2030, from 5.3 bcm in 2010, its share in total production will fall to 16% in 2030. Production in new fields, at about 43 bcm, is projected to come on line in the mid-2020s (Figure 14.7) and will account for 78% of total gas production by the end of the projection period.

Figure 14.7: **Libya's Natural Gas Production in the Reference Scenario**

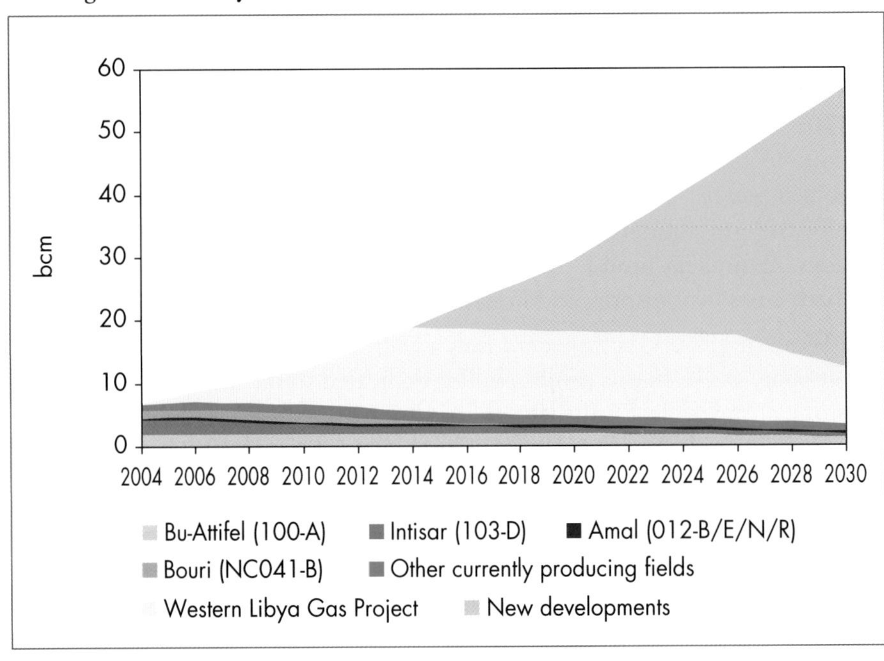

There is no need for new discoveries to be made to add the 50 bcm needed to meet the projected gas production level of 57 bcm in 2030. However, 43 bcm — out of the 45 bcm needed from 2010 to 2030 — will have to come from new fields. Thus, Libya will need to mobilise heavy investment and modern technologies in the very near future. Attracting investors will be essential to ensuring the expected production levels from 2015 to the end of the projection period.

Exports

In 2003, domestic gas consumption absorbed some 88% of Libya's natural gas production. Libya has been exporting gas in liquefied form since 1971 from the Exxon-built liquefaction plant at Marsa al-Brega, but lack of processing equipment to strip natural gas liquids out of the gas streams before liquefaction resulted in "wet" LNG, with higher heat content, which only a few receiving terminals could handle, limiting the marketability of Libyan LNG. The Marsa al-Brega plant has a nominal capacity of 3.9 bcm per year of LNG and 20 kb/d of naphtha, but it is now operating at less than one-quarter of its capacity due to technical constraints.

Libya's LNG exports fell from 1.5 bcm a year in the mid-1990s to under 1 bcm in 2003. NOC is now carrying out a major upgrade of the Marsa al-Brega plant to enable it to produce both low and normal grade LNG. LNG exports are projected to reach around 10 bcm by 2030.

The Western Libya Gas Project will deliver exports of some 8 bcm per year by pipeline to Italy. Edison Gas will lift 4 bcm per year, Gaz de France 2 bcm and Energia Gas 2 bcm under 24-year contracts. In a second phase, the total volume will increase to 11 bcm per year.

By the end of the *Outlook* period, total gas exports will be just under 34 bcm per year, representing 59% of total gas production in Libya.

Investment Needs and Financing

Libyan overall energy investment over the *Outlook* period will amount to $80 billion (in year-2004 dollars). Over half of this investment will be needed for the oil sector and the remaining will be shared between the electricity sector and the gas sector. Libya will need $3 billion per year on average over the next 25 years to sustain its growing domestic energy demand and, more importantly, to develop its oil and gas-export potential.

14

Table 14.6: **Libya's Cumulative Energy Investment**
($ billion in year-2004 dollars)

	2004-2010	2011-2020	2021-2030	2004-2030
Oil	5.3	15.4	20.0	40.7
Gas	1.3	7.7	12.3	21.3
Electricity	3.5	6.4	8.1	18.0
Total	**10.0**	**29.5**	**40.4**	**80.0**

Oil

Total cumulative oil investment needs over the projection period are expected to be $41 billion. Over 85% of total oil investment will be directed to upstream oil exploration and development. Production costs in Libya are believed to range from as low as $1 to $5 a barrel (among the lowest in North Africa and well below world averages, but similar to average costs in Middle Eastern countries).The remaining $4 billion of oil investment will go to refining capacity upgrades and conversions to meet product specifications.

Figure 14.8: **Libya's Annual Oil Investment and Oil Production in the Reference Sceanrio**

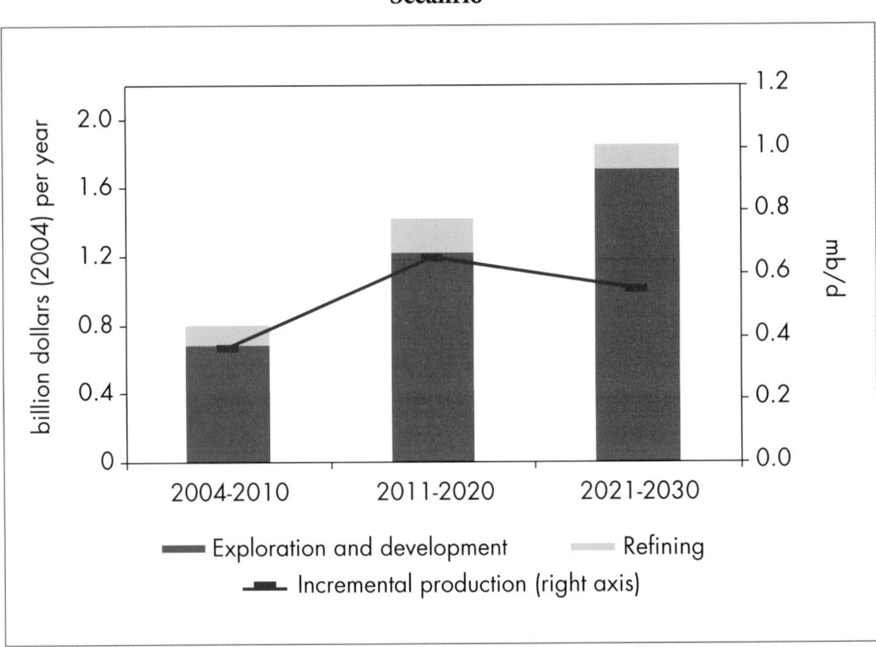

In the Reference Scenario, over $20 billion will be needed by 2020 to increase oil production to 3 mb/d. The investment requirements in upstream oil are expected to be some $1.3 billion per year on average over the projection period (Figure 14.8). Up to 2010, over $5 billion (around $0.8 billion per year) will be needed to finance enhanced oil recovery in order to reach the projected 1.9 mb/d from existing producing fields. From 2010 to 2020, an average of $1.3 billion dollars per year will be needed to increase oil production by 0.7 mb/d. In the last decade of the *Outlook* period, as existing fields mature and new smaller fields are brought into production, upstream investment costs will increase to $1.8 billion dollars per year, achieving net incremental production of about 0.6 mb/d. Libya's upstream investment requirements will account for 7% of total MENA investments, and will produce around 7% of total incremental MENA oil production over the projection period.

In the near term, refinery investments are urgently needed. Investments in this sector are expected to be nearly $3 billion in the period to 2020, nearly $1 billion of which will be spent in the period to 2010. Over the projection period, investment needs in the refinery sector will represent 11% of total oil investment. Timely investment will be crucial for Libya to meet growing domestic demand.

Natural Gas

Total gas investment needs will amount to $21 billion from 2004 to 2030, most of it concentrated in the last decade. Some $13 billion – nearly 58% of total gas investment – will be needed from 2020 to 2030. In the short term, production will come from existing fields. But as the use of gas expands, especially in domestic power generation and for export, investment will be needed for exploration and development and for enhanced recovery in producing fields. $180 million will be required each year from now to 2010, rising to $1.2 billion annually from 2020 to 2030. Expenditures for exploration and development will account for 59% of total cumulative gas investments.

Electricity

Total cumulative investment in Libya's power sector in 2004-2030 is projected to amount to $18 billion, of which $8 billion will be needed in power generation, $3 billion in transmission and $7 billion in distribution. Libya needs to improve the financial health of its power company in order to garner the necessary investment. The greatest challenges are to tackle the non-payment problem and eventually to make tariffs cost-reflective.

Deferred Investment Scenario[8]

While prospects are good and international oil companies are competing to enter Libya, a lack of transparency and uncertain licensing terms, especially in EPSA-IV agreements, have caused delays. If this confusion were to last or

8. See Chapter 7 for a detailed discussion of the assumptions and methodology underlying the Deferred Investment Scenario.

worsen, or if Libya were unable to achieve a stable regulatory framework and other economic reforms, investment in Libyan oil and gas could be severely compromised. Such a possibility is examined in the Deferred Investment Scenario.

Energy Demand

In this scenario, total primary energy demand in Libya grows at an average annual rate of 3.3% over the projection period, reaching 43 Mtoe in 2030. Demand at that time is 7% lower than in the Reference Scenario. Gas demand is nearly 7% lower and oil demand is 7.4% lower. Libya's efforts to diversify its economy are hampered and energy demand in the industry and transformation sectors are lower. Energy demand for both oil and gas, but primarily gas, falls in the petrochemical sector. Over the *Outlook* period, gas demand in the industry sector grows by 7 percentage points less than in the Reference Scenario.

In the Deferred Investment Scenario, lower economic growth results in decreased demand for energy also in the residential and transport sectors. In 2030, energy demand is projected to be around 3 Mtoe lower than in the Reference Scenario, with over 80% of this reduction in final consumption and 40% in industry.

Oil and Gas Production

The assumption adopted in the Differed Investment Scenario that the ratio of upstream investment to GDP remains constant at the level of the last ten years results in an average level of investment some 40% lower than in the Reference Scenario. Growth in production of both crude oil and natural gas declines, with crude oil production 1.95 mb/d, or about 37%, lower in 2030 compared with the Reference Scenario. Natural gas production in 2030 is 33 bcm, 42% lower than in the Reference Scenario.

The slower growth in domestic oil demand reduces the need to expand refining capacity in the Deferred Investment Scenario. Capacity grows by only 50% over 2003-2030, reaching 0.6 mb/d, compared with 0.7 mb/d in the Reference Scenario.

Oil and Gas Exports

In the Deferred Investment Scenario, net oil exports are 1 mb/d lower, almost 40% less than their level in the Reference Scenario, reaching 1.6 mb/d in 2030. Natural gas exports are also markedly lower, because of slower demand growth in the net importing regions of Europe, North America and Asia. Exports are 11.4 bcm in 2030, 22.4 bcm less than in the Reference Scenario.

QATAR

HIGHLIGHTS

- Qatar's GDP is expected to grow by 3.9% per year on average over the *Outlook* period, on the back of a large-scale expansion of the natural gas and petrochemical sectors. Oil and gas will continue to play a major role in the economy in the next 25 years.
- Total primary energy demand will grow by almost 6% per year in 2003-2030, almost twice the average rate of other MENA countries. Energy demand in the LNG and GTL transformation sectors will account for about two-thirds of the growth in total demand. Final energy consumption will increase by 3.7% per year, driven mainly by strong growth in the petrochemical industry.
- Qatar has significant oil reserves, two-thirds of which are in the form of condensates. Around 90% of the country's 1 mb/d of oil production was exported in 2004. In the Reference Scenario, oil production is projected to rise to 1.25 mb/d in 2030. Condensates will make up most of the increase. Cumulative oil investment requirements will amount to $50 billion in 2004-2030.
- Qatar has the third-largest proven gas reserves in the world, mostly in the North Field, the largest non-associated gas field in the world. Gas production is projected to increase sixfold from 2004 through to 2030. Exports, mainly as LNG but also by pipeline to neighbouring countries, will increase along with gas production, reaching almost 17% of global gas trade in 2030. Cumulative investment needs in the gas sector are close to $100 billion in 2003-2030.
- Three major gas-to-liquids projects are under way. Total capacity is projected to reach over 100 kb/d in 2010 and, with additional projects, almost 650 kb/d in 2030, making Qatar the world's largest GTL producer.
- Per capita electricity generation, at more than 16 500 kWh, is among the highest in the world. Consumption will continue to grow briskly over the projection period, reaching 22 500 kWh per capita in 2030 – more than twice the OECD average. Two-thirds of additional electricity-generating capacity will be new combined water and power plants with desalination units, which will supply almost all the country's drinking water.
- In the Deferred Investment Scenario, primary energy demand in Qatar in 2030 is about 10% lower than in the Reference Scenario. In 2030, oil exports are 13% lower and gas exports 30% lower.

Overview of the Qatari Energy Sector

Oil is the cornerstone of Qatar's economy, accounting for more than 70% of total government revenue and for nearly 40% of GDP in 2004. Oil exports increased from just under 500 kb/d in 1990 to nearly 820 kb/d in 2003, fuelling rapid economic growth. But gas is becoming increasingly important. With proven gas reserves of nearly 26 trillion cubic metres, Qatar's natural gas reserves rank third in size after Russia and Iran (Cedigaz, 2005). It has the largest known non-associated gas field in the world, the North Field. Qatar is on track to becoming the world's largest LNG and gas-to-liquids (GTL) producer over the next decade.

State-owned Qatar Petroleum (QP) was established in 1974 and is responsible for all aspects of the oil and gas industry in Qatar, including exploration and drilling for oil and natural gas, production and refining. QP conducts exploration and new development through production-sharing agreements with international oil and gas companies and through agreements with the international oil companies for LNG development via joint ventures using a tax/royalty scheme.

Total primary energy demand grew by an average of over 9% per year from 1971 to 2003 (Table 15.1). Per capita energy demand is three times that of the United States and Canada. High incomes, air-conditioning, desalination of sea-water and large industrial energy needs explain the high energy intensity of the Qatari economy. Oil accounts for some one-quarter of energy demand and gas for the rest. CO_2 emissions were 26.7 million tonnes in 2003, nearly three-

Table 15.1: **Key Energy Indicators for Qatar**

	1971	2003	1971-2003*
Total primary energy demand (Mtoe)	0.9	15.2	9.1%
Total primary energy demand per capita (toe)	7.6	20.8	3.2%
Total primary energy demand/GDP**	0.08	0.61	6.5%
Share of oil in total primary energy demand (%)	11	26	–
Net oil exports (mb/d)	0.52	0.82	1.5%
Share of oil exports in production (%)	99.6	89	–
Share of gas in total primary energy demand (%)	89	74	–
Net gas exports (bcm)	–	19.5	–
Share of gas exports in production (%)	–	59	–
CO_2 emissions (Mt)	2.2	26.7	8.1%

* Average annual growth rate.

** Toe/thousand dollars of GDP in year-2004 dollars and PPPs.

quarters of which came from gas – the main fuel in power generation, water desalination, LNG production and other industries.

Political and Economic Situation

Political Developments

Qatar gained independence from the United Kingdom in 1971 and Emir Khalifa bin Hamad al-Thani became ruler. In 1995, his son, the deputy ruler, Sheikh Hamad bin Khalifa al-Thani, deposed him. Sheikh Hamad has introduced gradual political, social and economic reforms. Two of the most important were the creation of an elected council and the extension to women of the right to vote. A new constitution, which was approved by almost 97% of voters in a public referendum in 2003 and which came into force in June 2005, slightly broadened the scope of political participation. In April 2003, elections were held for the Central Municipal Council, which advises the government on domestic issues but does not have the authority to change policy.

Like other oil-rich Persian Gulf states, Qatar needs to continue to diversify its economy away from dependence on oil exports and to scale back state subsidies. Thanks to the rapid growth of exports of LNG, and to its very small population, per capita GDP in Qatar held up better than in other Persian Gulf oil exporters in the 1990s. As in other Gulf countries, higher oil and gas prices in the last few years have led to an economic boom. Qatar's policy of economic diversification has led to a surge in investment in LNG and petrochemical projects (Box 15.1). Qatar has a relatively large foreign debt, at around $17 billion as of 2004 (EIA, 2005). This debt was largely accumulated for infrastructure investments in oil and gas projects. But surging oil and gas export revenues have enabled Qatar to maintain a large budget surplus despite sharply increased government spending.

Box 15.1: **Economic Diversification in Qatar**

Qatar has sought to broaden its economic base by developing industries unrelated to oil, as well as downstream activities using hydrocarbons as fuel or feedstock. A number of state-owned industrial enterprises, such as Qatar Steel Company and Qatar National Cement Company, were set up in the 1960s and 1970s. State fertilizer and petrochemical companies were also established. The development of the North Field in the 1990s gave rise to a fresh wave of industrial ventures based on natural gas. The oil and gas sector accounts for over 60% of total GDP and fuelled two-thirds of the growth from 2000 to 2004 (Table 15.2). But virtually every sector of the economy has been growing rapidly. The manufacturing, electricity and water, and construction sectors saw output more than double from 2000 to 2004.

Economic Developments

Oil production and revenues increased sharply in the late 1970s, leading to average economic growth of over 3% per annum from 1975 to 1980. In the 1980s OPEC quotas on crude oil production and lower oil prices led to reduced export earnings and an economic downturn, with GDP declining on average by 1.8% per year from 1980 to 1990.

Table 15.2: **Qatar's Nominal GDP by Sector** (QR million)

	2000	2001	2002	2003	2004*
Oil and gas sectors	39 065	36 812	40 717	50 551	64 365
Non-oil sector	25 581	27 767	31 016	35 367	39 198
Total	**64 646**	**64 579**	**71 733**	**85 918**	**103 563**
Oil and gas sectors	60%	57%	57%	59%	62%

* Preliminary data.
Source: Qatar National Bank (2005).

Since 1990, the economy has recovered strongly, growing by an average 6.4% per year between 1990 and 2003. Real GDP growth in 2004 was very strong, around 10%. Qatar's per capita GDP is among the highest in the world, averaging $34 000 per capita in 2003. GDP is projected to grow by 3.9% per year on average in 2003-2030 (Table 15.3). The oil and gas sectors will continue to be the major drivers for GDP growth.

In 2003, the population of Qatar was some 730 000. Population growth averaged 5.8% per year from 1971 to 2003, but slowed down considerably from 1990 to 2003, growing on average by about 3% per year. Foreign nationals make up over three-quarters of the population. Population is assumed to grow by 1.7% per year over the projection period.

Table 15.3: **GDP and Population Growth Rates in Qatar in the Reference Scenario** (average annual rate of change in %)

	1971-2003	1990-2003	2003-2010	2010-2020	2020-2030	2003-2030
GDP	2.5	6.4	6.6	3.7	2.4	3.9
Population	5.8	3.2	2.9	1.5	1.1	1.7
Active labour force	6.0	2.0	2.9	1.7	1.5	1.9
GDP per capita	–3.1	3.1	3.5	2.2	1.2	2.2

The oil and gas sectors will account for nearly two-thirds of GDP in 2030. Given the expected expansion in production of LNG and GTL, the gas share of hydrocarbon output will grow throughout the projection period (Figure 15.1). If all proposed GTL projects are realised, the share of GTL production in total GDP will be about 8% in 2030.

Figure 15.1: **Shares of Oil and Gas in GDP in Qatar**

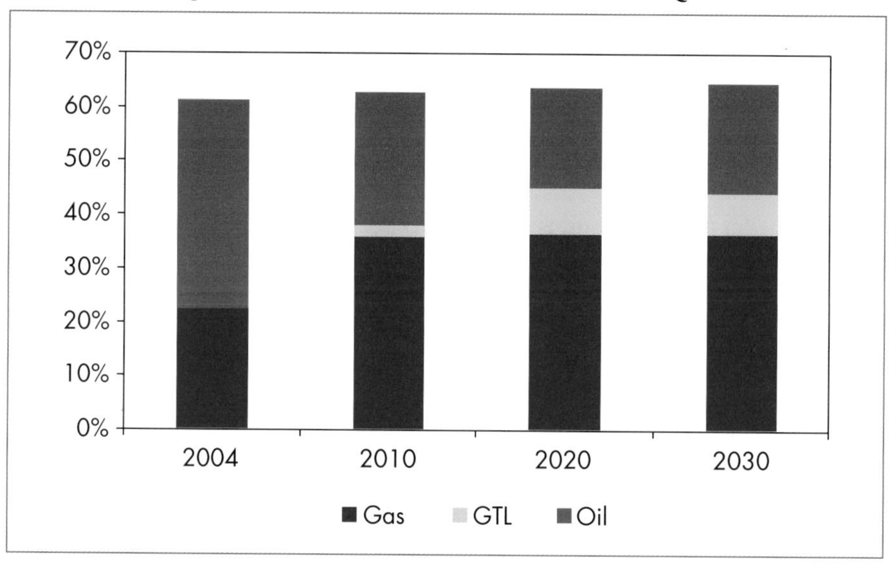

Energy Policy

In the late 1990s, Qatar initiated new exploration and production policies aimed at increasing oil reserves, raising oil production and encouraging the use of advanced oil recovery techniques to extend the life of existing fields. The government recently further improved the terms of exploration and production-sharing agreements (EPSAs) and development and production-sharing agreements (DPSAs) to encourage foreign oil companies to improve oil recovery in producing fields and to explore for new oil deposits. Foreign companies now account for more than one-third of Qatar's oil production capacity.

Qatar does not have comprehensive petroleum legislation or a unified oil policy under which the conditions governing the granting of exploration and development permits are fixed and known in advance. The legal framework is still determined by the terms of individually negotiated agreements between the government and foreign operators.

Currently, Qatar has two major joint ventures in the LNG industry: the Qatar Liquefied Company (Qatargas) and Ras Laffan Natural Gas Company I and II

(RasGas). LNG expansion projects under way have created Qatargas II, 3 and 4 as well as RasGas III ventures, all of which will go on stream in 2008 to 2010. In addition, pipeline gas projects Dolphin and Al-Khaleej Gas are scheduled to start up in 2006 and will serve domestic and regional markets. Qatar's first GTL project, with Sasol, will go on stream in 2006 and two additional projects, with Shell and ExxonMobil, have recently been approved by the government.

Energy Demand

Primary Fuel Mix

Total primary energy demand[1] in Qatar is projected to increase from 15 Mtoe in 2003 to 67 Mtoe in 2030, an average rate of growth of nearly 6% per year (Table 15.4). The share of oil is expected to decline from 26% to 18% over the period. The share of natural gas will increase, mainly because of increased use in LNG and GTL projects. Gas use in these transformation sectors will account for about two-thirds of the increase in energy demand over the *Outlook* period (Figure 15.2). Gas use in the petrochemical industry will also increase significantly. In Qatar, the fuel requirements for water desalination will account for 30% of the increase in energy demand in the power and water sector from 2003 to 2030. CO_2 emissions are expected to rise by 4.2% per annum, reaching about 80 million tonnes in 2030. Gas use will continue to account for about three-quarters of total emissions.

Table 15.4: **Qatar's Primary Energy Demand in the Reference Scenario** (Mtoe)

	1990	2003	2010	2020	2030	2003-2030 *
Oil	1.3	4.0	6.4	9.2	12.0	4.2%
Gas	5.6	11.2	25.4	50.2	55.2	6.1%
Total**	**6.9**	**15.2**	**31.9**	**59.4**	**67.3**	**5.7%**

* Average annual growth rate.
** Renewable energy demand is negligible.

The efficiency of the natural gas liquefaction will increase throughout the projection period, but as LNG exports are projected to rise almost sevenfold, consumption for gas liquefaction will rise rapidly. Gas consumption for GTL production will rise enormously after 2010 and will account for 40% of primary gas demand in 2030. Losses in the GTL conversion process are

1. Primary gas demand in the energy balance includes the energy input but excludes gas used as a feedstock in GTL plants.

currently at about 45% and will decline to about 35% at the end of the projection period. Losses in the liquefaction process are also significant. Gas use for GTL and LNG will rise by over 10% per year over the projection period, while gas for power generation will rise by 2.3% per year.

Figure 15.2: **Qatar's Total Primary Energy Demand by Sector** (Mtoe)

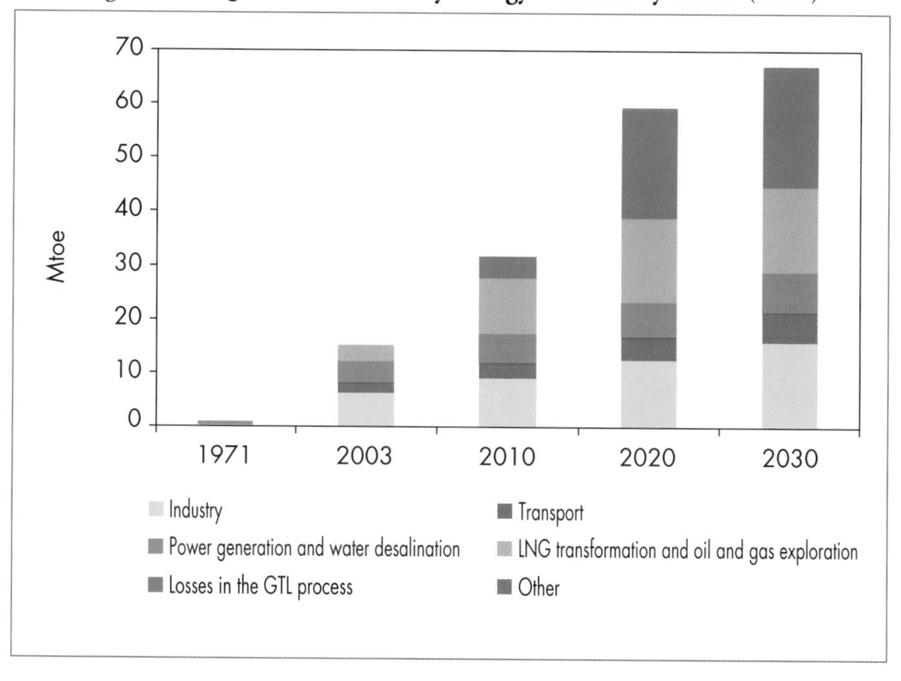

In 2003, oil accounted for about 46% of total final consumption, with gas representing 43% and electricity accounting for the remainder. By 2030, the share of oil will rise to 50%, while the shares of gas and electricity will fall by about two percentage points each in the Reference Scenario. Energy demand in the transport sector will grow fastest, at 4.3% per year on average, due to rapid growth in the use of aviation fuel and automotive fuel. Energy demand is projected to grow by 3.6% per year in industry and by 2.8% per year in the residential and services sector. The share of industrial energy demand in total final consumption will remain almost constant at about 70% throughout the projection period.

Sectoral Trends

Industry

The industry sector in Qatar is comprised of ammonia, fertilizers, petrochemicals, steel, cement and commercial ship repair. The petrochemical industry is by far the largest energy consumer, accounting for some 92% of industrial energy demand in 2003. This share will fall slightly to 90% in 2030. The petrochemical sector will account for over 60% of final energy consumption in 2030.

Four petrochemical projects are planned, all starting in 2008 (Arab Petroleum Research Centre, 2005):

- Qatofin, a joint venture between QAPCO (63%), Atofina (36%) and Qatar Petroleum (1%) which will produce 450 000 tonnes per year of linear low-density polyethylene.
- Q-Chem II, a joint venture between Qatar Petroleum (51%) and Chevron (49%), with a design capacity of 350 000 tonnes per year of high-density polyethylene and 350 000 tonnes per year of normal alfa olefins.
- A joint venture between Q-Chem II (53.3%), Qatofin (45.7%) and QP (1%) to build a steam cracker at Ras Laffan with design capacity of 1.3 million tonnes per year of ethylene.
- A letter of intent was signed in June 2003 with Mitsubishi Gas Chemicals and Itochu for the production of 1.7 million tonnes per year of dimethyl-ether at Ras Laffan. An ethylene pipeline will be built from Ras Laffan to Mesaieed to supply ethylene to the Q-Chem II and Qatofin plants.

Transport

Energy demand in the transport sector was 1.8 Mtoe in 2003. Aviation demand accounted for about a third. There were an estimated 340 000 vehicles on the road in Qatar in 2002, of which 230 000 were cars. Road fuel demand will grow by 3.2% in 2003-2030 in the Reference Scenario. Aviation fuel demand is expected to grow quite rapidly over the *Outlook* period. Qatar Airways recently signed a $5.1 billion agreement with Airbus Industries to purchase 32 aircraft, increasing the fleet size to 56 aircraft by 2008. Aviation fuel demand is projected to grow by 5.8% per year between 2003 and 2030, by which time it will account for more than half of transport energy demand.

Residential and Services

Energy demand in the residential and services sector was 0.8 Mtoe in 2003. Electricity accounted for more than 90%, with oil accounting for the remainder. Strong growth in electricity demand in the services sub-sector will drive energy demand for the entire sector. Electricity consumption in the services sector will increase by 3.9% per year, compared with a growth of 2.9% for electricity in total final consumption.

Electricity Supply and Desalinated Water Production

Overview

Qatar's power plants had an installed capacity of 3.4 GW in 2003 and produced 12 TWh (Table 15.5). Electricity generation per capita, at about 16 500 kWh, is among the highest in the world. Despite this very high level of consumption, demand for electricity continues to grow rapidly. The average annual growth rate over the five-year period 1998-2003 was 8.1%.

Almost all Qatar's power generation is based on open-cycle gas turbines and all power plants are fired on natural gas. Some of Qatar's power plants are integrated with water desalination units. Natural gas consumption in Qatar's power and water plants represents 36% of total primary gas consumption and 15% of total production in 2003. The electricity-generation efficiency was 26% in 2003, partly because gas turbines have relatively low efficiency and partly because some of the fuel consumed goes into the desalination process in combined water and power facilities.

Table 15.5: **Power Plants in Qatar, 2003**

Company	Installed capacity (MW)	Owner
Utilities	**2 419**	
Ras Abu Fontas B	1 105	QEWC
Ras Abu Fontas A*	618	QEWC
Al Wajbah	301	QEWC
Ras Abu Aboud	150	QEWC
Saliyah	134	QEWC
Doha South Super	67	QEWC
Dukhan ACPG	44	QEWC
Industry	**962**	
Ras Laffan RasGas	330	Ras Laffan LNG
Ras Laffan Qatargas	187	Qatar Liquefied Gas (Qatargas)
Mesaieed Works	156	Qatar Vinyl Co
Mesaeeid QAFCO	141	Qatar Fertilizer Co (QAFCO)
Umm Said Refinery	128	QP Refinery
Maersk Qatar	20	Maersk Oil Qatar As
Total capacity	**3 380**	

* Ras Abu Fontas A was owned by Kahramaa until 2004.
Source: Platts (2003).

The Ras Abu Fontas A and B plants – combined water and power facilities – are Qatar's major power plants. Over 70% of installed capacity is operated by Qatar Electricity and Water Company (QEWC), a joint-stock company in which the government holds a 42.7% stake. The remaining power plants are used in industrial sites and are operated by various government-owned companies. The electricity and water sectors are controlled by Qatar General Electricity & Water Corporation, known as Kharamaa, which was established in 2000. Kharamaa operates as an independent corporation and, having transferred its power plants to QEWC, it remains responsible for transmission and distribution.

In April 1999, the Ras Abu Fontas B power plant was transferred to Qatar Electricity and Water Company. This event marked the beginning of private-sector participation in the market. In 2003, more power plants owned by Kahramaa were transferred to QEWC, which also acquired the Dukhan power plant from Qatar Petroleum. In 2004, QEWC acquired Ras Abu Fontas A. QEWC sells the electricity and water produced in its plants to Kahramaa. The output of the Dukhan power plant is sold to Qatar Petroleum. QEWC's revenues from electricity sales amounted to QR 874 million (about $240 million) in 2003. This corresponds roughly to 2 US cents per kWh generated, which is below the full cost of electricity. Qatari nationals do not pay for their electricity; foreigners do.

Electricity Production

Qatar's electricity generation is projected to increase at an annual rate of 2.9% in the period 2003-2030. Installed capacity is projected to rise to 8 GW by 2030. Electricity generation will continue to be based exclusively on natural gas. While traditionally Qatar's power generation was based on open-cycle gas turbines, recent projects have marked the beginning of use of combined-cycle gas-turbine (CCGT) plants. This technology is far more fuel-efficient. It is expected that most new capacity in the future will be CCGT, so raising the average thermal efficiency of power generation.

Qatar has taken steps to promote private investment in independent power and water projects. In 2001, the US-based AES, QEWC, Qatar Petroleum and the Gulf Investment Corporation formed a joint venture – the Ras Laffan Power Company – to build an integrated water and power plant at Ras Laffan. AES is the majority shareholder with an equity stake of 55%. The company signed a 25-year purchasing agreement with Kahramaa. The facility began operation in November 2004. It is a combined-cycle gas-turbine plant with an installed capacity of 756 MW and uses multi-stage flash evaporators to produce up to 190 thousand cubic metres of water a day.

A second water and power project in Ras Laffan is underway. A contract to build a 1 025 MW facility was awarded in late 2004 to an international consortium comprising the UK's International Power, QEWC and Japan's Chubu Electric Power. The first phase of the project is expected to be operational in 2006 and will be fully commissioned in 2008.

There are also plans to expand the power generation and water production capacity of existing power stations. The government is also considering plans to privatise the transmission and distribution network, although no concrete steps have been taken.

Water Production

Since it has virtually no permanent water resources, Qatar relies almost entirely on energy-intensive desalination plants for its drinking water supply. Residential water consumption was 178 million cubic metres in 2003, accounting for over 45% of total water demand. Water production from desalination plants was 142 million cubic metres in 2003.

By 2030, residential water consumption is projected to rise to 292 million cubic metres, growth of 1.9% per year, and to account for about 60% of total water demand. The share of desalinated water in total water consumption will increase to 60% in 2030.

Qatari citizens are exempt from municipal water charges and tariff rates are too low to recover costs. The water sector is struggling with inadequate metering systems and water leakages from distribution networks.

To meet the rising demand for water, desalination capacity will need to increase by nearly 200 million cubic metres. Desalinated water production is expected to increase to over 300 mcm in 2030, growth of 2.8% per year. Two-thirds of the total electricity capacity additions in Qatar will be for new combined water and power plants with desalination units. Gas is the only fuel used in such plants in Qatar. Fuel requirements for desalination will rise from 1.2 Mtoe in 2003 to 2.3 Mtoe in 2030 (Table 15.6), accounting for 3% of total primary energy demand.

15

Table 15.6: **Water Consumption and Desalination Capacity in Qatar**

	2003	2010	2020	2030
Water consumption (mcm)	375	420	462	501
Desalination capacity (mcm)	206	282	336	401
Oil and gas use for desalination (Mtoe)	1.2	1.6	1.9	2.3

Oil Supply

Overview

Crude oil and condensate production was just over 1 mb/d in 2004, up from 930 kb/d in 2003. Exports increased from 820 kb/d in 2003 to 920 kb/d in 2004. The share of condensates was almost one-quarter of total oil production in 2004. Crude oil and condensate production in Qatar is projected to rise to 1.25 mb/d in 2030 in the Reference Scenario, growth of 1.1% per year on average over the *Outlook* period. In 2030, oil production in Qatar will represent only 3% of total Middle East oil production – down from 4% in 2004. But the share will be higher with the inclusion of GTL production of refined products. GTL plants are projected to produce over 600 kb/d in 2030. Adding this production to crude oil and condensates, total oil production will be nearly 1.9 mb/d in 2030 (Figure 15.3).

Figure 15.3: **Qatar's Oil Balance in the Reference Scenario**

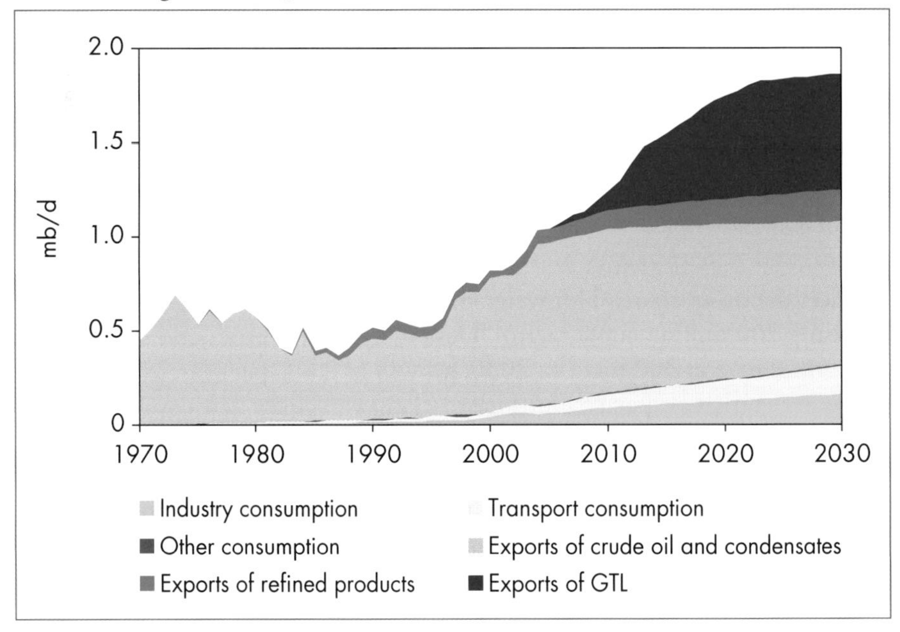

Resources and Reserves

According to the *Oil and Gas Journal*, proven oil reserves in Qatar on 1 January 2005 were 15 billion barrels, of which more than two-thirds were condensates.[2] Proven reserves are equivalent to 40 years of production at current rates and

2. *Oil and Gas Journal*, 20 December 2004. Qatar Petroleum puts proven oil reserves at 25.5 billion barrels as of 1 January 2004, nearly 90% of which are condensates in the North Field.

Figure 15.4: **Main Oil and Gas Infrastructure in Qatar**

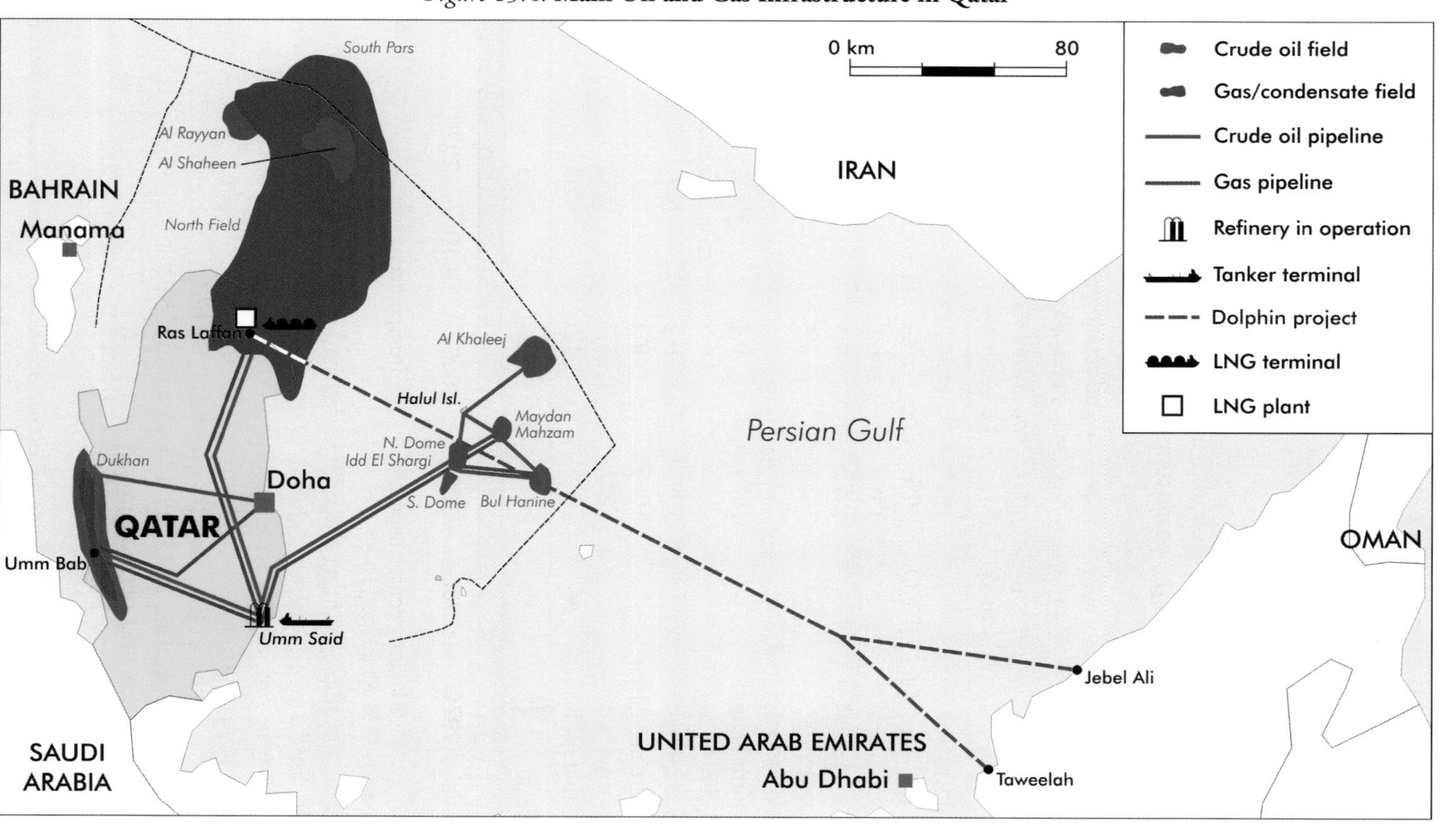

15

have increased substantially over the past five years – in 1999 they were less than 4 billion barrels. IHS Energy estimates that remaining proven and probable reserves are 33 billion barrels (Table 15.7), of which around 80% are condensates.

Table 15.7: **Oilfields in Qatar**

Field	Year of first production	Remaining proven and probable oil reserves at end-2004 (billion barrels)	Cumulative production to 2004 (billion barrels)	API gravity
Al Khaleej	1997	0.3	0.03	27.5
Al Rayyan	1996	0.3	0.03	24.0
Al Shaheen	1995	0.8	0.45	30.6
Bul Hanine	1972	0.5	1.30	35.0
Dukhan	1940	2.4	3.75	41.1
Idd El Shargi North Dome	1964	0.4	0.51	31.0
Idd El Shargi South Dome	1999	0.2	0.01	29.0
Maydan Mahzam	1965	0.3	0.92	38.0
North Field	1991	25.7	0.32	55.3
Other fields		1.9	0.00	
Total		**32.8**	**7.33**	

Note: Includes NGLs and condensates.
Sources: IHS Energy and IEA databases.

Crude Oil Production[3]

Qatar, an OPEC member since 1961, has had a crude oil production quota of 700 kb/d since November 2004. Qatar Petroleum has embarked upon an investment programme with the intention of expanding crude oil production capacity (excluding NGLs) from its onshore and offshore fields from the current capacity of 800 kb/d to around 875 kb/d by the end of 2006. In its five-year plan starting 2005, it budgeted about $51 billion for projects in crude oil, natural gas and petrochemicals, with 3.2% allocated specifically for oil-related projects.

3. All the oil-production figures cited in this section include crude oil, natural gas liquids (NGLs) and condensates.

Qatar Petroleum's oil production capacity accounted for 56% of Qatar's total capacity as of September 2004 (Arab Petroleum Research Centre, 2005). QP produces crude oil, associated gas, condensate and non-associated gas from the onshore Dukhan field, Qatar's oldest and largest field. Dukhan has four hydrocarbon reservoirs, three of which produce oil and the fourth non-associated gas. The company also produces crude oil from the offshore Maydan Mahzam and Bul Hanine fields. Oil is piped to Halul Island for storage and export. Halul Island is a major international oil terminal, located north-east of Doha. The island has 11 large crude oil storage tanks, with a total capacity of 5 million barrels, crude oil pumping facilities and power generation and water desalination plants.

Table 15.8: **Qatar's Crude Oil Production in the Reference Scenario** (mb/d)

	2004	2010	2020	2030
Currently producing fields	**1.03**	**1.14**	**0.90**	**0.77**
Al Khaleej	0.05	0.04	0.02	0.00
Al Rayyan	0.02	0.01	0.00	0.00
Al Shaheen	0.20	0.14	0.03	0.00
Bul Hanine	0.10	0.07	0.00	0.00
Dukhan	0.35	0.33	0.17	0.05
Idd El Shargi North Dome	0.10	0.07	0.00	0.00
Idd El Shargi South Dome	0.01	0.02	0.00	0.00
Maydan Mahzam	0.05	0.05	0.02	0.00
North Field	0.11	0.31	0.41	0.46
Other fields	0.04	0.12	0.25	0.25
New developments	**0.00**	**0.00**	**0.30**	**0.48**
Fields awaiting development	0.00	0.00	0.02	0.00
Reserve additions and new discoveries	0.00	0.00	0.28	0.48
Total	**1.03**	**1.14**	**1.20**	**1.25**

Note: Includes NGLs and condensates.
Sources: IHS Energy databases; IEA analysis.

Production capacity at the Al Khaleej offshore field, operated by Total, will increase from 38 kb/d to 60 kb/d in 2006 (Qatar National Bank, 2005). Capacity at the Al Rayyan offshore field, operated by Anadarko, will reach 20 kb/d in 2006. Qatar Petroleum and Maersk are jointly developing the Al Shaheen offshore field using horizontal drilling and water injection, at a cost of over $2 billion. Qatar Petroleum is also developing the north and south domes of Idd El Shargi with Occidental Petroleum; total capacity of the field will reach 120 kb/d in 2006. But, with the exception of the North Field, all of the currently producing fields are expected to decline later in the projection period. In 2030, currently producing fields will produce 770 kb/d of crude oil and condensates (Table 15.8). With output from new discoveries and fields awaiting development, total oil production is projected to reach 1.25 mb/d in 2030 (Figure 15.5).

Figure 15.5: **Qatar's Crude Oil Production by Source in the Reference Scenario**

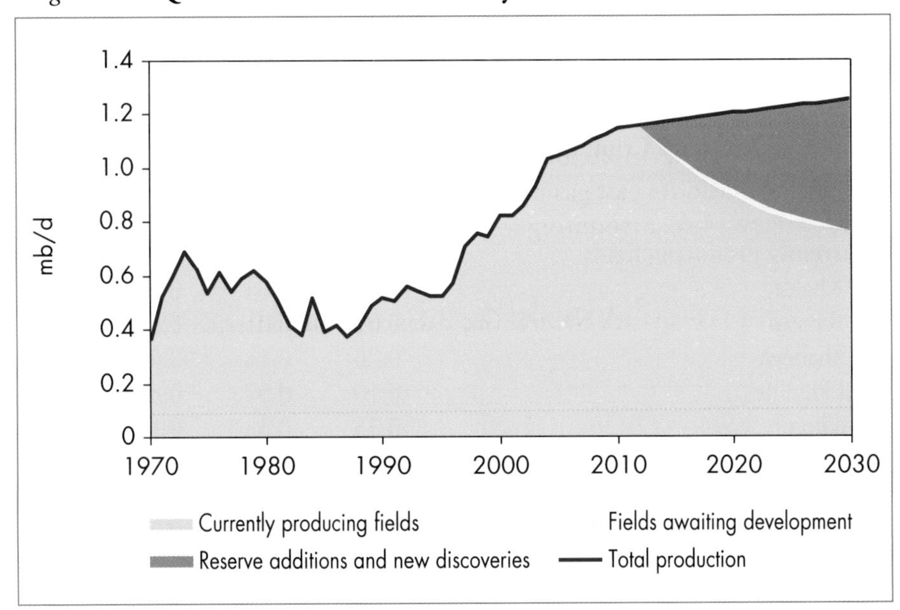

Note: Includes NGLs and condensates, but excludes GTL production.

Refining

Qatar Petroleum owns and operates the only refinery in the country at Umm Said (Mesaieed Industrial City), south of Doha. It was recently upgraded and capacity was increased to 136 kb/d. Qatar's refining capacity is expected to increase to 320 kb/d by 2030, with most of the increase coming from a new refinery to be built in Ras Laffan, north of Doha. The refinery will produce initially 140 kb/d of LPG, naphtha, kerosene and diesel.

Exports

In 2004, Qatar's exports of crude oil, condensates and refined products were 920 kb/d, equal to less than 5% of total Middle East exports. The largest importers of Qatari oil are Japan, Singapore, South Korea and Thailand. Exports are projected to grow only marginally, to 930 kb/d in 2030. The share of refined products (excluding GTL) in these exports will rise from 8% today to 18% by 2030. Qatar will also export over 600 kb/d of oil products from new GTL plants, bringing total oil-related exports to over 1.5 mb/d in 2030.

Natural Gas Supply

Overview

Natural gas production jumped by an estimated 8 bcm to 41 bcm in 2004, as new LNG capacity came on stream. LNG exports have increased rapidly over the last few years, from 15 bcm in 2000 to 26 bcm in 2004. Qatar's share in total MENA gas exports approached one-quarter. Gas production in Qatar is projected to rise to 255 bcm in 2030, an annual growth rate of almost 8% over the *Outlook* period. In 2030, gas production in Qatar will represent almost 30% of total Middle East gas production. Gas exports will rise to 152 bcm in 2030 (Figure 15.6), accounting for almost 17% of global gas exports.

Figure 15.6: **Qatar's Natural Gas Balance in the Reference Scenario**

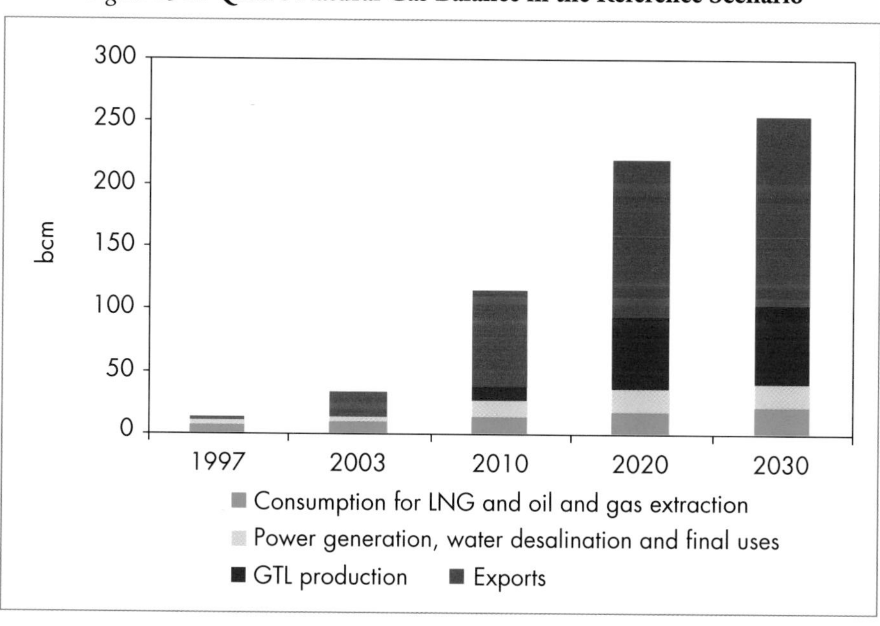

Resources and Reserves

As of 1 January 2005, Qatar's proven natural gas reserves were some 26 tcm, the third-largest in the world after Russia and Iran, and 14% of the world total (Cedigaz, 2005). They could last several hundred years at current production levels. In the Reference Scenario, remaining reserves in 2030 will still be equivalent to 80 years of production at that time. All but 1% of Qatar's gas reserves are in the offshore North Field – the largest non-associated gas field in the world (Box 15.2). The onshore Dukhan field contains an estimated 152 bcm of associated and 65 bcm of non-associated gas. There are smaller associated gas reserves in the Idd El Shargi, Maydan Mahzam, Bul Hanine, and Al Rayyan offshore oilfields. Total associated gas is estimated to be some 303 bcm.[4]

Box 15.2: **Qatar's North Field**

Qatar's North Gas Field is part of a super-giant gas condensate accumulation extending over an area of more than 6 000 square kilometres off the north-east coast of Qatar and southern Iran. The Iranian part of the reservoir is called South Pars. Qatar Petroleum began developing the North Field primarily for LNG exports, but also for domestic gas requirements. Currently 8.3 bcm per year of raw gas is being produced from North Field Alpha and processed to produce 7.7 bcm of lean gas for domestic consumption. A further 28 bcm per year is produced from the North Field for LNG and condensate exports. Qatar will soon export gas by pipeline to Abu Dhabi. The initial stage of the $3.5 billion Dolphin Project, which involves the construction of a pipeline linking the North Field to Abu Dhabi and Dubai, is expected to be completed by the end of 2006. Kuwait and Bahrain have also held discussions with Qatar about importing gas from the North Field.

Production and Distribution

Qatar's natural gas production has increased rapidly since the first phase of the development of the North Field was completed in 1991. North Field gas is processed in four NGL recovery plants in Mesaieed Industrial City. Qatar Liquefied Natural Gas Company (Qatargas) and Ras Laffan Liquefied Natural Gas Company (RasGas) produce LNG (see Exports section below). All the natural gas feedstock for Qatar's LNG plants comes from the North Field.

Other North Field projects include the Al Khaleej Gas Project, which will provide gas for the domestic and export market and the Dolphin Project, which calls for a gas pipeline to carry Qatari gas to Abu Dhabi, Dubai and Oman. Under the Al Khaleej Gas Project, 21.9 bcm a year of North Field gas will be developed by RasGas. The gas will be supplied to power and water facilities at Ras Laffan, the Oryx GTL Project and to end-users located in

4. Qatar Petroleum website, http://www.qp.com.

Mesaieed Industrial City, south of Doha. Qatar Petroleum and ExxonMobil plan to extend the Al Khaleej Gas Project through incremental pipeline exports as well as Qatari industrial sale contracts. Gas sales agreements covering the entire output of the project have been concluded and first gas is scheduled for November 2005. The Al Khaleej Gas Project could be expanded to allow gas exports to Kuwait through a 590-km sub-sea gas pipeline from the North Field.

The Dolphin Project is the first export-oriented pipeline project in the GCC region. In 1999, the UAE government's Offsets Group (UOG) announced the formation of a company, Dolphin Energy, to manage the project. The company is a joint venture with Mubadala Development Company (51%), Total (24.5%) and Occidental Petroleum (24.5%). The first phase of the project will involve production and distribution of gas through a sub-sea pipeline from the North Field to Taweelah in Abu Dhabi and Jebel Ali in Dubai. Deliveries are due to start at the end of 2006. The second phase of the project involves increased volumes of piped gas to the UAE. The pipeline will continue to Oman, with a possible submarine extension to Pakistan.

In the Reference Scenario, gas production is projected to grow on average by almost 8% per year over the *Outlook* period, reaching 255 bcm in 2030 (Figure 15.7). The majority of the increase will be from existing and future additions to reserves in the North Field.

Figure 15.7: **Qatar's Natural Gas Production in the Reference Scenario**

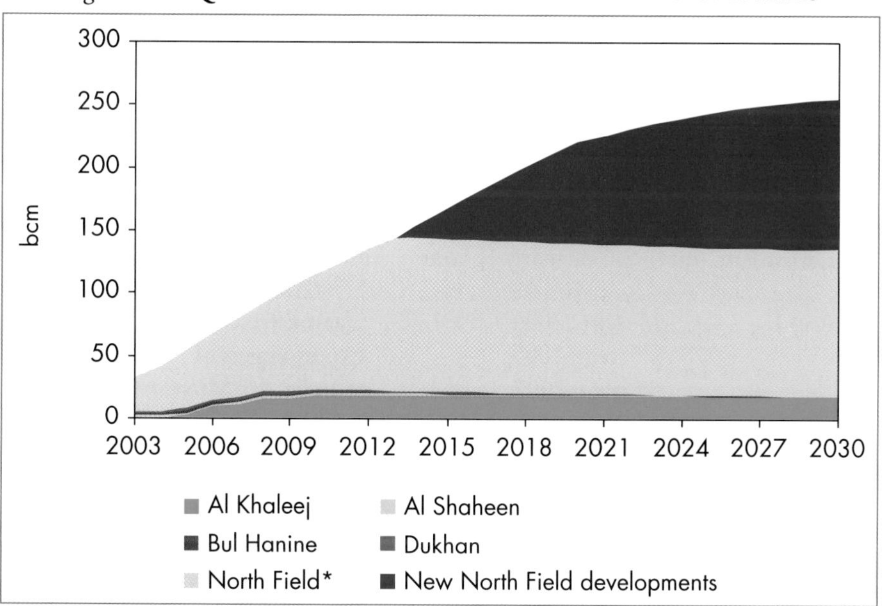

*Already planned or implemented projects.

Exports

Liquefied Natural Gas

Qatar Petroleum has developed two major LNG projects in partnership with international oil companies to exploit North Field gas reserves. Qatargas is a joint venture with Qatar Petroleum, Total, ExxonMobil, Mitsui and Marubeni. The Qatargas plant has three trains with a combined capacity of 9.5 million tonnes per year and was commissioned in 1997. Qatar Petroleum owns RasGas. ExxonMobil, Itochu and Nissho Iwai are joint venture partners. The first train began operation in 1999, and the second train came on stream in 2000, bringing total capacity to 6.6 Mt/year. RasGas-II, which started up in 2004, can produce 4.8 Mt/year (Table 15.9). Qatar Petroleum holds a 70% stake and ExxonMobil 30%. Most of the LNG from the Qatargas and RasGas plants is shipped to Japan, South Korea, Spain and India. There are also occasional spot shipments to the United States.

Table 15.9: **Planned LNG Export Projects in Qatar**

	Expected completion date	Export capacity (Mt/year)	Notes
Qatargas II	1st train: 2008 2nd train: 2009	2 x 7.8	Qatar Petroleum/ExxonMobil; 15 to 16 LNG carriers to be built to ship LNG to Europe and the United States.
Qatargas 3	2009/2010	7.8	Qatar Petroleum/ConocoPhillips. To be built at Ras Laffan. Exports to the US.
Qatargas 4	2010/2011	7.8	Qatar Petroleum/Shell. Heads of agreement signed 2005. Exports to US and Europe.
RasGas II	1st train : 2004 2nd train: 2005 3rd train: 2007	3 x 4.78	Qatar Petroleum/ExxonMobil. Export targets of 7.5 Mt/year to India, 4.7 Mt/year to Italy, 3.4 Mt/year to Belgium and 3 Mt/year to Chinese Taipei.
RasGas III	2008/2009	2 x 7.8	Qatar Petroleum /ExxonMobil. Export target of up to 15.6 Mt/year to the US.

Sources: Arab Petroleum Research Centre (2005); Qatar Petroleum (2005).

LNG facilities are being expanded with five new projects: RasGas-II (expansion), RasGas-III, Qatargas II, Qatargas 3 and Qatargas 4. Sales and purchase agreements have been reached with a number of countries. Several Heads of Agreement have also been signed and, should they turn into confirmed sales agreements, total LNG exports would reach about 77 Mt/year by 2012 (Table 15.10). Qatar Petroleum has allocated $33 billion in its five-year plan, starting 2005, for LNG and piped natural gas projects.

Table 15.10: **Qatar's Contracted LNG Exports** (million tonnes)

		2004	2005	2006	2008	2010	2012
Sales agreements		**16.4**	**20.1**	**22.9**	**38.6**	**46.4**	**46.4**
Japan	(Qatargas)	6.7	6.7	6.7	6.7	6.7	6.7
Korea	(RasGas)	4.8	4.8	4.8	4.8	4.8	4.8
India	(RasGas II)	2.5	5.1	7.5	7.5	7.5	7.5
Italy	(RasGas II)	-	-	-	4.7	4.7	4.7
Spain	(Qatargas)	1.4	1.4	1.4	1.4	1.4	1.4
Spain	(Qatargas)	0.8	0.8	0.4	-	-	-
Spain	(RasGas II)	-	0.6	0.8	0.8	0.8	0.8
Spain	(Qatargas)	0.2	0.7	0.7	0.7	0.7	0.7
Spain	(Qatargas)	-	-	0.6	0.8	0.8	0.8
UK	(Qatargas II)	-	-	-	7.8	15.6	15.6
Belgium	(RasGas II)	-	-	-	3.4	3.4	3.4
Heads of agreement		-	-	-	**1.7**	**23.1**	**30.9**
Chinese Taipei	(RasGas)	-	-	-	1.7	3.0	3.0
US	(Qatargas 3)	-	-	-	-	7.8	7.8
US	(RasGas III)	-	-	-	-	12.3	12.3
US	(Qatargas 4)	-	-	-	-	-	7.8
Total		**16.4**	**20.1**	**22.9**	**40.3**	**69.5**	**77.3**

Sources: Qatargas Marketing Plans (August 2005) and Qatar Petroleum.

In the Reference Scenario, LNG exports are projected to be somewhat lower than contracted exports in 2010 (Figure 15.8). Although planned projects will add 57 Mt/year of production capacity between 2004 and 2012, a fourfold increase, simultaneously building several LNG trains could result in bottlenecks in engineering, procurement and construction (EPC). There are a limited number of EPC companies in the international market that are able to undertake this work. There may also be delays in getting the gas production capacity on stream due to capacity constraints in building LNG carriers, holdups in approving receiving terminals and hiring carrier crews. Total capacity in the Reference Scenario will reach more than 100 Mt/year by the end of the projection period.

Figure 15.8: **Qatar's LNG Capacity and Projected Exports in the Reference Scenario**

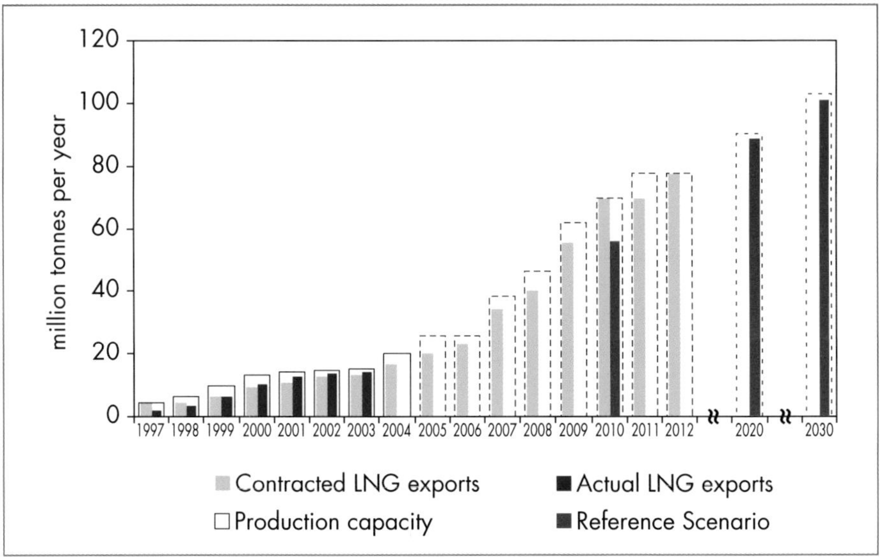

Sources: Qatar National Bank (2005); Qatar Petroleum (2004); IEA analysis.

Gas to Liquids (GTL)[5]

Qatar is aiming to make gas to liquids the centre-piece of its downstream efforts once various LNG projects are completed. The projects are all integrated with offshore developments to supply the large amounts of gas needed. The country's first plant, with a capacity of 34 kb/d, is under construction and is expected to be commissioned in early 2006 (Table 15.11). The plant, which will take gas from the North Field, is a joint venture between Qatar Petroleum (51%) and South Africa's Sasol and Chevron. Two other projects – Shell's 140 kb/d Pearl project and ExxonMobil's 154 kb/d plant – are at an advanced planning stage. If these projects proceed to schedule, total Qatari capacity will reach 104 kb/d in 2009, when Phase 1 of the Pearl project is commissioned, and 330 kb/d in 2011 when Pearl Phase 2 and the ExxonMobil plants are completed. These plants, when operating at full capacity, will together consume about 33 bcm a year of gas – equal to more than 20% of Qatar's entire output of natural gas.

5. The use of natural gas in GTL plants includes volumes used as a feedstock and as fuel.

Table 15.11: **Planned GTL Projects in Qatar**

	Expected completion date	Notes
Oryx GTL (expansion)	2006 (2009)	Qatar Petroleum (51%), Sasol-Chevron (49%). MoU between QP and Sasol Chevron (on hold).
Pearl GTL	2009 (Phase 1) 2011 (Phase 2)	D/PSA with Shell.
ExxonMobil GTL	2011	Heads of agreement.
ConocoPhillips GTL	delayed	PSA with ConocoPhillips for 2 phases with 80 kb/d each.
Sasol/Chevron GTL	delayed	Sasol Chevron submitted a proposal to Qatar Petroleum in July 2002.
Marathon GTL	delayed	

In May 2005, Qatar declared a three-year moratorium on new GTL projects using gas from the North Field. The projects affected by this moratorium are those under production-sharing agreements with ConocoPhillips, Marathon and, the second-phase Oryx development led by Sasol/Chevron. This is partly because of spiralling EPC costs, but the Qatari government also wants to allow more time for a careful evaluation of the impact of a rapid expansion of production on the recovery rate of the North Field.

We expect that Qatari consumption of gas for GTL production will reach 10 bcm in 2010 and 63 bcm in 2030, when the total capacity of GTL plants will reach about 640 kb/d.

Energy Investment

Cumulative investment needs in the Qatari energy sector over the projection period will amount to $155 billion (in year-2004 dollars), or almost $6 billion per year (Figure 15.9). Investments in the gas sector will account for more than 60% of total investments. Oil sector investments will account for almost one-third and the electricity sector for the remaining 4%.

Figure 15.9: **Qatar's Energy Investment by Sector in the Reference Scenario**

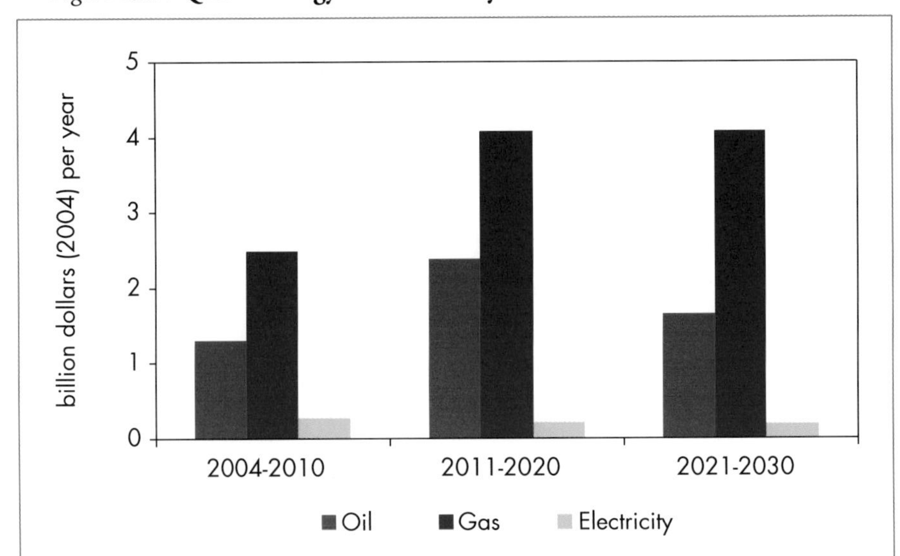

Oil and Gas

Cumulative oil investments over the projection period are $50 billion in the period 2004-2030 in the Reference Scenario, of which more than 60% in the upstream oil sector, almost one-third in GTL projects and the remaining 6% in the refining sector. Oil investments in Qatar will account for 8% of the total oil investment in the MENA region over the *Outlook* period.

Gas investments will amount to almost $100 billion in 2004-2030, with two-thirds going to upstream developments, about one-fifth to liquefaction facilities and the rest to transmission and distribution systems. Projected investments in the gas sector from 2004 to 2030 in Qatar account for almost one-quarter of total cumulative gas investment in the MENA region.

Electricity

Total investment in Qatar's power sector over the *Outlook* period is projected to be $6 billion. Power plants will require nearly $3 billion. Transmission will need $1 billion and distribution $2 billion. Qatar is likely to rely increasingly on independent power producers to meet its future needs. Financing new projects is not likely to be a problem, although they are becoming larger and more complex.[6]

6. This is the case in other Gulf countries, including the UAE.

Deferred Investment Scenario[7]

The Deferred Investment Scenario assumes that upstream oil investment in 2004-2030 is constant as a proportion of GDP at the level of the past decade. This leads to a 10% reduction in oil investment in Qatar compared with the Reference Scenario. Though the same proportionate cut has not been applied initially to gas investment, it also falls as the higher oil prices resulting from lower investment and production would push up gas prices, curbing global gas demand and the call on Qatar gas supply. As a result, the call on Qatari LNG exports is markedly lower in the Deferred Investment Scenario. Gas production in Qatar is about 20% lower in 2030 than in the Reference Scenario, while cumulative gas investment is almost 30% lower. Lower gas production leads to a substantial fall in associated condensate output.

Energy Demand

Qatar's primary energy demand in the Deferred Investment Scenario grows at a more moderate pace in 2003-2030, averaging 5.2% per year over the projection period. As a result, demand reaches 60 Mtoe in 2030, compared with 67 Mtoe in the Reference Scenario. Primary demand for both oil and gas falls, with gas accounting for most of the decline. Most of the decline in demand is caused by slower growth in energy consumption in LNG production (because of lower exports), in oil and gas extraction and in the petrochemical sector. The fall in demand in the petrochemical sector is the result of slightly slower global economic growth, which depresses export demand.

Oil Production and Exports

Unlike in most other MENA countries, the fall in oil production in Qatar in the Deferred Investment Scenario is due mainly to lower gas production rather than lower oil investment and production, because a large and rising portion of total oil output is condensates associated with gas production. Production of crude oil and condensates, taken together, rises by 0.7% per year over the projection period compared with 1.1% in the Reference Scenario. Production reaches 1.1 mb/d in 2030, compared with 1.25 mb/d in the Reference Scenario.

Exports of crude oil, condensates and refined products in 2030 are 13% lower, at 810 kb/d, in the Deferred Investment Scenario. GTL production and exports are assumed to remain constant and their share in total oil production

7. See Chapter 7 for a detailed discussion of the assumptions and methodology underlying the Deferred Investment Scenario.

in 2030 increases from 33% in the Reference Scenario to 35%. We do not expect GTL production from new plants already being built to decline, given the large capital outlays involved.

Gas Production and Exports

Gas production in the Deferred Investment Scenario is lower than in the Reference Scenario because LNG export sales fall and because internal demand is lower. Total gas production in 2030 reaches 202 bcm in the Deferred Investment Scenario, a difference of about 20% with the Reference Scenario. Because Qatar is by far the largest LNG exporter in the Reference Scenario, the fall in the call on Qatar gas resulting from lower global demand is larger than for other MENA exporters in volume terms. Gas exports reach 106 bcm in 2030, 45 bcm, or 30%, less than in the Reference Scenario (Figure 15.10).

Figure 15.10: **Qatar's Natural Gas Balance in the Reference and Deferred Investment Scenarios**

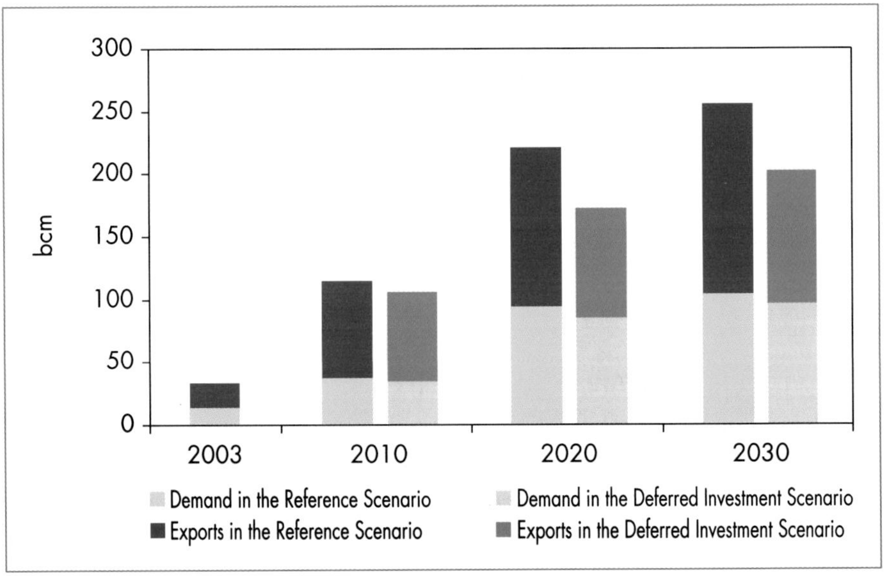

CHAPTER 16

SAUDI ARABIA

HIGHLIGHTS

- The Kingdom of Saudi Arabia is the largest nation in the Middle East geographically, and has the third-largest population after Iran and Iraq. Its social and economic record is intimately linked to the development of the country's exceptional endowment of hydrocarbon resources. The country has made substantial progress in diversifying its economy, but economic growth remains closely linked to the value of oil earnings.
- Saudi Arabia's primary energy demand is projected to grow at an average rate of 3% per year from 2003 to 2030, down from 5.5% since 1990. Future demand growth will be tempered by a less rapid expansion in population and a slow-down in the growth of gross domestic product and industrial output.
- Saudi Arabia has the largest proven reserves of oil in the world, at 262 billion barrels at the start of 2005. More than half are located in eight super-giant fields, including the onshore Ghawar field, the largest oilfield in the world, and Safaniyah, the world's largest offshore field. In our Reference Scenario, oil production (including natural gas liquids) grows from 10.4 mb/d in 2004 to 11.9 mb/d in 2010 and 18.2 mb/d in 2030. Saudi Arabia will remain the world's largest oil exporter, with exports rising from 8.3 mb/d in 2004 to 14.4 mb/d in 2030. These trends call for cumulative oil investment of $174 billion (in year-2004 dollars) over 2004-2030, more than 80% for upstream projects.
- Production of natural gas will grow even faster than that of crude oil, as more gas will come from non-associated fields. We project marketed production to rise from 60 bcm in 2003 to 86 bcm in 2010 and then surge to 155 bcm in 2030. All of this gas will be consumed domestically.
- Saudi Arabia's power generation capacity is projected to rise from 35 GW in 2003 to 102 GW by 2030. Power-sector investment will reach $110 billion, about a third of total energy investment. Just over half of new power capacity will be integrated with desalination. Fuel requirements for desalination will rise from 11 Mtoe in 2003 to 31 Mtoe in 2030, when they will account for 27% of the increase in primary energy demand in the power and water sector.
- In the Deferred Investment Scenario, crude oil production reaches 14.1 mb/d in 2030 – 4.1 mb/d, or 23%, lower than in the Reference Scenario. Internal oil demand is little changed, and oil exports are reduced by 27% in 2030, reaching 10.5 mb/d. Natural gas production reaches 150 bcm in 2030.

Overview of the Saudi Energy Sector

The Kingdom of Saudi Arabia is the largest nation in the Middle East geographically and has the third-largest population after Iran and Iraq. Saudi Arabia's social and economic record is closely linked to the development of the country's exceptional endowment of hydrocarbon resources. Saudi Arabia holds the world's largest proven oil reserves and the fourth-largest gas reserves. It is the world's leading oil producer alongside Russia, providing nearly 13% of global crude oil and NGL supply, and is the largest exporter.

Table 16.1: **Key Energy Indicators for Saudi Arabia**

	1971	2003	1971-2003*
Total primary energy demand (Mtoe)	6.1	130.8	10.0%
Total primary energy demand per capita (toe)	1.0	5.8	5.6%
Total primary energy demand /GDP**	0.08	0.43	5.4%
Share of oil in total primary energy demand (%)	81%	62%	–
Net oil exports (mb/d)	4.9	8.1	1.6%
Share of oil exports in production (%)	98%	81%	–
CO_2 emissions (Mt)	13.0	306.5	10.4%

* Average annual growth rate.
** Toe/thousand dollars of GDP in year-2004 dollars and PPPs.

Saudi Arabia plays a central role in OPEC and in balancing the world market, because of the size of its production and its spare capacity. In mid-2005, it was the only country in the world with any appreciable amount of sustainable capacity in reserve.

Political and Economic Situation

Political Developments

Saudi Arabia is a hereditary monarchy, but is slowly moving to a more participatory system of government. Abdullah bin Abdul Aziz became king in August 2005 on the death of his brother, King Fahd bin Abdul Aziz. Abdullah had effectively ruled in place of Fahd since 1995, due to Fahd's ill health, so no abrupt change in policy direction is expected. The country's first-ever elections, covering half of the members of consultative municipal councils in

178 towns across the country, were held in February 2005 in the Riyadh region and one month later in the eastern region. The third and final phase took place in the second half of April in the western and northern provinces of the country. The government plans to extend the elections to all members by 2010. Voter registration and turnout, however, were both low, and women were not allowed to vote. The government has indicated that the next round of municipal elections in 2009 will include women. Half of the members of regional councils may be elected in 2006. The government also plans to extend elections to the "majlis", the national consultative council, though no time-frame has been announced.

Terrorism has emerged as a challenge to the stability of the country. A series of terrorist attacks in recent years has prompted vigorous efforts on the part of the Saudi government to counter domestic terrorism and religious extremism. The security of oil facilities and personnel is a particular concern. Saudi Aramco employs 5 000 security guards to protect oil facilities, with the support of the Saudi National Guard, regular Saudi military forces and Interior Ministry officers. The government is reported to have substantially boosted spending on oil security. The maintenance of duplicate facilities for key installations is thought to provide a large degree of flexibility in the event of any one facility being disabled.

Economic Trends

Recent Economic Performance

Saudi Arabia has made substantial progress in diversifying its economy away from oil extraction, though economic growth remains linked to the value of oil earnings. The economy grew strongly in 2003 and 2004, after a weak performance during the previous two years. Real GDP surged by 7.7% in 2003 and 5.2% in 2004, thanks in part to the stimulus provided by increased oil-export revenues, resulting from high oil prices and rising output. Nonetheless, private non-oil sector growth is becoming more independent of fluctuations in oil-export earnings.

Oil contributed 40% of GDP, around 90% of total export earnings and three-quarters of the central government budget in 2004. Oil exports generated $106 billion in revenues in 2004 alone – an increase of 30%. These revenues, which are likely to reach at least $150 billion in 2005, are being channelled into the non-oil sector through higher government spending and into the reduction of the country's large debt, which has contributed to increased private-sector liquidity. The share of oil in GDP and export earnings has risen sharply in recent years with higher oil prices, but is still significantly lower than in the 1970s (Figure 16.1).

Figure 16.1: **Share of Oil in GDP and Exports by Value in Saudi Arabia**

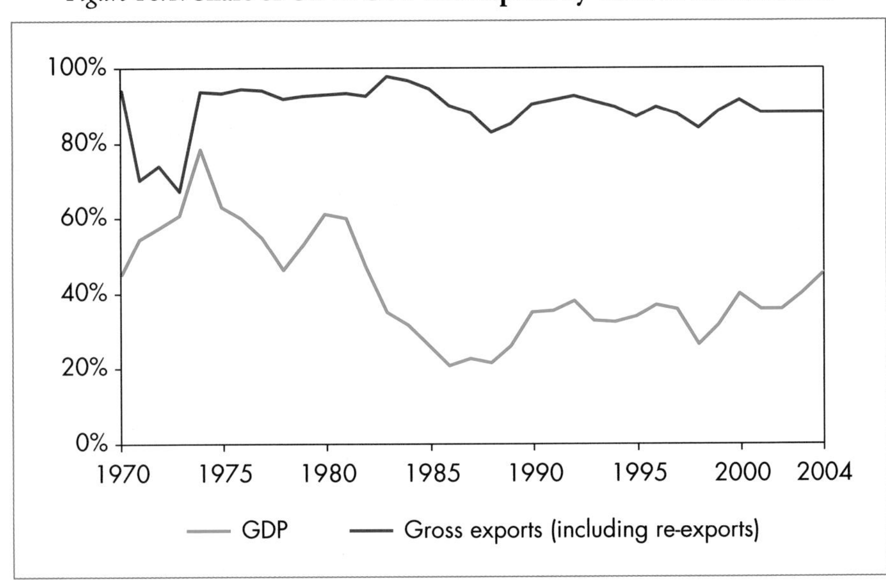

Source: SAMA (2005).

Higher oil revenues in recent years have turned around government finances. The central government ran a large fiscal surplus in 2004, amounting to $26 billion, or 10.5% of GDP, up from only 1% in 2003. This was despite spending 28% more than budgeted. Before 2003, Saudi Arabia experienced two decades of large fiscal deficits. Total public debt in 2002 was around 97% of annual GDP, but has since fallen to around 65%. The jump in oil revenues has encouraged the government to increase capital spending and, to a lesser extent, current spending, as well as to repay domestic debt. Higher oil revenues have also transformed the trade balance, from a deficit of $13 billion in 1998 to a surplus of $51.5 billion in 2004, equal to 20.6% of GDP. Per capita income jumped by 16% between 2002 and 2004. It fell during much of the 1980s and 1990s (SAMA, 2005).

Structural Economic Factors and Policy Reforms

The government continues to liberalise the economy and privatise state-owned companies, in line with some other Gulf Cooperation Council (GCC) countries. Private firms have been allowed to invest in some sectors and opportunities for foreign investors have been increased. Improving the investment climate and increasing the legal protection to foreign investors are central elements in the government's reform programme.

In May 2000, a new law aimed at attracting more foreign investment came into effect. The law permits full foreign ownership of Saudi property and licensed projects. It also established the Saudi Arabian General Investment Authority (SAGIA) to streamline applications by foreign investors, and lowered taxes on company profits from 45% to 30%. Previously, foreign companies were not allowed to own more than 49% of any Saudi enterprise. However, several sectors remain closed to full foreign ownership, including oil exploration and production, oil pipelines, media and publishing, insurance, telecommunications, defence and security, health services, and wholesale and retail trade. Fully foreign-owned companies are not eligible for the same financial incentives as Saudi-controlled firms. Nonetheless, in January 2004, the Saudi cabinet approved a reduction in taxes on the profits of foreign-owned businesses to 20% in most sectors and 30% in the natural gas sector.

Some state assets have been sold to private buyers, including stakes in the National Company for Cooperative Insurance and the Saudi Telecommunications Company. Large state corporations, including Saudi Aramco, which has a monopoly on Saudi upstream oil development, and the Saudi Basic Industries Corporation (Sabic), still dominate the Saudi economy. But there has been some private investment in petrochemicals, the refining sector and gas exploration. Some 30% of Sabic's shares and 19% of the Saudi Electric Company's shares are publicly listed The regulation of capital markets has been overhauled and markets have been opened up to foreign investors.

Saudi Arabia's application for membership of the World Trade Organization (WTO), which will require the opening-up of domestic markets and the removal of price controls, is giving additional impetus to economic reform. Saudi Arabia, the only GCC country that is not yet a WTO member, is expected to complete negotiations on joining by late 2005 or early 2006. The country's membership bid was brought one step closer in September 2005, when it signed a bilateral trade agreement with the United States. It had earlier signed WTO-related bilateral agreements with the European Union, China and several other countries. Membership is likely to benefit the economy, through increased competition and efficiency and through increased confidence in the country's legal framework and dispute resolution procedures. However, the oil sector – unlike petrochemicals – will remain outside the scope of the agreement with the WTO.

Saudi Arabia continues to face serious long-term economic-policy challenges. These include high rates of unemployment and a growing demand for increased government spending on social welfare programmes in response to a rapidly growing population. Saudi Arabia's per capita oil-export revenues have fallen considerably in the past two-and-a-half decades, from $21 652 (in year-2004 dollars) in 1980 to only $4 685 in 2004 (SAMA, 2005). This is mainly the result of a tripling of the population. Subsidies, notably in water and electricity, are growing as demand rises.

The unemployment rate among Saudi nationals is officially about 9.7% , but the actual level is reported to be much higher. Some unofficial estimates put male unemployment at 25% or more. Whatever the true rate, unemployment is most prevalent among young males. There is nonetheless a severe local shortage of skilled Saudi workers and a dislike of doing menial work. As a result, just under half of the workforce is foreign. The government is tackling high unemployment among Saudi nationals by encouraging the replacement of foreign workers with Saudis. All companies in the country are obliged by law to increase the numbers of Saudi nationals in their workforce by 5% a year. There are signs that the authorities are beginning to get tougher in enforcing this policy. The government has also introduced a training scheme for Saudi workers. As in other GCC countries, the government places restrictions on the employment of foreign nationals, through a quota system, and offers tax incentives for companies employing Saudis.

Macroeconomic Prospects and Assumptions

Efforts to expand the role of foreign private investment in the economy are expected to continue. Foreign private investment will most likely be focused on the upstream gas industry and the power, refining, petrochemicals, water and telecommunications sectors. Further steps will probably also be taken towards encouraging the development of capital markets. Nonetheless, there is a danger that the recent surge in oil revenues will reduce the incentive for the government to push ahead with economic reforms. And economic reform will remain constrained by consensus-building traditions. The government's approach to privatisation and to the expansion of the role of the private sector in general is expected to remain cautious.

Saudi Arabia's GDP is expected to grow at an average annual rate of 4.2% between 2003 and 2010, slowing slightly to 3.5% in 2010-2020 and then to 3.0% in 2020-2030 (Table 16.2). The hydrocarbon sector is expected to remain the primary driving force behind economic growth over the *Outlook* period. The assumed fall in oil prices from recent peaks will depress growth in the second half of the current decade. But rising oil output, together with rising prices from early in the second decade, will push up oil revenues in the long term. Growth in the non-oil sector will continue to depend to some degree on the health of the hydrocarbon market, which will affect how much the government has to spend on public infrastructure projects. The development of upstream gas and power and water projects – bolstered by rising foreign direct investment – will help to support economic development, particularly after 2010. Rapid growth in the labour force and an assumed modest increase in labour productivity will also boost non-hydrocarbon GDP. Non-hydrocarbon productivity is assumed to rise from an average of $30 700 per year per worker (in year-2004 prices) in 2004-2010 to $33 500 in 2020-2030.

1. In 2004, the last year for which official data are available (SAMA, 2005).

Table 16.2: **GDP and Population Growth Rates in Saudi Arabia in the Reference Scenario**
(average annual rate of change)

	1971-2003	1990-2003	2003-2010	2010-2020	2020-2030	2003-2030
GDP	4.4%	2.8%	4.2%	3.5%	3.0%	3.5%
Population	4.2%	2.8%	2.7%	2.2%	1.8%	2.2%
Active labour force	4.5%	2.6%	3.5%	2.9%	2.4%	2.9%
GDP per capita	0.1%	0.0%	1.5%	1.3%	1.2%	1.3%

Demographic Trends and Assumptions

The Saudi population is growing rapidly – 23 million in 2003. Over the 1990 to 2003 period, population growth averaged 2.8% per year. The annual population growth rate is assumed to slow slightly over the projection period, averaging 2.2% per year (UNDESA, 2004). This assumes little or no growth in the number of foreign workers. If this projection proves accurate, Saudi Arabia's population will reach 40 million by 2030 (Figure 16.2). Almost 40% of Saudi population is aged 15 years-old or younger, and 73% at or under 29 years-old. As a result, the number of people entering the labour force will surge in the next few years. Combined with the projected growth of GDP, these trends imply that per capita income will grow at 1.3% per year. Per capita income in Saudi Arabia will reach 44% of that in the OECD countries in 2030.

Figure 16.2: **Saudi Arabia's Population and GDP Per Capita in PPPs**

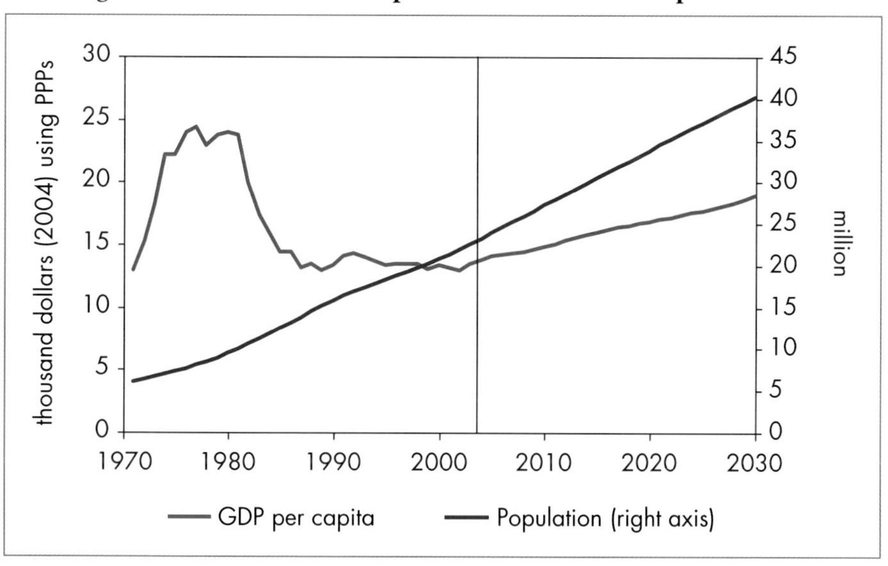

Energy Demand

Primary Energy Mix

Saudi Arabia's primary energy demand is projected to grow at an average rate of 3% per year from 2003 to 2030 (Table 16.3). This is a slower rate than that observed since 1990, when annual demand growth averaged 5.5%. Demand will nonetheless more than double over the projection period. Demand growth will slow progressively over the *Outlook* period, from 4.8% per year in 2003-2010 to 1.6% in 2020-2030, as economic growth slows. Slower growth in population and industrial output will also contribute to lower energy demand growth over 2003-2030.

Table 16.3: **Total Primary Energy Demand in Saudi Arabia in the Reference Scenario** (Mtoe)

	2000	2003	2010	2020	2030	2003-2030*
Oil	73.7	81.7	110.5	141.1	161.9	2.6%
Gas	40.6	49.0	70.5	105.6	126.9	3.6%
Other	0.0	0.0	0.1	0.1	0.3	16.1%
Total	**114.3**	**130.8**	**181.0**	**246.8**	**289.1**	**3.0%**

* Average annual growth rate.

Today, Saudi Arabia's primary energy needs are met entirely by oil and natural gas, almost all of which is produced in association with oil. Primary oil consumption in the country reached 1.9 mb/d in 2003, absorbing 19% of the country's total oil production (including NGLs). Oil accounted for 62% of the country's total primary energy demand. Domestic demand has grown steadily with the rise in population. Oil demand is projected to reach 3.7 mb/d in 2030. Oil's share of primary energy demand will nonetheless decline over the projection period, from 62% in 2003 to 56% in 2030 as demand for gas rises more rapidly than that for oil (Figure 16.3).

Natural gas demand is projected to grow at an average annual rate of 3.6% over 2003-2030, driven by power generation and, to a lesser extent, by industry. Gas recently overtook oil as the leading fuel in power and water production and its share in power plant fuel use will reach two-thirds by 2030. In total, gas use will rise from 60 bcm in 2003 to 155 bcm in 2030. The contribution of renewable energy sources – essentially solar – will remain negligible. The share of power generation in total primary energy demand will increase, from 33% in 2003 to 40% in 2030. The fuel requirements for water desalination will account for 27% of the increase in energy demand in the power and water sector from 2003 to 2030.

Figure 16.3: **Shares of Oil and Natural Gas in Primary Energy Demand in Saudi Arabia in the Reference Scenario**

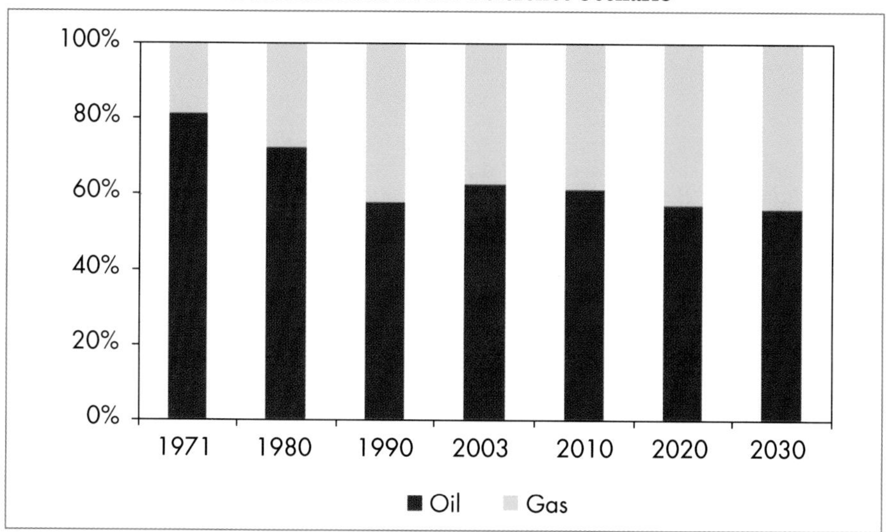

Primary energy intensity – the amount of energy needed to produce a unit of GDP – will continue to rise in the short term, with rapidly rising demand from energy-intensive manufacturing industry, but will turn down in the second decade (Figure 16.4). Primary energy use per capita will continue to rise over the first half of the projection period at more or less the same pace as during the last ten years. It will begin to level off thereafter, largely as a result of energy efficiency improvements.

Figure 16.4: **Primary Energy Intensity and Per Capita Energy Use in Saudi Arabia in the Reference Scenario**

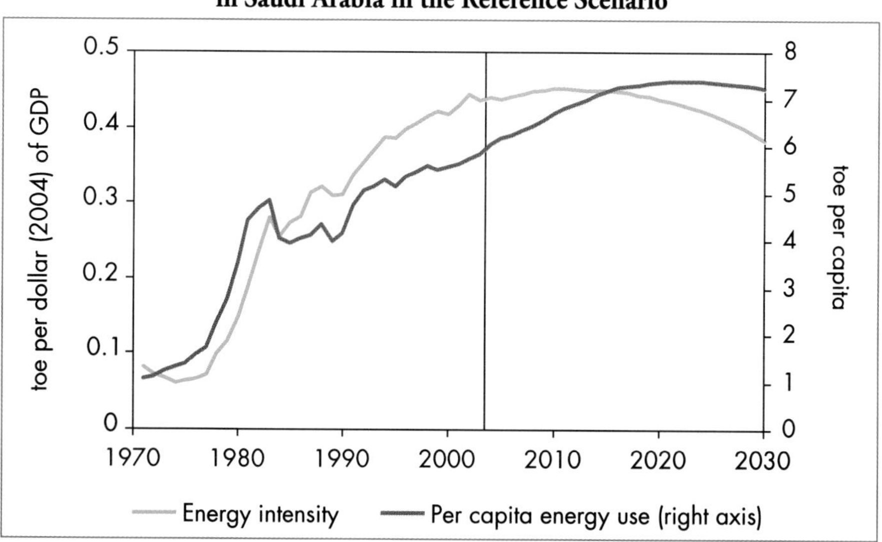

16

Total final energy consumption is projected to grow on average by 2.8% over 2003-2030, a marginally slower pace than primary demand. Electricity consumption will grow fastest, by 4% per annum, in line with growing population and rising household incomes. Saudi Arabia already has one of the highest levels of per capita electricity consumption in the world, because of its extreme climate. The wide variations in temperature between seasons, between day and night and even during the day result in surges in demand for electricity for both air-conditioning and space-heating. Air-conditioning load is particularly strong from June to September. Per capita electricity use is projected to increase from 6 790 kWh per year in 2003 to 10 600 kWh in 2030.

Sectoral Trends

Industry

Industry will remain the largest end-use sector, although its share of total final energy consumption will drop by five percentage points to 49% in 2030 (Figure 16.5). In 2003, chemicals, made up almost entirely of petrochemicals, accounted for 58% of total industrial energy consumption.

Figure 16.5: **Total Final Energy Consumption by Sector in Saudi Arabia**

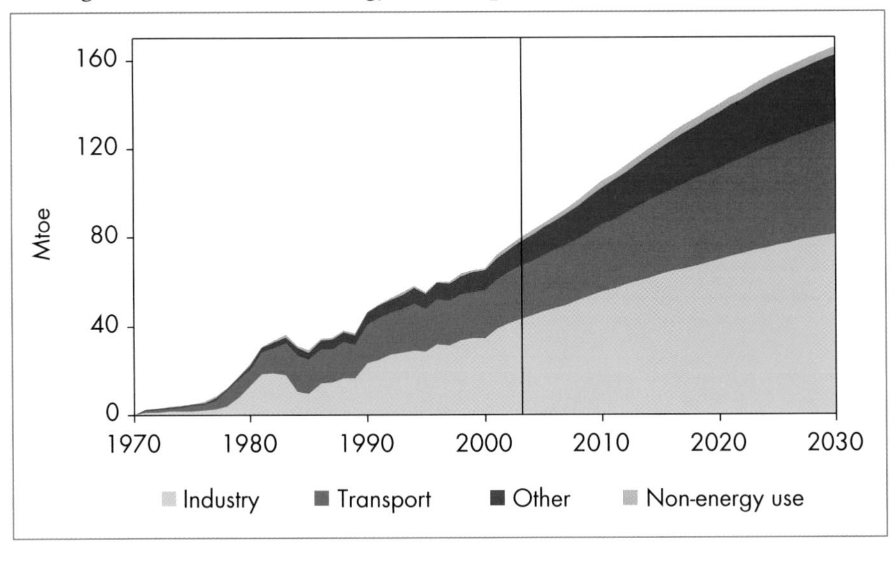

Saudi Arabia is a major petrochemical producer. The sector consumed 11.5 Mtoe of natural gas in 2003, equal to 23.5% of total marketed gas supply. Sabic, which is 70% owned by the state, is the Middle East's largest

Table 16.4: **Saudi Arabian Petrochemical Plants**

Location	**Owner**	**Capacity** (kt/year)
Jubail	Saudi Methanol (Al-Razi) – a 50:50 Sabic/Mitsubishi joint venture.	2 980 methanol
Jubail	Kemya - a 50:50 Sabic joint venture with ExxonMobil	700 ethylene 200 polypropylene
Jubail	Arabian Petrochemicals (Petrokemya) – a Sabic affiliate	2 250 ethylene
Jubail	Saudi Petrochemicals (Sadaf) – a 50:50 Sabic/Pecten (Shell affiliate) joint venture	485 propylene 1 050 ethylene 700 MTBE
Jubail	Saudi European Petrochemicals (Ibn Zahr) – Sabic (70%); Neste Oy, Finland (10%); Ecofuel, Italy (10%); Arab Petroleum Investment Corporation, Apicorp (10%)	1 300 MTBE
Jubail/ Dammam	Saudi Fertilizers (Safco) – Sabic (41%), Safco Employees (10%), private Saudi Arabian shareholders (49%)	700 ammonia
Jubail	National Fertilizer (Ibn Baytar) – a 50:50 Sabic joint venture with Safco	500 ammonia
Jubail	Jubail Fertilizers (Samad) – a 50:50 Sabic/ Taiwan Fertilizer Company joint venture	350 ammonia
Jubail	National Methanol (Ibn Sina) – Sabic (50%); Hoechst/Celanese, US (25%); Pan Energy, US (25%)	1 000 methanol 850 MTBE
Yanbu	Arabian Industrial Fibres (Ibn Rushd) – Sabic (70%) and 15 Saudi Arabian and regional private-sector partners	730 benzene-toluene-xylene
Yanbu	Yanpet – a 50:50 Sabic/ExxonMobil joint venture	1 480 ethylene-glycol, polyethylene, polypropylene
Jubail	Saudi Chevron Petrochemical Co. – 50:50 Chevron Chemical/Saudi Industrial Venture Capital Group (private) joint venture	482 benzene 220 cyclohexane

MTBE: methyl tertiary butyl ether.
Source: Global Insight Markets Research Centre database.

non-oil industrial company, contributing around 10% of world petrochemical production. With the completion of a new petrochemicals plant in the eastern Saudi Arabian industrial city of Jubail in the second half of 2004, Saudi Arabia became the fifth-largest ethylene producer in the world.

There are ambitious plans for further expanding petrochemical production using natural gas as a feedstock. The last two years have seen huge investments in petrochemicals, Saudi Arabia's fastest-growing industrial sector. Sabic continues to dominate, but new players have emerged. However, this boom has created some bottlenecks, notably a scarcity of ethane, the most widely-used feedstock in the country's projects. Ethane is still priced lower than other feedstock, despite the current shortage.

Saudi Aramco has already allocated the majority of its feedstock supply and no new round of supplies is expected before 2007. Liquids extracted from any non-associated gas discovered by international oil companies in partnership with Saudi Aramco under recent deals will not be available before the end of the decade. Associated gas from the planned development of oilfields in the eastern region will similarly not become available until the next decade.

Ten additional petrochemical plants are under construction, all of which are designed to use ethane as feedstock in cracking units. Because of the ethane shortage, a number of crackers are being converted to be able to use other feedstocks. Doing so will produce new by-products, which will create new opportunities for further downstream diversification. Total production capacity is expected to increase by more than 9 Mt/year when all these projects are completed. Other projects are expected later in the *Outlook* period, but the rate of expansion of petrochemical production is assumed to slow.

We project natural gas consumption in the chemical industry, as feedstock and energy input, to grow at an average annual rate of 2.5%, reaching 16 Mtoe in 2010 and 22 Mtoe in 2030 – nearly twice the level of 2003.

Transport

As a result of its relatively large population and heavily-subsidised fuel prices, Saudi Arabia has the largest vehicle fleet in the GCC area and the second-largest in the MENA region. There were an estimated 6.5 million vehicles in use in 2003 (National Commercial Bank, 2003). Given the size of Saudi families and the low cost of fuel, the most popular vehicles are multi-utility vehicles and vans. Nearly all Saudi households own at least two cars, one of which can typically accommodate a large family.

Oil demand for transport is expected to grow at an average rate of 2.8% per year over the projection period, in line with the growth of GDP (Figure 16.6). Car ownership in the country, currently 185 vehicles per 1 000 inhabitants, will continue to grow strongly as the country's young population reaches adulthood and household incomes remain buoyant. Around 300 000 new vehicles, equal to almost 5% of the total vehicle fleet, were imported in 2003. The country has no car-making industry. Imports are expected to be steady through to the end of the decade.

Figure 16.6: **Transport Energy Consumption and GDP in Saudi Arabia**

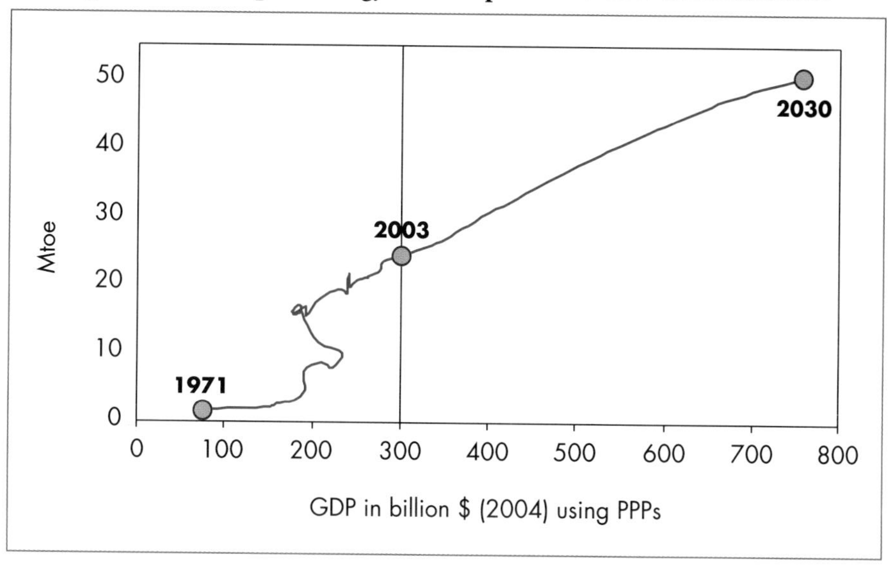

The existing Saudi railway network connects the capital, Riyadh, to the eastern port city of Dammam. Two new projects are planned: the Saudi Landbridge Project, which would run from Jeddah in the west to Dammam, transporting freight and passengers; and the Makkah-Madinah Rail Link Project, a high-speed mass-transit system for carrying 7.5 million pilgrims each year between the country's two major holy places in the western provinces.

Residential and Services

Rapid population growth, rising per capita incomes and a structural economic shift towards services will push up energy consumption in the residential and services sectors. Household demand is projected to grow by 4% per year and services demand by 3.8% per year from 2003 to 2030. Residential demand for oil, which will grow by 2.5% per year, and electricity, 4.2% per year, will also be boosted by the maintenance of large subsidies, which result in retail prices that are far below the true cost of supply.

Electricity Supply and Desalinated Water Production

Overview

Saudi Arabia's electricity generation was 153 TWh in 2003, equal to 6 790 kWh per capita – close to the EU average. Preliminary estimates for 2004 indicate that electricity generation reached 160 TWh. Saudi Arabia is one of the few MENA countries to still rely significantly on oil, which meets over half its total generation needs (the average for MENA countries is 33%). However, the share of natural gas, now 46%, is growing. The use of natural gas in the power sector is constrained by supply-related factors, as several power stations, particularly in the west, are far from gas production centres.

Total installed generating capacity was 35 GW in 2003, about 10% of which was in combined water and electricity production facilities. Most electricity is produced in open-cycle gas turbines or steam boilers. More efficient combined-cycle power plants accounted for only 6% of the installed capacity in 2003.

The country's rapid industrialisation and rapid population growth will continue to push electricity consumption up strongly. Electricity tariffs, particularly for households, are heavily subsidised. At less than 3 cents per kWh on average, these tariffs do not reflect costs and do not encourage efficient electricity use. Because of the rapid growth in electricity demand, investment in new power generation projects has become a major challenge. The government has even had to delay several industrial projects in recent years because of fears of inadequate power supply. It has also had to ration supplies to various areas at times of peak summer demand.

Electricity and water supply in Saudi Arabia are controlled by the Ministry of Water and Electricity. The electricity sector is dominated by the Saudi Electricity Company (SEC). SEC is a private joint-stock company, 74% owned by the government, 19% by Saudi Aramco and 7% by other shareholders. The government expects to start selling off its stake in SEC once the company is commercially viable. Power plants producing desalinated water are controlled by the Saline Water Conversion Corporation (SWCC).

Over the past six years, the Saudi government has taken steps towards restructuring the power industry and towards attracting private investment with the aim of increasing economic efficiency. Saudi Arabia is now opening electricity generation, combined electricity and water production, and electricity transmission to domestic or foreign private investors. The government has also set up the Electricity Services Regulatory Authority as an independent body to oversee power sector-related activities. In June 2004, its remit was extended to include desalination and was renamed the Electricity and Dual Production Organizing Authority. The authority reports to the Minister of Water and Electricity. One of its key responsibilities is to recommend electricity tariffs. It is now holding discussions with SEC about unbundling generation, transmission and distribution, and about creating an independent national transmission company, initially as a subsidiary of SEC (SAMA, 2005). There are also plans to reform tariffs to make them more cost-reflective, particularly in industry. A new law setting out rules and procedures for regulating the electricity sector is expected to be approved by the end of 2005.

Early in 2004, SEC invited the private sector to invest in power generation and transmission projects with a total value of about $12 billion. Table 16.5 lists Saudi Arabia's planned power generation projects. These projects have a total capacity of about 20 GW and most of them are combined power and water production plants. The first four projects are now in various stages of tendering under the control of SEC and SWCC.

Table 16.5: **SEC's Generating Projects Open to the Private Sector**

	Capacity (MW)	**Notes**
Al Shouaiba-3	700	IWPP, top-bidder selected in June 2005
Al Shuqaiq-2	700	IWPP, competitive tender issued
Ras Alzor	2500	IWPP, to be tendered in 2006
Jubail-3	Up to 2 500	IWPP, request for proposal
Muzahimiyah	1725	Planned for 2011
Rabigh-2	2400	Planned for 2012, electricity and water
Qurayyah	3600	Planned for 2012, electricity and water
Sulbokh	1725	Planned for 2014
Yanbu-2	2400	Planned for 2015, electricity and water
Al Shuqaiq-3	600	Planned for 2016, electricity and water
Power plant-10	1725	Planned for 2017

IWPP: Independent water and power producer.
Sources: SEC (2004) and IEA.

The country's first independent power producer project came on line in July 2005. The 250-MW facility is a co-generation plant producing electricity and steam at the Saudi Petrochemical Company complex in Jubail. The first IWPP project is also under development at Al Shouaiba, on a build-own-operate basis. It is expected to be completed by 2007. Saudi Aramco has signed a contract for a 25-year build-own-operate-transfer project at its Rabigh site. The plant will produce electricity, steam and water.

Saudi Arabia's transmission network now extends to over 200 000 kilometres. Only two of the country's four power regions are interconnected. SEC plans to connect them all eventually. This is a major challenge, requiring an additional 30 000 kilometres of lines. In the near term, SEC expects to have two major transmission "islands" – the eastern-central region (accounting for two-thirds of the country's total electrical consumption) and the western-southern region (accounting for 30% of consumption). The sparsely-populated northern region will account for the remaining 5%. Saudi Arabia also plans to be part of the Gulf electricity grid (see Box 6.4). The first phase of the regional grid is expected to be operational by 2008.

Electricity Production

Saudi Arabia's electricity generation is projected to increase by 3.9% per year between 2003 and 2030. Growth will be strongest in the decade to 2010, averaging 6.3% per year. Installed capacity is projected to increase from 35 GW in 2003 to 102 GW in 2030. This will make Saudi Arabia the largest electricity market in the entire MENA region. Most new capacity is expected to be natural gas-fired.[2] The increase in natural gas use in the power sector will depend critically on Saudi Arabia's ability to increase gas production and to develop the infrastructure to bring gas to areas that currently do not have access to it. The share of natural gas in the country's generation mix is projected to rise from 46% in 2003 to 66% in 2030. Some 39 GW, or 51% of additions to generating capacity will be integrated with water desalination (Figure 16.7).

Figure 16.7: **Electricity-Only and CWP Capacity Additions in Saudi Arabia**

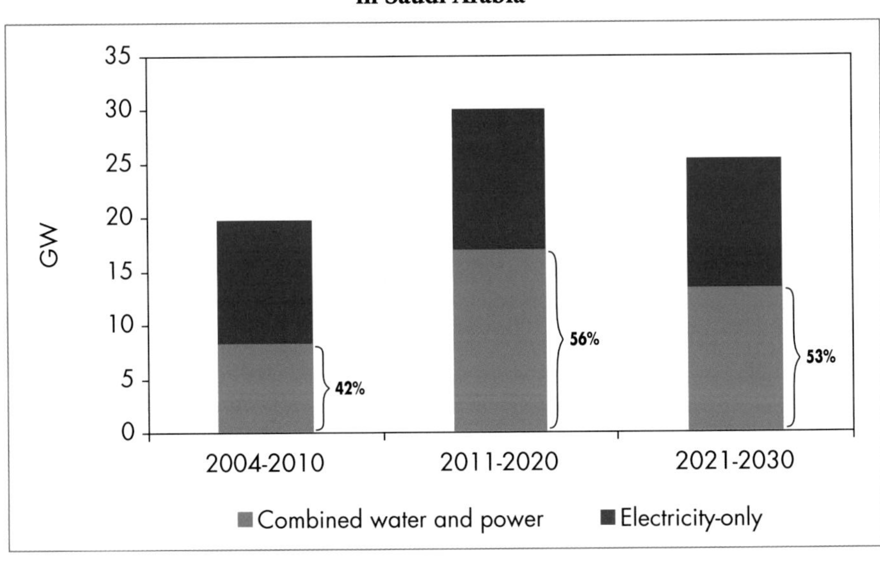

2. See Chapter 6 for an analysis of the economics of oil- and gas-based electricity generation.

Oil-fired generation will increase in absolute terms, but its share in total generation will fall from 54% to 34%. A number of new power projects open to the private sector are oil-fired. Some older oil-fired boilers will be retired before 2030 and some oil-fired power plants could be converted to burning gas. A small increase in solar power is projected (to just 0.5 TWh in 2030), but this will not be big enough to make any significant contribution to Saudi Arabia's electricity needs.

Water Production

Saudi Arabia is the largest producer of desalinated water in the world. Production amounted to 1.75 billion cubic metres in 2003 – equal to 8% of the country's total water needs.[3] It is home to the largest desalination plant in the world, the combined power and water Jubail facility, with capacity of 290 million cubic metres per year of water and 1 500 MW of power. Desalination is expected to increase rapidly over the projection period, greatly boosting energy needs in combined power and water plants. Fuel requirements for desalination will rise from 11 Mtoe in 2003 to 31 Mtoe in 2030, when they will account for 11% of total primary energy supply.

Saudi Arabia has limited groundwater resources. One-third of the country's non-renewable reserves are estimated to have been depleted already, though there are doubts about the reliability of reserve estimates. Some estimates indicate that the aquifers may not last more than 15 to 20 years at current rates of extraction. Water consumption has been rising quickly, but will slow sharply with falling availability of aquifer water. Total water consumption is projected to increase from 22.5 bcm in 2003 to over 25 bcm in 2030, an average annual rate of growth of only 0.5% (Table 16.6). Because of the country's rapidly declining supply of freshwater, almost all residential and industrial water and a quarter of the country's total water supply will come from desalination plants by 2030. As a result, the cost of water supply is set to rise sharply.

Water losses, caused by poor metering and billing practices as well as physical losses in the distribution network, are large in Saudi Arabia. In addition, water prices are heavily subsidised. There is a progressive block tariff system: tariffs for the first two consumption categories are extremely low – $0.03 per cubic metre for 1-50 cubic metres/month and $0.04 per cubic metre for 50-100 cubic metres/month. Total water subsidies reached an estimated $3.2 billion in 2000, equal to 7% of total oil-export earnings (World Bank, 2005). The government plans to phase out subsidies and introduce a country-wide metering system and a public awareness campaign about water conservation.

3. Excludes water used by the upstream oil industry for injection in reservoirs to maintain oil-flow pressure. This unprocessed water is extracted from salty aquifers or from the sea.

Table 16.6: **Saudi Arabia's Water Desalination**

	2003	2010	2020	2030
Water consumption (bcm)	22.5	23.1	24.0	25.1
Desalination capacity (bcm)	2.2	3.5	5.6	7.8
Oil and gas use for desalination (Mtoe)	11	17	24	31

Agriculture is a major user of water, accounting for almost 90% of the country's total water consumption. During the 1980s, the government adopted a strategy of national food self-sufficiency, involving heavy subsidies. By the 1990s, Saudi Arabia was a large net exporter of wheat. Rapid expansion of irrigated areas led to rapid depletion of deep non-renewable aquifers, large water losses and high costs. The Saudi government is seeking to curb water use for agriculture through better crop selection and the adoption of more efficient irrigation techniques.

More than 40% of the country's desalination plants are due to be retired by 2015. Planned capacity additions amount to more than 1.8 bcm per year by 2010, more than offsetting retired capacity. The share of reverse osmosis technology in total desalination capacity was 35% in 2003 and is projected to rise to 39% in 2030. Such plants provide potable water to citizens living in outlying populated areas that have no other source but brackish groundwater. The unit production cost is much less than that of sea-water desalination due to lower salinity ranges. New combined water and power plants (CWP) will rely on both oil and gas, with the share of gas-fired plants increasing over the projection period. About half of the total electricity capacity additions will be for new CWP plants with desalination units. Investment in new desalination capacity in Saudi Arabia, at almost $21 billion over the projection period, is equal to over half of total MENA desalination investment and almost one-third of the total investment in the power and water generation sector in Saudi Arabia.

Oil Supply

Overview

Saudi Arabia is the world's largest oil exporter and the largest producer. It holds one-quarter of global proven reserves, amounting to around 262 billion barrels. Since the 1980s, the country has often assumed the role of swing producer, adjusting its production in line with movements in global oil demand in order to stabilise prices. Saudi Arabia currently has around 12 mb/d of sustainable production capacity,[4] around 1 to 1.5 mb/d more than actual production. Most

4. Includes estimated NGL capacity of 1.3 mb/d and 50% of Neutral Zone crude production capacity, but excludes the Abu Sa'fah field jointly owned by Bahrain. According to the September 2005 edition of the IEA's *Oil Market Report*, Saudi crude oil production capacity currently stands at 10.5 to 11 mb/d. "Sustainable production capacity" is the level of production that can be reached within 30 days and sustained for 90 days.

of this capacity is heavy crude, which is less in demand than lighter grades. More than half of the country's production comes from Ghawar – the largest onshore oilfield in the world, which has been in production for more than 50 years. Prospects for development of Saudi reserves will have a major bearing on the evolution of the world oil market. Saudi Arabia's willingness and ability to invest in expanding and maintaining oil-production capacity will be a major determinant of future trends in international energy prices.

Saudi Arabia has a long history of oil exploration and production. Oil was first discovered in the 1930s and commercial production began at the Abqaiq field in 1946. Production was initially dominated by Aramco, a consortium of four international oil companies: Chevron, Texaco, Exxon and Mobil (now merged into Chevron and ExxonMobil). The assets of Aramco and other companies operating in the country were nationalised in 1981. Aramco became Saudi Aramco when it was formally incorporated as a Saudi enterprise in 1988. The only remaining foreign ownership in the upstream sector is Chevron's share of Saudi Arabian Texaco, which has a concession in the Partitioned Neutral Zone.[5] This concession is due to expire in 2009, but negotiations are underway to extend it.

Saudi crude oil production has fluctuated markedly in recent decades in response to changes in the global balance of supply and demand. Total production (including the country's share of output in the Neutral Zone and natural gas liquids) soared from 3.6 mb/d in 1970 to almost 11 mb/d in 1980, but fell back to only 3.8 mb/d in 1985 as a result of the country's policy of lowering its production to support official OPEC fixed prices. Saudi Arabia abandoned the OPEC pricing system at the end of 1985 and adopted a netback-pricing approach, whereby crude oil export prices were adjusted regularly according to changes in spot product prices in the main consuming markets. Production[6] rebounded to 5.5 mb/d in 1986, but fell back again the following year with the introduction of an OPEC production-quota system. Production since then has more or less followed the country's quota, rising steadily in the late 1980s and 1990s. It spiked in early 1991 in response to the loss of Iraqi and Kuwaiti exports following the invasion of Kuwait. Annual production peaked again at 9.4 mb/d in 1998, but was then lowered again as part of an OPEC deal to cut production. This was intended to force up prices, which had fallen to below $10 a barrel following the Asian financial crisis. Output dipped to 8.7 mb/d in 2002, but surged in 2003 in response to

5. A disputed area of coastal land lying between Saudi Arabia and Kuwait. The exploitation of onshore and offshore petroleum resources are shared by the two countries. In 2000, Saudi Arabian Texaco took over a concession in the Neutral Zone held since the late 1950s by the Japanese-owned company, the Arab Oil Company, in 2000.

6. Unless otherwise mentioned, all oil production figures quoted in this chapter include NGLs and condensates.

reduced supplies from Venezuela, Nigeria and Iraq. Output reached 10.4 mb/d in 2004, including 1.3 mb/d of NGLs and 0.3 mb/d of Neutral Zone output. On current trends, in 2005, it will approach its all-time record average annual level of 10.7 mb/d.

Saudi Aramco is proceeding with several projects aimed at maintaining output at existing fields and developing new fields. There is considerable potential to raise output in the coming decades given the country's large proven reserves – even without any reserve additions. In our Reference Scenario, oil production is projected to reach 11.9 mb/d in 2010 and close to 18.2 mb/d in 2030. This projection assumes investment is forthcoming. Total oil net exports (crude and refined products) grow from 8.3 mb/d in 2004 to 9.3 mb/d in 2010 and 14.4 mb/d in 2030. Doubts have been raised about whether actual capacity could increase this much because of production problems at the country's main fields, including Ghawar. The reliability of reserve estimates has also been questioned. If reserves are indeed as large as official estimates show, production increases will depend critically on the pace of investment, which is largely a matter of government policy. In the Deferred Investment Scenario, production rises more slowly, reaching only 14.1 mb/d in 2030.

Policy Framework

Saudi oil and gas policy is formulated by the Supreme Council of Petroleum and Mineral Affairs, established in January 2000. It is chaired by King Abdullah and its members are drawn from the government and the private sector. The Ministry of Petroleum and Mineral Resources is responsible for executing general policy related to oil, gas and minerals. The ministry supervises Saudi Aramco, which has a monopoly over upstream oil activities in all parts of the country except the Neutral Zone and a dominant position in the downstream oil sector. The ministry is responsible for all high-level oil-sector planning, budgeting and project decisions. It also monitors and regulates the activities of all other companies working in oil exploration, development, production, refining, transportation and distribution. These include Saudi Arabian Texaco, which is jointly owned by Saudi Aramco and Chevron. Other companies active in the Saudi oil sector include the Petroleum Services Company, a partially state-owned enterprise set up by the government in 2002 to carry out seismic and geological surveys, petroleum engineering and related services in the country. Some foreign companies have interests in joint-venture refineries and lubricant plants in Saudi Arabia.

Oil Minister Al-Naimi has indicated that additional projects to develop proven reserves have been identified that can be given the green light when market conditions are appropriate. These projects, he has claimed, would allow production of 15 mb/d to be sustained for at least 50 years.

Saudi Arabia has invested heavily in domestic refining capacity for export and in overseas refineries, in order to extract more value from its crude oil output. Product exports accounted for around 17%, or 1.4 mb/d, of total oil exports of 8.3 mb/d in 2004. All eight refineries in Saudi Arabia are wholly or partly owned by Saudi Aramco. The company also holds equity stakes in several refineries in the United States, Europe and the Far East, with total capacity of 1.8 mb/d. Saudi Aramco is expanding capacity at its domestic refineries and is considering building new export refineries. The company is also pursuing investments in downstream activities in China, India and other Asian countries.

Despite the implementation of economic reforms aimed at opening up markets to private domestic and foreign investors, including upstream gas activities, Saudi Arabia has made it clear that it does not intend to allow foreign participation in crude oil production (beyond the small existing Chevron stake in Saudi Aramco Texaco in the Neutral Zone) for the time being. Foreign involvement in the oil industry is a sensitive issue for ideological reasons. Moreover, the government contends that Saudi Aramco has no particular need for financial or technical help.

Resources and Reserves

Saudi Arabia has the largest proven reserves of oil in the world, estimated by Saudi Aramco at 262 billion barrels at the start of 2005, including a half share of the 5-billion barrel reserves of the Neutral Zone (*Oil and Gas Journal*, 2004; BP, 2005).[7] More than one-half are located in eight super-giant fields, including the onshore Ghawar field – the largest oilfield in the world. Ghawar's remaining proven and probable oil reserves are estimated at 86 billion barrels (at 1 January 2005) and original oil in place at around 300 billion barrels. Around 90% of proven reserves are in fields that are already in production, reflecting the fact that non-producing discovered fields have not been appraised in detail. All eight super-giants are located in the central and eastern provinces (Figure 16.8). Around two-thirds of reserves are classified by Saudi Arabia as light or extra-light grades of oil, while the rest are either medium or heavy. By the end of 2004, 109 billion barrels had already been produced.

Recoverable reserves are likely to be significantly higher. IHS Energy estimates remaining proven and probable (2P) reserves at 289 billion barrels. The US Geological Survey estimates ultimately recoverable resources (crude oil and NGLs), including undiscovered oil, remaining proven and probable reserves and cumulative production, at 511 billion barrels[8] (USGS, 2000). The Saudi oil minister recently stated that proven reserves will grow to around 420 billion

7. Throughout all 2005, and recently at the 18th World Petroleum Congress in Johannesburg, South Africa, Saudi officials reiterated that Saudi Arabia reserves may be substantially increased.
8. Including NGLs and a half share of Neutral Zone resources.

Figure 16.8: **Oil and Gas Infrastructure in Saudi Arabia**

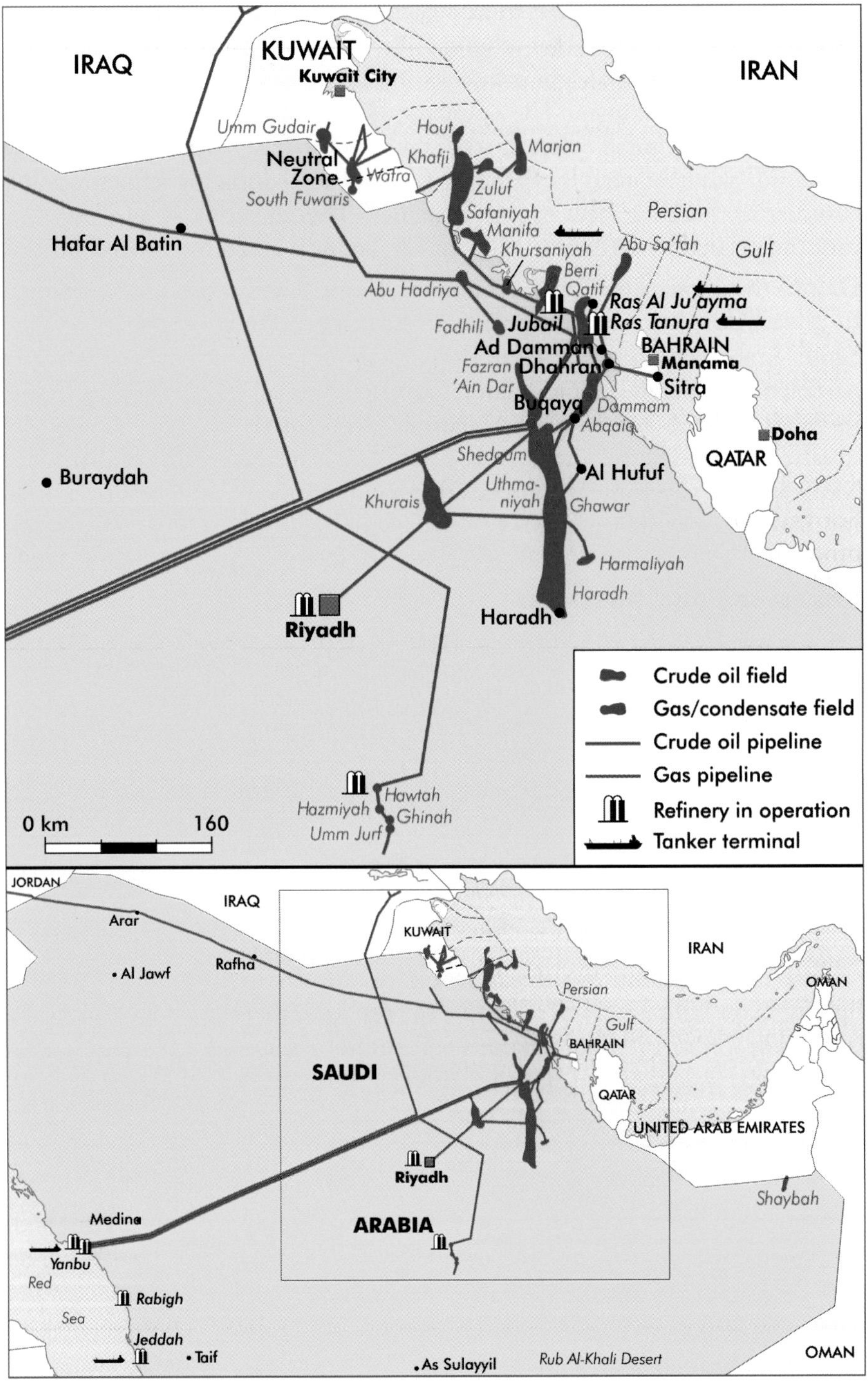

barrels by 2025 as a result of exploration and appraisal drilling. Parts of the country, including the region close to the border with Iraq, the Red Sea region and the vast Rub Al-Khali desert region (the Empty Quarter) in the south-east, have yet to be thoroughly explored.

There are two main geological zones in Saudi Arabia:

- The western third of the country, the Arabian Shield, which is made up of old basement rocks with no hydrocarbon potential, except for a narrow strip of sedimentary deposits in the Red Sea.
- The eastern two-thirds of the country, the Arabian platform, which is sedimentary, with the thickness of sedimentary rocks increasing towards the Persian Gulf. The entire region has high hydrocarbon prospectivity.

The Jurassic and Cretaceous horizons in the eastern part of the country are well explored, with the exception of Rub Al-Khali. Some deeper reservoirs, such as the Khuff gas reservoir in the Ghawar area, have been discovered, but generally the deeper (Permian) horizons are less well explored. Consequently, the Red Sea and the Rub Al-Khali regions, as well as deeper reservoirs in the main producing areas, still have potential for new discoveries. The eastern region holds a number of giant and super-giant fields, some of which are in production. Saudi Aramco reports that only half of proven reservoirs are currently developed. Half of current production comes from a single field, the Ghawar reservoir (Box 16.1).

Saudi Arabia has stepped up drilling since the late 1990s. The number of active rigs grew from 20 in 1999 to 33 in 2002 and, reportedly, as many as 60 in late 2004. Saudi Aramco has announced plans to increase the number of rigs to over 100 in 2006. The company drilled a total of 718 km of exploration and development wells in 2003, an increase of 16% over 2002. The number of wells drilled increased from 250 in 2002 to 290 in 2003. Five oilfields were discovered in the three years to 2003, compared with none at all in the period 1998-2000. Recent discoveries include two oil/gas fields south-east of Riyadh, one in the Khurais region and the other at Takhman (just south of the Ghawar field). In October 2004, Saudi Aramco announced that it had discovered a deposit, yielding 54° API super-light crude oil, in the Abu Sidr region south of Riyadh. Saudi Aramco has also commissioned extensive seismic and geophysical surveys to identify new structures and obtain better data on known formations.

Crude Oil Production

There are currently less than 30 fields in production in Saudi Arabia (Table 16.7), the largest of which (in terms of production) are Ghawar, Safaniyah, Shaybah, Abqaiq, Zuluf, Marjan, and Berri. Only about one-third of the initial reserves of these fields have been depleted. More than 80% of the oil ever produced in Saudi Arabia has come from Ghawar, Abqaiq and Safaniyah.

Table 16.7: **Oilfields in Saudi Arabia**

Field	Year of first production	Remaining proven and probable oil reserves at end-2004 (billion barrels)	Cumulative production to 2004 (billion barrels)	API gravity (degrees)
Abqaiq	1946	5.5	13.0	37.0
Abu Hadriyah	1961	1.2	0.6	34.7
Abu Sa'fah	1966	6.8	1.7	30.0
Berri	1967	15.3	3.1	33.0
Fadhili	1963	0.6	0.3	39.0
Ghawar	1951	86.3	60.7	34.0
Ghinah	1994	0.3	0.0	52.0
Harmaliyah	1973	1.9	0.2	35.0
Hawtah	1994	2.0	0.0	49.0
Hazmiyah	1994	0.5	0.0	47.5
Khurais	1963	16.8	0.2	36.0
Khursaniyah	1960	3.3	1.0	30.0
Manifa	1966	22.8	0.3	29.0
Marjan	1973	9.3	0.7	33.0
Qatif	1946	9.2	0.8	33.5
Safaniyah	1957	39.6	15.4	27.0
Shaybah	1998	20.7	0.8	42.5
Umm Jurf	1994	0.2	0.0	53.0
Zuluf	1973	18.2	1.8	32.0
Other fields	–	29.0	5.0	–
Total Saudi Arabia	–	**289.4**	**105.6**	–
Khafji	1961	1.7	1.8	28.2
Umm Gudair South	1968	0.2	0.2	24.3
Wafra	1953	0.5	1.2	21.2
Other Neutral Zone fields	–	0.3	0.6	–
Saudi Arabia share of Neutral Zone	–	**2.6**	**3.7**	–
Total Saudi Arabia including share of Neutral Zone	–	**292.1**	**109.3**	–

Sources: IHS Energy and IEA databases.

Sustainable production capacity (including NGLs) amounted to 11.5 to 12 mb/d in mid-2005, of which around 1 to 1.5 mb/d was unused. Most of the country's spare capacity is at three offshore fields – Safaniyah, Zuluf and Marjan – all of which produce medium or heavy crude oil. According to the *Oil and Gas Journal*, oil is currently produced from over 1 500 wells, nearly two-thirds of which are at Ghawar.

Ghawar is by far the most important field, currently producing 5.8 mb/d (Box 16.1). Safaniyah is the world's largest offshore oilfield, with estimated proven and probable reserves of 40 billion barrels and current production of 1.7 mb/d. Shaybah, a much more recent discovery located in the Rub Al-Khali bordering the United Arab Emirates with 21 billion barrels of proven and probable reserves, was brought into production in 1998 at a cost of $2.5 billion. The Shaybah complex comprises three gas/oil separation plants and a 630-km pipeline that connects the field to a gathering centre at Abqaiq. Output from Shaybah, which has a capacity of about 600 kb/d, helped to offset most of a drop in production at Ghawar and Safaniyah in the late 1990s. Shaybah crude is blended into Arab extra light from the Berri and Abqaiq fields. Saudi Aramco has managed its oilfields very conservatively over several decades in order to extend their plateau production for as long as possible – a long-standing objective of Saudi oil policy.

The quality of Saudi Arabian crude oil varies widely, ranging from 28° to 50° API and from 4% to less than 1% sulphur. Output is blended into five export grades (Table 16.8). About 80% of current production capacity is Arab light and extra light oil. Most light grades are produced onshore, while medium and heavy grades come mainly from offshore fields. The Ghawar field is the main producer of Arabian light crude, the main export grade. Abqaiq, along with Shaybah, is the principal source of Arab extra light. Since 1994, the Najd fields in the Hawtah Trend south of Riyadh have been producing in excess of 200 kb/d of 45-50° API, 0.06% sulphur Arab super light crude. Arab heavy crude comes mostly from the offshore Safaniyah field. Most Saudi oil production, with the notable exception of Arab super light, is sour, containing relatively high levels of sulphur.

Saudi Aramco's technical capabilities are widely respected, and it actively uses and develops new technology. For example, it was an early user of multilateral wells, pioneering what it calls "maximum reservoir contact" wells as a way to maximise production per dollar of capital invested. It has a very conservative approach to managing its reservoirs, generally aiming for fairly low depletion rates – less than 5% per year – in an attempt to maximise long-term recovery factors. This involves maximising field life to benefit from the impact of implementation of successive generations of technology.

Box 16.1: **The Ghawar Field**

The Ghawar field – the world's largest – was discovered in 1948 and started producing in 1951. The field area is partitioned into six geographical areas, from north to south: Ain Dar, Shedgum, Farzan, Hawiyah, Uthmaniyah and Haradh. Oil is produced from the Jurassic formations, namely Arab, Dhruma and Hanifa, while gas and condensate are extracted from the deeper and more ancient reservoirs, in the Khuff, Unayzah and Jauf formations. Ghawar is a large anticline[9] structure, 280 km long by 25 km wide, with about 50 metres of net oil pay. Initial oil in place is believed to amount to at least 300 billion barrels. Cumulative production is 61 billion barrels and, according to Saudi Aramco, remaining proven reserves are about 65 billion barrels. Saudi Aramco has not provided sufficient data for a proper assessment, but the recovery rate could be between 40% and 60%. Ghawar produced 5.2 mb/d of crude oil (excluding NGLs and condensates) in 2004, down from a peak of 5.7 mb/d in 1981 and a recent peak of 5.5 mb/d in 1997.

Reservoir pressure is maintained through the use of peripheral water flooding, whereby sea-water is injected into the reservoir in the oil layer just above the tar mat that separates the oil layer from the aquifer. The water pushes the oil inwards and upwards towards the producing wells. This technique, first used at Ghawar in 1965, results in lower flow rates than the more commonly used pattern water flooding, but tends to result in a higher recovery rate and allows for plateau production to be maintained for longer. As the field matures, maintaining reservoir pressure and sustaining production has become more difficult and costs have risen. The water cut – the share of water in the liquids extracted – increased sharply in the 1990s, reaching 37% in 2000. However, recent work has succeeded in reducing the water cut to 33% in 2003 and 31% in 2004.

Ghawar's current maximum sustainable crude oil (excluding NGLs and condensates) capacity is about 5.3 mb/d. The second phase of a project to develop the Haradh area in the southern part of Ghawar was recently completed, doubling capacity there to about 600 kb/d. A third phase is underway. Once a new gas/oil separation plant is completed in 2006, production from the zone could rise to about 900 kb/d. The project will also boost production of non-associated gas, natural gas condensates (by up to 170 kb/d), and sulphur. The first phase was completed in 1996.

Ghawar produces most of the country's Arab light crude. Like most oil produced in Saudi Arabia, Ghawar crude, which has a gravity of 34° API and sulphur content of 1.8%, is piped to the Abqaiq processing facility and then on to one of the country's ports at Ras Tanura, Al Juaymah and Yanbu for export. The Petroline pipeline system, which runs from Abqaiq in the east to Yanbu on the Red Sea, is 1 200 km long and has a capacity of 4.8 million b/d.

Ghawar also contains considerable volumes of natural gas, both associated and non-associated. The field currently produces most of the country's output of natural gas. Four gas-processing plants located on the Ghawar field handle all the gas produced in Saudi Arabia and transported through the Master Gas System.

Table 16.8: **Saudi Arabia's Crude Oil Export Streams**

Blend	**Average API gravity**	**Sulphur content** (%)	**Field**
Arab heavy	28°	2.90	Manifa, Safaniyah
Arab medium	30°	2.50	Abu Hadriyah, Abu Jifan, Abu Sa'fah, Khurais, Khursaniyah, Marjan, Zuluf
Arab light	33°	1.95	Ghawar, Harmaliyah, Qatif
Arab extra light	38°	1.20	Abqaiq, Berri, Qatif, Shaybah
Arab super light	45-50°	0.06	Ghinah, Hazmiyah, Hawtah, Umm Jurf

Source: IEA databases.

In the last two years, there has been growing speculation about the ability of Saudi Aramco to sustain output from its mature fields because of technical difficulties, notably falling pressures and increasing water cuts. For example, production from the offshore Safaniyah field has been partially curtailed due to water encroachment. The company estimates that the natural decline rate – the rate at which production would decline in the absence of any maintenance or new drilling – at existing fields is around 6%. If this is the case, about 600 kb/d of capacity has to be added each year just to maintain overall capacity. Studies by Matthew Simmons, a US investment banker, released in 2002 and 2005, question the ability of Saudi Aramco to prevent further rapid declines in output from its mature fields. He claims that Saudi oil production could be close to reaching its peak, if it has not already done so.

9. An arch-shaped fold in rock in which the rock layers are upwardly convex.

Saudi Aramco claims that the problems highlighted by Simmons's studies are exaggerated and that it is capable of substantially raising production, even without the need to prove up more reserves. It is addressing the problem of production decline and the goal of restoring spare capacity through a dual strategy of enhanced development of existing fields and development of new fields. The water cut at Ghawar and other fields, which is already low compared with giant oilfields in most other parts of the world, is being reduced through water management (Box 16.2). Recently completed projects to bolster output from fields already in production include further development of the Qatif and Abu Sa'fah fields, and the second phase of the Haradh project at Ghawar.

Box 16.2: **Water Management in a Giant Reservoir**

Since water injection begun in 1983, the water cut – the share of water produced together with the oil – at the Ghawar field gradually increased, reaching 37% in 2000. In other words, for every 63 barrels of oil produced, 37 barrels of water are also produced. This has prompted some commentators to forecast the imminent demise of this giant field, a prediction that Saudi Aramco disputes. The purpose of this box is to provide a non-technical introduction to water management in a giant reservoir of similar type to Ghawar to enable the reader to understand this controversy.

In a giant reservoir in a simple anticline structure, oil is found at the top and water at the bottom (Figure 16.9). Where the water does not provide sufficient pressure to maintain adequate oil-flow rates as oil is progressively extracted, additional water can be injected into the reservoir to compensate for the volume of oil extracted. This technique can help maintain production rates by keeping pressure above the bubble point, at which some of the oil turns into free gas, reducing the output of oil. It is often classified as a secondary recovery technique when used for water flood and is frequently practised from the start of production.

In order to maximise recovery, it is often best to locate the water injector wells at the periphery of the reservoir, very close to the point of oil/water contact. As oil is removed from the reservoir, water slowly rises, pushing the oil upwards, since oil with an API gravity of more than 10 floats on water. This "peripheral water flooding" approach differs from the common "five-spot pattern" approach, whereby a checkerboard of alternating injector and producer wells is drilled throughout the reservoir. The latter technique tends to yield higher production rates, but lower ultimate oil recovery. With peripheral water flooding, the producing wells are initially drilled at the periphery too, but on the side closest to the centre of the reservoir. Indeed, drilling producing wells too close to the top of the structure would increase the risk of water moving too quickly all the way to the top through high permeability paths, such as fractures or fissures, and bypassing the oil. In time, the injected water rises to the point where it reaches the producing wells, causing the water cut to increase and the volume of oil produced to fall.

An obvious solution to this problem, deployed by Saudi Aramco and the Abu Dhabi National Oil Company, is to turn off the initial circle of producing wells or convert them into injector wells and drill new producer wells higher up the structure, which will produce water-free oil. Another approach is simply to increase total fluid production, in order to maintain oil production in spite of raising water content. The arrival of water in the producer wells does not mean that there is no oil left between injectors and producers. In general, neither solution is optimal. One has to balance the costs of drilling new wells against the cost of handling more water. The latter can be expensive, especially if surface facilities have not been built to anticipate rising water cut. Typically, the best solution is a combination of letting the water cut increase slowly until it reaches the maximum existing handling capacity while slowly adding producing wells higher up in the structure. Data published by Saudi Aramco suggest that this is the strategy adopted at Ghawar over the past 10 years. However, Saudi Aramco has released no information on well positions, the capacity of surface facilities or drilling costs. Saudi Aramco has also drilled horizontal wells, as a way of increasing oil production at lowest cost: horizontal wells typically produce more oil per dollar invested than traditional vertical wells.

Most fields in the world produce a lot of water together with the oil. On average worldwide, the industry produces four barrels of water for every barrel of oil. In mature provinces such as the US and the North Sea, many old fields have a water cut of more than 90%. Water, as it is being circulated through the reservoir, continues to dislodge small amounts of oil. In an old field with depreciated facilities, as long as the cost of pumping and processing the water is less than the value of the oil produced, it is economic to keep producing even with a very high water cut. A water cut of 35% is far from signalling the end of production. But careful optimisation of the further development of the reservoir is needed in order to maintain production and minimise cost.

The Qatif and Abu Sa'fah developments, involving the addition of three gas/oil separation plants, new gas gathering lines and facilities for water injection, gas-processing and sulphur recovery, were completed in late 2004. Capacity

Fig 16.9: **Peripheral Water Flooding in a Simple Anticline Oil Reservoir Structure**

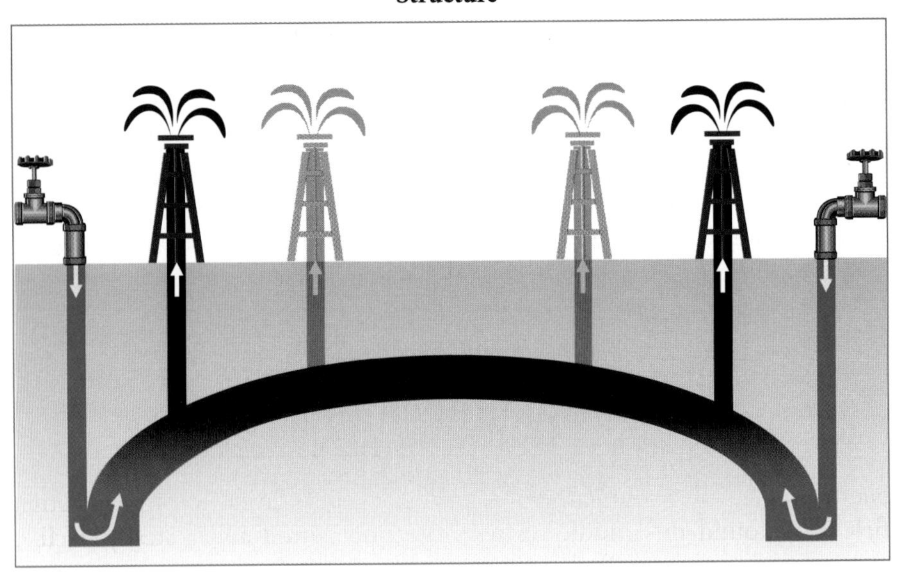

was boosted to 500 kb/d at Qatif, which produces Arab light, and by about 100 kb/d to 300 kb/d at Abu Sa'fah, which produces Arab medium. Output from Abu Sa'fah is shared with Bahrain.

Four major projects are currently underway, which together will add approximately 2.3 mb/d of capacity over 2006-2010 (Table 16.9). Part of this increase will compensate for declining capacity elsewhere. The third phase of

Table 16.9: **Current Upstream Oil Projects in Saudi Arabia**

Project	**Capacity addition** (kb/d)	**Completion date**
Haradh-3	300	2006
Khursaniyah	500	2007
Shaybah	300	2008
Khurais/Manifa	1 200	2009 - 2010
Total	**2 300**	

Source: Saudi Aramco (2005).

the Haradh development will be commissioned first, in 2006. The demothballing of the Khursaniyah field is expected to be completed in 2007 and the expansion of the Shaybah field in 2008. The largest increment, of 1.2 mb/d, will come from the Khurais field – one of five onshore fields mothballed by Saudi Aramco in the early 1990s. Khurais, a satellite of Ghawar, will be developed in parallel with the offshore Manifa field. The project is reported to be technically complex. Khurais will involve a major water-flood programme, while the heavy crude from Manifa has high metals and hydrogen sulphide content, requiring special processing. Combined proven reserves in Khurais and Manifa amount to 40 billion barrels. Several other mothballed fields – Abu Hadriyah (with 1.2 billion barrels of reserves), Abu Jifan (0.56 billion barrels), Fadhili (0.6 billion barrels) and Harmaliyah (2 billion barrels) – could be brought back on line at relatively low cost.

As the mature fields age, it will become harder and more expensive to maintain production. Our field-by-field production analysis, which takes account of Saudi Aramco's development plans, points to a peak in output from existing fields by around the middle of the projection period and a steady decline thereafter (Figure 16.10). Future production rates in large fields are very difficult to forecast as they basically depend on expectations about the global oil market and the call on MENA supply. These projections should, therefore, be treated as indicative.

The Abqaiq field, one of the world's largest oilfields with slightly less than 19 billion barrels of proven plus probable recoverable reserves, started production in 1946 and peaked at more than 1 mb/d in 1973. It maintained output at 0.7-0.8 mb/d in the 1970s and then dropped to less than 0.2 mb/d in the mid-1980s, as Saudi Arabia cut its output to help support oil prices. Output recovered in the 1990s, reaching an estimated 0.43 mb/d in 2004. The field is thought to have a maximum sustainable capacity of more than 0.45 mb/d at present. We project Abqaiq production to remain flat at around 0.44 mb/d through to 2010 and fall off gradually thereafter.

Berri, another super-giant with 18 billion barrels of recoverable reserves, increased its production rapidly from 1967, when it came on stream, to 1976, when it peaked at 0.77 mb/d. A low of 0.1 mb/d was recorded in 1985, but production then started to rise again. In 2004, it produced slightly more than 0.2 mb/d, even though its maximum sustainable capacity is as high as 0.45 mb/d. We see the field peaking again in 2020 at 0.43 mb/d and then gradually declining.

Ghawar (see Box 16.1) is projected to produce around 6 mb/d (including NGLs and condensates) over the next few years after the completion of the Haradh Phase-3 project in 2006, and then start to decline progressively, dropping to 3.6 mb/d in 2030. However, given its size, it is likely that future developments will increase sustainable capacity further. These projections do not take account of any new reserve additions that may be made, or the impact of enhanced recovery techniques that may be deployed in the future.

The maximum sustainable capacity of the Manifa field, which produces heavy crude oil, was raised to 0.3 mb/d in the 1990s, though production is currently lower. We believe that capacity will be further increased in the next ten years,

Table 16.10: **Saudi Arabia's Crude Oil Production in the Reference Scenario** (mb/d)

	2000	2004	2010	2020	2030
Currently producing fields	**9.286**	**10.354**	**11.850**	**13.142**	**10.081**
Abqaiq	0.443	0.434	0.438	0.362	0.359
Abu Hadriyah	0.000	0.000	0.070	0.198	0.014
Abu Sa'fah	0.125	0.189	0.284	0.278	0.253
Berri	0.222	0.213	0.351	0.425	0.387
Fadhili	0.040	0.050	0.050	0.041	0.017
Ghawar	4.898	5.772	6.048	5.519	3.654
Ghinah	0.041	0.041	0.039	0.000	0.000
Harmaliyah	0.000	0.028	0.054	0.152	0.109
Hawtah	0.029	0.026	0.029	0.031	0.031
Hazmiyah	0.060	0.059	0.051	0.005	0.000
Khurais	0.000	0.000	0.260	1.183	1.183
Khursaniyah	0.000	0.000	0.000	0.250	0.221
Manifa	0.040	0.050	0.100	0.360	0.360
Marjan	0.216	0.223	0.250	0.320	0.026
Qatif	0.000	0.100	0.375	0.500	0.413
Safaniyah	2.011	1.728	1.627	1.472	1.331
Shaybah	0.432	0.492	0.560	0.642	0.581
Umm Jurf	0.010	0.010	0.010	0.003	0.000
Zuluf	0.385	0.407	0.440	0.712	0.586
Other fields*	0.335	0.530	0.815	0.687	0.557
New developments	**0.000**	**0.000**	**0.000**	**2.258**	**8.069**
Fields awaiting development	0.000	0.000	0.000	1.129	4.639
Reserve additions and new discoveries	0.000	0.000	0.000	1.129	3.429
Total	**9.286**	**10.354**	**11.850**	**15.400**	**18.150**

* Includes Saudi Arabia's share of the Neutral Zone.

Note: Includes NGLs and condensates.

Sources: IHS Energy databases; IEA analysis.

when the development of new export-oriented refineries able to process heavy blends will be completed. The field has enough reserves for capacity to be increased much more, possibly to as much as 3 mb/d according to Saudi Aramco executives. This possibility is not reflected in our projections, as demand for heavy, sour crude oil on the international market is not expected to be strong enough to absorb a large increase in output from Manifa.

Safaniyah started to produce in 1957, peaked in 1980 at 1.56 mb/d, but quickly fell back to around 0.5 mb/d. According to IHS Energy, production recovered and peaked again in 1998, exceeding 2 mb/d. Today, the field is believed to have a maximum sustainable capacity of around 1.7 b/d. Given the size of the field's reserves, it is likely to maintain production at this level for many years.

Shaybah has been operating close to its maximum capacity level ever since it came on stream in 1998. It produces 0.5 mb/d of light, sweet crude. Although an additional 300 kb/d of capacity will be added by 2008, this will partly offset natural declines from existing wells. There is, nonetheless, scope for boosting output further, but this will require the capacity of the existing export pipeline to be expanded. Assuming that the necessary investment in this line and in drilling new wells is forthcoming, production could increase by up to 50% in the next ten years. Associated gas production is currently reinjected to sustain reservoir pressure.

Zuluf, discovered in 1965 and brought on stream in 1973, is currently producing just over 0.4 mb/d. Production could be boosted to 0.7 mb/d, a rate that we believe will be reached in the late 2010s and maintained for several years. The Hawtah Trend fields, brought into production in 1994, are currently producing around 0.026 mb/d of crude oil and NGLs. Output is projected to remain at a plateau of around 0.03 mb/d through to the end of the projection period.

Total production from currently producing fields is projected to peak in the middle of the next decade and then to begin to decline, reaching 10.1 mb/d by 2030. Enhanced oil recovery is expected to temper these declines somewhat, though it will become progressively more economic to develop new fields.

At present, about 70 fields, holding more than 10% of the country's proven and probable reserves, await development. This will, however, require large investments in new processing facilities, as well as gathering and trunk pipelines to connect to export terminals. Emphasis is expected to be placed on developing fields with light crude oil reserves. Nuayyim, in the centre of the country, may be the next virgin field to be developed. It could produce up to 100 kb/d of Arab super light grade oil. New field developments are projected to account for more than one-quarter of total Saudi oil production in 2030 (Figure 16.10).

Figure 16.10: **Saudi Arabia's Crude Oil Production by Source in the Reference Scenario**

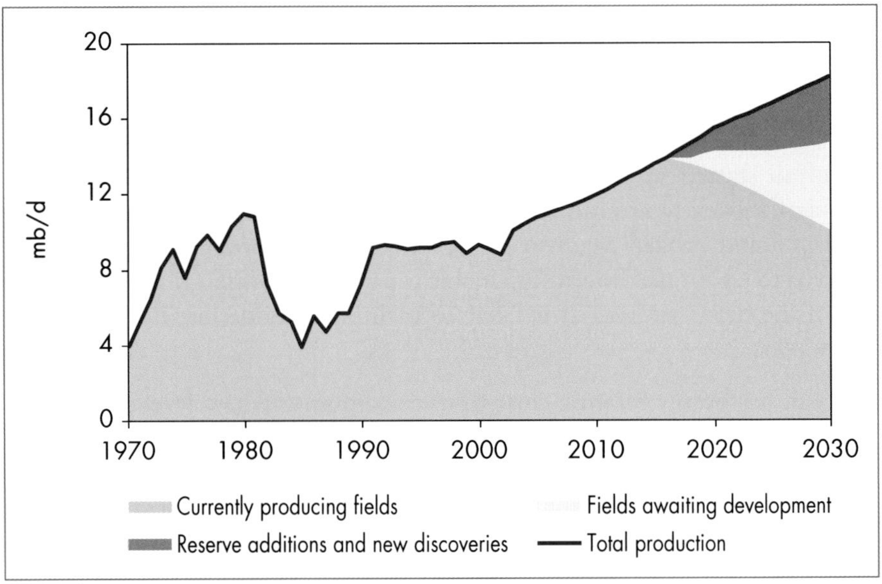

Note: Includes NGLs and condensates.

Oil Refining

The country has eight refineries, most of them built in the 1980s, with a combined crude throughput capacity of around 2.1 mb/d at the beginning of 2005 (Table 16.11). Saudi Aramco envisages a further rapid increase in capacity, through long-planned improvements at existing refineries and the construction of new export plants. These projects were put on hold after oil prices collapsed in 1998 and Saudi Aramco's revenues dropped. In the Reference Scenario, distillation capacity is projected to reach 2.6 mb/d in 2010, based on current plans, 3.4 mb/d in 2020 and 4.5 mb/d in 2030.

Saudi Aramco currently has a combined capacity of more than 1.3 mb/d at its five refineries at Ras Tanura, Rabigh, Yanbu (which serves the domestic market), Riyadh and Jeddah. It also has a 50% stake in two joint-venture export-oriented plants: Samref at Yanbu with ExxonMobil, and Sasref at Jubail with Petromin and Shell. Together, they account for 0.67 mb/d of refining capacity. This capacity was reached after a new 200-kb/d condensate splitter at the Ras Tanura refinery was completed in August 2003.

Saudi Aramco plans to upgrade its Rabigh refinery on the Red Sea, to reduce output of heavy products and boost that of gasoline and kerosene. The expansion project includes the addition of an ethane cracker and other

Table 16.11: **Saudi Arabia's Oil Refineries, at 1 January 2005**

Location	**Owner**	**Crude oil distillation capacity** (kb/d)
Ras Tanura	Saudi Aramco	525
Rabigh	Saudi Aramco	425
Yanbu (Samref)	Saudi Aramco/Mobil	400
Jubail (Sasref)	Petromin/Shell	305
Yanbu	Saudi Aramco	190
Riyadh	Saudi Aramco	120
Jeddah	Saudi Aramco	60
Ras Al-Khafji	Aramco Gulf Operations Company	30
Total		**2 055**

Sources: Saudi Aramco (2004); Arab Oil and Gas Research Centre (2005).

petrochemical facilities (under a joint-venture deal with Japan's Sumitomo Chemical). Upgrades are also underway at Saudi Aramco's Riyadh and Yanbu plants, in order to boost the company's total refining capacity to 2.5 mb/d by the end of 2006. The addition of hydrotreating units at both these plants will enable them to produce products, in particular gasoil, with a significantly lower sulphur content. A similar addition is planned soon for Ras Tanura. In April 2005, Saudi Aramco unveiled plans to build a new refining complex to be located in the Red Sea city of Yanbu.

Box 16.3: **New Export Refinery in Yanbu**

Saudi Aramco announced, at an energy conference in Dubai in April 2005, its intention to build a new export-oriented refinery in Yanbu. The new plant, with a capacity of 400 kb/d, will require $4-5 billion of capital expenditure and will be built as a joint venture with one or more international partners. This will be Saudi Aramco's third joint venture in the downstream oil sector. The main purpose of this project is to process heavy crude oil to produce light products, notably high-quality gasoline for export to the east coast of the United States, naphtha for supply to the Far East, and low-sulphur gasoil for Europe.

The Indian Oil Corporation and the Hindustan Petroleum Corporation are two of the potential investors in this venture. Saudi Aramco may consider swapping equity in the new Yanbu refinery with an interest in HPCL's Vizag refinery or IOC's Paradip refinery, both in India.

Exports

In 2004, Saudi Arabia exported 8.3 mb/d of oil, of which 6.9 mb/d was crude. About 46% of all the oil exported was shipped to Far East markets and the rest mainly to the United States (20%) and Europe/Mediterranean (16%). Exports are still below the peak reached in 1981. Exports are projected to grow strongly over the projection period in line with higher production. In the Reference Scenario, total oil exports will increase to 9.3 mb/d in 2010 and will then climb to 14.4 mb/d in 2030 (Figure 16.11). By 2030, approximately 20% of exports will be in the form of oil products.

Figure 16.11: **Saudi Arabia's Oil Balance in the Reference Scenario**

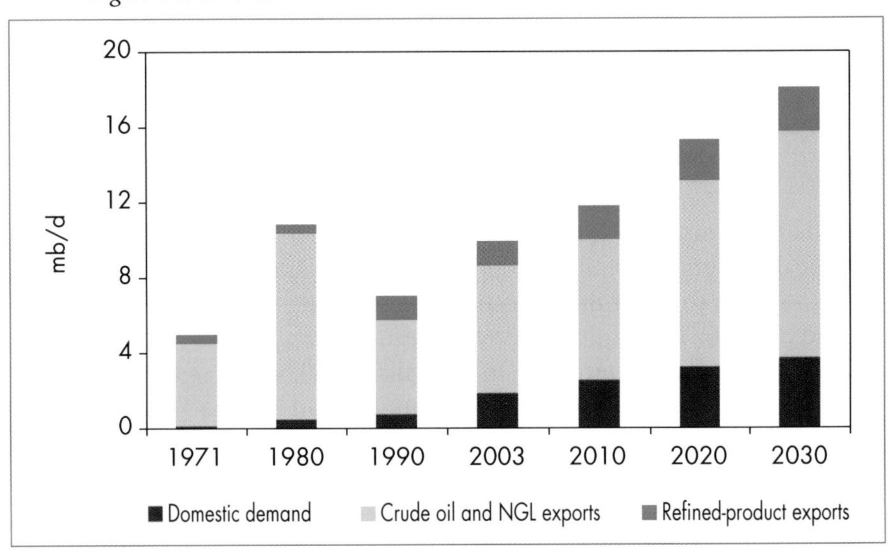

Most crude oil exports pass through the huge Abqaiq processing facility, which handles around two-thirds of the country's oil output. The main export terminals are at Ras Tanura and Ras Al Ju'aymah on the Persian Gulf and Yanbu on the Red Sea. Ras Tanura, the world's largest oil-loading terminal, has a capacity of 6 mb/d. Ras Al Ju'aymah can handle up to 3 mb/d and Yanbu up to 5 mb/d. There is a smaller Gulf port at Az Zuluf and two others at Ras Al Khafji and Mina Saud in the Kuwaiti sector of the Neutral Zone. Total export capacity of over 14 mb/d is sufficient to handle the projected growth in output, at least for the next decade. There is also ample spare capacity in the internal pipeline system to transport crude oil and NGLs from processing plants to domestic refineries and export terminals. The Saudi authorities are nonetheless planning to conduct a feasibility study on the construction of an oil pipeline from the Empty Quarter of south-eastern Saudi Arabia to the port of Hamaradam on the Arabian Sea in Yemen.

Saudi Aramco's shipping subsidiary, Vela, has the world's largest fleet of oil tankers, including 19 VLCCs (very large crude carriers) and 4 ULCCs (ultra large crude carriers). Overall, Vela carries around half of Saudi oil exports. In September 2004, the Saudis placed a $200 million order for two VLCCs from Hyundai Heavy Industry, with delivery expected in 2007. In addition to tankers, Saudi Aramco owns or leases oil storage facilities around the world, in Rotterdam, Sidi Kerir (the Sumed pipeline terminal on Egypt's Mediterranean coast), South Korea, the Philippines, the Caribbean, the United States and elsewhere.

Natural Gas Supply

Resources and Reserves

Saudi Arabia has vast resources of natural gas. Proven natural gas reserves are estimated at 6.7 trillion cubic metres at the beginning of 2005 – the fourth-largest in the world after Russia, with 48 tcm, Iran, 28 tcm, and Qatar, 26 tcm (Cedigaz, 2005). About 60% of these reserves consist of associated gas, mainly from the onshore Ghawar field and the offshore Safaniyah and Zuluf fields. The Ghawar oilfield alone holds one-third of the country's total gas reserves. Most non-associated gas reserves are located in the deep Khuff and pre-Khuff formations, underlying the Ghawar oilfield. Non-associated Khuff gas has also been found in the north-west, at Midyan, and in the Rub Al-Khali region. Saudi Aramco believes Rub Al-Khali alone could contain as much as 8 tcm of recoverable gas resources.

There has been very little exploration for non-associated gas until recently, so the ultimate gas production potential is likely to be considerably higher than current reserve estimates suggest. The US Geological Survey estimates that the country may contain 19 trillion cubic metres of undiscovered gas (USGS, 2000). Saudi Aramco has made a series of discoveries of non-associated fields as a result of a number of test wells drilled in the vicinity of Ghawar underneath oil reservoirs. Other discoveries have been made recently in the eastern region in the northern dome of the Abqaiq field and in the Ghazal zone of the Marjan field. In 2004, a natural gas and condensate reservoir was found at the Shaybah oilfield. In the last ten years, natural gas reserves in Saudi Arabia have grown by nearly 30%, or 2.5% per annum.

Gas Production and Distribution

Natural gas is expected to make a growing contribution to Saudi Arabia's energy mix over the projection period. Production is projected to rise from

60 bcm in 2003 to 86 bcm in 2010 and 155 bcm in 2030. All of this gas will be used domestically. The share of gas in the country's total primary energy demand will rise from 38% in 2003 to 44% in 2030 (Figure 16.12). This will reverse the trend of the past decade or so, when gas-production capacity trailed the growth in domestic demand, primarily because gas output was constrained by oil-production rates. The government is now giving priority to the rapid development of the country's large non-associated natural gas resources.

Figure 16.12: **Natural Gas Supply and Share of Primary Energy Demand in Saudi Arabia in the Reference Scenario**

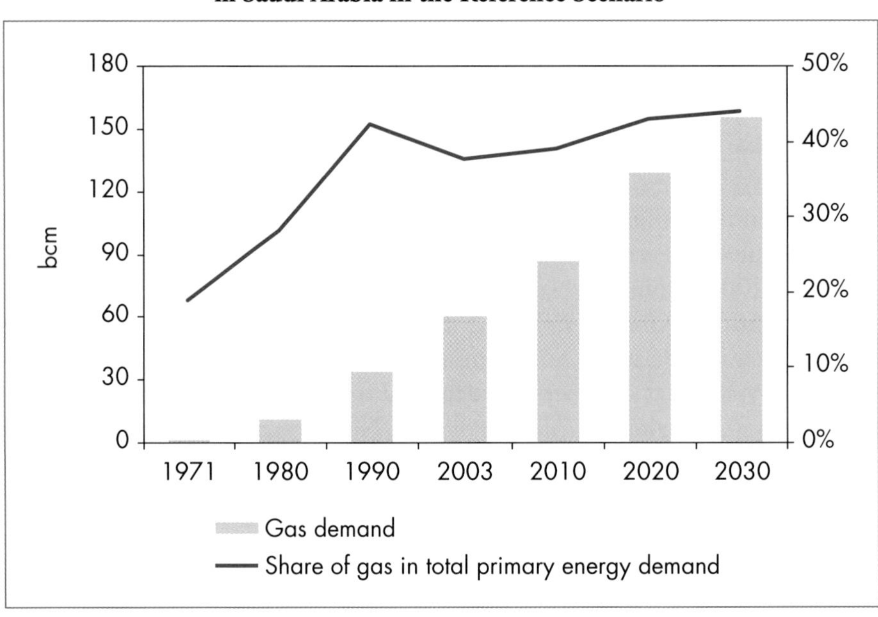

The production of natural gas in Saudi Arabia has always been closely linked to oil production. Almost all of the country's gas output is associated with oil production. For many years, virtually all the gas produced was flared. In 1975, the government directed Saudi Aramco to build an integrated gas-gathering network for the associated gas produced at the largest fields. The network, known as the Master Gas System (MGS), came on stream in 1982. Initially, the MGS handled only associated gas from Ghawar, but the system was expanded in the mid-1980s to take non-associated gas from the Khuff reservoirs, the gas caps at Abqaiq and Ghawar, and associated gas from Safaniyah and Zuluf. The MGS feeds gas to petrochemical and other industrial plants at Yanbu on the Red Sea and Jubail on the east coast. Demand for gas as an industrial feedstock and, increasingly, as a fuel for power generation and water desalination has since grown strongly, outstripping associated production. Very little associated gas is now flared.

Growing industrial and power-generation needs, in part linked to water demand and the need to expand desalination capacity, will require a major increase in gas-production capacity. Much of that additional capacity will come from non-associated gas fields, the development of which has been prioritised by the government. In 1997, the government drew up plans to invite international oil companies to develop new large-scale integrated gas projects, known as the Saudi Gas Initiative (SGI). It was abandoned in 2003 because of a failure to reach agreement on investment terms.

The government subsequently repackaged the project as a series of smaller, stand-alone upstream projects, with more attractive financial terms. Sabic and Saudi Aramco are expected to take on most of the downstream water, power, and petrochemical-related components that had initially formed part of the SGI.

The Saudi government has signed four deals with foreign oil companies – the first since Aramco was nationalised – to explore for and develop non-associated gas resources. In late 2003, a deal was negotiated with Shell and Total covering the Shaybah and Kidan areas of the Rub Al-Khali region. This area had originally been part of the SGI Core Venture 3. Shell has a 40% share, Total 30% and Saudi Aramco the remaining 30% share in the consortium, the South Rub Al-Khali Company, which is developing the project. It involves investment of about $2 billion.

In 2004, Russia's Lukoil signed a deal to explore for and develop non-associated gas in the northern part of the Rub Al-Khali desert south of Ghawar. It holds an 80% share in the Luksar project, with Saudi Aramco taking the remaining 20%. In the same year, China's Sinopec and a consortium made up by Italy's ENI and Spain's Repsol were each granted exploration and production licences. The acreage in all three of the 2004 deals, which were reached following an open tender procedure, had formed part of SGI Core Venture 1.

In all these agreements, Saudi Aramco will buy dry gas on a take-or-pay basis for $0.75 per million Btu. The condensates and NGLs extracted from the gas will be sold at international market rates. The Saudi government has agreed to pay the costs of connecting the fields to the MGS.

Current gas processing capacity in the country is around 90 bcm/year. There are plans to expand it to 100 bcm/year by 2007. Saudi Aramco intends to build a further 3 000 km of gas pipelines by 2006 as part of the continued expansion of the MGS.

Abqaiq produces associated gas, which is injected into the MGS, and has large non-associated high-sulphur gas reserves that, being sour, are currently not considered commercial. The gas to oil ratio at Abqaiq is estimated at a relatively high 15 cubic metres per barrel produced.

The super-giant Berri field contains huge volumes of gas, both associated and non-associated. At present, production of 1.5 bcm/year is limited to associated gas. One of Saudi Arabia's gas processing plants, which also receives gas from Abqaiq, is located at Berri. It processes large volumes of NGLs which are then sent to the refineries. All the remaining dry gas is transmitted to the MGS.

A huge amount of associated and non-associated gas – more than 60 bcm in 2003, according to IHS Energy – is produced in Ghawar and then transferred almost entirely to the MGS. Four gas-processing plants are installed at the field site. Most of the associated gas produced in Safaniyah (4 bcm/year) and Zuluf (around 3 bcm/year) is shipped through the MGS and processed in Berri, though some of the Safaniyah gas is reinjected to support offshore oil production.

Energy Investment

The amount of investment needed to support the growth in Saudi energy supply projected in our Reference Scenario will total $332 billion (in year-2004 dollars) from 2004 to 2030, or $12.3 billion per year. Oil will require the largest amounts of capital, averaging $6.5 billion per year, or over half of the total needs. It will be higher than the sum of gas and electricity investment. Gas investment needs will total $1.8 billion per year, while electricity investment requirements will amount to $4.1 billion per year.

Projected investment in the energy sector is equal to 2.2% of Saudi GDP. Finance will come from various sources: Saudi Aramco's cash flow; borrowing from commercial banks; initial public offerings, which are becoming more common in the local stock market; and foreign direct investment (FDI) for projects in petrochemicals, power and water. However, Saudi Arabia does not have the same need for foreign direct investment as many other developing countries, thanks mainly to its oil exports, which on average are worth three times the value of imported goods and services, and to the ample liquidity of the local economy.

The share in Saudi Arabia's GDP of total investment in fixed assets increased sharply in 2005, having been broadly constant during the last 15 years at around 18%. This level has just been sufficient to replace the naturally depleting capital stock and was accompanied by little or no economic growth. Now the picture is changing. Historically, in countries experiencing rapid economic expansion, investment can reach as much as 50% of GDP. An increase in investment to around 25% of GDP in Saudi Arabia would certainly be manageable and would allow existing capital stock to be replaced, boosting economic growth and job creation. However, not all capital expenditures play an equal role in advancing growth: it is crucial to have a balanced mix of investment among the public sector, the private sector and the oil sector. Samba, a local bank, reckons that an efficient mix of investment would assign 20% of total investment to the oil sector (Samba, 2005).

Table 16.12: **Saudi Arabia's Investment Mix** (% of total)

	1990	**1995**	**2000**	**2005 (target)**
Government	56.8	26.9	13.2	20.0
Non-oil private sector	37.5	57.3	75.2	60.0
Oil sector	5.7	15.8	11.5	20.0

Sources: Samba Financial Group (2005); Central Department of Statistics, Saudi Ministry of Economy and Planning (2004).

Oil

Cumulative investment needs in the Saudi oil industry over 2004-2030 are estimated at $174 billion (in year-2004 dollars). The upstream sector will continue to absorb most oil investment: more than 80%, or $141 billion, over the projection period. Investment needs will be much larger in the second half of the projection period, when crude oil production capacity is expected to grow most rapidly. New field developments will be increasingly preferred over projects to boost output at existing fields, as the cost of the latter rises progressively.

Average annual investment needs in the upstream oil sector will amount to $5.2 billion over the projection period (Figure 16.13). Over the rest of the current decade, around $3.4 billion per year will be needed to reach the 12.5 mb/d of production capacity projected for 2010. This is in line with Saudi Aramco's own estimate of its upstream investment needs. In spring 2005, the company revised its plan for upstream expenditures in the next five years. It now projects that investment needs will amount to $2.5 billion in 2007, $2.4 billion in 2008, $2.1 billion in 2009 and $1.9 billion in 2010.

In the decade after 2010, we expect upstream investment needs to increase sharply along with the projected increase in production of 3.6 mb/d between 2010 and 2020. They will average $5 billion per year. Investment needs will go up to $6.7 billion a year in the last decade of the *Outlook* period. Depending on the government's dividend policy, future Saudi Aramco investments can be financed solely out of the company's own cash flow.

Figure 16.13: **Saudi Arabia's Oil and Gas Investment in the Reference Scenario**

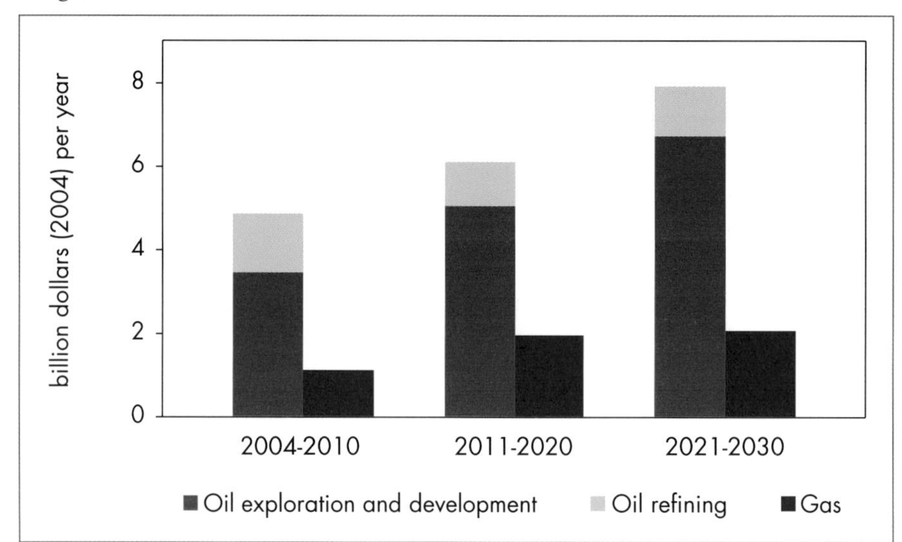

Natural Gas

The projected trends in gas production imply a need for cumulative investment of $48 billion over 2004-2030, most of it in the upstream sector. Annual expenditures will grow over time. The Saudi government is giving priority to exploring for and developing non-associated gas. Much of the capital needed for upstream gas projects is expected to come from foreign investors.

Electricity

The power sector in Saudi Arabia will need to invest a total of $110 billion in generation, transmission and distribution over 2004-2030. Generation will account for 41%, transmission for 18% and distribution for 41%. Saudi Arabia expects the private sector to provide much of the capital needed in the power sector. While efforts are being made to establish an attractive investment climate for private investors, it is not clear yet how quickly the private sector will respond. Some large projects may be difficult to finance because of their complexity.

Deferred Investment Scenario[10]

Saudi Arabia will not have a problem in funding the capital expenditures needed in the energy sector to boost supply as projected in our Reference Scenario. But it is possible that investment will, in fact, turn out to be lower.

10. See Chapter 7 for a detailed discussion of the assumptions and methodology underlying the Deferred Investment Scenario.

This may occur because of delays in the investment decision-making process or because of a deliberate policy of holding back the development of the country's resources to slow the rate of depletion. This policy may be justified by the goal of boosting international oil prices and export revenues, or by the aim of conserving resources for future generations. In 2001, the latest year for which firm data are available, investment in the oil sector accounted for 11% of total investment in the economy.

In our Deferred Investment Scenario, we analyse what would happen if the ratio of upstream oil investment to GDP were to remain constant over the projection period at the average level of the last ten years. This translates into a 20% reduction in cumulative upstream spending for the 2004-2030 period compared with the Reference Scenario. The cumulative investment in the oil upstream sector would fall from $141 billion to $115 billion.

Energy Demand

In the Deferred Investment Scenario, total primary energy demand in Saudi Arabia grows at an average annual rate of 2.8% per year over the projection period – 0.3 percentage points lower than in the Reference Scenario. Demand reaches 277 Mtoe in 2030, compared with 289 Mtoe in the Reference Scenario (Figure 16.14). Slightly lower GDP growth rates explain most of the divergence in energy demand trends between the two scenarios. Higher end-use prices in Saudi Arabia also contribute marginally to slower demand growth. The shares of oil and gas in primary demand are similar in the two scenarios.

Figure 16.14: **Saudi Arabia's Primary Energy Demand in the Reference and Deferred Investment Scenarios**

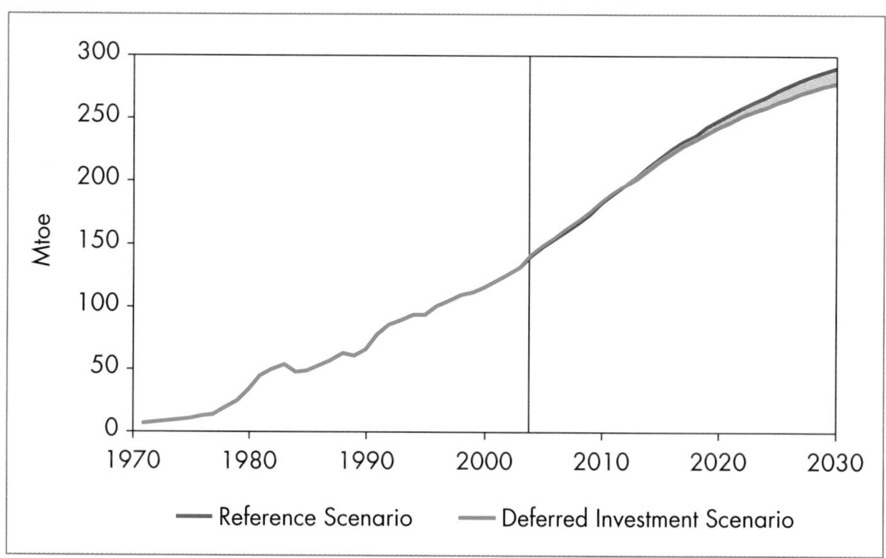

16

Oil Production and Exports

The assumption that upstream investment over the projection period is constant in relation to GDP at the level of the last ten years results in an average level of investment some 20% lower than in the Reference Scenario. Unsurprisingly, this leads to much slower growth in production of crude oil and NGLs. As a result, oil production is 14.1 mb/d, or 23%, lower in 2030 compared with the Reference Scenario (Figure 16.15).

Figure 16.15: **Saudi Arabia's Oil Production by Source in the Deferred Investment Scenario**

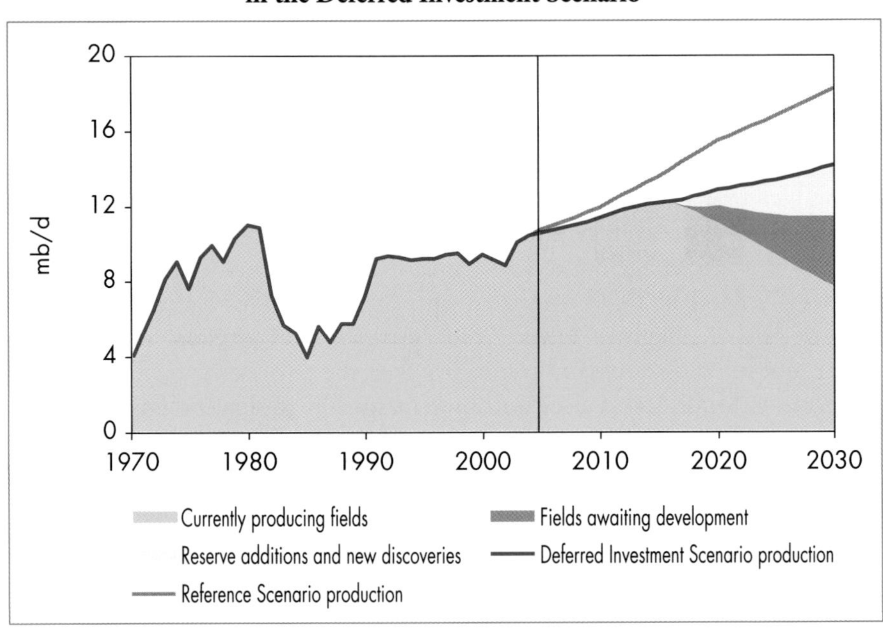

New production from fields awaiting development is 0.9 mb/d lower than in the Reference Scenario. Output from existing fields falls more than total production in percentage terms. It grows slightly to about 2015 and then begins to decline gradually, reaching just over 7.6 mb/d compared with nearly 10.1 mb/d in the Reference Scenario. The share of production from new developments in total Saudi oil production in 2030 is stable, at 19% in the Deferred Investment Scenario compared to the Reference Scenario.

The slower growth in domestic oil demand in the Deferred Investment Scenario leads to a less rapid expansion of refining capacity, as lower global demand for oil reduces the opportunity for Saudi Arabia to export oil products. Consequently, capacity grows by 78% over 2004-2030, reaching 3.7 mb/d, compared with more than doubling to 4.5 mb/d in the Reference Scenario.

Oil exports fall even more than production in percentage terms, but slightly less in volume terms – because lower inland consumption helps to compensate for the fall in production. Net oil exports are 3.9 mb/d, or 27%, lower in 2030 in the Deferred Investment Scenario, reaching 10.5 mb/d in 2030 (Figure 16.16).

Figure 16.16: **Saudi Arabia's Oil Balance in the Reference and Deferred Investment Scenarios**

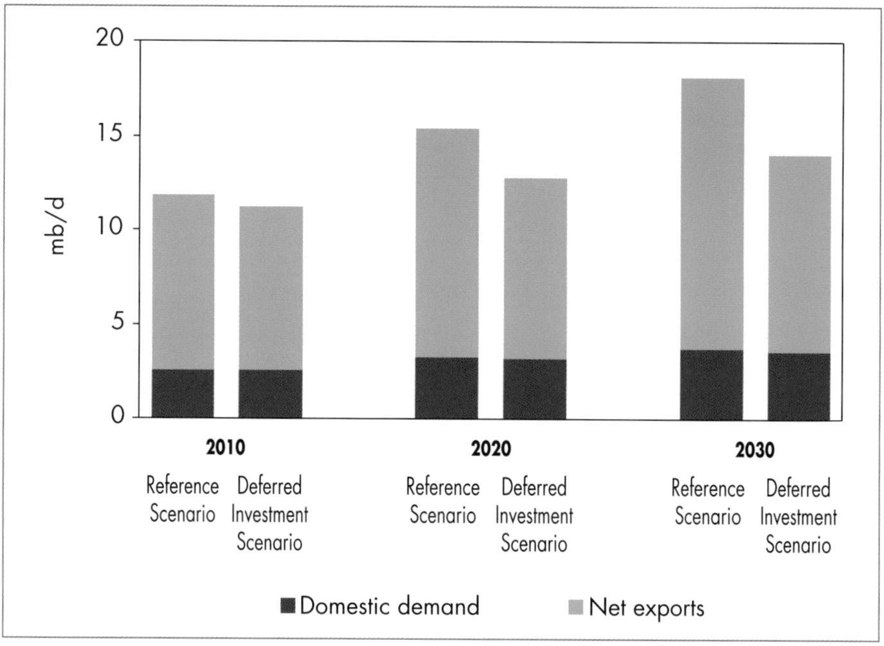

Natural Gas Production

Lower oil output results in lower natural gas production from associated oil/gas fields. Gas production from non-associated fields also increases slightly less in the Deferred Investment Scenario, since domestic demand is lower because of slower economic growth. Total gas production is 5 bcm, or 3%, lower in 2030 than in the Reference Scenario.

UNITED ARAB EMIRATES

HIGHLIGHTS

- The United Arab Emirates, a federation of seven emirates, has achieved both strong growth and extensive diversification of its economy over the past three decades, though the hydrocarbon sector remains a key component of the economy. The country's stable political environment and economic policies have contributed to these trends.
- In the Reference Scenario, the UAE's total primary energy demand, which is already high in per capita terms, is projected to grow by 2.9% per year over the projection period, reaching 84 Mtoe in 2030 – more than double the current level. Natural gas will remain the dominant fuel, meeting more than 80% of energy needs in 2030.
- Power generation will expand rapidly in response to burgeoning demand in the residential and the services sectors. Electricity output will rise from 50 TWh in 2003 to 128 TWh in 2030, requiring $35 billion in power plants and networks. The UAE is MENA's second-largest producer of desalinated water. Fuel requirements for desalination will rise from 9 Mtoe in 2003 to 16 Mtoe, or nearly one-fifth of total primary energy demand, in 2030.
- The UAE's abundant proven oil reserves, at 98 billion barrels or 8% of the world total, will allow for an expansion of oil-production capacity, assuming continued participation of foreign companies. In the Reference Scenario, oil production climbs from 2.7 mb/d in 2004 to 3.2 mb/d in 2010 and 5.1 mb/d in 2030. Exports rise to 2.9 mb/d in 2010 and 4.7 mb/d in 2030. In the Deferred Investment Scenario, output is more than a third lower in 2030 and exports correspondingly grow much less rapidly.
- The increasing need for reinjection in oilfields will partly limit the expansion of marketed gas production, which will rise from 44 bcm in 2003 to 59 bcm in 2010 and 75 bcm in 2030. Demand grows more rapidly, making the country a net gas importer after 2020. Gas imports will come from Qatar through the Dolphin Project. Gas production is around a third lower in 2030 in the Deferred Investment Scenario, and the country becomes a net gas importer much earlier than in the Reference Scenario.
- Cumulative energy investment needs in the Reference Scenario in 2004-2030 will amount to $115 billion, around 40% of which will go to the oil sector. Rising upstream oil investment needs will push up total energy investment towards the end of the projection period.

Overview of the UAE's Energy Sector

The United Arab Emirates (UAE) is one of the key suppliers in the world energy market. The country's oil reserves account for around 8% of the world total and the gas reserves are the world's fifth-largest. The UAE is the ninth-largest crude oil producer in the world and the third-largest among MENA countries. Most of the oil produced is exported to Asia, in particular to Japan.

Oil production, which declined in the early 1980s, rebounded to around 2.5 mb/d in the early 1990s and has remained close to that level since. Natural gas production has increased steadily since the 1980s to reach 44 bcm in 2003, more than four-fifths of which goes to domestic consumption. Although national companies remain key players in the energy sector in each emirate, foreign companies are participating in the development of hydrocarbon resources and privatisation has begun in Abu Dhabi's power sector.

The UAE's primary energy demand grew strongly, at an average annual rate of 12.1% between 1971 and 2003 (Table 17.1), driven by strong economic and population growth and low domestic energy prices. It reached 39 Mtoe in 2003, or 9.8 toe per capita – one of the highest levels in the region and more than twice that of the OECD average. The country's high energy consumption is also reflected in high energy intensity, which is far above the MENA and OECD averages. Government policy aims to maximise the export of oil, with the result that the country depends on natural gas for a large part of its domestic energy needs, especially for power generation and desalination, and industry. Natural gas accounted for more than three-quarters of total energy demand in 2003.

Table 17.1: **Key Energy Indicators for the UAE**

	1971	2003	1971-2003*
Total primary energy demand (Mtoe)	1.0	39.2	12.1%
Total primary energy demand per capita (toe)	4.0	9.8	2.8%
Total primary energy demand/GDP**	0.12	0.50	4.6%
Share of oil in total primary energy demand (%)	14	23	–
Net oil exports (mb/d)	1.18	2.47	2.3%
Share of oil exports in production (%)	100	93	–
Share of gas in total primary energy demand (%)	86	77	–
Net gas exports (bcm)	–	7.0	–
Share of gas exports in production (%)	–	16	–
CO_2 emissions (Mt)	2.4	96.1	12.2%

*Average annual growth rate.

** Toe/thousand dollars of GDP in year-2004 dollars and PPPs.

Political and Economic Situation

Political Developments

The UAE consists of seven emirates: Abu Dhabi, Dubai, Sharjah, Ajman, Fujairah, Ras Al Khaimah and Umm Al Qaiwain. The federation was established in 1971, three years after the United Kingdom announced the withdrawal of its troops from the region. Sheikh Zayed bin Sultan Al Nahyan, the then ruler of Abu Dhabi, became president of the unified country. Abu Dhabi dominates the UAE politically and economically, accounting for more than 90% of the UAE's hydrocarbon resources, nearly 60% of GDP and around 40% of population. On the death of Sheikh Zayed in November 2004, his son, Sheikh Khalifa, took over as president of the UAE and ruler of Abu Dhabi. The succession was carried out smoothly.

The Federal Supreme Council, which consists of the rulers of the seven emirates, determines the country's broad policy. However, the constitution limits the legislative and administrative power of the federation and each emirate exercises considerable autonomy in judicial, economic and energy matters.

Economic and Demographic Trends

The UAE's economy has expanded at a brisk 7.2% per year over 1971-2003. Although the economy stagnated in the 1980s due to lower oil prices and production, it recovered in the 1990s, thanks to growth of the non-hydrocarbon sector. Higher oil prices and production have driven an economic boom since 2003. Real GDP growth reached 11.9% in 2003 and is estimated at 7.4% in 2004 (Ministry of Economy and Planning, 2005). GDP per capita is around $20 000, which is the second-highest in the MENA region after that of Qatar.

Though the hydrocarbon sector remains a key component of the economy, the UAE has significantly reduced its dependence on the sector over the last two decades (Figure 17.1). Reflecting the growing contribution to the economy of manufacturing, trade, tourism, real estate, transport, communication and finance, the share of the hydrocarbon sector in GDP declined from around 60% in 1980 to around 20% in 2003, and that in total exports declined from nearly 90% to less than 50%. Government economic policies have played a vital role in the country's diversification (Box. 17.1).

The UAE keeps sound fiscal conditions, with the ratio of government debt to GDP being less than 10%. Abu Dhabi has been the main source of finance for the federal government and other emirates, based on the revenues from its abundant hydrocarbon resources. Corporate income tax is levied only on foreign banks and oil and gas companies in the UAE, with revenues from current hydrocarbon sales and from accumulated financial assets being used to

Figure 17.1: **Share of GDP by Sector in the UAE**

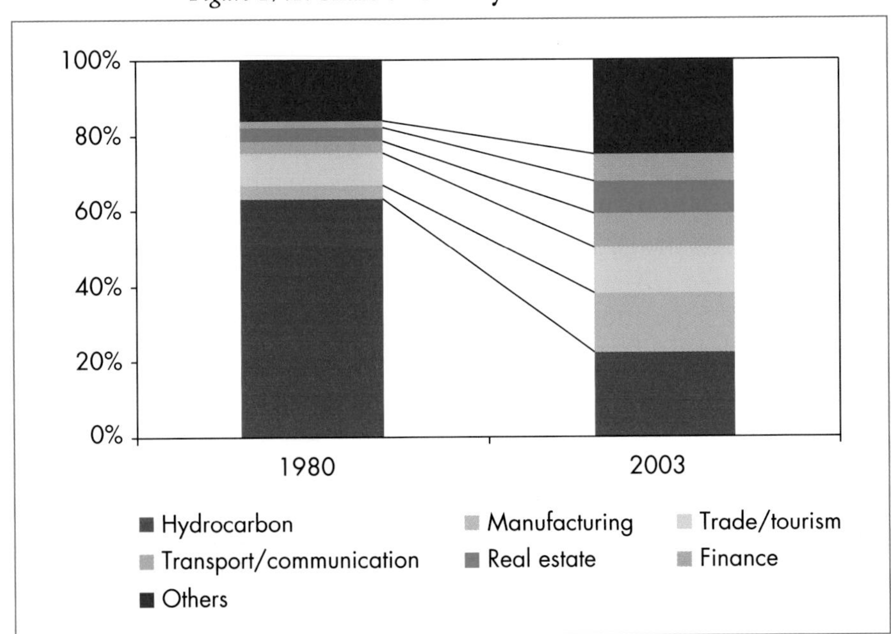

Sources: Ghanem (2001); Central Bank of the UAE (2004).

balance the federal budget. Subsidies and transfers account for 13% of total government spending, or 3% of GDP in 2004 (IMF, 2005). Abu Dhabi's government has reduced agricultural subsidies already and is also discussing the reduction of water and electricity subsidies.

The population of the UAE grew on average by 9.0% per year over the three decades to 2003. This rapid growth reflects a massive influx of foreign workers, mostly from South Asia. Foreign nationals now account for about 80% of the country's total population. While most nationals are employed in the government sector, foreign workers are predominantly employed in labour-intensive, low-wage private-sector industries and services. Though the UAE's overall unemployment rate is still low at around 3% in 2004, increasing unemployment among nationals is becoming a key policy issue.

In the Reference Scenario, the UAE's GDP is assumed to grow on average by 4.6% per year in 2003-2010, slowing to 3.0% in 2010-2020 and 2.6% in 2020-2030 (Table 17.2). The non-hydrocarbon sector is expected to drive the growth of the country's diversified economy, while the hydrocarbon sector is assumed to grow in line with oil and gas production expansion. With slower growth of the labour force, the rates of growth of the non-hydrocarbon sector and the overall economy are assumed to decline towards the end of the *Outlook* period.

Box 17.1: The UAE's Economic Diversification Policy

Economic diversification has been promoted since the 1970s, in particular in Dubai, which is endowed with limited hydrocarbon resources. In addition to establishing state-owned companies in heavy industries such as petrochemicals and aluminium, the government of each emirate has developed infrastructure to encourage private business and foreign direct investment.

Free zones, like Dubai's Jebel Ali, have provided an opportunity for foreign companies to expand business in the region. In free zones, 100% foreign ownership of a company is allowed, while it is limited to 49% outside. Other incentives, such as duty-free trade and unlimited capital transfer, are also available. Currently more than 4 500 companies from 100 countries are located in Jebel Ali Free Zone.[1] Foreign direct investment in the UAE between 2001 and 2003 reached $2.5 billion, by far the largest in the Middle East (UNCTAD, 2004). Dubai's government is now seeking to expand its industrial base by setting up new types of free zone, such as the Dubai Internet City and Dubai International Financial Centre.

Although the business environment of the UAE is generally favourable compared to other MENA countries, further legislative and administrative reform will be needed. A relaxation of limits on foreign ownership of property in Abu Dhabi, announced in August 2005, is expected to boost real estate business and to attract further foreign investment in non-hydrocarbon sectors.

Table 17.2: **GDP and Population Growth Rates in the UAE in the Reference Scenario** (average annual rate of change in %)

	1971-2003	1990-2003	2003-2010	2010-2020	2020-2030	2003-2030
GDP	7.2	4.3	4.6	3.0	2.6	3.3
Population	9.0	6.5	3.4	2.0	1.6	2.2
Active labour force	9.7	6.6	4.0	2.4	1.7	2.6
GDP per capita	–1.7	–2.0	1.2	1.0	0.9	1.0

1. *Khaleej Times*, 5 August 2005.

Energy Policy

Under the UAE constitution, hydrocarbon resources belong to each emirate, not to the federation. Accordingly, the government of each emirate formulates its own energy policy, with the federal government playing only a limited role. In Abu Dhabi, the Supreme Petroleum Council (SPC), which was established in 1988, formulates and monitors the emirate's oil and gas policy. Under the authority of the SPC, the Abu Dhabi National Oil Company (ADNOC), established in 1971, is responsible for the development, production and distribution of hydrocarbon resources. Most of the other emirates also have a council or department for the oil sector, under the ruler's control. In November 2004, the federal Ministry of Petroleum and Mineral Resources and the federal Ministry of Electricity and Water merged to form the Ministry of Energy, with Mohammed bin Dhaen Al Hamili, Marketing and Refinery Director of ADNOC and former OPEC governor, becoming the minister.

The Abu Dhabi's energy policy seeks to maximise oil exports and this is reflected in heavy investment in oilfield development.[2] Unlike many other oil-producing countries, Abu Dhabi and the other emirates grant upstream concession agreements to foreign companies, even to the extent of 100% in some fields. Natural gas development is undertaken by ADNOC on behalf of the Abu Dhabi government, with foreign companies permitted up to 49% participation.

Private investment in the electricity and water sector is being promoted in Abu Dhabi. The Abu Dhabi Water and Electricity Authority (ADWEA) is responsible for formulating and implementing the emirate's electricity and water policy, including privatisation. Four independent power and water projects have so far been set up, with ADWEA holding a 60% share in each of them. The remaining 40% shares are held by foreign private investors.

Energy Demand

Primary Energy Mix

Primary energy demand has grown strongly since the 1970s, even in the 1980s when economic growth was stagnant. Demand grew faster than the economy and population, at more than 20% per year in the 1970s and around 10% per year in the 1980s. Energy-demand growth slowed to 5.5% per year between 1990 and 2003, closer to the average GDP growth rate of 4.3% per year and below the population-growth rate of 6.5%. Per capita energy consumption and energy intensity are much higher than OECD averages, because of the strong needs for air-conditioning and water desalination, and low energy prices.

2. Development wells drilled by ADNOC in the past five years numbered about 1 000 (ADNOC, 2005).

Domestic energy prices are set far below international levels. For example, the price of regular gasoline averaged $0.28 per litre and that of electricity for households around $0.02 per kWh in 2004 (Ministry of Economy and Planning, 2005).

In the Reference Scenario, total primary energy demand is projected to grow by 2.9% per year over the projection period, reaching 84 Mtoe in 2030 – more than double the current level (Table 17.3). Growth of 4.7% per year is expected in the rest of the current decade, then slowing to 2.2% per year until 2030 as economic activity and population growth slow. Per capita energy consumption will reach 11.7 toe in 2030. Total CO_2 emissions increase from 96 million tonnes in 2003 to 202 Mt in 2030.

Table 17.3: **UAE's Primary Energy Demand in the Reference Scenario** (Mtoe)

	1990	2003	2010	2020	2030	2003-2030 *
Oil	6.1	9.1	11.2	14.0	16.2	2.2%
Gas	13.5	30.1	43.0	57.8	68.2	3.1%
Total**	**19.6**	**39.2**	**54.3**	**71.8**	**84.4**	**2.9%**

*Average annual growth rate.
** Renewable energy demand is negligible.

Figure 17.2: **UAE's Primary Energy Demand by Fuel in the Reference Scenario**

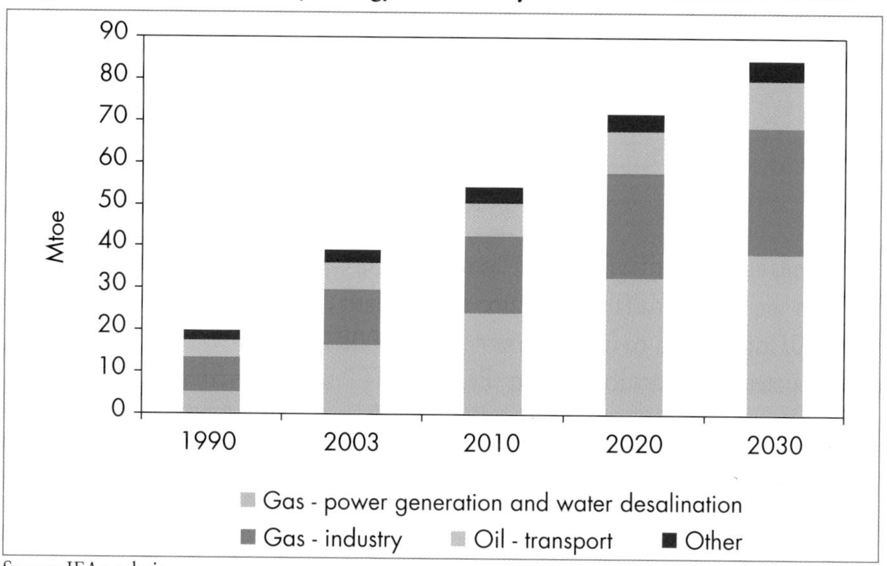

Source: IEA analysis.

The UAE will depend on gas for the bulk of its primary energy needs until 2030 (Figure 17.2). With strong demand from power generation and desalination and industry, gas use is projected to rise by 3.1% per year on average, from 30 Mtoe in 2003 to 68 Mtoe in 2030. The high share of gas in primary energy demand will increase yet further to more than 80% in 2030. In order to meet its growing domestic demand, the UAE has launched the Dolphin Project – a regional gas production and transportation network – to pave the way for imports from Qatar (Box 17.2). Oil demand, around 70% of which comes from the transport sector, will grow at 2.2% per year on average over the projection period. Total fuel requirements for water desalination will account for 30% of the increase in energy demand in the power and water sector from 2003 to 2030.

Box 17.2: **The Dolphin Project**

The Dolphin project will connect the prolific North Field in Qatar to the national gas grids of the UAE and Oman. The project was initiated by the UAE Offsets Group and Qatar Petroleum. Dolphin Energy Limited (DEL) was established to implement the project in 1999. The current shareholders are Mubadala Development Company (51%), a wholly-owned subsidiary of the Abu Dhabi government, Total (24.5%) and Occidental Petroleum (24.5%).

A 370-km offshore pipeline will be built from Ras Laffan in Qatar to Taweelah in Abu Dhabi, with an extension to Dubai and the northern part of Oman. The UAE will import about 20 bcm/year from Qatar in the initial phase. A plan to increase imports to more than 30 bcm/year is under discussion.

Upstream development in Qatar began in 2004 and initial deliveries of gas from Qatar to the UAE are planned by the end of 2006. DEL made 25-year gas supply commitments to ADWEA and Union Water and Electricity Company in Fujairah in 2003, and to Dubai Supply Authority in May 2005. DEL has been supplying gas from Oman to Fujairah since January 2004 – the first ever cross-border gas sales among countries of the Gulf Co-operation Council (GCC). Supplies will be continued until gas from Qatar becomes available.

The project will cost an estimated $3.5 billion – $2.5 billion for the construction of the processing plant at Ras Laffan and $1 billion for the pipeline (Al Yabhouni, 2005). DEL secured financing from 25 local and international financial institutions in July 2005.

Final energy consumption is expected to grow by 2.8% per year on average over the projection period in the Reference Scenario. Energy demand in the industry and transport sectors will account for more than 80% of total final consumption throughout the period, while that in the residential and services sectors will grow faster, driving the strong growth of 3.6% per year in total final electricity consumption.

Sectoral Trends

The *industry sector* is the largest end-use energy consumer in the UAE. Energy consumption in the industry sector grew by 3.3% per year on average between 1990 and 2003, reaching 15 Mtoe, or nearly 60% of total final energy demand. The petrochemical industry accounts for a large share of total industry demand. The main petrochemical company is Abu Dhabi Polymers Company (Borouge), a joint venture between ADNOC (60%) and Danish Borealis (40%). It plans to expand its ethylene and polyethylene production capacity significantly this decade. Other heavy industries will also contribute to growth in industrial demand. Dubai Aluminium (Dubal), Dubai's leading manufacturing company, set up in 1979, has continuously expanded its production capacity and has now become one of the world's largest smelters. The growth of petrochemical and other industries, backed by the government's diversification policy, will boost industrial energy consumption in the Reference Scenario by 3% per year on average between 2003 and 2030, when it will reach 34 Mtoe.

Energy consumption in the *transport sector* rose by 3.4% per year on average between 1990 and 2003. The vehicle stock has increased slightly faster than population growth since the mid-1990s, with the passenger car ownership rate rising from around 135 per thousand people in the late 1990s to 170 in the early 2000s. This is high among MENA countries but much lower than in OECD countries. Increases in vehicle ownership will drive up oil consumption, though growth will be tempered by the decline in population growth and by efficiency improvements. Transport energy consumption will grow by 2.1% per year over 2003-2030.

Residential energy consumption grew rapidly between 1990 and 2003, by 10.9% per year on average. The growth rate will moderate over the projection period, as the efficiency of household appliances improves and population growth slows. The sector's energy use will grow by 3.1% per year on average until 2030. The energy consumption of the *services sector* has grown by 9.5% per year between 1990 and 2003. The sector has continuously increased its share in GDP and is expected to grow further in response to strong government promotional efforts to boost financial services and tourism. For example, Dubai aims to attract 15 million visitors in 2010 and 40 million by 2015 (Ministry of Information and Culture, 2005). Energy consumption in the sector will grow by 3.3% per year until 2030.

Electricity Supply and Desalinated Water Production

Overview

The United Arab Emirates has one of the fastest growing electricity industries in the world. Electricity generation doubled between 1995 and 2003, reaching 50 TWh. Demand growth was strong in the services sector, which now accounts for more than 40% of total demand for electricity.[3] Per capita electricity generation, in excess of 12 000 kWh, is higher than the OECD average. The government has made substantial investments in electricity infrastructure to underpin economic growth. Preliminary estimates indicate that electricity generation reached 53 TWh in 2004.

Natural gas fuels over 99% of total electricity generation, the remainder being based on oil. Installed capacity was 17 GW in 2003, including nearly 4 GW in various industrial, commercial and other sites. The emirates of Abu Dhabi, Dubai and Sharjah account for 90% of the country's installed capacity (Figure 17.3). Abu Dhabi – the most populous of the emirates – dominates electricity production. It had an installed capacity of 8 GW at the end of 2003. ADWEA owns about 60% of Abu Dhabi's installed capacity, while the remainder is owned principally by the oil and gas industry. Most power plants in the UAE are combined electricity and water desalination facilities.

Figure 17.3: **UAE's Installed Power Generating Capacity, End-2003**

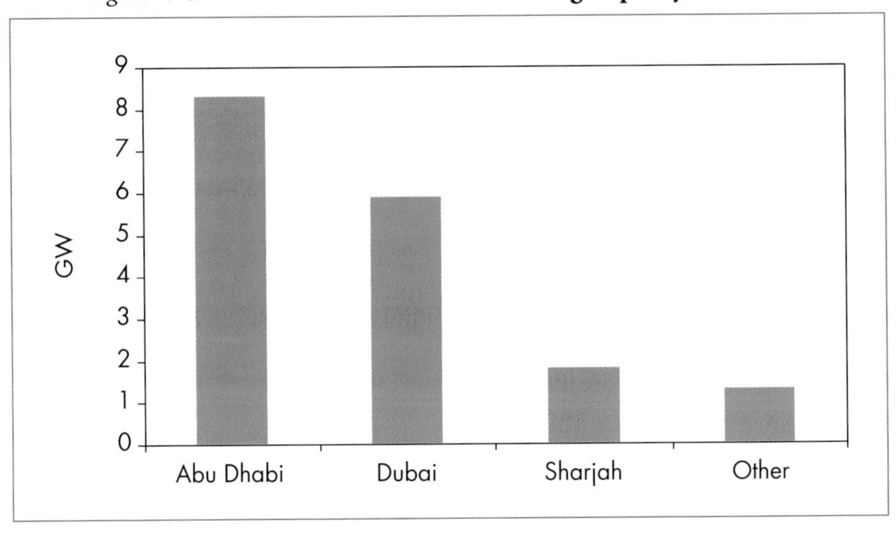

3. This share compares with 25% in the European Union and 34% in North America.

The power plant mix is based on open-cycle gas turbines, steam boilers and combined-cycle gas-turbine (CCGT) power plants, with a few diesel engines. CCGT capacity has increased substantially in recent years and now accounts for more than a third of installed capacity – the highest share in the region.

The power sector is organised in different ways in each of the seven emirates. Only Abu Dhabi has embraced privatisation and has one of the most advanced privatisation programmes in the MENA region. Abu Dhabi's power generation is controlled by the ADWEA – the single buyer of independently produced water and power. There are currently four such projects, developed on a build-own-operate basis through joint ventures (Table 17.4). ADWEA has a 60% share in all four projects. Abu Dhabi is also the only emirate to have separated network activities from power generation. Transmission is handled by TRANSCO, while distribution is divided between the Abu Dhabi Distribution Company and the Al Ain Distribution Company.

In January 2005, ADWEA signed a contract with a consortium headed by Japan's Marubeni Corporation for the purchase and expansion of the Taweelah B power and water plant, Abu Dhabi's fifth independent water and power project. By 2008, this plant's generating capacity will be expanded to 2 000 MW and its desalination capacity to 263 million cubic metres per year.

Table 17.4: **Independent Water and Power Projects in Abu Dhabi, 2004**

Project	Power generation capacity (MW)	Water production capacity (mcm/year)	Owner
Al Taweelah A-2	777	83	Emirates CMS Power Company
Al Taweelah A-1	1 431	139	Gulf Total Tractebel Power Company
Shuweihat 1	1 610	166	Shuweihat CMS International Power Company
Umm Al Nar	1 038	269	Arabian Power Company

Dubai's power sector is controlled by the Dubai Electricity and Water Authority (DEWA), which owns about 4 GW of capacity. In Sharjah, the responsibility for supplying power lies with the Sharjah Electricity and Water Authority. The Federal Electricity and Water Authority is responsible for the power sector in the remaining emirates. Electricity prices are heavily subsidised in all emirates. Figure 17.4 shows current electricity tariffs in Abu Dhabi, compared to household and industry tariffs in the United States. There have been announcements in the past that Abu Dhabi would eliminate subsidies as privatisation proceeded, but there are no concrete steps in this direction yet.

Figure 17.4: **Electricity Tariffs in Abu Dhabi**

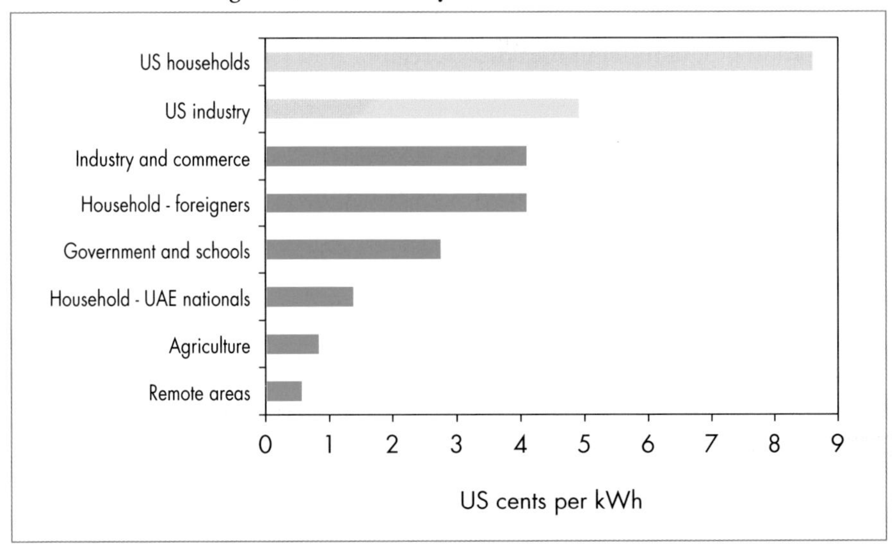

The UAE's electricity demand shows extreme seasonal variation, with demand for power doubling in the summer, primarily because of extensive use of air-conditioning. Low electricity prices and a lack of demand-side efficiency policies accentuate the problem, while power plants and the transmission grid operate under extreme conditions in the summer because of both high temperatures and sand pollution from the desert. All these factors put enormous stress on the system. This was clearly demonstrated in a major power failure in Dubai on 9 June 2005. According to DEWA, the six-hour power cut resulted from a major failure in one of its main substations at Jebel Ali. Strong demand during the hot months of the year means that UAE power plants operate at low capacity factors during the winter. Uneven demand makes it more difficult to match power and water production, since demand for water is fairly constant throughout the year.

The UAE is one of the world's largest per capita water consumers and MENA's second-largest producer of desalinated water after Saudi Arabia. Residential water consumption in the UAE was 760 million cubic metres in 2003, accounting for about 28% of total water demand. Desalinated water production was 1 140 mcm in 2003, accounting for more than 40% of total water demand.

The groundwater aquifers of the UAE are recharged from limited surface flows. The total renewable amount is estimated to be 190 mcm (World Bank, 2005). The Environmental Research and Wildlife Development Agency of Abu Dhabi estimates that the UAE has about 235 billion cubic metres of non-renewable groundwater reserves, about 80% of which is brackish. Groundwater in the UAE is used primarily for irrigation purposes, and the rate of extraction exceeds the estimated safe yield of the aquifers, resulting in saline intrusion, abandonment of wells and other adverse effects. Both national and emirate-level strategies for water resources management are currently being developed.[4]

Water tariffs in the UAE vary from one emirate to another, but metering and pricing policies effectively exempt nationals from water charges. No water-distribution companies have introduced progressive block tariff systems.

Electricity Production

Electricity generation in the UAE will continue to increase rapidly over the current decade, at 5.6% per year on average. Growth is projected to slow in the following two decades, falling to 3.7% per year in 2010-2020 and to 2.1% per year in 2020-2030. Total electricity generation is projected to increase from 50 TWh in 2003 to 128 TWh in 2030.

The UAE will need 31 GW of new generating capacity between 2004 and 2030. About 60% of the total electricity capacity additions will be for new combined water and power plants with desalination units. The country is expected to install more CCGT power plants for base-load to mid-load operation and in desalination plants. Almost all new power plants are expected to be gas-fired. A wind turbine was installed in Dubai in 2004, using a German grant, but it is unlikely that there will be substantial increases in wind power in the future.

Water Production

Residential water demand is projected to increase to nearly 1 500 mcm in 2030, a growth of 2.5% per year. The residential sector will account for 40% of total water demand in 2030. Most of the installed desalination capacity will continue to use the distillation process (see Box 6.6), but the share of reverse osmosis will increase from 5% in 2003 to over 15% in 2030. Fuel requirements for desalination will rise from 9 Mtoe in 2003 to 16 Mtoe in 2030 (Table 17.5), and they will account for nearly one-fifth of total primary energy demand at the end of the projection period.

4. The Environmental Research and Wildlife Development Agency in Abu Dhabi is planning to have its water conservation policy in force by end-2007 (ERWDA, 2005).

Table 17.5: **UAE's Water Consumption and Desalination Capacity**

	2003	2010	2020	2030
Water consumption (mcm)	2 694	2 979	3 312	3 656
Desalination capacity (mcm)	1 465	2 482	2 684	2 948
Oil and gas use for desalination (Mtoe)	9	13	15	16

Oil Supply

Policy Framework

In order to increase oil production and exports, the UAE has welcomed foreign companies' participation in upstream operations. In Abu Dhabi, ADNOC has three operating subsidiaries to develop the emirate's major oilfields in partnership with foreign companies (Table 17.6). Some small fields are operated entirely by foreign operators. In Dubai, Dubai Petroleum Company, a 100% subsidiary of the US firm, ConocoPhillips, operates major fields of the emirate on behalf of a consortium of foreign companies. The involvement of foreign operators has made it possible for the UAE to gain access to the latest technologies and additional financial resources for field development and enhanced recovery.

Table 17.6: **ADNOC's Upstream Operating Subsidiaries, End-2004**

Company	Production capacity (mb/d)	Shareholders	Major fields
Abu Dhabi Company for Onshore Oil Operations (ADCO)	1.30	ADNOC 60%, BP/ExxonMobil//Royal Dutch Shell/Total 9.5% each, Partex 2%	Bab, Bu Hasa, Asab, Al Dabb'iya, Sahil, Shah
Abu Dhabi Marine Operating Company (ADMA-OPCO)	0.55	ADNOC 60%, BP 14.67%, Total 13.33%, JODCO* 12%	Lower Zakum, Umm Shaif
Zakum Development Company (ZADCO)	0.60	ADNOC 88%, JODCO* 12%	Upper Zakum

* Japan Oil Development Company.
Sources: Al Yabhouni (2005); Arab Petroleum Research Centre (2005).

Abu Dhabi started production and exports of oil in 1962 and became a member of OPEC in 1966. Its membership was transferred to the UAE when the country was formed in 1971. Abu Dhabi has generally taken responsibility for adjusting its oil production to ensure that the UAE's total production is in line with the OPEC production target. The UAE has sought to raise NGL production, which is not restricted by OPEC quotas. ADNOC has two subsidiaries dealing with NGLs. Abu Dhabi Gas Industries (GASCO), a joint venture between ADNOC (68%), Total (15%), Shell (15%) and Partex (2%), processes associated gas from ADCO's onshore fields at the plants at Habshan-Bab, Bu Hasa, Asab and Ruwais. Abu Dhabi Gas Liquefaction Company (ADGAS), a joint venture between ADNOC (70%), Mitsui & Co., Ltd. (15%), BP (10%) and Total (5%), handles associated and non-associated gas from offshore fields at its plant on Das Island.

Resources and Reserves

According to the *Oil and Gas Journal* (2004) and the Ministry of Information and Culture (2005), the UAE's proven oil reserves at the end of 2004 were 98 billion barrels, or 8% of the world total. The UAE's undiscovered recoverable resources are estimated by the US Geological Survey at 10 billion barrels (USGS, 2000).

Proven oil reserves were revised sharply upward in the 1980s, on the basis of an improvement in recovery rates. Remaining exploration opportunities are limited to smaller reservoirs in stratigraphic traps. Most of the large oilfields are in carbonate structures, at times heavily faulted, which present particular technology challenges: heterogeneities, wettability and the need for improved monitoring of fluid movements between wells (see Chapter 4).

Abu Dhabi's proven oil reserves are estimated to amount to 92 billion barrels, or 94% of the UAE's total reserves. The emirate's reserves-to-production (R/P) ratio is around 100 years at the 2004 rate of output. Dubai's proven oil reserves are officially estimated at 4 billion barrels, but the figure has not been revised for many years and the American Association of Petroleum Geologists estimates a much lower figure, 1.6-2 billion barrels (Arab Petroleum Research Centre, 2005). Sharjah's oil and condensate reserves are officially estimated at 1.5 billion barrels, but the estimate has not been revised for many years. Ras Al Khaimah has 100 million barrels of oil reserves. There are negligible reserves in Fujairah, Ajman and Umm Al Qaiwain.

Crude Oil Production

About 30 fields are currently in production in the UAE. The largest is the offshore Zakum field, which has remaining proven reserves of 16 billion barrels (Table 17.7) and produced around 770 kb/d in 2004. The other major fields are, onshore, Bab, Bu Hasa, Asab, and offshore, Umm Shaif. All these fields are located in Abu Dhabi. The producing fields in the other emirates are relatively small and most of them are in decline. Crude oil produced in the UAE is typically light: the average API gravity for Murban, a blend from Abu Dhabi's onshore fields, is 39°. Murban is sweet, with 0.8% sulfur content, while crude oil from the offshore Upper Zakum and Fateh fields is more sour, with around 2% sulphur content.

Table 17.7: **Oilfields in the UAE**

Field	**Year of first production**	**Remaining proven and probable oil reserves at end-2004** (billion barrels)	**Cumulative production to 2004** (billion barrels)	**API gravity**
Zakum	1967	16.0	5.4	39.0
Bab	1963	9.9	1.6	44.0
Bu Hasa	1964	5.5	6.0	39.0
Asab	1973	5.3	3.1	41.0
Umm Shaif	1962	3.6	2.7	37.0
Al Dabb'iya	1996	1.7	0.0	39.3
Fateh	1969	1.0	2.9	32.4
Sahil	1975	1.0	0.2	39.7
Shah	1983	0.7	0.2	30.0
Other		10.4	2.8	
Total		**55.1***	**24.9**	

* This figure is derived from IHS Energy's field-by-field production data and differs from estimates in the *Oil and Gas Journal* and other sources.
Note: Includes NGLs and condensates.
Sources: IHS Energy and IEA databases.

The Zakum field, which was discovered in 1964, is one of the biggest offshore fields in the world. The Upper Zakum field, operated by ZADCO, contains an estimated 50 billion barrels of initial oil in place. It started production in 1967 and large-scale development work, including water injection, increased its production capacity to over 600 kb/d in the late 1990s. However, production then declined owing to water encroachment and falling reservoir pressure. The Abu Dhabi government sought a strategic partner to acquire a 28% stake in ZADCO in order to assist in boosting the production from this complex field from around 550 kb/d to around 750 kb/d. After nearly five years of

Figure 17.5: **UAE's Oil and Gas Resources and Supply Infrastructure**

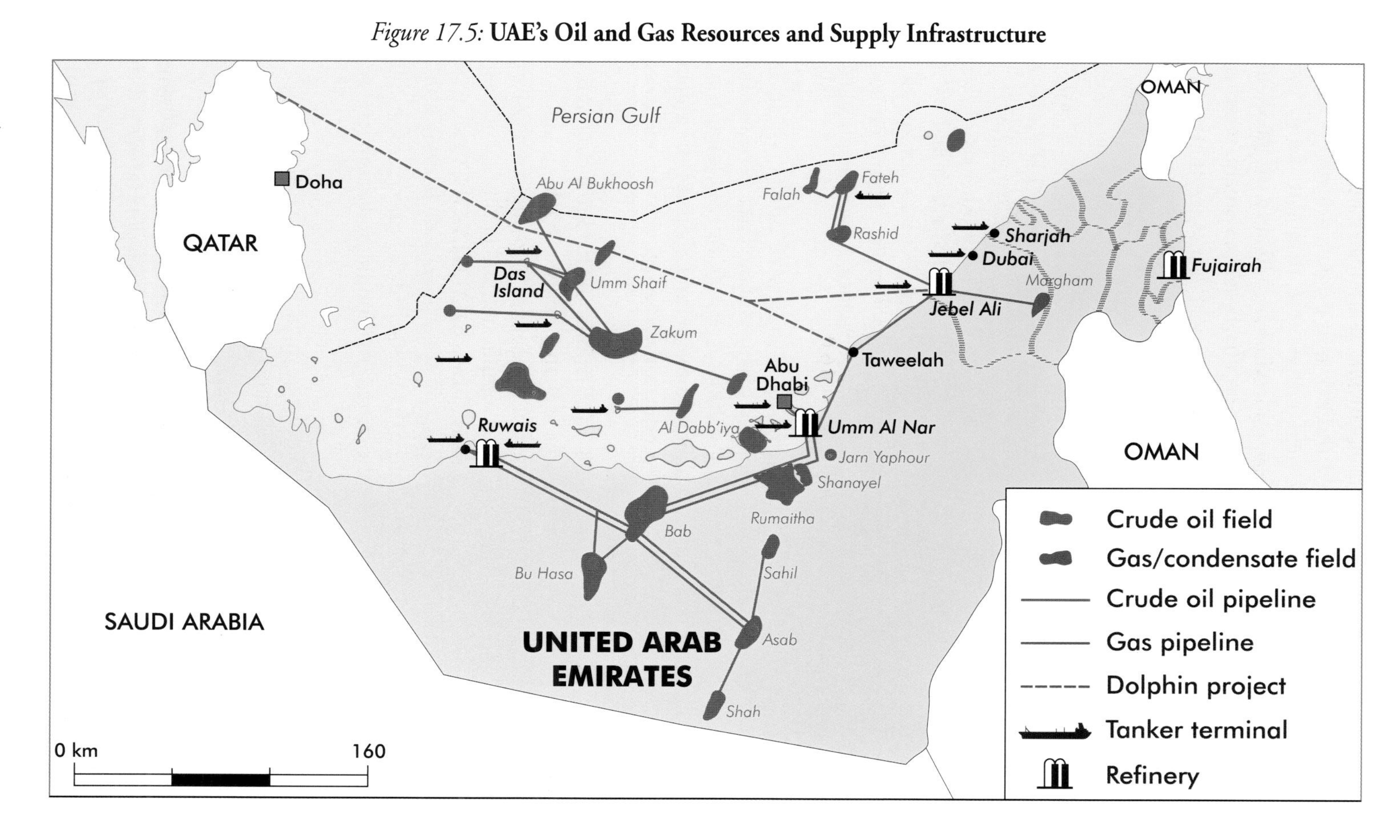

consultations, SPC announced in April 2005 that it had chosen ExxonMobil for final negotiations to become the strategic partner. Current sustainable capacity of the Lower Zakum field is no more than 240 kb/d, though maximum production capacity is considered to be 320 kb/d. ADMA-OPCO is expanding gas injection facilities to lift the field's sustainable production capacity.

The Bab field, which is located 85 km south-west of Abu Dhabi city, was the first onshore oilfield to be discovered, in 1958. Production started five years later. Capacity reached over 300 kb/d in the late 1990s but has since declined to 250 kb/d. ADCO started work in 2003 to increase capacity to 350 kb/d by installing processing trains, gas separators, a degassing station and associated pipelines.

Bu Hasa is ADCO's other large onshore field, located south-west of the Bab field. Bu Hasa was discovered in 1962 and started production in 1964. The production capacity of the field is 550 kb/d and a project is in hand to increase it to 730 kb/d by the end of 2006. The project will replace ageing production facilities and enhance pressure maintenance schemes through the installation of gas- and water-injection facilities.

The Asab field, which was discovered in 1965, started production in 1973. ADCO is planning further development of this onshore field, to boost capacity by 30 kb/d to 310 kb/d by the end of 2006. ADCO's four fields north-east of Abu Dhabi (Al Dabb'iya, Jarn Yaphour, Rumaitha and Shanayel) are also under development. Their combined production capacity will be increased to 110 kb/d.

Umm Shaif was discovered in 1958, the first offshore field to be found in Abu Dhabi. The field was the first of the emirate's fields to begin production, in 1962. The field is located north-west of the Zakum field and covers an area of 360 square kilometres. The structure is complicated and horizontal drilling and gas injection are carried out on a large scale by ADMA-OPCO to enhance recovery. Maximum production capacity is estimated to be 280 kb/d but sustainable capacity is put at no more than 220 kb/d. A project aimed at lifting the field's sustainable capacity to the current maximum production level will be completed in 2006.

Dubai has three major offshore oilfields in production: Fateh, Falah and Rashid, all of which are operated by the Dubai Petroleum Company. The emirate's oil and condensate production peaked at more than 400 kb/d in 1991. Output in 2004 was estimated to be below 200 kb/d.

The UAE's NGL and condensate production in 2004 is estimated at 0.4 mb/d. GASCO is expanding its NGL and condensate production capacity through projects, called OGD-3 and AGD-2, at the Bab and Asab fields. These projects involve the construction of new gas-processing plants and NGL recovery units that will produce around 0.3 mb/d of NGLs and condensates. According to official projections, Abu Dhabi's LPG production will increase from around 160 kb/d at present to around 260 kb/d in 2007-2008 (Al Yabhouni, 2005). This will make the UAE one of the largest LPG exporters in the world.

The UAE's total crude oil production (including NGLs and condensates), which declined in the early 1980s mainly in response to sluggish world oil demand, rebounded to around 2.5 mb/d in the early 1990s and has remained close to that level since then. According to the official plans, the UAE's crude oil production capacity will increase from around 2.7 mb/d at present to over 3 mb/d between 2008 and 2010 (Al Yabhouni, 2005). In the Reference Scenario, total UAE oil production (including NGLs and condensates) is projected to grow from 2.75 mb/d in 2004 to 3.15 mb/d in 2010 and 5.05 mb/d in 2030 (Table 17.8).

Table 17.8: **UAE's Crude Oil Production in the Reference Scenario** (mb/d)

Field	2004	2010	2020	2030
Currently producing fields	**2.75**	**3.09**	**3.08**	**2.75**
Zakum	0.77	0.80	0.97	0.97
Bab	0.36	0.42	0.42	0.42
Bu Hasa	0.44	0.67	0.60	0.35
Asab	0.32	0.38	0.38	0.37
Umm Shaif	0.22	0.29	0.29	0.28
Other	0.64	0.53	0.42	0.36
New developments	**0.00**	**0.06**	**0.87**	**2.30**
Fields awaiting development	0.00	0.04	0.61	1.73
Reserve additions and new discoveries	0.00	0.02	0.26	0.57
Total	**2.75**	**3.15**	**3.95**	**5.05**

Note: Includes NGLs and condensates.
Sources: IHS Energy database; IEA analysis.

The Zakum field will remain the most important contributor to the country's production throughout the projection period. Bu Hasa, from which a large amount of oil has already been extracted, will produce progressively less after 2010. Production from the Bab, Asab and Umm Shaif fields will be expanded to reach a plateau in the next decade. While total production from current producing fields declines after 2010, production from new fields and output from reserve additions will lift the country's overall production level, accounting for around 45% of total production by 2030 (Figure 17.6).

Figure 17.6: **UAE's Crude Oil Production by Source in the Reference Scenario**

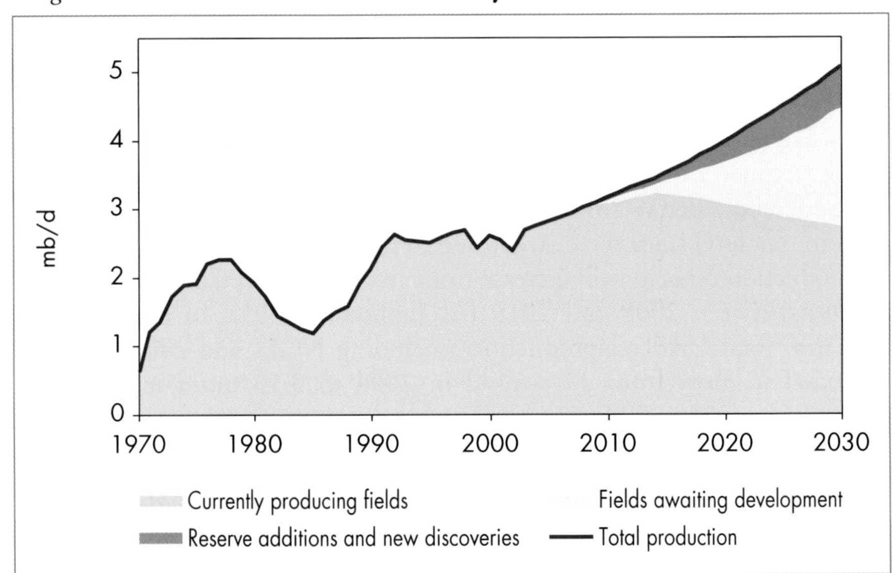

Note: Includes NGLs and condensates.

Oil Refining

There are four major oil refineries in the UAE, with a total capacity of 708 kb/d (Table 17.9). Two of them, at Ruwais and Umm Al Nar, are operated by the Abu Dhabi Oil Refining Company (Takreer), which was established in 1999 to take over refining operations from ADNOC. The Emirates National Oil Company (ENOC), Dubai's state-owned company, operates the refinery in Jebel Ali. The refinery plant in Fujairah was developed by a Greek company but was temporarily closed in March 2003 because of a hike in feedstock

Table 17.9: **UAE's Crude Oil Distillation Capacity, End-2004**

Owner	**Location**	**Start-up year**	**Capacity** (kb/d)
Takreer	Ruwais	1981	420
	Umm Al Nar	1976	88
ENOC	Jebel Ali	1999	120
Fujairah government	Fujairah	1993	80
Total			**708**

Source: Ministry of Information and Culture (2005).

prices. The Fujairah government now owns the plant and plans to bring it back into production (Ministry of Information and Culture, 2005).

The capacity of the Ruwais plant was expanded in 2000 with the installation of two 140-kb/d condensate-processing trains. The plant produces light products for export, mainly to Japan and India, and fuel oil for local power generation. ENOC's plant in Jebel Ali has two 60-kb/d condensate trains, one for sweet and the other for sour condensate. The plant produces LPG, jet fuel, gas oil and bunker fuel for domestic use, as well as naphtha for export. With relatively slow growth of domestic petroleum product demand, the increase in the UAE's refining capacity is expected to be moderate. Total distillation capacity is projected to rise to 800 kb/d in 2010 and to 1.1 mb/d in 2030.

Oil Exports

Almost all of the crude oil exported by the UAE goes to Asia and more than half to Japan. The UAE has been the main exporter to Japan since the mid-1980s, accounting for around a quarter of Japan's total crude oil imports. Quality and supply stability of Abu Dhabi's crude oil have helped the UAE to become the major supplier to Japan. Dubai crude oil is mainly sold in the Far East on the spot market, and, despite its very limited volume, serves as a price marker for the Asian market. Abu Dhabi's refined products are exported by ADNOC Distribution – mainly to India and Japan.

In the Reference Scenario, exports of crude oil, NGLs and condensates are projected to rise steadily over the projection period, in line with the growth in production. They increase from 2.5 mb/d in 2004 to 2.9 mb/d in 2010 and 4.7 mb/d in 2030.

Natural Gas Supply

Resources and Reserves

The UAE has abundant natural gas resources. Cedigaz estimates that the UAE's proven gas reserves are 6 tcm, making the country the fifth-largest gas reserve-holder in the world (Cedigaz, 2005). Reserves are equal to more than 100 years production at current rates. The USGS puts undiscovered resources at 1.26 tcm (USGS, 2000). Most of the UAE's proven reserves are located in Abu Dhabi, which has 5.6 tcm, or 93%, of the UAE's total. Sharjah has 0.3 tcm and Dubai 0.1 tcm. Abu Dhabi revised its estimate of gas reserves upward in the late 1990s after discovery of the large Khuff gas reservoirs and other non-associated gas structures. There have been no revisions since then.

Gas Production and Distribution

Production of marketed gas has increased steadily in recent years, more than doubling from 20 bcm in 1990 to 44 bcm in 2003. These numbers do not include gas reinjected into reservoirs or flared gas, which amounted to some 20 bcm in total in 2003. Abu Dhabi accounts for more than 80% of the UAE's gas production. The rapid production expansion came from the completion of the onshore development projects, OGD-1 and OGD-2 at the Bab field and AGD-1 at the Asab field. Ongoing OGD-3 project is expected to expand GASCO's natural gas production capacity by around 12 bcm by 2008. In 2004, an offshore gas production expansion project at the Khuff reservoirs under the Abu Al Bukhoosh field was completed, increasing the gas recovered from the reservoir by some 2 bcm per year.

Dubai's gas production at the Margham field, the emirate's only gas field, has been declining, falling below 3 bcm in 2004. With a huge increase in domestic gas demand, Dubai is importing gas from neighbouring emirates. Imports from Sharjah commenced in 1986 and amounted to some 3 bcm in 2004. Imports from Abu Dhabi started in 2001, after the completion of the OGD-2 project, and reached 9 bcm in 2004. Dubai also plans to acquire some 7 bcm of gas from the Dolphin Project. Production in Sharjah also has been declining and was less than 5 bcm in 2004.

Total UAE gas production is set to grow steadily as existing field facilities are expanded through to the end of the decade. However, proximity to the massive North Field in Qatar will make it more economic to import Qatari gas rather than developing the smaller and more complex domestic gas fields. The increasing need for reinjection in oilfields will also limit the expansion of marketed gas. We project production of marketed gas to rise from 44 bcm in 2003 to 59 bcm in 2010 and 75 bcm in 2030.

Gas Exports

Since the commissioning of the Das Island plant in 1977, ADGAS has sold LNG to Japan's Tokyo Electric Power Company (TEPCO) under long-term contracts. Under the current contract, ADGAS is to export LNG to TEPCO until 2019. ADGAS has also sold LNG on the spot market to Europe, the United States, and South Korea.

In the Reference Scenario, we expect the UAE's net gas exports to decline from 7 bcm in 2003 to 6.5 bcm in 2010. The country will become a net gas importer after 2020, because domestic demand will grow faster than production. Net imports are expected to reach 8.4 bcm in 2030 (Figure 17.7).

Figure 17.7: **UAE's Gas Balance in the Reference Scenario**

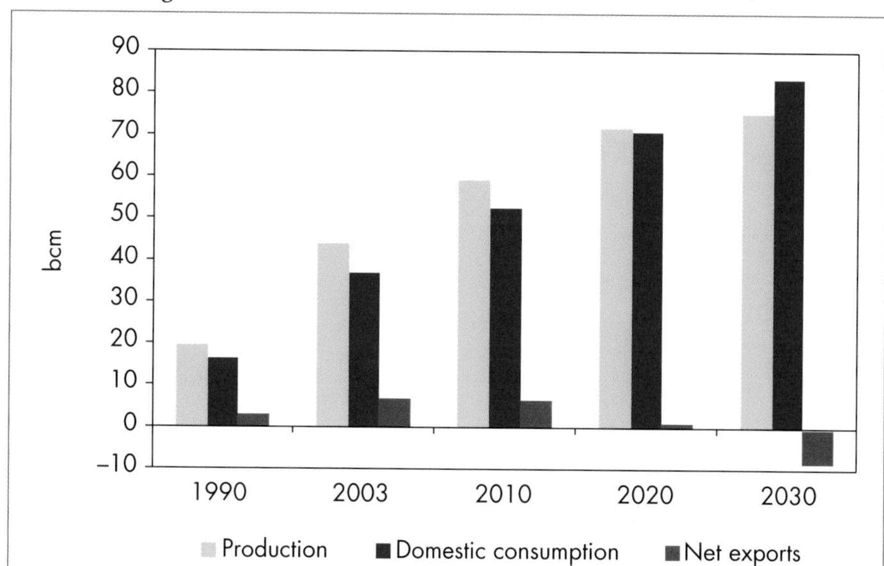

Energy Investment

The cumulative energy investment needs in the UAE between 2004 and 2030 are projected to reach $115 billion (in year-2004 dollars), or $4.3 billion per year. Investment needs in the oil sector amount to $46 billion, or around 40% of the total. The gas sector will need $34 billion and the electricity sector $35 billion. Rising oil upstream investment needs will push up total energy investment towards the end of the projection period (Figure 17.8).

The upstream sector accounts for almost 90% of the total cumulative oil investment over 2004-2030. Most of the upstream investment to 2010 will be for maintenance and expansion of the production capacity of fields already in production. The increase in average costs, as the easiest reserves are depleted, and the increase in production will raise investment needs towards the end of the *Outlook* period, to $24 billion between 2021 and 2030. Investment needs in refining will grow in accordance with capacity expansion, but its proportion of total investment will remain small. In the gas sector, $21 billion of cumulative investment will be needed in the upstream and $14 billion in downstream over 2004-2030. Imports from Qatar will slow the need to develop indigenous gas reserves, which will limit the increase in investment in the last decade. The UAE has involved a number of foreign companies in the investment in the oil and gas sectors and is expected to continue to seek foreign companies' participation in order to obtain both technology and financial resources.

In the electricity sector, required investment will reach $16 billion in power generation and $19 billion in transmission and distribution over 2004-2030.

Funding new projects is not expected to be difficult, although some of the projects in the country are becoming larger and more complex. Private-sector participation is likely to increase as more emirates seek to restructure and privatise their industries.

Figure 17.8: **UAE's Energy Investment in the Reference Scenario**

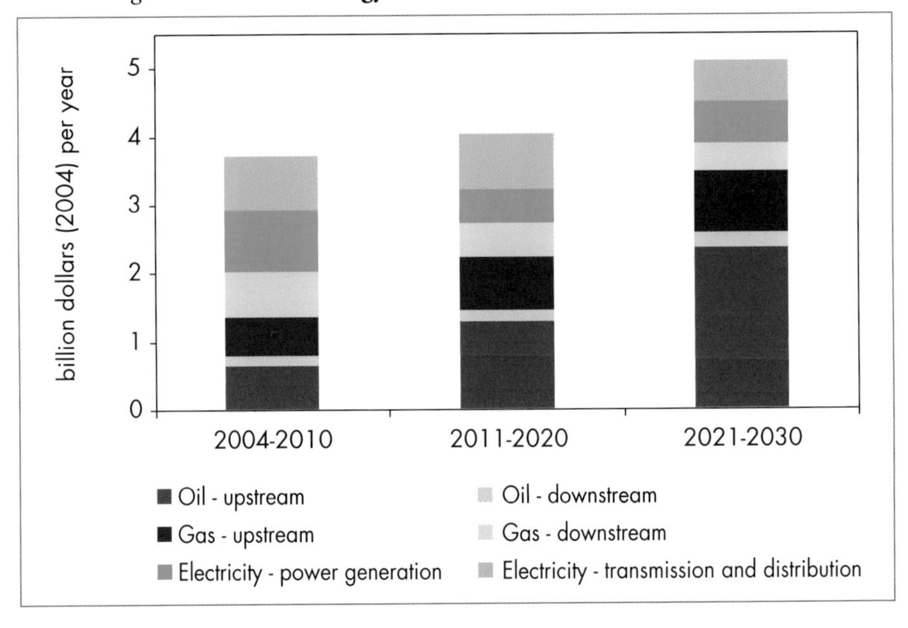

Deferred Investment Scenario[5]

Background

Large investments have been made in the oil and gas sectors of the UAE in recent years, but there are uncertainties about future investment. First, the government may impose a slower production expansion path than we project in the Reference Scenario, if favourable prices permit the country to constrain development without unacceptable budgetary loss. Second, political and economic obstacles may impede investment, for example through delays in negotiating agreements on the role of the international oil companies.[6] In addition, ADNOC may face operational difficulties associated with the emiratisation programme, which aims to increase the share of nationals in the workforce from 40% in 2004 to 75% by 2009. Investment could be hampered if ADNOC is unable to recruit qualified managers and incurs higher operating costs. Strong demand for government spending in other sectors, including

5. See Chapter 7 for a detailed discussion of the assumptions and methodology underlying the Deferred Investment Scenario.
6. The selection of the strategic partner for the Zakum field project took nearly five years.

infrastructure investment, could limit the availability of funds for upstream development.

In the Deferred Investment Scenario, we assume that the ratio of upstream investment to GDP over the projection period remains constant at the level of the last ten years. This results in an average level of investment some 27% lower than in the Reference Scenario. The slower development of the oil and gas industry leads to slower economic growth which, combined with higher end-use energy prices, dampens energy demand growth. Total primary energy demand in the UAE grows at an average annual rate of 2.7% per year over the projection period – 0.2 percentage points lower than in the Reference Scenario. Demand reaches 81 Mtoe in 2030, compared with 84 Mtoe in the Reference Scenario.

Oil and Gas Production

In the Deferred Investment Scenario, production of both crude oil and natural gas, much of which is associated with oil, will grow at a much slower pace. Production of crude oil (including NGLs and condensates) will reach 3.35 mb/d, 1.7 mb/d or 34% lower in 2030 compared with the Reference Scenario (Figure 17.9).

Lower investment has a large impact on production from fields already in production. Production from those fields begins to decline after 2010, reaching 2.0 mb/d in 2030 compared with 2.7 mb/d in the Reference Scenario. Output

Figure 17.9: **UAE's Crude Oil Production in the Deferred Investment Scenario**

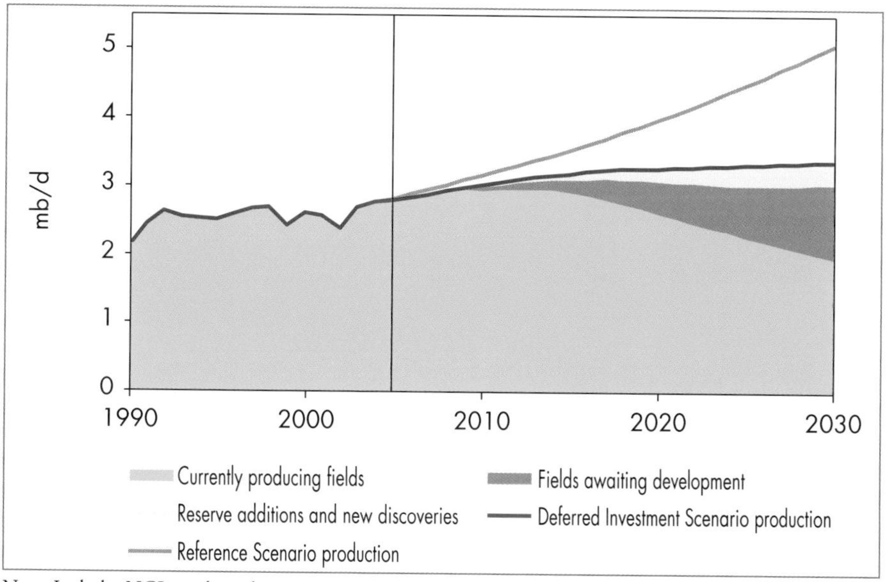

Note: Includes NGLs and condensates.

17

from fields awaiting development will be 0.6 mb/d lower in 2030 than in the Reference Scenario, and that from reserve additions and new discoveries will be 0.3 mb/d lower.

Slower growth in domestic oil demand reduces the need to expand refining capacity in the Deferred Investment Scenario. Capacity grows by 0.9% per year over 2004-2030, reaching 0.9 mb/d, around 18% lower than in the Reference Scenario.

Lower oil output results in lower natural gas production from associated oil/gas fields. Total gas production will begin to decline in the mid-2010s to reach 51 bcm in 2030, 24 bcm or 33%, lower than in the Reference Scenario.

Oil and Gas Exports

The impact of deferred investment on oil and gas trade will be significant, especially towards the end of the projection period (Figure 17.10). Net oil exports are 1.7 mb/d, or 36%, lower in 2030 in the Deferred Investment Scenario, reaching 3.0 mb/d. In the Deferred Investment Scenario, the UAE becomes a net gas importer around 2015, much earlier than in the Reference Scenario. Net imports reach 29.3 bcm in 2030, 21 bcm more than in the Reference Scenario.

Figure 17.10: **UAE's Oil and Gas Net Exports in the Reference and Deferred Investment Scenarios**

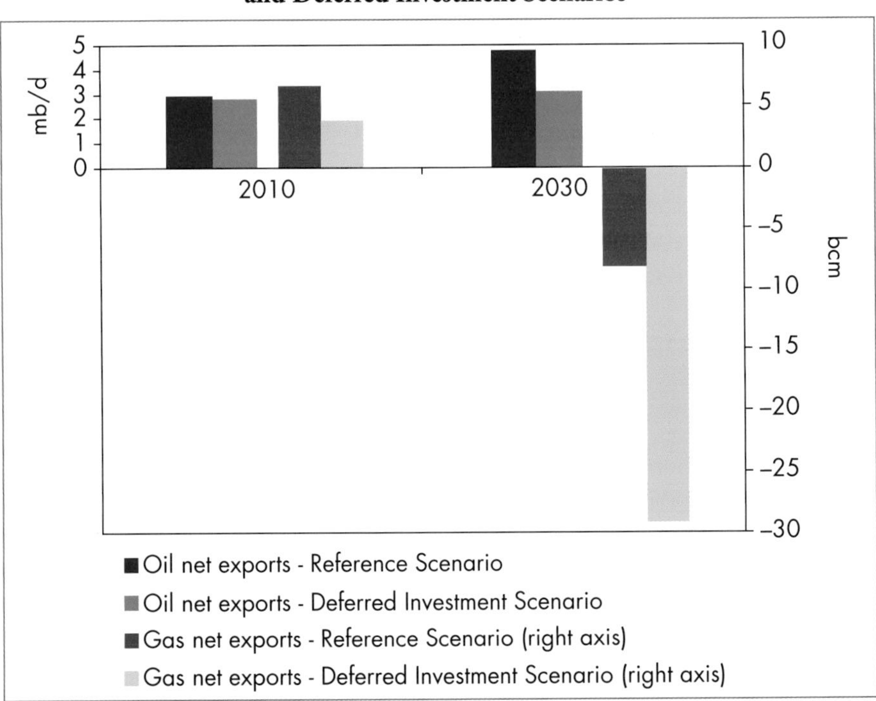

Note: Oil includes NGLs and condensates.

TABLES FOR REFERENCE SCENARIO PROJECTIONS

General Note to the Tables

The energy demand, supply and infrastructure, and investment tables show projections for the following countries/regions:

- MENA
- Middle East
- Iran
- Iraq
- Kuwait
- Qatar
- Saudi Arabia
- United Arab Emirates
- North Africa
- Algeria
- Egypt
- Libya

The definitions for fuels and regions are given in Annex B. In all Annex A tables, biomass is included in other renewables.

For MENA countries, the analysis of energy demand is based on data up to 2003, published in mid-2005 in *Energy Balances of Non-OECD Countries* and obtained from national sources. Supply and infrastructure tables present 2004 data for oil, while gas, electricity and energy indicators data are for 2003. Investment numbers are cumulative for the period indicated. All monetary values are in billion year-2004 dollars.

Oil production includes crude oil, condensates and NGLs. Primary gas demand in the energy balances in this annex includes the energy used in the GTL production process, but excludes gas used as a feedstock in GTL plants.

For net trade, negative values are exports and positive values imports. Both in the text of this book and in the tables, rounding may cause some differences between the total and the sum of the individual components. A hyphen represents a nil figure, while the figure zero appears due to rounding.

Reference Scenario: MENA

	Energy Demand (Mtoe)					Shares (%)					Growth Rates (% per annum)			
	1971	2003	2010	2020	2030	1971	2003	2010	2020	2030	1971-2003	2003-2010	2003-2020	2003-2030
Total Primary Energy Demand	**69**	**570**	**757**	**1 020**	**1 225**	**100**	**100**	**100**	**100**	**100**	**6.8**	**4.2**	**3.5**	**2.9**
Coal	1	14	18	22	25	2	2	2	2	2	8.1	3.7	2.7	2.3
Oil	52	309	393	491	576	75	54	52	48	47	5.7	3.5	2.8	2.3
Gas	13	240	335	491	601	19	42	44	48	49	9.5	4.9	4.3	3.5
Nuclear	–	–	2	2	2	–	–	0	0	0	–	–	–	–
Hydro	1	3	4	5	6	1	0	1	1	0	3.3	5.7	4.2	3.1
Other renewables	2	5	7	9	15	3	1	1	1	1	2.8	4.1	3.8	4.2
Power Generation and Water Desalination	**11**	**180**	**246**	**337**	**401**	**100**	**100**	**100**	**100**	**100**	**9.2**	**4.6**	**3.8**	**3.0**
Coal	0	11	14	17	20	1	6	6	5	5	15.9	4.2	3.0	2.5
Oil	8	60	77	89	92	71	33	31	26	23	6.7	3.6	2.4	1.6
Gas	2	107	150	222	277	19	59	61	66	69	13.2	4.9	4.4	3.6
Nuclear	–	–	2	2	2	–	–	1	0	0	–	–	–	–
Hydro	1	3	4	5	6	9	1	2	2	1	3.3	5.7	4.2	3.1
Other renewables	0	0	0	1	4	0	0	0	0	1	9.3	32.6	20.3	17.5
Other Transformation, Own Use and Losses	**15**	**76**	**112**	**168**	**196**						**5.3**	**5.7**	**4.8**	**3.6**
Total Final Consumption	**47**	**376**	**488**	**639**	**783**	**100**	**100**	**100**	**100**	**100**	**6.7**	**3.8**	**3.2**	**2.8**
Coal	1	2	2	2	3	1	0	0	0	0	3.3	2.2	2.1	1.8
Oil	35	224	281	357	431	74	60	58	56	55	6.0	3.3	2.8	2.4
Gas	8	96	128	171	212	16	25	26	27	27	8.2	4.2	3.5	3.0
Electricity	3	50	72	101	127	6	13	15	16	16	9.2	5.3	4.3	3.5
Renewables	1	5	6	8	10	3	1	1	1	1	3.9	3.6	3.2	3.1

Industry	**17**	**139**	**180**	**232**	**281**	**100**	**100**	**100**	**100**	**100**	**6.8**	**3.8**	**3.1**	**2.6**
Coal	1	2	2	2	3	3	1	1	1	1	3.4	2.2	2.1	1.8
Oil	8	58	70	85	99	49	42	39	37	35	6.3	2.6	2.3	2.0
Gas	6	65	89	118	146	37	47	50	51	52	7.7	4.5	3.5	3.0
Electricity	1	13	18	25	32	9	9	10	11	11	7.0	5.1	4.0	3.4
Renewables	0	1	1	1	2	2	1	1	1	1	2.9	2.1	2.0	1.9
Transport	**16**	**113**	**149**	**198**	**248**	**100**	**100**	**100**	**100**	**100**	**6.4**	**4.0**	**3.3**	**2.9**
Oil	16	111	146	194	243	100	98	98	98	98	6.3	4.0	3.3	2.9
Other fuels	0	2	3	3	4	0	2	2	2	2	18.2	5.8	4.1	3.3
Other Sectors	**13**	**114**	**146**	**192**	**234**	**100**	**100**	**100**	**100**	**100**	**7.0**	**3.6**	**3.1**	**2.7**
Coal	0	–	–	–	–	0	–	–	–	–	–	–	–	–
Oil	9	44	51	60	68	69	39	35	31	29	5.1	2.0	1.8	1.6
Gas	1	29	37	50	63	11	25	25	26	27	9.7	3.5	3.4	3.0
Electricity	2	37	54	76	95	12	33	37	40	41	10.4	5.4	4.3	3.6
Renewables	1	4	4	6	8	7	3	3	3	3	4.2	2.7	2.8	2.9
Non–Energy Use	**2**	**10**	**13**	**17**	**21**						**5.6**	**4.1**	**3.1**	**2.6**
Electricity Generation (TWh)	**41**	**724**	**1 028**	**1 443**	**1 799**	**100**	**100**	**100**	**100**	**100**	**9.4**	**5.1**	**4.1**	**3.4**
Coal	0	48	65	80	96	1	7	6	6	5	17.1	4.2	3.0	2.6
Oil	25	239	309	352	366	61	33	30	24	20	7.3	3.8	2.3	1.6
Gas	5	406	598	931	1 234	12	56	58	65	69	14.8	5.7	5.0	4.2
Nuclear	–	–	6	6	6	–	–	1	0	0	–	–	–	–
Hydro	11	31	45	61	70	26	4	4	4	4	3.3	5.7	4.2	3.1
Other renewables	0	1	4	11	27	0	0	0	1	2	9.3	32.0	18.6	15.2

Reference Scenario: MENA

Supply and Infrastructure		2004	2010	2020	2030
Oil	Oil production (mb/d)	29.0	33.0	41.8	50.5
	Net trade (mb/d)	–22.3	–25.0	–31.8	–38.7
	GTL capacity (mb/d)	–	0.10	0.69	0.80
	Refinery distillation capacity (mb/d)	8.8	10.9	13.4	16.0

		2003	2010	2020	2030
Gas	Gas production (bcm)	385	596	956	1 211
	Net trade (bcm)	–97	–188	–327	–444
Electricity	Generating capacity (GW)	178	256	359	447
	Coal	*7*	*9*	*12*	*15*
	Oil	*66*	*85*	*105*	*112*
	Gas	*92*	*142*	*213*	*282*
	Nuclear	–	*1*	*1*	*1*
	Hydro	*13*	*18*	*23*	*27*
	Other renewables	*0*	*2*	*4*	*10*
Indicators	GDP (billion dollars 2004 using PPPs)	1 970	2 640	3 661	4 870
	Population (million)	323	370	441	505
	Per capita energy demand (toe)	1.8	2.0	2.3	2.4
	Energy intensity*	0.29	0.29	0.28	0.25
	CO_2 emissions (Mt)	1 397	1 783	2 345	2 795

* Toe/thousand dollars of GDP in year-2004 dollars and PPPs.

Reference Scenario: MENA

Investment – billion dollars (2004)		2004-2010	2011-2020	2021-2030	2004-2030
Total Regional Investment		**314**	**558**	**637**	**1 508**
Oil	**Total**	**112**	**223**	**279**	**614**
	Exploration and development	72	170	242	484
	GTL	3	15	3	20
	Refining	37	39	34	110
	Additions	*25*	*27*	*27*	*78*
	Conversion	*12*	*12*	*8*	*32*
Gas	**Total**	**70**	**171**	**196**	**436**
	Exploration and development	38	96	135	269
	Downstream*	32	74	61	168
Electricity	**Total**	**132**	**164**	**162**	**458**
	Generating capacity	60	67	76	203
	Transmission	22	30	27	79
	Distribution	49	67	59	176

* Includes transmission, distribution, LNG and storage.

Reference Scenario: Middle East

	Energy Demand (Mtoe)					Shares (%)					Growth Rates (% per annum)			
	1971	2003	2010	2020	2030	1971	2003	2010	2020	2030	1971-2003	2003-2010	2003-2020	2003-2030
Total Primary Energy Demand	**52**	**446**	**597**	**807**	**963**	**100**	**100**	**100**	**100**	**100**	**7.0**	**4.3**	**3.6**	**2.9**
Coal	0	9	12	15	18	0	2	2	2	2	12.2	4.1	3.0	2.5
Oil	39	247	316	393	457	76	55	53	49	47	5.9	3.6	2.8	2.3
Gas	11	187	262	389	473	21	42	44	48	49	9.2	5.0	4.4	3.5
Nuclear	–	–	2	2	2	–	–	0	0	0	–	–	–	–
Hydro	0	1	2	4	4	1	0	0	0	0	4.6	8.5	6.0	4.3
Other renewables	1	2	3	5	9	1	0	0	1	1	4.0	5.9	5.5	6.0
Power Generation and Water Desalination	**8**	**141**	**194**	**268**	**319**	**100**	**100**	**100**	**100**	**100**	**9.4**	**4.7**	**3.9**	**3.1**
Coal	–	8	11	13	16	–	6	5	5	5	–	4.3	3.1	2.6
Oil	6	52	67	78	81	74	37	35	29	25	7.0	3.9	2.5	1.7
Gas	2	80	112	171	213	22	57	58	64	67	12.8	4.9	4.6	3.7
Nuclear	–	–	2	2	2	–	–	1	1	1	–	–	–	–
Hydro	0	1	2	4	4	4	1	1	1	1	4.6	8.5	6.0	4.3
Other renewables	0	0	0	1	3	0	0	0	0	1	2.6	68.9	32.8	25.7
Other Transformation, Own Use and Losses	**12**	**57**	**88**	**134**	**153**						**5.1**	**6.3**	**5.1**	**3.7**
Total Final Consumption	**34**	**295**	**384**	**501**	**610**	**100**	**100**	**100**	**100**	**100**	**7.0**	**3.8**	**3.2**	**2.7**
Coal	0	1	1	1	1	0	0	0	0	0	5.8	2.5	2.4	1.9
Oil	25	176	221	279	334	72	60	57	56	55	6.3	3.3	2.8	2.4
Gas	7	79	105	139	171	20	27	27	28	28	7.9	4.1	3.4	2.9
Electricity	2	38	55	79	99	6	13	14	16	16	9.6	5.5	4.4	3.6
Renewables	0	2	2	3	5	1	1	1	1	1	4.2	4.5	4.4	4.4

Industry	**12**	**112**	**146**	**190**	**230**	**100**	**100**	**100**	**100**	**100**	**7.3**	**3.9**	**3.2**	**2.7**
Coal	0	1	1	1	1	1	1	1	1	1	5.7	2.5	2.4	2.0
Oil	5	49	60	74	87	43	44	41	39	38	7.3	2.8	2.5	2.1
Gas	6	53	73	97	120	48	48	50	51	52	7.3	4.6	3.6	3.0
Electricity	1	8	12	17	22	8	7	8	9	10	7.3	5.7	4.3	3.7
Renewables	0	0	0	0	0	0	0	0	0	0	4.3	4.2	3.6	3.1
Transport	**12**	**89**	**117**	**153**	**188**	**100**	**100**	**100**	**100**	**100**	**6.4**	**4.0**	**3.2**	**2.8**
Oil	12	89	117	152	187	100	100	100	100	100	6.4	3.9	3.2	2.8
Other fuels	–	0	0	0	0	–	0	0	0	0	–	27.3	16.6	11.9
Other Sectors	**9**	**87**	**111**	**146**	**177**	**100**	**100**	**100**	**100**	**100**	**7.2**	**3.5**	**3.1**	**2.7**
Coal	–	–	–	–	–	–	–	–	–	–	–	–	–	–
Oil	6	31	35	40	45	69	35	31	28	25	5.0	1.7	1.6	1.4
Gas	1	25	31	41	51	15	29	28	28	29	9.5	3.1	2.9	2.6
Electricity	1	30	43	61	77	12	34	39	42	43	10.8	5.5	4.4	3.6
Renewables	0	1	2	3	4	4	2	2	2	3	4.2	4.0	4.1	4.2
Non–Energy Use	**1**	**7**	**10**	**12**	**15**						**5.7**	**4.5**	**3.2**	**2.7**
Electricity Generation (TWh)	**27**	**553**	**794**	**1 122**	**1 397**	**100**	**100**	**100**	**100**	**100**	**9.8**	**5.3**	**4.2**	**3.5**
Coal	–	36	49	61	75	–	7	6	5	5	–	4.3	3.1	2.7
Oil	20	212	280	317	329	71	38	35	28	24	7.7	4.0	2.4	1.6
Gas	4	289	429	688	919	15	52	54	61	66	14.3	5.8	5.2	4.4
Nuclear	–	–	6	6	6	–	–	1	1	0	–	–	–	–
Hydro	4	16	28	42	49	14	3	4	4	4	4.6	8.5	6.0	4.3
Other renewables	0	0	3	7	19	0	0	0	1	1	2.6	67.8	30.0	22.4

Reference Scenario: Middle East

Supply and Infrastructure		2004	2010	2020	2030
Oil	Oil production (mb/d)	24.6	28.3	36.8	45.3
	Net trade (mb/d)	–19.3	–21.8	–28.7	–36.0
	GTL capacity (mb/d)	–	0.10	0.65	0.77
	Refinery distillation capacity (mb/d)	7.0	8.8	10.7	12.9

		2003	2010	2020	2030
Gas	Gas production (bcm)	259	425	692	860
	Net trade (bcm)	–34	–102	–185	–244
Electricity	Generating capacity (GW)	140	204	286	355
	Coal	*5*	*7*	*9*	*11*
	Oil	*56*	*74*	*91*	*98*
	Gas	*69*	*108*	*164*	*217*
	Nuclear	*–*	*1*	*1*	*1*
	Hydro	*9*	*13*	*18*	*21*
	Other renewables	*0*	*1*	*3*	*7*
Indicators	GDP (billion dollars 2004 using PPPs)	1 268	1 707	2 385	3 218
	Population (million)	177	207	252	295
	Per capita energy demand (toe)	2.5	2.9	3.2	3.3
	Energy intensity*	0.35	0.35	0.34	0.30
	CO_2 emissions (Mt)	1 102	1 411	1 852	2 191

* Toe/thousand dollars of GDP in year-2004 dollars and PPPs.

Reference Scenario: Middle East

Investment – billion dollars (2004)		**2004-2010**	**2011-2020**	**2021-2030**	**2004-2030**
Total Regional Investment		**255**	**443**	**506**	**1 203**
Oil	**Total**	**92**	**186**	**239**	**517**
	Exploration and development	58	142	209	409
	GTL	3	14	3	19
	Refining	31	31	27	89
	Additions	*21*	*21*	*21*	*62*
	Conversions	*10*	*10*	*6*	*26*
Gas	**Total**	**53**	**125**	**140**	**318**
	Exploration and development	28	71	98	197
	Downstream*	25	54	41	121
Electricity	**Total**	**109**	**132**	**127**	**368**
	Generating capacity	51	54	62	166
	Transmission	18	24	20	63
	Distribution	40	54	45	139

* Includes transmission, distribution, LNG and storage.

Reference Scenario: Iran

	Energy Demand (Mtoe)					Shares (%)					Growth Rates (% per annum)			
	1971	2003	2010	2020	2030	1971	2003	2010	2020	2030	1971-2003	2003-2010	2003-2020	2003-2030
Total Primary Energy Demand	**19**	**136**	**173**	**225**	**271**	**100**	**100**	**100**	**100**	**100**	**6.3**	**3.4**	**3.0**	**2.6**
Coal	0	1	1	2	2	1	1	1	1	1	5.1	3.2	2.8	2.3
Oil	16	66	79	96	111	84	48	46	43	41	4.5	2.6	2.2	2.0
Gas	2	68	88	120	149	12	50	51	54	55	11.1	3.8	3.5	3.0
Nuclear	–	–	2	2	2	–	–	1	1	1	–	–	–	–
Hydro	0	1	2	2	3	1	1	1	1	1	4.5	8.6	5.4	3.7
Other renewables	0	1	1	2	5	2	1	1	1	2	2.8	6.8	6.8	7.0
Power Generation and Water Desalination	**2**	**33**	**43**	**56**	**66**	**100**	**100**	**100**	**100**	**100**	**9.5**	**4.1**	**3.2**	**2.6**
Coal	–	–	–	–	–	–	–	–	–	–	–	–	–	–
Oil	1	6	7	7	6	82	19	16	13	10	4.6	1.2	1.0	0.1
Gas	0	26	33	44	54	6	78	76	79	81	18.9	3.8	3.2	2.8
Nuclear	–	–	2	2	2	–	–	4	3	3	–	–	–	–
Hydro	0	1	2	2	3	13	3	4	4	4	4.5	8.6	5.4	3.7
Other renewables	–	0	0	1	2	–	0	0	1	3	–	75.5	37.3	27.2
Other Transformation, Own Use and Losses	**4**	**8**	**13**	**23**	**29**						**2.3**	**7.3**	**6.4**	**4.9**
Total Final Consumption	**14**	**109**	**135**	**171**	**207**	**100**	**100**	**100**	**100**	**100**	**6.6**	**3.1**	**2.7**	**2.4**
Coal	0	1	1	1	1	1	1	1	1	1	5.4	2.9	2.8	2.2
Oil	12	60	71	86	101	82	55	53	51	49	5.3	2.5	2.2	2.0
Gas	1	38	48	62	77	11	35	35	36	37	10.6	3.4	3.0	2.7
Electricity	1	10	14	19	24	4	9	11	11	12	9.2	5.1	3.8	3.3
Renewables	0	1	1	2	3	2	1	1	1	2	2.8	5.4	5.4	5.6

Industry	**4**	**27**	**34**	**42**	**51**	**100**	**100**	**100**	**100**	**100**	**5.8**	**3.3**	**2.6**	**2.4**
Coal	0	1	1	1	1	2	2	2	2	2	5.3	3.0	2.8	2.3
Oil	2	10	11	12	13	54	37	31	28	25	4.6	1.0	1.0	1.0
Gas	1	13	17	22	27	34	48	50	52	54	6.9	4.1	3.2	2.8
Electricity	0	3	5	7	9	9	13	16	17	18	7.0	6.1	4.2	3.6
Renewables	0	0	0	0	0	1	1	1	1	1	4.8	4.2	3.6	3.1
Transport	**5**	**30**	**40**	**52**	**65**	**100**	**100**	**100**	**100**	**100**	**5.9**	**4.1**	**3.2**	**2.8**
Oil	5	30	40	52	64	100	100	100	100	99	5.9	4.0	3.2	2.8
Other fuels	–	0	0	0	0	–	0	0	0	1	–	27.3	16.6	11.9
Other Sectors	**4**	**48**	**57**	**72**	**87**	**100**	**100**	**100**	**100**	**100**	**7.8**	**2.5**	**2.4**	**2.2**
Coal	–	–	–	–	–	–	–	–	–	–	–	–	–	–
Oil	4	16	17	18	19	89	34	30	25	22	4.6	0.5	0.6	0.6
Gas	0	25	30	40	50	0	51	53	56	58	58.4	3.0	2.9	2.7
Electricity	0	7	9	12	15	5	14	16	17	18	11.4	4.6	3.6	3.1
Renewables	0	1	1	1	2	6	1	1	2	3	2.2	4.6	5.1	5.7
Non–Energy Use	**0**	**3**	**3**	**4**	**5**						**6.3**	**2.1**	**2.0**	**1.9**
Electricity Generation (TWh)	**8**	**153**	**214**	**286**	**359**	**100**	**100**	**100**	**100**	**100**	**9.6**	**4.9**	**3.8**	**3.2**
Coal	–	–	–	–	–	–	–	–	–	–	–	–	–	–
Oil	5	24	27	30	26	62	16	13	10	7	5.1	1.4	1.1	0.2
Gas	0	117	159	220	289	5	77	74	77	80	19.2	4.5	3.8	3.4
Nuclear	–	–	6	6	6	–	–	3	2	2	–	–	–	–
Hydro	3	11	20	27	30	33	7	9	9	8	4.5	8.6	5.4	3.7
Other renewables	–	0	1	4	9	–	0	1	1	2	–	74.4	32.5	23.3

Reference Scenario: Iran

Supply and Infrastructure		2004	2010	2020	2030
Oil	Oil production (mb/d)	4.1	4.5	5.5	6.8
	Net trade (mb/d)	–2.7	–2.8	–3.6	–4.4
	GTL capacity (mb/d)	–	–	0.08	0.13
	Refinery distillation capacity (mb/d)	1.5	1.7	2.2	2.6

		2003	2010	2020	2030
Gas	Gas production (bcm)	78	109	177	240
	Net trade (bcm)	2	–5	–31	–57
Electricity	Generating capacity (GW)	33	47	64	80
	Coal	–	–	–	–
	Oil	*10*	*11*	*13*	*12*
	Gas	*19*	*27*	*38*	*52*
	Nuclear	–	*1*	*1*	*1*
	Hydro	*4*	*8*	*10*	*11*
	Other renewables	*0*	*1*	*1*	*3*
Indicators	GDP (billion dollars 2004 using PPPs)	487	664	925	1 249
	Population (million)	66	73	84	91
	Per capita energy demand (toe)	2.1	2.4	2.7	3.0
	Energy intensity*	0.28	0.26	0.24	0.22
	CO_2 emissions (Mt)	349	427	545	648

* Toe/thousand dollars of GDP in year-2004 dollars and PPPs.

Reference Scenario: Iran

Investment – billion dollars (2004)		2004-2010	2011-2020	2021-2030	2004-2030
Total Country Investment		**51**	**93**	**110**	**254**
Oil	**Total**	**11**	**30**	**36**	**77**
	Exploration and development	8	21	30	59
	GTL	–	2	1	3
	Refining	3	8	5	16
	Additions	*2*	*6*	*3*	*11*
	Conversions	*1*	*2*	*1*	*5*
Gas	**Total**	**13**	**32**	**40**	**85**
	Exploration and development	6	18	28	52
	Downstream*	6	14	12	32
Electricity	**Total**	**28**	**30**	**34**	**92**
	Generating capacity	14	13	16	43
	Transmission	4	5	5	15
	Distribution	10	12	12	34

* Includes transmission, distribution, LNG and storage.

Reference Scenario: Iraq

	Energy Demand (Mtoe)					Shares (%)					Growth Rates (% per annum)			
	1971	2003	2010	2020	2030	1971	2003	2010	2020	2030	1971-2003	2003-2010	2003-2020	2003-2030
Total Primary Energy Demand	**5**	**26**	**35**	**47**	**62**	**100**	**100**	**100**	**100**	**100**	**5.6**	**4.6**	**3.7**	**3.3**
Coal	–	–	–	–	–	–	–	–	–	–	–	–	–	–
Oil	4	24	33	40	49	82	95	92	85	79	6.1	4.2	3.0	2.6
Gas	1	1	3	7	12	17	5	7	14	20	1.6	10.9	10.2	8.8
Nuclear	–	–	–	–	–	–	–	–	–	–	–	–	–	–
Hydro	0	0	0	0	1	0	0	0	1	1	2.4	20.9	16.4	11.9
Other renewables	0	0	0	0	0	1	0	0	0	0	–0.0	1.0	1.0	1.0
Power Generation and Water Desalination	**1**	**7**	**8**	**11**	**14**	**100**	**100**	**100**	**100**	**100**	**5.1**	**2.0**	**2.6**	**2.5**
Coal	–	–	–	–	–	–	–	–	–	–	–	–	–	–
Oil	1	7	8	8	6	99	99	98	70	45	5.2	1.8	0.5	–0.4
Gas	–	–	0	3	7	–	–	0	26	50	–	–	–	–
Nuclear	–	–	–	–	–	–	–	–	–	–	–	–	–	–
Hydro	0	0	0	0	1	1	1	2	4	5	2.4	20.9	16.4	11.9
Other renewables	–	–	–	–	–	–	–	–	–	–	–	–	–	–
Other Transformation, Own Use and Losses	**0**	**3**	**4**	**5**	**6**						**8.5**	**1.0**	**2.1**	**2.0**
Total Final Consumption	**3**	**18**	**27**	**36**	**49**	**100**	**100**	**100**	**100**	**100**	**5.6**	**6.4**	**4.4**	**3.9**
Coal	–	–	–	–	–	–	–	–	–	–	–	–	–	–
Oil	2	14	21	28	37	67	80	78	76	75	6.2	6.0	4.1	3.7
Gas	1	1	3	4	5	25	7	10	10	11	1.6	10.9	6.6	5.5
Electricity	0	2	3	5	7	8	13	13	13	14	7.5	5.8	4.4	4.0
Renewables	0	0	0	0	0	1	0	0	0	0	–1.1	1.0	1.0	1.0

Industry	**1**	**5**	**7**	**9**	**12**	**100**	**100**	**100**	**100**	**100**	**4.7**	**4.5**	**3.7**	**3.5**
Coal	–	–	–	–	–	–	–	–	–	–	–	–	–	–
Oil	0	4	4	5	7	23	74	60	58	56	8.6	1.5	2.2	2.4
Gas	1	1	3	4	5	69	26	40	42	44	1.6	10.9	6.6	5.5
Electricity	0	–	–	–	–	7	–	–	–	–	–	–	–	–
Renewables	0	–	–	–	–	1	–	–	–	–	–	–	–	–
Transport	**1**	**7**	**13**	**17**	**22**	**100**	**100**	**100**	**100**	**100**	**6.8**	**7.9**	**4.9**	**4.2**
Oil	1	7	13	17	22	100	100	100	100	100	6.8	7.9	4.9	4.2
Other fuels	–	–	–	–	–	–	–	–	–	–	–	–	–	–
Other Sectors	**1**	**5**	**7**	**9**	**12**	**100**	**100**	**100**	**100**	**100**	**5.2**	**5.5**	**4.1**	**3.7**
Coal	–	–	–	–	–	–	–	–	–	–	–	–	–	–
Oil	1	2	3	4	6	81	49	48	46	46	3.6	5.1	3.7	3.4
Gas	0	–	–	–	–	0	–	–	–	–	–	–	–	–
Electricity	0	2	3	5	7	17	50	52	53	54	8.9	5.8	4.4	4.0
Renewables	0	0	0	0	0	2	0	0	0	0	0.1	1.0	1.0	1.0
Non–Energy Use	**0**	**1**	**1**	**2**	**2**						**5.2**	**7.6**	**5.3**	**4.7**
Electricity Generation (TWh)	**3**	**28**	**40**	**59**	**82**	**100**	**100**	**100**	**100**	**100**	**7.5**	**4.9**	**4.4**	**4.0**
Coal	–	–	–	–	–	–	–	–	–	–	–	–	–	–
Oil	3	28	38	37	30	93	98	96	62	36	7.7	4.5	1.6	0.2
Gas	–	–	0	17	43	–	–	0	29	53	–	–	–	–
Nuclear	–	–	–	–	–	–	–	–	–	–	–	–	–	–
Hydro	0	0	2	6	9	7	2	4	10	11	2.4	20.9	16.4	11.9
Other renewables	–	–	–	–	–	–	–	–	–	–	–	–	–	–

Reference Scenario: Iraq

Supply and Infrastructure		2004	2010	2020	2030
Oil	Oil production (mb/d)	2.0	3.2	5.4	7.9
	Net trade (mb/d)	–1.4	–2.5	–4.6	–6.9
	GTL capacity (mb/d)	–	–	–	–
	Refinery distillation capacity (mb/d)	0.6	0.8	1.0	1.2
		2003	**2010**	**2020**	**2030**
Gas	Gas production (bcm)	2	4	15	32
	Net trade (bcm)	–	–1	–7	–17
Electricity	Generating capacity (GW)	11	12	16	22
	Coal	–	–	–	–
	Oil	*8*	*10*	*10*	*9*
	Gas	–	*0*	*3*	*8*
	Nuclear	–	–	–	–
	Hydro	*3*	*3*	*3*	*5*
	Other renewables	–	–	–	–
Indicators	GDP (billion dollars 2004 using PPPs)	27	55	96	164
	Population (million)	25	30	37	45
	Per capita energy demand (toe)	1.0	1.2	1.3	1.4
	Energy intensity*	0.96	0.64	0.50	0.38
	CO_2 emissions (Mt)	69	90	110	132

* Toe/thousand dollars of GDP in year-2004 dollars and PPPs.

Reference Scenario: Iraq

Investment – billion dollars (2004)		2004-2010	2011-2020	2021-2030	2004-2030
Total Country Investment		**16**	**31**	**50**	**96**
Oil	**Total**	**7**	**20**	**33**	**59**
	Exploration and development	4	16	30	51
	GTL	–	–	–	–
	Refining	3	3	3	8
	Additions	*2*	*2*	*2*	*6*
	Conversions	*1*	*1*	*0*	*2*
Gas	**Total**	**0**	**4**	**6**	**11**
	Exploration and development	0	1	3	5
	Downstream*	0	3	3	6
Electricity	**Total**	**9**	**7**	**10**	**26**
	Generating capacity	5	3	5	12
	Transmission	1	1	2	4
	Distribution	3	3	4	9

* Includes transmission, distribution, LNG and storage.

A

Reference Scenario: Kuwait

	Energy Demand (Mtoe)					Shares (%)					Growth Rates (% per annum)			
	1971	2003	2010	2020	2030	1971	2003	2010	2020	2030	1971-2003	2003-2010	2003-2020	2003-2030
Total Primary Energy Demand	**6**	**23**	**30**	**41**	**49**	**100**	**100**	**100**	**100**	**100**	**4.3**	**4.0**	**3.5**	**2.8**
Coal	0	–	–	–	–	0	–	–	–	–	–	–	–	–
Oil	2	15	21	24	27	28	66	68	59	56	7.1	4.5	2.8	2.2
Gas	4	8	10	17	21	71	34	32	41	44	1.9	3.1	4.6	3.8
Nuclear	–	–	–	–	–	–	–	–	–	–	–	–	–	–
Hydro	–	–	–	–	–	–	–	–	–	–	–	–	–	–
Other renewables	0	–	–	–	0	0	–	–	–	0	–	–	–	–
Power Generation and Water Desalination	**1**	**9**	**12**	**16**	**18**	**100**	**100**	**100**	**100**	**100**	**7.3**	**4.1**	**3.6**	**2.7**
Coal	–	–	–	–	–	–	–	–	–	–	–	–	–	–
Oil	0	7	9	10	10	0	80	73	60	57	27.6	2.9	1.8	1.5
Gas	1	2	3	6	8	100	20	27	40	43	2.0	8.4	8.0	5.7
Nuclear	–	–	–	–	–	–	–	–	–	–	–	–	–	–
Hydro	–	–	–	–	–	–	–	–	–	–	–	–	–	–
Other renewables	–	–	–	–	0	–	–	–	–	0	–	–	–	–
Other Transformation, Own Use and Losses	**1**	**8**	**11**	**14**	**17**						**7.7**	**3.8**	**3.3**	**2.7**
Total Final Consumption	**4**	**9**	**12**	**17**	**20**	**100**	**100**	**100**	**100**	**100**	**2.3**	**4.3**	**3.5**	**3.0**
Coal	–	–	–	–	–	–	–	–	–	–	–	–	–	–
Oil	1	5	7	9	10	22	56	56	53	51	5.3	4.4	3.2	2.6
Gas	3	2	3	4	6	74	22	21	25	28	–1.5	3.6	4.2	3.9
Electricity	0	2	3	4	4	3	23	23	22	22	8.6	4.6	3.4	2.8
Renewables	0	–	–	–	–	0	–	–	–	–	–	–	–	–

Industry	**2**	**3**	**3**	**5**	**7**	**100**	**100**	**100**	**100**	**100**	**0.9**	**4.0**	**4.1**	**3.7**
Coal	–	–	–	–	–	–	–	–	–	–	–	–	–	–
Oil	–	0	0	0	1	–	9	9	8	7	–	3.6	3.2	2.8
Gas	2	2	3	4	6	98	78	76	79	81	0.1	3.6	4.2	3.9
Electricity	0	0	1	1	1	2	13	15	13	12	7.3	6.1	4.3	3.2
Renewables	–	–	–	–	–	–	–	–	–	–	–	–	–	–
Transport	**1**	**3**	**4**	**6**	**7**	**100**	**100**	**100**	**100**	**100**	**4.1**	**4.8**	**3.4**	**2.8**
Oil	1	3	4	6	7	100	100	100	100	100	4.1	4.8	3.4	2.8
Other fuels	–	–	–	–	–	–	–	–	–	–	–	–	–	–
Other Sectors	**2**	**3**	**4**	**4**	**5**	**100**	**100**	**100**	**100**	**100**	**2.1**	**2.5**	**2.0**	**1.8**
Coal	–	–	–	–	–	–	–	–	–	–	–	–	–	–
Oil	0	1	1	1	1	5	43	36	31	28	9.3	–0.2	0.1	0.1
Gas	1	–	–	–	–	87	–	–	–	–	–	–	–	–
Electricity	0	2	2	3	4	7	57	64	69	72	9.0	4.3	3.2	2.7
Renewables	0	–	–	–	–	1	–	–	–	–	–	–	–	–
Non–Energy Use	**0**	**0**	**1**	**2**	**2**						**7.2**	**11.9**	**7.2**	**5.3**
Electricity Generation (TWh)	**3**	**40**	**54**	**68**	**82**	**100**	**100**	**100**	**100**	**100**	**8.9**	**4.3**	**3.2**	**2.7**
Coal	–	–	–	–	–	–	–	–	–	–	–	–	–	–
Oil	0	32	40	44	48	0	80	75	65	59	30.1	3.4	2.0	1.5
Gas	3	8	13	24	34	100	20	25	35	41	3.5	7.8	6.7	5.5
Nuclear	–	–	–	–	–	–	–	–	–	–	–	–	–	–
Hydro	–	–	–	–	–	–	–	–	–	–	–	–	–	–
Other renewables	–	–	–	–	0	–	–	–	–	0	–	–	–	–

Reference Scenario: Kuwait

Supply and Infrastructure		2004	2010	2020	2030
Oil	Oil production (mb/d)	2.5	2.9	3.8	4.9
	Net trade (mb/d)	–2.2	–2.5	–3.4	–4.4
	GTL capacity (mb/d)	–	–	–	–
	Refinery distillation capacity (mb/d)	0.9	1.3	1.4	1.4

		2003	2010	2020	2030
Gas	Gas production (bcm)	10	11	18	21
	Net trade (bcm)	–	1	3	6
Electricity	Generating capacity (GW)	10	14	19	21
	Coal	–	–	–	–
	Oil	*6*	*8*	*9*	*10*
	Gas	*4*	*6*	*9*	*11*
	Nuclear	–	–	–	–
	Hydro	–	–	–	–
	Other renewables	–	–	–	*0*
Indicators	GDP (billion dollars 2004 using PPPs)	40	51	69	90
	Population (million)	2	3	3	4
	Per capita energy demand (toe)	9.5	10.4	11.8	12.2
	Energy intensity*	0.58	0.59	0.59	0.54
	CO_2 emissions (Mt)	58	68	92	109

* Toe/thousand dollars of GDP in year-2004 dollars and PPPs.

Reference Scenario: Kuwait

Investment – billion dollars (2004)		**2004-2010**	**2011-2020**	**2021-2030**	**2004-2030**
Total Country Investment		**21**	**28**	**37**	**86**
Oil	**Total**	**13**	**18**	**27**	**59**
	Exploration and development	5	16	26	47
	GTL	–	–	–	–
	Refining	8	3	1	12
	Additions	*5*	*1*	*1*	*6*
	Conversions	*3*	*2*	*0*	*6*
Gas	**Total**	**1**	**3**	**4**	**8**
	Exploration and development	1	2	2	5
	Downstream*	1	2	1	4
Electricity	**Total**	**7**	**6**	**6**	**19**
	Generating capacity	3	3	3	9
	Transmission	1	1	1	3
	Distribution	2	2	2	7

* Includes transmission, distribution, LNG and storage.

Reference Scenario: Qatar

	Energy Demand (Mtoe)					Shares (%)					Growth Rates (% per annum)			
	1971	2003	2010	2020	2030	1971	2003	2010	2020	2030	1971-2003	2003-2010	2003-2020	2003-2030
Total Primary Energy Demand	**1**	**15**	**32**	**59**	**67**	**100**	**100**	**100**	**100**	**100**	**9.1**	**11.2**	**8.3**	**5.7**
Coal	–	–	–	–	–	–	–	–	–	–	–	–	–	–
Oil	0	4	6	9	12	11	26	20	15	18	12.2	7.1	5.1	4.2
Gas	1	11	25	50	55	89	74	80	85	82	8.5	12.4	9.2	6.1
Nuclear	–	–	–	–	–	–	–	–	–	–	–	–	–	–
Hydro	–	–	–	–	–	–	–	–	–	–	–	–	–	–
Other renewables	–	–	0	0	0	–	–	0	0	0	–	–	–	–
Power Generation and Water Desalination	**0**	**4**	**5**	**7**	**7**	**100**	**100**	**100**	**100**	**100**	**6.8**	**4.3**	**3.1**	**2.3**
Coal	–	–	–	–	–	–	–	–	–	–	–	–	–	–
Oil	0	–	–	–	–	4	–	–	–	–	–	–	–	–
Gas	0	4	5	7	7	96	100	100	100	100	6.9	4.3	3.1	2.3
Nuclear	–	–	–	–	–	–	–	–	–	–	–	–	–	–
Hydro	–	–	–	–	–	–	–	–	–	–	–	–	–	–
Other renewables	–	–	0	0	0	–	–	0	0	0	–	–	–	–
Other Transformation, Own Use and Losses	**0**	**3**	**15**	**36**	**38**						**27.0**	**23.9**	**15.2**	**9.6**
Total Final Consumption	**0**	**9**	**13**	**18**	**24**	**100**	**100**	**100**	**100**	**100**	**9.7**	**5.7**	**4.4**	**3.7**
Coal	–	–	–	–	–	–	–	–	–	–	–	–	–	–
Oil	0	4	6	9	12	18	46	48	49	50	13.0	6.1	4.7	3.9
Gas	0	4	6	8	10	76	43	42	42	41	7.8	5.5	4.2	3.5
Electricity	0	1	1	2	2	6	11	10	9	9	11.8	4.7	3.5	2.9
Renewables	–	–	–	–	–	–	–	–	–	–	–	–	–	–

Industry	**0**	**6**	**9**	**13**	**16**	**100**	**100**	**100**	**100**	**100**	**9.5**	**5.8**	**4.3**	**3.6**
Coal	–	–	–	–	–	–	–	–	–	–	–	–	–	–
Oil	–	2	4	5	6	–	37	38	37	37	–	6.2	4.3	3.7
Gas	0	4	6	8	10	100	60	59	60	59	7.8	5.5	4.2	3.5
Electricity	–	0	0	0	1	–	3	3	3	3	–	5.5	4.0	3.4
Renewables	–	–	–	–	–	–	–	–	–	–	–	–	–	–
Transport	**0**	**2**	**3**	**4**	**6**	**100**	**100**	**100**	**100**	**100**	**10.2**	**6.2**	**5.1**	**4.3**
Oil	0	2	3	4	6	100	100	100	100	100	10.2	6.2	5.1	4.3
Other fuels	–	–	–	–	–	–	–	–	–	–	–	–	–	–
Other Sectors	**0**	**1**	**1**	**1**	**2**	**100**	**100**	**100**	**100**	**100**	**10.4**	**4.4**	**3.3**	**2.8**
Coal	–	–	–	–	–	–	–	–	–	–	–	–	–	–
Oil	0	0	0	0	0	12	7	6	6	6	8.3	3.4	3.0	2.5
Gas	0	–	–	–	–	0	–	–	–	–	–	–	–	–
Electricity	0	1	1	1	2	79	93	94	94	94	10.9	4.4	3.4	2.8
Renewables	–	–	–	–	–	–	–	–	–	–	–	–	–	–
Non–Energy Use	**–**	**–**	**–**	**–**	**–**						**–**	**–**	**–**	**–**
Electricity Generation (TWh)	**0**	**12**	**17**	**22**	**26**	**100**	**100**	**100**	**100**	**100**	**12.1**	**4.7**	**3.5**	**2.9**
Coal	–	–	–	–	–	–	–	–	–	–	–	–	–	–
Oil	0	–	–	–	–	10	–	–	–	–	–	–	–	–
Gas	0	12	16	22	26	90	100	100	100	100	12.4	4.6	3.5	2.9
Nuclear	–	–	–	–	–	–	–	–	–	–	–	–	–	–
Hydro	–	–	–	–	–	–	–	–	–	–	–	–	–	–
Other renewables	–	–	0	0	0	–	–	0	0	0	–	–	–	–

Reference Scenario: Qatar

Supply and Infrastructure		2004	2010	2020	2030
Oil	Oil production (mb/d)	1.0	1.1	1.2	1.2
	Net trade (mb/d)	–0.9	–1.0	–1.0	–0.9
	GTL capacity (mb/d)	–	0.10	0.58	0.64
	Refinery distillation capacity (mb/d)	0.1	0.2	0.3	0.3

		2003	2010	2020	2030
Gas	Gas production (bcm)	33	115	220	255
	Net trade (bcm)	–19	–78	–126	–152
Electricity	Generating capacity (GW)	3	5	6	8
	Coal	–	–	–	–
	Oil	*0*	*0*	*0*	*0*
	Gas	*3*	*5*	*6*	*8*
	Nuclear	–	–	–	–
	Hydro	–	–	–	–
	Other renewables	–	*0*	*0*	*0*
Indicators	GDP (billion dollars 2004 using PPPs)	25	39	55	70
	Population (million)	1	1	1	1
	Per capita energy demand (toe)	20.8	35.7	57.3	58.1
	Energy intensity*	0.61	0.82	1.07	0.96
	CO_2 emissions (Mt)	27	45	69	81

* Toe/thousand dollars of GDP in year-2004 dollars and PPPs.

Reference Scenario: Qatar

Investment – billion dollars (2004)		**2004-2010**	**2011-2020**	**2021-2030**	**2004-2030**
Total Country Investment		**29**	**67**	**59**	**155**
Oil	**Total**	**9**	**24**	**17**	**50**
	Exploration and development	5	11	14	31
	GTL	3	12	2	16
	Refining	1	1	1	3
	Additions	*1*	*1*	*1*	*2*
	Conversions	*0*	*0*	*0*	*1*
Gas	**Total**	**17**	**41**	**41**	**99**
	Exploration and development	10	25	31	65
	Downstream*	8	16	10	34
Electricity	**Total**	**2**	**2**	**2**	**6**
	Generating capacity	1	1	1	3
	Transmission	0	0	0	1
	Distribution	1	1	1	2

* Includes transmission, distribution, LNG and storage.

Reference Scenario: Saudi Arabia

	Energy Demand (Mtoe)					Shares (%)					Growth Rates (% per annum)			
	1971	2003	2010	2020	2030	1971	2003	2010	2020	2030	1971-2003	2003-2010	2003-2020	2003-2030
Total Primary Energy Demand	**6**	**131**	**181**	**247**	**289**	**100**	**100**	**100**	**100**	**100**	**10.0**	**4.8**	**3.8**	**3.0**
Coal	–	–	–	–	–	–	–	–	–	–	–	–	–	–
Oil	5	82	110	141	162	81	62	61	57	56	9.1	4.4	3.3	2.6
Gas	1	49	70	106	127	19	38	39	43	44	12.4	5.3	4.6	3.6
Nuclear	–	–	–	–	–	–	–	–	–	–	–	–	–	–
Hydro	–	–	–	–	–	–	–	–	–	–	–	–	–	–
Other renewables	0	0	0	0	0	0	0	0	0	0	3.5	44.0	22.1	16.1
Power Generation and Water Desalination	**1**	**43**	**65**	**98**	**115**	**100**	**100**	**100**	**100**	**100**	**14.6**	**6.1**	**5.0**	**3.7**
Coal	–	–	–	–	–	–	–	–	–	–	–	–	–	–
Oil	1	20	30	37	39	100	46	46	38	34	11.9	6.0	3.7	2.6
Gas	–	23	35	61	76	–	54	54	62	66	–	6.2	6.0	4.5
Nuclear	–	–	–	–	–	–	–	–	–	–	–	–	–	–
Hydro	–	–	–	–	–	–	–	–	–	–	–	–	–	–
Other renewables	–	–	0	0	0	–	–	0	0	0	–	–	–	–
Other Transformation, Own Use and Losses	**3**	**22**	**32**	**40**	**45**						**6.0**	**5.3**	**3.6**	**2.7**
Total Final Consumption	**2**	**79**	**105**	**139**	**165**	**100**	**100**	**100**	**100**	**100**	**11.5**	**4.1**	**3.4**	**2.8**
Coal	–	–	–	–	–	–	–	–	–	–	–	–	–	–
Oil	2	52	66	85	101	93	65	63	61	61	10.3	3.4	3.0	2.5
Gas	0	17	22	28	33	0	21	21	20	20	56.4	4.4	3.1	2.5
Electricity	0	11	17	26	31	7	14	16	19	19	14.0	6.5	5.2	4.0
Renewables	0	0	0	0	0	0	0	0	0	0	3.5	44.0	22.1	15.3

Industry	**1**	**43**	**55**	**69**	**81**	**100**	**100**	**100**	**100**	**100**	**13.6**	**3.8**	**2.9**	**2.4**
Coal	–	–	–	–	–	–	–	–	–	–	–	–	–	–
Oil	1	24	31	38	44	85	57	56	55	55	12.2	3.3	2.7	2.2
Gas	–	17	22	28	33	–	39	41	40	40	–	4.4	3.1	2.5
Electricity	0	1	2	3	4	15	4	4	5	5	8.5	5.3	4.5	3.8
Renewables	–	–	–	–	–	–	–	–	–	–	–	–	–	–
Transport	**1**	**24**	**30**	**41**	**50**	**100**	**100**	**100**	**100**	**100**	**9.6**	**3.3**	**3.2**	**2.8**
Oil	1	24	30	41	50	100	100	100	100	100	9.6	3.3	3.2	2.8
Other fuels	–	–	–	–	–	–	–	–	–	–	–	–	–	–
Other Sectors	**0**	**11**	**16**	**25**	**30**	**100**	**100**	**100**	**100**	**100**	**12.4**	**6.3**	**5.2**	**3.9**
Coal	–	–	–	–	–	–	–	–	–	–	–	–	–	–
Oil	0	1	2	2	3	78	12	10	9	9	6.1	3.5	3.3	2.5
Gas	–	–	–	–	–	–	–	–	–	–	–	–	–	–
Electricity	0	9	15	23	27	22	88	89	90	91	17.4	6.7	5.3	4.0
Renewables	0	0	0	0	0	1	0	0	1	1	3.5	44.0	22.1	15.3
Non–Energy Use	**0**	**2**	**3**	**4**	**4**						**7.8**	**5.9**	**3.6**	**2.6**
Electricity Generation (TWh)	**2**	**153**	**235**	**356**	**426**	**100**	**100**	**100**	**100**	**100**	**14.4**	**6.3**	**5.1**	**3.9**
Coal	–	–	–	–	–	–	–	–	–	–	–	–	–	–
Oil	2	82	116	137	145	100	54	49	38	34	12.2	5.1	3.1	2.1
Gas	–	71	119	219	280	–	46	51	62	66	–	7.6	6.9	5.2
Nuclear	–	–	–	–	–	–	–	–	–	–	–	–	–	–
Hydro	–	–	–	–	–	–	–	–	–	–	–	–	–	–
Other renewables	–	–	0	0	1	–	–	0	0	0	–	–	–	–

Reference Scenario: Saudi Arabia

	Supply and Infrastructure	2004	2010	2020	2030
Oil	Oil production (mb/d)	10.4	11.9	15.4	18.2
	Net trade (mb/d)	–8.3	–9.3	–12.1	–14.4
	GTL capacity (mb/d)	–	–	–	–
	Refinery distillation capacity (mb/d)	2.1	2.6	3.4	4.5
		2003	**2010**	**2020**	**2030**
Gas	Gas production (bcm)	60	86	129	155
	Net trade (bcm)	–	–	–	–
Electricity	Generating capacity (GW)	35	55	84	102
	Coal	*0*	*0*	*0*	*0*
	Oil	*20*	*28*	*38*	*41*
	Gas	*15*	*27*	*47*	*61*
	Nuclear	–	–	–	–
	Hydro	–	–	–	–
	Other renewables	–	*0*	*0*	*0*
Indicators	GDP (billion dollars 2004 using PPPs)	302	402	566	758
	Population (million)	23	27	34	40
	Per capita energy demand (toe)	5.8	6.7	7.3	7.2
	Energy intensity*	0.43	0.45	0.44	0.38
	CO_2 emissions (Mt)	306	416	568	664

* Toe/thousand dollars of GDP in year-2004 dollars and PPPs.

Reference Scenario: Saudi Arabia

Investment – billion dollars (2004)		**2004-2010**	**2011-2020**	**2021-2030**	**2004-2030**
Total Country Investment		**74**	**127**	**131**	**332**
Oil	**Total**	**34**	**61**	**79**	**174**
	Exploration and development	24	50	67	141
	GTL	–	–	–	–
	Refining	10	11	12	33
	Additions	*7*	*8*	*11*	*26*
	Conversions	*3*	*3*	*1*	*7*
Gas	**Total**	**8**	**20**	**21**	**48**
	Exploration and development	4	10	14	28
	Downstream*	4	9	7	20
Electricity	**Total**	**32**	**46**	**32**	**110**
	Generating capacity	12	18	15	45
	Transmission	6	9	5	20
	Distribution	13	20	11	45

* Includes transmission, distribution, LNG and storage.

Reference Scenario: United Arab Emirates

	Energy Demand (Mtoe)					Shares (%)					Growth Rates (% per annum)			
	1971	2003	2010	2020	2030	1971	2003	2010	2020	2030	1971-2003	2003-2010	2003-2020	2003-2030
Total Primary Energy Demand	**1**	**39**	**54**	**72**	**84**	**100**	**100**	**100**	**100**	**100**	**12.1**	**4.7**	**3.6**	**2.9**
Coal	–	–	–	–	–	–	–	–	–	–	–	–	–	–
Oil	0	9	11	14	16	14	23	21	19	19	13.8	3.1	2.6	2.2
Gas	1	30	43	58	68	86	77	79	80	81	11.7	5.2	3.9	3.1
Nuclear	–	–	–	–	–	–	–	–	–	–	–	–	–	–
Hydro	–	–	–	–	–	–	–	–	–	–	–	–	–	–
Other renewables	–	–	0	0	0	–	–	0	0	0	–	–	–	–
Power Generation and Water Desalination	**0**	**17**	**24**	**33**	**39**	**100**	**100**	**100**	**100**	**100**	**19.0**	**5.7**	**4.1**	**3.2**
Coal	–	–	–	–	–	–	–	–	–	–	–	–	–	–
Oil	–	0	0	0	0	–	1	1	1	1	–	8.4	5.9	5.4
Gas	0	16	24	33	38	100	99	99	99	99	18.9	5.7	4.1	3.2
Nuclear	–	–	–	–	–	–	–	–	–	–	–	–	–	–
Hydro	–	–	–	–	–	–	–	–	–	–	–	–	–	–
Other renewables	–	–	0	0	0	–	–	0	0	0	–	–	–	–
Other Transformation, Own Use and Losses	**0**	**1**	**1**	**1**	**1**						**18.9**	**7.8**	**2.9**	**2.8**
Total Final Consumption	**1**	**26**	**35**	**47**	**56**	**100**	**100**	**100**	**100**	**100**	**10.9**	**4.2**	**3.4**	**2.8**
Coal	–	–	–	–	–	–	–	–	–	–	–	–	–	–
Oil	0	10	12	14	16	15	37	33	30	29	14.1	2.8	2.3	1.9
Gas	1	13	18	25	30	84	50	52	53	54	9.1	4.7	3.8	3.1
Electricity	0	4	5	8	9	2	14	15	16	17	18.6	5.6	4.5	3.6
Renewables	–	–	–	–	–	–	–	–	–	–	–	–	–	–

Industry	**1**	**15**	**21**	**29**	**34**	**100**	**100**	**100**	**100**	**100**	**9.6**	**4.6**	**3.7**	**3.0**
Coal	–	–	–	–	–	–	–	–	–	–	–	–	–	–
Oil	–	2	2	2	2	–	11	9	8	7	–	1.5	1.2	1.0
Gas	1	13	18	25	30	100	85	86	88	88	9.1	4.7	3.8	3.1
Electricity	0	1	1	1	2	0	3	5	5	5	21.3	9.6	5.8	4.4
Renewables	–	–	–	–	–	–	–	–	–	–	–	–	–	–
Transport	**0**	**6**	**8**	**10**	**11**	**100**	**100**	**100**	**100**	**100**	**12.6**	**3.0**	**2.6**	**2.1**
Oil	0	6	8	10	11	100	100	100	100	100	12.6	3.0	2.6	2.1
Other fuels	–	–	–	–	–	–	–	–	–	–	–	–	–	–
Other Sectors	**0**	**4**	**6**	**8**	**10**	**100**	**100**	**100**	**100**	**100**	**19.7**	**4.3**	**3.8**	**3.1**
Coal	–	–	–	–	–	–	–	–	–	–	–	–	–	–
Oil	–	1	2	2	3	–	31	29	25	24	–	2.9	2.5	2.2
Gas	–	–	–	–	–	–	–	–	–	–	–	–	–	–
Electricity	0	3	4	6	8	99	68	71	75	75	18.3	4.9	4.3	3.5
Renewables	–	–	–	–	–	–	–	–	–	–	–	–	–	–
Non–Energy Use	**–**	**0**	**0**	**0**	**0**						**–**	**3.5**	**2.9**	**2.2**
Electricity Generation (TWh)	**0**	**50**	**72**	**105**	**128**	**100**	**100**	**100**	**100**	**100**	**18.7**	**5.6**	**4.5**	**3.6**
Coal	–	–	–	–	–	–	–	–	–	–	–	–	–	–
Oil	–	0	1	1	1	–	1	1	1	1	–	10.0	6.7	5.5
Gas	0	49	72	104	127	100	99	99	99	99	18.7	5.6	4.5	3.6
Nuclear	–	–	–	–	–	–	–	–	–	–	–	–	–	–
Hydro	–	–	–	–	–	–	–	–	–	–	–	–	–	–
Other renewables	–	–	0	0	0	–	–	0	0	0	–	–	–	–

Reference Scenario: United Arab Emirates

Supply and Infrastructure		2004	2010	2020	2030
Oil	Oil production (mb/d)	2.7	3.2	4.0	5.1
	Net trade (mb/d)	–2.5	–2.9	–3.7	–4.7
	GTL capacity (mb/d)	–	–	–	–
	Refinery distillation capacity (mb/d)	0.7	0.8	0.9	1.1

		2003	2010	2020	2030
Gas	Gas production (bcm)	44	59	72	75
	Net trade (bcm)	–7	–7	–1	8
Electricity	Generating capacity (GW)	17	28	37	45
	Coal	–	–	–	–
	Oil	*0*	*0*	*0*	*0*
	Gas	*17*	*28*	*37*	*44*
	Nuclear	–	–	–	–
	Hydro	–	–	–	–
	Other renewables	–	*0*	*0*	*0*
Indicators	GDP (billion dollars 2004 using PPPs)	79	107	144	187
	Population (million)	4	5	6	7
	Per capita energy demand (toe)	9.8	10.8	11.7	11.7
	Energy intensity*	0.50	0.50	0.50	0.45
	CO_2 emissions (Mt)	96	131	172	202

* Toe/thousand dollars of GDP in year-2004 dollars and PPPs.

Reference Scenario: United Arab Emirates

Investment – billion dollars (2004)		2004-2010	2011-2020	2021-2030	2004-2030
Total Country Investment		**26**	**40**	**50**	**115**
Oil	**Total**	**6**	**15**	**26**	**46**
	Exploration and development	5	13	24	41
	GTL	–	–	–	–
	Refining	1	2	2	5
	Additions	*1*	*1*	*1*	*3*
	Conversions	*0*	*1*	*1*	*2*
Gas	**Total**	**9**	**13**	**13**	**34**
	Exploration and development	4	8	9	21
	Downstream*	5	5	4	14
Electricity	**Total**	**12**	**12**	**11**	**35**
	Generating capacity	6	5	6	16
	Transmission	2	2	2	6
	Distribution	4	5	4	13

* Includes transmission, distribution, LNG and storage.

Reference Scenario: North Africa

	Energy Demand (Mtoe)					Shares (%)					Growth Rates (% per annum)			
	1971	2003	2010	2020	2030	1971	2003	2010	2020	2030	1971-2003	2003-2010	2003-2020	2003-2030
Total Primary Energy Demand	**18**	**124**	**160**	**213**	**262**	**100**	**100**	**100**	**100**	**100**	**6.3**	**3.7**	**3.2**	**2.8**
Coal	1	4	5	6	7	5	4	3	3	3	5.1	2.9	2.1	1.7
Oil	12	62	77	98	119	70	50	48	46	45	5.2	3.1	2.7	2.5
Gas	2	53	73	102	128	12	43	45	48	49	10.5	4.6	3.9	3.3
Nuclear	–	–	–	–	–	–	–	–	–	–	–	–	–	–
Hydro	1	1	1	2	2	3	1	1	1	1	2.4	2.3	1.5	1.2
Other renewables	2	3	4	5	6	9	3	2	2	2	2.2	3.0	2.6	2.5
Power Generation and Water Desalination	**3**	**39**	**52**	**68**	**81**	**100**	**100**	**100**	**100**	**100**	**8.7**	**4.3**	**3.4**	**2.8**
Coal	0	3	3	4	4	4	7	7	6	5	10.9	4.0	2.7	2.0
Oil	2	8	9	11	11	62	21	18	16	13	5.2	1.9	1.7	0.9
Gas	0	26	37	51	64	12	69	72	75	79	15.0	5.0	4.0	3.3
Nuclear	–	–	–	–	–	–	–	–	–	–	–	–	–	–
Hydro	1	1	1	2	2	23	3	3	2	2	2.4	2.3	1.5	1.2
Other renewables	–	0	0	0	1	–	0	0	1	1	–	12.9	12.8	10.9
Other Transformation, Own Use and Losses	**3**	**19**	**25**	**34**	**42**						**5.8**	**3.7**	**3.5**	**3.0**
Total Final Consumption	**13**	**81**	**104**	**138**	**173**	**100**	**100**	**100**	**100**	**100**	**5.9**	**3.6**	**3.2**	**2.9**
Coal	0	1	1	1	1	4	1	1	1	1	2.2	2.0	1.9	1.7
Oil	10	48	60	78	97	78	60	58	56	56	5.0	3.1	2.8	2.6
Gas	1	17	23	32	41	5	21	22	24	24	11.1	4.5	3.9	3.3
Electricity	1	12	16	22	28	8	15	16	16	16	8.0	4.6	3.9	3.3
Renewables	1	3	4	4	5	7	4	3	3	3	3.8	3.0	2.3	2.1

Industry	**5**	**27**	**33**	**42**	**51**	**100**	**100**	**100**	**100**	**100**	**5.5**	**3.0**	**2.7**	**2.4**
Coal	0	1	1	1	1	9	3	3	3	3	2.4	2.0	1.9	1.7
Oil	3	9	10	11	13	63	34	30	27	25	3.4	1.4	1.3	1.2
Gas	1	12	16	21	26	10	44	47	50	51	10.4	3.9	3.4	2.9
Electricity	1	4	6	8	9	11	16	17	18	19	6.6	3.8	3.4	3.0
Renewables	0	1	1	1	1	7	3	3	2	2	2.7	1.5	1.5	1.5
Transport	**4**	**24**	**32**	**45**	**60**	**100**	**100**	**100**	**100**	**100**	**6.0**	**4.2**	**3.7**	**3.4**
Oil	4	22	30	42	56	100	93	92	93	94	5.8	4.1	3.7	3.4
Other fuels	0	2	2	3	4	0	7	8	7	6	18.2	5.5	3.7	3.0
Other Sectors	**4**	**27**	**35**	**46**	**57**	**100**	**100**	**100**	**100**	**100**	**6.3**	**3.9**	**3.3**	**2.8**
Coal	0	-	-	-	-	-	-	-	-	-	-	-	-	-
Oil	3	14	17	20	23	71	52	48	43	40	5.2	2.7	2.2	1.9
Gas	0	3	5	9	12	2	13	15	19	21	12.0	6.5	5.8	4.8
Electricity	0	7	10	15	19	11	28	30	32	33	9.3	5.1	4.1	3.5
Renewables	1	2	2	3	3	14	8	7	6	6	4.3	1.8	1.7	1.7
Non-Energy Use	**1**	**3**	**4**	**5**	**6**						**5.2**	**2.9**	**2.7**	**2.5**
Electricity Generation (TWh)	**14**	**171**	**233**	**321**	**402**	**100**	**100**	**100**	**100**	**100**	**8.1**	**4.5**	**3.8**	**3.2**
Coal	0	12	16	19	21	2	7	7	6	5	12.2	4.0	2.7	2.0
Oil	6	26	29	35	37	42	15	13	11	9	4.8	1.5	1.8	1.2
Gas	1	117	169	243	315	6	68	73	76	78	16.6	5.4	4.4	3.7
Nuclear	-	-	-	-	-	-	-	-	-	-	-	-	-	-
Hydro	7	15	17	19	21	50	9	7	6	5	2.4	2.3	1.5	1.2
Other renewables	-	1	1	4	8	-	0	1	1	2	-	12.9	12.8	10.9

Reference Scenario: North Africa

Supply and Infrastructure		2004	2010	2020	2030
Oil	Oil production (mb/d)	4.3	4.7	5.0	5.1
	Net trade (mb/d)	–3.0	–3.2	–3.0	–2.7
	GTL capacity (mb/d)	–	–	0.04	0.04
	Refinery distillation capacity (mb/d)	1.8	2.1	2.6	3.1

		2003	2010	2020	2030
Gas	Gas production (bcm)	125	171	264	352
	Net trade (bcm)	–63	–86	–143	–200
Electricity	Generating capacity (GW)	39	52	72	92
	Coal	*2*	*2*	*3*	*4*
	Oil	*10*	*11*	*14*	*14*
	Gas	*23*	*33*	*49*	*65*
	Nuclear	–	–	–	–
	Hydro	*4*	*5*	*5*	*6*
	Other renewables	*0*	*0*	*2*	*3*
Indicators	GDP (billion dollars 2004 using PPPs)	703	933	1 276	1 652
	Population (million)	145	164	189	209
	Per capita energy demand (toe)	0.9	1.0	1.1	1.3
	Energy intensity*	0.18	0.17	0.17	0.16
	CO_2 emissions (Mt)	295	372	493	604

* Toe/thousand dollars of GDP in year-2004 dollars and PPPs.

Reference Scenario: North Africa

	Investment – billion dollars (2004)	2004-2010	2011-2020	2021-2030	2004-2030
Total Regional Investment		**59**	**115**	**131**	**305**
Oil	**Total**	**19**	**37**	**40**	**96**
	Exploration and development	14	28	32	74
	GTL	–	1	–	1
	Refining	6	8	8	21
	Additions	*4*	*6*	*6*	*16*
	Conversions	*2*	*2*	*2*	*6*
Gas	**Total**	**17**	**45**	**56**	**118**
	Exploration and development	10	25	37	71
	Downstream*	7	21	19	47
Electricity	**Total**	**23**	**32**	**35**	**90**
	Generating capacity	9	13	15	37
	Transmission	4	6	6	17
	Distribution	9	13	14	37

* Includes transmission, distribution, LNG and storage.

Reference Scenario: Algeria

	Energy Demand (Mtoe)					Shares (%)					Growth Rates (% per annum)			
	1971	2003	2010	2020	2030	1971	2003	2010	2020	2030	1971-2003	2003-2010	2003-2020	2003-2030
Total Primary Energy Demand	**4**	**33**	**43**	**58**	**70**	**100**	**100**	**100**	**100**	**100**	**7.1**	**3.7**	**3.4**	**2.8**
Coal	0	1	1	1	1	5	2	2	2	1	3.9	1.7	1.6	1.5
Oil	2	10	13	17	21	62	32	32	30	30	4.8	3.7	3.0	2.6
Gas	1	22	28	39	47	31	66	66	68	68	9.6	3.7	3.5	2.9
Nuclear	–	–	–	–	–	–	–	–	–	–	–	–	–	–
Hydro	0	0	0	0	0	1	0	0	0	0	–0.7	–	–	–
Other renewables	0	0	0	0	0	0	0	0	0	1	7.0	4.9	6.1	5.9
Power Generation and Water Desalination	**1**	**9**	**12**	**16**	**18**	**100**	**100**	**100**	**100**	**100**	**8.5**	**4.5**	**3.6**	**2.8**
Coal	–	–	–	–	–	–	–	–	–	–	–	–	–	–
Oil	0	0	0	0	0	48	2	1	1	1	–1.5	–1.4	–0.6	–1.9
Gas	0	9	12	16	18	47	98	98	98	98	11.0	4.6	3.6	2.8
Nuclear	–	–	–	–	–	–	–	–	–	–	–	–	–	–
Hydro	0	0	0	0	0	4	0	0	0	0	–0.7	–	–	–
Other renewables	–	–	0	0	0	–	–	0	1	1	–	–	–	–
Other Transformation, Own Use and Losses	**1**	**7**	**9**	**13**	**16**						**6.3**	**2.7**	**3.4**	**2.9**
Total Final Consumption	**2**	**19**	**25**	**34**	**41**	**100**	**100**	**100**	**100**	**100**	**7.1**	**3.9**	**3.4**	**2.9**
Coal	0	0	0	0	0	2	1	1	1	1	3.0	2.9	2.7	2.4
Oil	2	10	12	16	19	78	51	49	47	47	5.6	3.3	2.9	2.6
Gas	0	7	10	13	17	12	38	39	40	40	10.9	4.4	3.7	3.1
Electricity	0	2	3	4	5	6	10	11	12	12	8.6	5.3	4.3	3.5
Renewables	0	0	0	0	0	0	0	0	0	0	7.0	3.2	3.1	3.0

Industry	**0**	**5**	**6**	**8**	**9**	**100**	**100**	**100**	**100**	**100**	**8.0**	**3.1**	**2.7**	**2.3**
Coal	0	0	0	0	0	13	3	3	3	3	3.0	2.9	2.7	2.4
Oil	0	1	1	1	2	26	21	20	19	20	7.2	2.7	2.3	2.1
Gas	0	3	4	5	5	45	62	62	61	60	9.1	3.1	2.6	2.1
Electricity	0	1	1	1	2	17	14	15	16	17	7.5	3.8	3.6	3.1
Renewables	-	-	-	-	-	-	-	-	-	-	-	-	-	-
Transport	**1**	**7**	**9**	**12**	**15**	**100**	**100**	**100**	**100**	**100**	**6.4**	**4.0**	**3.5**	**3.0**
Oil	1	5	7	9	12	100	78	77	79	80	5.6	4.0	3.5	3.1
Other fuels	0	1	2	3	3	0	22	23	21	20	23.7	4.1	3.2	2.6
Other Sectors	**1**	**7**	**10**	**14**	**17**	**100**	**100**	**100**	**100**	**100**	**7.5**	**4.3**	**3.7**	**3.1**
Coal	-	-	-	-	-	-	-	-	-	-	-	-	-	-
Oil	1	3	4	4	5	78	44	38	33	29	5.7	2.1	1.9	1.6
Gas	0	3	4	6	8	11	38	42	47	50	11.7	6.0	4.9	4.1
Electricity	0	1	2	3	3	10	17	19	20	20	9.3	6.1	4.7	3.8
Renewables	0	0	0	0	0	1	1	1	1	1	7.0	2.4	2.3	2.2
Non-Energy Use	**0**	**0**	**1**	**1**	**1**						**3.8**	**4.1**	**3.5**	**3.1**
Electricity Generation (TWh)	**2**	**30**	**42**	**60**	**74**	**100**	**100**	**100**	**100**	**100**	**8.4**	**5.2**	**4.2**	**3.4**
Coal	-	-	-	-	-	-	-	-	-	-	-	-	-	-
Oil	1	1	1	1	0	46	2	2	1	1	-1.3	-0.2	0.1	-1.2
Gas	1	29	41	58	71	39	97	97	97	96	11.6	5.3	4.2	3.4
Nuclear	-	-	-	-	-	-	-	-	-	-	-	-	-	-
Hydro	0	0	0	0	0	15	1	1	0	0	-0.7	-	-	-
Other renewables	-	-	0	1	2	-	-	0	2	3	-	-	-	-

Reference Scenario: Algeria

Supply and Infrastructure		2004	2010	2020	2030
Oil	Oil production (mb/d)	1.9	2.2	1.9	1.6
	Net trade (mb/d)	–1.7	–1.8	–1.5	–1.1
	GTL capacity (mb/d)	–	–	0.04	0.04
	Refinery distillation capacity (mb/d)	0.5	0.7	0.9	1.0

		2003	2010	2020	2030
Gas	Gas production (bcm)	88	107	160	198
	Net trade (bcm)	–64	–76	–114	–144
Electricity	Generating capacity (GW)	7	10	15	18
	Coal	–	–	–	–
	Oil	*0*	*0*	*0*	*0*
	Gas	*7*	*10*	*14*	*17*
	Nuclear	–	–	–	–
	Hydro	*0*	*0*	*0*	*0*
	Other renewables	–	*0*	*0*	*1*
Indicators	GDP (billion dollars 2004 using PPPs)	202	273	356	446
	Population (million)	32	36	41	44
	Per capita energy demand (toe)	1.0	1.2	1.4	1.6
	Energy intensity*	0.16	0.16	0.16	0.16
	CO_2 emissions (Mt)	78	98	133	161

* Toe/thousand dollars of GDP in year-2004 dollars and PPPs.

Reference Scenario: Algeria

Investment – billion dollars (2004)		2004-2010	2011-2020	2021-2030	2004-2030
Total Country Investment		**24**	**45**	**45**	**114**
Oil	**Total**	**11**	**15**	**12**	**38**
	Exploration and development	7	11	10	27
	GTL	–	1	–	1
	Refining	4	3	3	10
	Additions	*3*	*3*	*2*	*9*
	Conversions	*0*	*1*	*0*	*1*
Gas	**Total**	**9**	**23**	**27**	**59**
	Exploration and development	5	13	18	35
	Downstream*	4	11	9	24
Electricity	**Total**	**5**	**7**	**6**	**17**
	Generating capacity	2	3	3	7
	Transmission	1	1	1	3
	Distribution	2	3	2	7

* Includes transmission, distribution, LNG and storage.

Reference Scenario: Egypt

	Energy Demand (Mtoe)					Shares (%)					Growth Rates (% per annum)			
	1971	2003	2010	2020	2030	1971	2003	2010	2020	2030	1971-2003	2003-2010	2003-2020	2003-2030
Total Primary Energy Demand	**8**	**54**	**68**	**88**	**109**	**100**	**100**	**100**	**100**	**100**	**6.2**	**3.4**	**2.9**	**2.6**
Coal	0	0	1	1	1	4	1	1	1	1	1.1	4.1	2.2	1.7
Oil	6	27	33	41	51	81	51	48	47	47	4.7	2.5	2.4	2.3
Gas	0	23	32	42	52	1	43	47	48	48	19.9	4.4	3.5	3.0
Nuclear	–	–	–	–	–	–	–	–	–	–	–	–	–	–
Hydro	0	1	1	1	2	6	2	2	2	1	3.0	2.4	1.5	1.2
Other renewables	1	1	2	2	3	8	3	3	3	3	2.4	4.4	3.2	2.9
Power Generation and Water Desalination	**1**	**17**	**22**	**27**	**32**	**100**	**100**	**100**	**100**	**100**	**8.7**	**3.7**	**2.8**	**2.4**
Coal	–	–	–	–	–	–	–	–	–	–	–	–	–	–
Oil	1	2	1	1	1	64	9	6	5	4	2.2	–1.3	–0.9	–1.0
Gas	–	14	19	24	29	–	84	87	89	91	–	4.2	3.2	2.7
Nuclear	–	–	–	–	–	–	–	–	–	–	–	–	–	–
Hydro	0	1	1	1	2	36	7	6	5	5	3.0	2.4	1.5	1.2
Other renewables	–	0	0	0	0	–	0	0	1	1	–	13.8	11.3	10.0
Other Transformation, Own Use and Losses	**1**	**8**	**11**	**13**	**17**						**8.0**	**4.1**	**3.1**	**2.8**
Total Final Consumption	**7**	**37**	**46**	**60**	**76**	**100**	**100**	**100**	**100**	**100**	**5.5**	**3.3**	**3.0**	**2.7**
Coal	0	0	0	0	0	3	1	1	1	0	1.7	0.9	0.9	0.8
Oil	5	22	27	34	44	79	60	58	57	58	4.6	2.8	2.7	2.6
Gas	0	6	9	12	15	0	18	18	20	20	28.2	4.1	3.8	3.2
Electricity	1	7	9	11	14	9	18	19	19	18	7.8	4.1	3.2	2.8
Renewables	1	1	2	2	3	10	4	4	4	4	2.4	4.2	2.9	2.6

Industry	**3**	**14**	**17**	**20**	**24**	**100**	**100**	**100**	**100**	**100**	**5.0**	**2.4**	**2.1**	**1.9**
Coal	0	0	0	0	0	5	2	2	2	2	1.9	0.9	0.9	0.8
Oil	2	5	5	6	6	72	36	32	27	23	2.7	0.5	0.4	0.3
Gas	0	6	7	10	12	0	41	44	47	50	32.5	3.6	3.0	2.7
Electricity	0	2	3	4	5	12	17	18	19	21	6.1	3.6	3.1	2.7
Renewables	0	1	1	1	1	11	5	5	4	4	2.4	1.5	1.5	1.5
Transport	**1**	**11**	**15**	**22**	**30**	**100**	**100**	**100**	**100**	**100**	**6.8**	**4.6**	**4.1**	**3.8**
Oil	1	11	15	21	29	100	98	97	97	97	6.7	4.3	4.0	3.7
Other fuels	–	0	0	1	1	–	2	3	3	3	–	13.4	7.0	5.4
Other Sectors	**2**	**9**	**12**	**15**	**18**	**100**	**100**	**100**	**100**	**100**	**5.0**	**3.3**	**2.9**	**2.5**
Coal	–	–	–	–	–	–	–	–	–	–	–	–	–	–
Oil	1	4	4	5	5	71	42	36	30	27	3.3	1.1	0.9	0.8
Gas	0	0	1	2	3	0	5	8	15	18	19.2	10.9	10.4	7.7
Electricity	0	4	6	7	9	12	46	49	48	49	9.4	4.4	3.2	2.8
Renewables	0	1	1	1	1	16	7	7	6	6	2.3	2.0	1.8	1.7
Non–Energy Use	**0**	**2**	**2**	**3**	**4**						**6.4**	**2.8**	**2.6**	**2.3**
Electricity Generation (TWh)	**8**	**92**	**121**	**154**	**188**	**100**	**100**	**100**	**100**	**100**	**7.9**	**4.0**	**3.1**	**2.7**
Coal	–	–	–	–	–	–	–	–	–	–	–	–	–	–
Oil	3	5	5	5	4	37	6	4	3	2	1.8	–1.3	–0.9	–1.0
Gas	–	73	100	132	163	–	80	83	85	86	–	4.5	3.5	3.0
Nuclear	–	–	–	–	–	–	–	–	–	–	–	–	–	–
Hydro	5	13	15	17	18	63	14	13	11	10	3.0	2.4	1.5	1.2
Other renewables	–	0	1	2	4	–	0	1	1	2	–	13.8	11.3	10.0

Reference Scenario: Egypt

Supply and Infrastructure		2004	2010	2020	2030
Oil	Oil production (mb/d)	0.7	0.7	0.5	0.5
	Net trade (mb/d)	–0.2	–0.1	0.2	0.5
	GTL capacity (mb/d)	–	–	–	–
	Refinery distillation capacity (mb/d)	0.8	0.8	0.9	1.1
		2003	**2010**	**2020**	**2030**
Gas	Gas production (bcm)	29	49	71	92
	Net trade (bcm)	–	–10	–19	–28
Electricity	Generating capacity (GW)	18	23	29	36
	Coal	–	–	–	–
	Oil	*3*	*3*	*3*	*2*
	Gas	*12*	*17*	*23*	*29*
	Nuclear	–	–	–	–
	Hydro	*3*	*3*	*3*	*4*
	Other renewables	*0*	*0*	*1*	*1*
Indicators	GDP (billion dollars 2004 using PPPs)	270	359	512	695
	Population (million)	68	78	91	103
	Per capita energy demand (toe)	0.8	0.9	1.0	1.1
	Energy intensity*	0.20	0.19	0.17	0.16
	CO_2 emissions (Mt)	122	151	195	242

* Toe/thousand dollars of GDP in year-2004 dollars and PPPs.

Reference Scenario: Egypt

Investment – billion dollars (2004)		2004-2010	2011-2020	2021-2030	2004-2030
Total Country Investment		**20**	**31**	**34**	**85**
Oil	**Total**	**3**	**5**	**6**	**14**
	Exploration and development	2	3	3	9
	GTL	–	–	–	–
	Refining	1	2	3	5
	Additions	*0*	*1*	*2*	*3*
	Conversions	*0*	*1*	*1*	*2*
Gas	**Total**	**7**	**13**	**15**	**35**
	Exploration and development	4	8	10	22
	Downstream*	3	5	5	13
Electricity	**Total**	**11**	**12**	**13**	**36**
	Generating capacity	4	4	5	13
	Transmission	2	2	3	7
	Distribution	5	5	6	16

* Includes transmission, distribution, LNG and storage.

Reference Scenario: Libya

	Energy Demand (Mtoe)					Shares (%)					Growth Rates (% per annum)			
	1971	2003	2010	2020	2030	1971	2003	2010	2020	2030	1971-2003	2003-2010	2003-2020	2003-2030
Total Primary Energy Demand	**2**	**18**	**25**	**36**	**46**	**100**	**100**	**100**	**100**	**100**	**7.7**	**5.0**	**4.2**	**3.6**
Coal	–	–	–	–	–	–	–	–	–	–	–	–	–	–
Oil	1	13	17	23	27	40	73	67	63	59	9.8	3.5	3.2	2.7
Gas	1	5	8	13	19	54	26	33	37	41	5.2	8.7	6.3	5.4
Nuclear	–	–	–	–	–	–	–	–	–	–	–	–	–	–
Hydro	–	–	–	–	–	–	–	–	–	–	–	–	–	–
Other renewables	0	0	0	0	0	6	1	1	1	1	1.3	2.0	2.4	2.4
Power Generation and Water Desalination	**0**	**6**	**9**	**13**	**16**	**100**	**100**	**100**	**100**	**100**	**11.9**	**5.8**	**4.4**	**3.6**
Coal	–	–	–	–	–	–	–	–	–	–	–	–	–	–
Oil	0	5	6	8	8	100	83	68	60	48	11.3	2.7	2.3	1.5
Gas	–	1	3	5	8	–	17	32	40	51	–	16.2	9.8	7.9
Nuclear	–	–	–	–	–	–	–	–	–	–	–	–	–	–
Hydro	–	–	–	–	–	–	–	–	–	–	–	–	–	–
Other renewables	–	–	0	0	0	–	–	0	0	0	–	–	–	–
Other Transformation, Own Use and Losses	**1**	**4**	**5**	**6**	**8**						**5.3**	**4.3**	**3.7**	**3.1**
Total Final Consumption	**1**	**10**	**14**	**20**	**27**	**100**	**100**	**100**	**100**	**100**	**7.9**	**4.7**	**4.3**	**3.9**
Coal	–	–	–	–	–	–	–	–	–	–	–	–	–	–
Oil	0	6	8	12	16	54	63	61	58	57	8.5	4.1	3.8	3.5
Gas	0	3	4	6	8	32	25	27	28	28	7.2	5.7	5.0	4.3
Electricity	0	1	1	3	4	3	10	11	13	14	11.6	6.2	5.8	5.2
Renewables	0	0	0	0	0	11	2	1	1	1	1.3	1.5	1.5	1.5

Industry	**0**	**4**	**6**	**8**	**11**	**100**	**100**	**100**	**100**	**100**	**8.7**	**4.9**	**4.5**	**3.9**
Coal	–	–	–	–	–	–	–	–	–	–	–	–	–	–
Oil	–	1	2	2	3	–	32	28	26	25	–	3.1	3.2	2.9
Gas	0	3	4	6	8	98	63	66	68	69	7.2	5.7	5.0	4.3
Electricity	0	0	0	0	1	2	5	6	6	6	11.9	5.7	4.9	4.4
Renewables	–	–	–	–	–	–	–	–	–	–	–	–	–	–
Transport	**0**	**4**	**5**	**8**	**10**	**100**	**100**	**100**	**100**	**100**	**8.1**	**4.4**	**4.1**	**3.7**
Oil	0	4	5	8	10	100	100	100	100	100	8.1	4.4	4.1	3.7
Other fuels	–	–	–	–	–	–	–	–	–	–	–	–	–	–
Other Sectors	**0**	**2**	**3**	**4**	**6**	**100**	**100**	**100**	**100**	**100**	**6.5**	**5.1**	**4.6**	**4.2**
Coal	–	–	–	–	–	–	–	–	–	–	–	–	–	–
Oil	0	1	1	2	2	51	51	49	43	40	6.5	4.6	3.7	3.2
Gas	–	–	–	–	–	–	–	–	–	–	–	–	–	–
Electricity	0	1	1	2	3	9	41	45	52	56	11.5	6.3	6.0	5.4
Renewables	0	0	0	0	0	39	8	6	5	4	1.3	1.5	1.5	1.5
Non–Energy Use	**0**	**0**	**0**	**0**	**0**						**6.9**	**1.1**	**2.3**	**2.0**
Electricity Generation (TWh)	**1**	**19**	**27**	**43**	**60**	**100**	**100**	**100**	**100**	**100**	**12.0**	**5.1**	**4.9**	**4.4**
Coal	–	–	–	–	–	–	–	–	–	–	–	–	–	–
Oil	1	15	17	23	26	100	80	64	53	42	11.2	2.0	2.5	2.0
Gas	–	4	9	20	34	–	20	35	46	56	–	13.7	10.0	8.4
Nuclear	–	–	–	–	–	–	–	–	–	–	–	–	–	–
Hydro	–	–	–	–	–	–	–	–	–	–	–	–	–	–
Other renewables	–	–	0	0	1	–	–	0	1	1	–	–	–	–

Reference Scenario: Libya

Supply and Infrastructure		2004	2010	2020	2030
Oil	Oil production (mb/d)	1.6	1.9	2.5	3.1
	Net trade (mb/d)	–1.4	–1.5	–2.1	–2.5
	GTL capacity (mb/d)	–	–	–	–
	Refinery distillation capacity (mb/d)	0.4	0.4	0.6	0.7

		2003	2010	2020	2030
Gas	Gas production (bcm)	6	12	29	57
	Net trade (bcm)	–1	–2	–13	–34
Electricity	Generating capacity (GW)	6	8	13	19
	Coal	–	–	–	–
	Oil	*4*	*5*	*6*	*7*
	Gas	*2*	*4*	*7*	*11*
	Nuclear	–	–	–	–
	Hydro	–	–	–	–
	Other renewables	–	*0*	*0*	*0*
Indicators	GDP (billion dollars 2004 using PPPs)	32	41	58	75
	Population (million)	6	6	7	8
	Per capita energy demand (toe)	3.2	4.0	4.9	5.7
	Energy intensity*	0.55	0.61	0.62	0.62
	CO_2 emissions (Mt)	43	58	81	104

* Toe/thousand dollars of GDP in year-2004 dollars and PPPs.

Reference Scenario: Libya

Investment – billion dollars (2004)		**2004-2010**	**2011-2020**	**2021-2030**	**2004-2030**
Total Country Investment		**10**	**30**	**40**	**80**
Oil	**Total**	**5**	**15**	**20**	**41**
	Exploration and development	4	13	18	36
	GTL	–	–	–	–
	Refining	1	2	2	4
	Additions	*0*	*2*	*1*	*3*
	Conversions	*0*	*1*	*0*	*1*
Gas	**Total**	**1**	**8**	**12**	**21**
	Exploration and development	1	4	8	13
	Downstream*	0	4	5	9
Electricity	**Total**	**4**	**6**	**8**	**18**
	Generating capacity	2	3	4	8
	Transmission	1	1	1	3
	Distribution	1	3	3	7

* Includes transmission, distribution, LNG and storage.

DEFINITIONS, ABBREVIATIONS AND ACRONYMS

This annex provides general information on definitions, abbreviations and acronyms used throughout *WEO-2005*. Readers interested in obtaining more detailed information should consult the annual IEA publications *Energy Balances of OECD Countries; Energy Balances of Non-OECD Countries; Coal Information; Oil Information; Gas Information and Renewables Information.*

Fuel and Process Definitions

Coal

Coal includes all coal: both primary coal (including hard coal and lignite) and derived fuels (including patent fuel, brown-coal briquettes, coke-oven coke, gas coke, coke-oven gas and blast-furnace gas). Peat is also included in this category.

Oil

Oil includes crude oil, condensates, natural gas liquids, refinery feedstocks and additives, other hydrocarbons and petroleum products (refinery gas, ethane, liquefied petroleum gas, aviation gasoline, motor gasoline, jet fuels, kerosene, gas/diesel oil, heavy fuel oil, naphtha, white spirit, lubricants, bitumen, paraffin waxes, petroleum coke and other petroleum products).

Gas

Gas includes natural gas (both associated and non-associated with petroleum deposits but excluding natural gas liquids) and gas-works gas. Marketed gas excludes gas used for re-injection, flared gas, and gas leakage and theft.

Biomass and Waste

Biomass includes solid biomass and animal products, gas and liquids derived from biomass, industrial waste and municipal waste.

Other Renewables

Other renewables include geothermal, solar, wind, tide and wave energy for electricity generation. Direct use of geothermal and solar heat is also included in this category.

Nuclear

Nuclear refers to the primary heat equivalent of the electricity produced by a nuclear plant with an average thermal efficiency of 33%.

Hydro

Hydro refers to the energy content of the electricity produced in hydropower plants, assuming 100% efficiency.

Condensates

Condensates are liquid hydrocarbon mixtures recovered from non-associated gas reservoirs. They are composed of C_4 and higher carbon number hydrocarbons and normally have an API between 50° and 85°.

Natural Gas Liquids

Natural gas liquids (NGLs) are the liquid or liquefied hydrocarbons produced in the manufacture, purification and stabilisation of natural gas. These are those portions of natural gas which are recovered as liquids in separators, field facilities, or gas-processing plants. NGLs include but are not limited to ethane, propane, butane, pentane, natural gasoline and condensates. They may also include small quantities of non-hydrocarbons.

Gas-to-Liquids

Fischer-Tropsch technology is used to convert natural gas into synthesis gas (syngas) and then, through catalytic reforming or synthesis, into very clean conventional oil products. The main fuel produced in most GTL plants is diesel.

Gasoil

Gasoil is an intermediate product in oil refining, used as a feedstock for lubricants or made lighter to make gasoline or kerosene through a process known as "cracking".

Light Petroleum Products

Light petroleum products include liquefied petroleum gas, naphtha and gasoline.

Middle Distillates

Middle distillates include jet fuel, diesel and heating oil.

Heavy Petroleum Products

Heavy petroleum products include heavy fuel oil.

Other Petroleum Products

Other petroleum products include refinery gas, ethane, lubricants, bitumen, petroleum coke and waxes.

Hard Coal

Coal of gross calorific value greater than 5 700 kcal/kg on an ash-free but moist basis and with a mean random reflectance of vitrinite of at least 0.6. Hard coal is further disaggregated into coking coal and steam coal.

Coking Coal

Coking coal is hard coal with a quality that allows the production of coke suitable to support a blast furnace charge.

Brown Coal

Sub-bituminous coal and lignite. Sub-bituminous coal is defined as non-agglomerating coal with a gross calorific value between 4 165 kcal/kg and 5 700 kcal/kg. Lignite is defined as non-agglomerating coal with a gross calorific value less than 4 165 kcal/kg.

Clean Coal Technologies

Clean coal technologies (CCTs) are designed to enhance the efficiency and the environmental acceptability of coal extraction, preparation and use.

Total Primary Energy Supply

Total primary energy supply is equivalent to primary energy demand. This represents domestic demand only and, except for world energy demand, excludes international marine bunkers.

International Marine Bunkers

International marine bunkers cover those quantities delivered to sea-going ships of all flags, including warships. Consumption by ships plying in inland and coastal waters is not included.

Power Generation

Power generation refers to fuel use in electricity plants, heat plants and combined heat and power (CHP) plants. Both public plants and small plants that produce fuel for their own use (autoproducers) are included.

Total Final Consumption

Total final consumption is the sum of consumption by the different end-use sectors. TFC is broken down into energy demand in the following sectors: industry, transport, other (includes agriculture, residential, commercial and public services) and non-energy use. Industry includes manufacturing, construction and mining industries. In final consumption, petrochemical feedstocks appear under *industry use*. Other non-energy uses are shown under *non-energy use*.

Other Transformation, Own Use and Losses

Other transformation, own use and losses covers the use of energy by transformation industries and the energy losses in converting primary energy into a form that can be used in the final consuming sectors. It includes energy use and loss by gas works, petroleum refineries, coal and gas transformation and liquefaction. It also includes energy used in coal mines, in oil and gas extraction and in electricity and heat production. Transfers and statistical differences are also included in this category

Electricity Generation

Electricity generation shows the total amount of electricity generated by power plants. It includes own use and transmission and distribution losses.

Other Sectors

Other sectors include the residential, commercial and public services and agriculture sectors.

Regional Definitions

OECD Europe

OECD Europe consists of Austria, Belgium, the Czech Republic, Denmark, Finland, France, Germany, Greece, Hungary, Iceland, Ireland, Italy, Luxembourg, the Netherlands, Norway, Poland, Portugal, the Slovak Republic, Spain, Sweden, Switzerland, Turkey and the United Kingdom.

OECD North America

OECD North America consists of the United States of America, Canada and Mexico.

OECD Pacific

OECD Pacific consists of Japan, Korea, Australia and New Zealand.

European Union

The EU consists of Austria, Belgium, Cyprus, the Czech Republic, Denmark, Estonia, Finland, France, Germany, Greece, Hungary, Ireland, Italy, Latvia, Lithuania, Luxembourg, Malta, the Netherlands, Poland, Portugal, the Slovak Republic, Slovenia, Spain, Sweden and the United Kingdom.

Transition Economies

The transition economies include: Albania, Armenia, Azerbaijan, Belarus, Bosnia-Herzegovina, Bulgaria, Croatia, Estonia, the Federal Republic of Yugoslavia, the former Yugoslav Republic of Macedonia, Georgia, Kazakhstan, Kyrgyzstan, Latvia, Lithuania, Moldova, Romania, Russia, Slovenia, Tajikistan, Turkmenistan, Ukraine and Uzbekistan. For statistical reasons, this region also includes Cyprus, Gibraltar and Malta.

China

China refers to the People's Republic of China, including Hong Kong.

East Asia

East Asia includes: Afghanistan, Bhutan, Brunei, Chinese Taipei, Fiji, French Polynesia, Indonesia, Kiribati, Democratic People's Republic of Korea, Malaysia, Maldives, Myanmar, New Caledonia, Papua New Guinea, the Philippines, Samoa, Singapore, Solomon Islands, Thailand, Vietnam and Vanuatu.

South Asia

South Asia consists of Bangladesh, India, Nepal, Pakistan and Sri Lanka.

Latin America

Latin America includes: Antigua and Barbuda, Argentina, Bahamas, Barbados, Belize, Bermuda, Bolivia, Brazil, Chile, Colombia, Costa Rica, Cuba, Dominica, the Dominican Republic, Ecuador, El Salvador, French Guiana, Grenada, Guadeloupe, Guatemala, Guyana, Haiti, Honduras, Jamaica, Martinique, Netherlands Antilles, Nicaragua, Panama, Paraguay, Peru, St.

Kitts-Nevis-Anguilla, Saint Lucia, St. Vincent-Grenadines and Suriname, Trinidad and Tobago, Uruguay and Venezuela.

Africa

Africa comprises Algeria, Angola, Benin, Botswana, Burkina Faso, Burundi, Cameroon, Cape Verde, the Central African Republic, Chad, Congo, the Democratic Republic of Congo, Côte d'Ivoire, Djibouti, Egypt, Equatorial Guinea, Eritrea, Ethiopia, Gabon, Gambia, Ghana, Guinea, Guinea-Bissau, Kenya, Lesotho, Liberia, Libya, Madagascar, Malawi, Mali, Mauritania, Mauritius, Morocco, Mozambique, Niger, Nigeria, Rwanda, Sao Tome and Principe, Senegal, Seychelles, Sierra Leone, Somalia, South Africa, Sudan, Swaziland, the United Republic of Tanzania, Togo, Tunisia, Uganda, Zambia and Zimbabwe.

Middle East

The Middle East is defined as Bahrain, Iran, Iraq, Israel, Jordan, Kuwait, Lebanon, Oman, Qatar, Saudi Arabia, Syria, the United Arab Emirates and Yemen. It includes the neutral zone between Saudi Arabia and Iraq.

North Africa

North Africa includes Algeria, Egypt, Libya, Morocco and Tunisia.

MENA

The MENA region is comprised of all Middle Eastern and North African countries.

In addition to the WEO regions, the following groupings are also referred to in the text.

Developing Asia

China, East Asia and South Asia.

Gulf Co-operation Council (GCC)

Bahrain, Kuwait, Qatar, Oman, Saudi Arabia and the United Arab Emirates.

Organization of Arab Petroleum Exporting Countries (OAPEC)

Algeria, Bahrain, Egypt, Iraq, Kuwait, Libya, Qatar, Saudi Arabia, Syria, and the United Arab Emirates.

Organization of the Petroleum Exporting Countries (OPEC)
Algeria, Indonesia, Iran, Iraq, Kuwait, Libya, Nigeria, Qatar, Saudi Arabia, the United Arab Emirates and Venezuela.

Abbreviations and Acronyms

bcm	billion cubic metres
b/d	barrels per day
boe	barrels of oil equivalent
BOO	build-operate-own
BOOT	build-own-operate-transfer
BOT	build-operate-transfer
CCGT	combined-cycle gas turbine
CDM	Clean Development Mechanism (under the Kyoto Protocol)
CDU	crude distillation unit
CHP	combined production of heat and power; when referring to industrial CHP, the term co-generation is sometimes used
CNG	compressed natural gas
CWP	combined water and power
CO_2	carbon dioxide
DIS	Deferred Investment Scenario
E&P	exploration and production
EOR	enhanced oil recovery
EPC	engineering, procurement and construction
EU	European Union
FDI	foreign direct investment
FSU	former Soviet Union
GCC	Gulf Co-operation Council
GDP	gross domestic product
GHG	greenhouse gas
Gt	gigatonne (1 tonne x 10^9)

GTL	gas-to-liquids
GW	gigawatt (1 watt x 10^9)
GWh	gigawatt-hour
IEA	International Energy Agency
IMF	International Monetary Fund
IOC	international oil company
IPP	independent power producer
IWPP	independent water and power producer
kb/d	thousand barrels per day
kt	kilotonne
kW	kilowatt (1 watt × 1 000)
kWh	kilowatt-hour
LNG	liquefied natural gas
LPG	liquefied petroleum gas
mb/d	million barrels per day
MBtu	million British thermal units
mcm	million cubic metres
MSC	multiple service contract
Mt	million tonnes
Mtoe	million tonnes of oil equivalent
MW	megawatt (1 watt × 10^6)
MWh	megawatt-hour
NGL	natural gas liquid
NOC	national oil company
OCGT	open-cycle gas turbine
OECD	Organisation for Economic Co-operation and Development
OAPEC	Organization of Arab Petroleum Exporting Countries
OPEC	Organization of the Petroleum Exporting Countries

PPP	purchasing power parity; the rate of currency conversion that equalises the purchasing power of different currencies. PPPs compare costs in different currencies of a fixed basket of traded and non-traded goods and services and yield a widely-based measure of standard of living
ppm	parts per million
PSA	production-sharing agreement
SOE	state-owned enterprise
tcf	thousand cubic feet
tcm	trillion cubic metres
TFC	total final consumption
toe	tonne of oil equivalent
tonne	metric ton
TPES	total primary energy supply
TWh	terawatt-hour
UNDESA	United Nations Department of Economic and Social Affairs
UNDP	United Nations Development Programme
UNFCCC	United Nations Framework Convention on Climate Change
USGS	United States Geological Survey
WAPS	World Alternative Policy Scenario
WEM	World Energy Model

REFERENCES

CHAPTER 1: The Context

BP (2005), *Statistical Review of World Energy 2005*, BP, London.

Cedigaz (2005), *Natural Gas in the World*, Institut Français du Pétrole, Rueil-Malmaison.

Cordesman, A. (2004), *Energy Developments in the Middle East,* Center for Strategic and International Studies, Washington, DC.

International Energy Agency (IEA) (2004), *World Energy Outlook 2004*, OECD/IEA, Paris.

Oil and Gas Journal (2004), PennWell Corporation, Oklahoma City, 20 December.

United Nations Department of Economics and Social Affairs (UNDESA), Population Division (2005), *World Population Prospects: The 2004 Revision*, United Nations, New York.

United Nations Development Programme (UNDP) (2005), *Arab Human Development Report: Towards Freedom in the Arab World*, UNDP, New York.

World Bank (2005), *Middle East and North Africa Economic Developments and Prospects 2005: Oil Booms and Revenue Management*, World Bank, Washington, DC.

CHAPTER 2: Global Energy Trends

International Energy Agency (IEA) (2004), *World Energy Outlook 2004*, OECD/IEA, Paris.

Goldman Sachs (2004), Energy Watch Commodities, *Infrastructure Constraints Will Likely Create a Cyclical Strong Market in 2005*, London, December.

CHAPTER 4: MENA Oil Outlook

Alazard N. and Mathieu Y. (2005), *Past Evolution of Tools, Technologies and Scientific Knowledge in the Field of Exploration*, Institut Français du Pétrole, Rueil-Malmaison.

BP (2005), *BP Statistical Review of World Energy 2005*, BP, London.

Economic and Social Commission for Western Asia (2005), *Statistical Abstract of the ESCWA Region*, United Nations, New York.

Energy Information Administration (2003), *Performance Profiles of Major Energy Producers*, US Department of Energy, Washington, DC.

Gately, Dermot (2004), "OPEC's Incentives for Faster Output Growth", *Energy Journal*, Vol. 25, No. 2, IAEE, Toronto.

Ghalib, Sharif and Knapp, David (2004), *High Oil Prices: Causes and Consequences*, Energy Intelligence Group, New York.

IEA (2003), *World Energy Investment Outlook 2003*, OECD/IEA, Paris.

IEA (2004), *World Energy Outlook 2004*, OECD/IEA, Paris.

IEA (forthcoming 2005), *Resources to Reserves: Oil and Gas Technologies for the Energy Markets of the Future*, OECD/IEA, Paris.

IHS (2004), IHS Energy databases.

Mahroos, F.A. (2005), *Future Challenges for Producing Middle East Oil Fields During Maturation Stage*, SPE paper 93708.

Marcel V. (2005a), *Good Governance of the National Oil Company*, Position paper for the Workshop on Good Governance of the National Petroleum Sector, London.

Marcel, V. (forthcoming 2005b), *Oil Titans: National Oil Companies in the Middle East*, Brooking Institution Press, Washington, DC.

Oil and Gas Journal, various issues, PennWell Corporation, Oklahoma City.

Organization of the Petroleum Exporting Countries (OPEC) (2003), *OPEC Annual Statistical Bulletin 2003*, OPEC, Vienna.

OPEC (2004), *Oil Outlook to 2025*, OPEC, Vienna

Pepperoni, T. (2005), *Middle East Oil Prospects*, Morgan Publishing, London.

Simmons, Matthew R. (2005), *Twilight in the Desert*, John Wiley & Sons, Inc., Hoboken, New Jersey.

United States Geological Survey (USGS) (2000), *World Petroleum Assessment 2000*, USGS, Washington.

Verma M.K., Ahlbrandt T.S. and Al-Gailani M. (2004), "Petroleum Reserves and Undiscovered Resources in the Total Petroleum Systems of Iraq", in *GeoArabia,* Vol. 9, No.3, Gulf Petrolink, Bahrain.

World Energy Council (2004), *2004 Survey of Energy Resources*, Elsevier, London.

World Energy Council (2005), http://www.worldenergy.org/wec-geis/publications/reports/ser/shale/shale.asp.

World Oil, various issues, Gulf Publishing Company, Houston.

CHAPTER 5: MENA Gas Outlook

Cedigaz (2005), *Natural Gas in the World*, Institut Francais du Pétrole, Rueil-Malmaison.

United States Geological Survey (USGS) (2000), *World Petroleum Assessment 2000*, USGS, Washington, DC.

CHAPTER 6: MENA Electricity and Water Outlook

Arab Union of Producers, Transporters and Distributors of Electricity (AUPTDE) (2004), *Statistical Bulletin 2003*, AUPTDE, Jordan.

Global Water Intelligence (2004), *Desalination Markets 2005-2015: A Global Assessment and Forecast*, Media Analytics Ltd, Oxford.

Global Water Intelligence (2005), *Water Market Middle East: Exploiting a Booming Market*, Media Analytics Ltd, Oxford.

International Energy Agency (IEA) (2003), *World Energy Investment Outlook 2003*, OECD/IEA, Paris.

IEA (2004), *World Energy Outlook 2004*, OECD/IEA, Paris.

IPCC (2001), *Climate Change 2001: Mitigation*, IPCC.

Khatib, H. (2005), *MENA Electrical Power Sector, Challenges and Opportunities*, IEA/OPEC Joint Symposium, Kuwait, 15 May.

Ministry of Energy and Mines (2004), "Energies et Mines – Review of the Energy and Mining Sector", No. 3, Algeria.

Pankratz, Tom and John Tonner (2003), *Desalination.com: an Environmental Primer*, Lone Oak Publishing, Houston, Texas.

Platts (2003), World Electric Power Plants Database, Platts.

Roudi-Fahimi, Farzaneh, Liz Creel and Roger-Mark De Souza (2002), "Finding the Balance: Population and Water Scarcity in the Middle East and North Africa", Population Reference Bureau, MENA Policy Brief.

United Nations Environment Programme (UNEP) (2004), *TEAP Chiller Task Force Report*, United Nations, Nairobi.

United Nations Food and Agricultural Organization (1997), "Irrigation in the Near East Region in Figures", *Water Reports*, No. 9, United Nations, Rome, pp. 55-61.

United Nations Food and Agricultural Organization (2003) "Review of World Water Resources by Country", United Nations, Rome.

United Nations Food and Agricultural Organization (2005), AQUASTAT: FAO's Information System on Water and Agriculture, http://www.fao.org/ag/agl/aglw/aquastat/main/.

Wangnick, Klaus (2002), *IDA Worldwide Desalting Plants, Inventory*, Report No. 17, Wangnick Consulting, Gnarrenburg, Germany.

Wangnick, Klaus (2004), *IDA Worldwide Desalting Plants, Inventory*, Report No. 18, Wangnick Consulting, Gnarrenburg, Germany.

World Bank (2001), *The Development of Electricity Markets in the Euro-Mediterranean Area*, World Bank, Washington, DC.

World Bank (2004), "Seawater and Brackish Water Desalination in the Middle East, North Africa and Central Asia, Final Report, Annex 1: Algeria", World Bank, Washington, DC, December.

World Bank (2005), "A Water Sector Assessment Report on the Countries of the Cooperation Council of the Arab States of the Gulf", Report No. 32539-MNA, World Bank, Washington, DC, 31 March.

CHAPTER 7: Deferred Investment Scenario

Gately, Dermot (2004), "OPEC's Incentives for Faster Output Growth", *Energy Journal*, Vol. 25, No. 2, IAEE, Toronto.

Gately, D. and Huntingdon, H. (2002), "The Asymmetric Effects of Changes in Price and Income on Energy Demand and Oil Demand", *Energy Journal*, 25(2), IAEE, Toronto.

International Energy Agency (IEA) (2004), *World Energy Outlook 2004*, OECD/IEA, Paris.

CHAPTER 8: Implications for World Energy Markets and Government Policy

British Petroleum (2005), *Statistical Review of World Energy 2005*, BP, London.

Cordesman, A. (2004), *US and Global Dependence of Middle East Energy Exports 2004-2030*, Center for Strategic and International Studies, Washington, DC.

European Commission (2005), *Green Paper on Energy Efficiency or Doing More with Less*, COM(2005)265 Final, June, European Commission, Brussels.

International Energy Agency (IEA) (2004), "The Impact of High Oil Prices on the Global Economy", Economic Analysis Division Working Paper, OECD/IEA, Paris, May.

Popp, D. (2002), "Induced Innovation and Energy Prices", in *American Economic Review*, 92(1), March, pp. 160-180.

CHAPTER 9: Algeria

Aissaoui, Ali (2001), *The Political Economy of Oil and Gas*, Oxford Univeristy Press, Oxford.

Arab Petroleum Research Center (APRC) (2004), *Arab Oil and Gas Directory 2004*, Arab Petroleum Research Center, Paris.

APRC (2005), *Arab Oil and Gas Directory 2005*, APRC, Paris.

Cedigaz (2005), *Natural Gas in the World*, Institut Français du Pétrole, Rueil-Malmaison.

EURELECTRIC and UCTE (2005), *European, CIS and Mediterranean Interconnection, State of Play 2004*, Brussels.

Global Water Intelligence (2004), *Desalination Markets 2005-2015: A Global Assessment and Forecast*, Media Analytics Ltd, Oxford.

Global Water Intelligence (2005a), *Water Market Middle East: Exploiting a Booming Market*, Media Analytics Ltd, Oxford.

Global Water Intelligence (2005b), "First Algerian Desalination Deal Gets Financed", Media Analytics Ltd, Oxford, July.

International Monetary Fund (IMF) (2005), Algeria, Selected Issues, IMF Country Report No. 05/51, IMF, Washington, DC.

Ministry of Energy and Mining (MEM) (2004), *Review of the Energy and Mining Sector*, No 3, MEM.

Oil and Gas Journal (2004), PennWell Corporation, Oklahoma City, 20 December.

Sonatrach (2003), *Annual Report 2002*, Sonatrach, Algiers.

Sonatrach (2004a), *Annual Report 2003*, Sonatrach, Algiers.

Sonatrach (2004b), "Efforts de Sonatrach dans la réduction des gaz à effet de serre", presentation by Rabah Nadir Allouani, Algiers, 26 September.

Sonatrach (2005), *Annual Report 2004*, Sonatrach, Algiers.

United Nations Development Programme (UNDP) (2005), *Human Development Report 2005. International cooperation at a crossroads: Aid, trade and security in an unequal world*, UNDP, New York.

United States Geological Survey (USGS) (2000), *World Petroleum Assessment 2000*, USGS, Washington, DC.

World Bank (2004), "Seawater and Brackish Water Desalination in the Middle East, North Africa and Central Asia, Final Report, Annex 1: Algeria", World Bank, Washington, DC, December.

CHAPTER 10: Egypt

African Development Bank (AfDB) and Organisation for Economic Co-operation and Development (OECD) (2004), *African Economic Outlook*, OECD, Paris.

Arab Petroleum Research Center (APRC) (2004), *Arab Oil and Gas Directory 2004*, APRC, Paris.

APRC (2005), *Arab Oil and Gas Directory 2005*, Arab Petroleum Research Center, Paris.

BG Group (2005a), "BG Group and its partners announce first LNG from Egyptian LNG Train 2", Stock Exchange Announcement, BG Group, 5 September.

BG Group (2005b), "BG Group announces the first LNG shipment from Egyptian LNG Train 1", Stock Exchange Announcement, BG Group, 29 May.

Cedigaz (2005), *Natural Gas in the World*, Institut Français du Pétrole, Rueil-Malmaison.

Central Bank of Egypt (2004), *Annual Report 2003-2004*, Central Bank of Egypt, Cairo.

Egyptian Environmental Affairs Agency (EEAA) *et al.* (2003), *Egypt's National Strategy Study on the Clean Development Mechanism*, EEAA, Cairo.

Egyptian General Petroleum Company (EGPC) (2004), "2004 Bid Round-1, II – Main Commercial Parameters", EGPC, May.

Egyptian Natural Gas Holding Company (EGAS) (2005), website www.egas.com.eg.

Egyptian Cabinet (2005), The Cabinet Information and Decision Support Center (IDSC), www.idsc.gov.eg (accessed at various dates across January-September 2005).

Egyptian Cabinet (various issues), *Monthly Economic Bulletin*, Information and Decision Support Centre, Cairo.

Egyptian Electricity Holding Company (EEHC) (2003), *Annual Report 2003*, EGHC, Cairo.

International Monetary Fund (IMF) (2005), *Arab Republic of Egypt: Selected Issues*, IMF Country Report No. 05/179, IMF, Washington, DC.

Nazif, Ahmed Mahmoud (2005), Speech delivered to World Economic Forum, in Jordan 2005, Dead Sea, 22 May.

Oil and Gas Journal (2004), 20 December, PennWell Corporation, Oklahoma City.

United Nations Development Programme (UNDP) (2005a), *Human Development Report 2005. International cooperation at a crossroads: Aid, trade and security in an unequal world*, UNDP, New York.

United Nations Development Programme (UNDP) (2005b), *Arab Human Development Report: Towards Freedom in the Arab World*, UNDP, New York.

World Markets Research Centre (WMRC) (2002), "Middle East Regional: Can Compressed Natural Gas Provide a Solution to Cairo's Air Pollution Problem?",WMRC, 4 January.

CHAPTER 11: Iran

Arab Petroleum Research Center (2004), *Arab Oil and Gas Directory*, Paris.

Cedigaz (2005), *Natural Gas in the World*, Institut Français du Pétrole, Rueil Malmaison.

Central Bank of Iran (CBI) (2004/2005), Report No 39, http://www.cbi.ir/publications/PDF/etno39.pdf

Government of Islamic Republic of Iran and UNDP (2005), *Removing Barriers to Large Scale Commercial Wind Energy Development*, United Nations, New York.

International Monetary Fund (IMF) (2004a), *Islamic Republic of Iran--Selected Issues*, Country Report No 04/308, IMF, Washington, DC.

IMF (2004b), *Islamic Republic of Iran--Statistical Appendix*, Country Report No 04/307, IMF, Washington, DC.

Institute for International Energy Studies (2004), *Iran Energy Report*, IIES, Teheran.

Majlis Research Center (2004), Bill of Inflation Containment, Articles collection, Report No. 26 (in Persian).

Meibodi, A. (1998), *Efficiency Considerations in the Electricity Supply Industry: The Case of Iran*, Department of Economics, University of Surrey.

Ministry of Road and Transport (2005), *Comprehensive Transportation Studies of Iran*, Ministry of Road and Transport, Teheran.

Tavanir Company (2005), *Electric Power Industry in Iran 2003-2004*, Ministry of Energy, Teheran.

United Kingdom House of Commons, Foreign Affairs Committee (2004), *Iran - Third Report of Session 2003-04*, The Stationery Office Limited, London.

United Nations Department of Economics and Social Affairs (UNDESA), Population Division (2004), *World Population Prospects: The 2004 Revision*, United Nations, New York.

United Nations Industrial Development Organization (UNIDO) (1999), *Islamic Republic of Iran Industrial Sector Survey on the Potential for Non-Oil Manufactured Exports*, NC/IRA/94/01D/08/37, United Nations, Vienna.

United States Geological Survey (USGS) (2000), *World Petroleum Assessment 2000*, USGS, Washington, DC.

Von Moltke, A, C. McKee and T. Morgan (2004), *Energy Subsidies*, Greenleaf Publishing/United Nations Environment Programme, Sheffield, United Kingdom.

World Bank (2003), "Iran Medium Term Framework for Transition", *Economic Report*, No. 25848, World Bank, Washington, DC.

World Energy Council (2004), *2004 Survey of Energy Resources*, Elsevier, London.

CHAPTER 12: Iraq

Arab Petroleum Research Center (2004), *Arab Oil and Gas Directory*, Paris.

British Petroleum (BP) (2005), *BP Statistical Review of World Energy 2005*, BP, London.

Cedigaz (2005), *Natural Gas in the World*, Institut Français du Pétrole, Rueil-Malmaison.

Energy Intelligence Group (2004), *Iraqi Oil & Gas: A Bonanza Still In Waiting*, EIG, New York.

Katzman (2004), *Iraq: US Regime Change Efforts and Post-Saddam Governance*, Congressional Research Service: Library of Congress, 22 December.

Oil and Gas Journal (2004), PennWell Corporation, Oklahoma City, 20 December.

Power Engineering International (PEi) (2005), *Turkey to up power exports to Iraq*, PennWell Corporation, Tulsa, Oklahoma, April.

Tarnoff (2004), *Iraq: Recent Developments in Reconstruction Assistance*, Congressional Research Service: Library of Congress, 20 December.

Transparency International (2004), "Transparency International Corruption Perceptions Index 2004", http://www.transparency.org, accessed 12 September 2005.

United Nations Development Programme (UNDP) (2004), *Arab Human Development Report 2004*, United Nations, New York.

United States Agency for International Development (USAID) (2003), *Fixing Iraq's Infrastructure*, USAID, Washington, DC.

United States Agency for International Development (USAID) (2004), *Restoring Power – USAID's Role in Restoring Power in Iraq*, USAID, Washington, DC.

United States Geological Survey (USGS) (2000), *World Petroleum Assessment 2000*, USGS, Washington, DC.

Verma, M., Ahlbrandt, T. and Al-Gailani, M. (2004), "Petroleum Reserves and Undiscovered Resources in the Total Petroleum Systems of Iraq: Reserve Growth and Production Implications", in *GeoArabia*, Vol. 9, No. 3, Gulf Petrolink, Bahrain.

World Bank (2004), International Bank for Reconstruction and Development, *Interim Strategy Note of the World Bank Group for Iraq*, World Bank, Washington, DC.

World Bank (2005), *Middle East and North Africa Economic Developments and Prospects 2005: Oil Booms and Revenue Management*, World Bank, Washington, DC.

World Energy Council (2004), *2004 Survey of Energy Resources*, Elsevier, London.

CHAPTER 13: Kuwait

Al-Attar, Abdulaziz E. (2004), "A review of upstream development policies in Kuwait", *OPEC Review*, Vol. XXV111, No. 4, Blackwell Publishers, Oxford.

Al Shatti, Mohammed (2005), "Kuwait's Production Plans", paper presented at the 3rd Joint OPEC-IEA Workshop, Kuwait City, 15 May.

Arab Petroleum Research Center (APRC) (2004), *Arab Oil and Gas Directory 2004*, Arab Petroleum Research Center, Paris.

British Petroleum (2005), *Statistical Review of World Energy 2005*, BP, London.

Cedigaz (2005), *Natural Gas in the World*, Institut Français du Pétrole, Rueil-Malmaison.

Global Water Intelligence (2004), *Desalination Markets 2005-2015: A Global Assessment and Forecast*, Media Analytics Ltd, Oxford, April.

Oil and Gas Journal (2004), PennWell Corporation, Oklahoma City, 20 December.

Organization of the Petroleum Exporting Countries (OPEC) (2004), *Annual Statistical Bulletin 2004*, Blackwell Publishers, Oxford.

Organization of Arab Petroleum Exporting Countries (OAPEC) (2004), *Annual Statistical Report 2004*, OAPEC, Safat.

Platts (2003), World Electric Power Plants Database, Platts.

United Nations Department of Economics and Social Affairs (UNDESA), Population Division (2005), *World Population Prospects: The 2004 Revision*, United Nations, New York.

World Energy Council (2004), *Survey of Energy Resources 2004*, Elsevier, London.

CHAPTER 14: Libya

AFP (2004), Interview of Prime Minister Shukri Ghanem: *Libya to Abolish State Subsidies in Liberalization Move*, France, November.

Arab Petroleum Research Center (APRC) (2004), *Arab Oil and Gas Directory 2004*, Arab Petroleum Research Center, Paris.

Cedigaz (2005), *Natural Gas in the World*, Institut Français du Pétrole, Rueil-Malmaison.

Energy Intelligence Group (2004), *Libya Oil & Gas: Back In Business*, EIG, New York.

Eurelectric and UCTE (2005), *European, CIS and Mediterranean Interconnection*, State of Play 2004, Brussels.

Khodadad Nabil L. (2005), "Libya Unveils Terms for Foreign Investors", *International Energy Law and Taxation Review*, UK.

Observatoire Méditerranéen de l'Energie (OME) (2004), *The role and Future Prospects of Natural Gas in the Mediterranean Region*, OME, Paris.

Organization of Arab Exporting Countries (OAPEC) (2003), *Annual Statistical Report 2003*, Kuwait.

UNESCO (2002), *Black and Blue, Libya's Liquid Legacy*, United Nations, Paris.

Wood Mackenzie (2005), *What's Next for Libya*, Wood Mackenzie, Scotland.

CHAPTER 15: Qatar

Arab Petroleum Research Center (APRC) (2004), *Arab Oil and Gas Directory 2004*, Arab Petroleum Research Center, Paris.

APRC (2005), *Arab Oil and Gas Directory*, APRC, Paris.

Cedigaz (2005), *Natural Gas in the World*, Institut Français du Pétrole, Rueil-Malmaison.

Energy Information Administration (EIA) (2005), "Qatar Country Analysis Brief", February, http://www.eia.doe.gov/emeu/cabs/qatar.html.

IHS (2004), IHS Energy databases, http://www.ihs.com/energy/index.html.

Oil and Gas Journal (2004), PennWell Corporation, Oklahoma City, 20 December.

Organisation for Economic Cooperation and Development (OECD) (2003), "Improving Water Management: Recent OECD Experience", *OECD Observer*, Policy Brief, OECD, Paris.

Platts (2003), World Electric Power Plants Database, Platts.

Qatar National Bank (2005), *Qatar Economic Review*, Doha, January.

Qatar Petroleum (2005), official website, http://www.qp.com.qa/qp.nsf.

Qatargas (2005), official website, http://www.qatargas.com.qa/default.htm.

United States Geological Survey (USGS) (2000), *World Petroleum Assessment 2000*, USGS, Washington, DC.

CHAPTER 16: Saudi Arabia

Arab Petroleum Research Center (APRC) (2004), *Arab Oil and Gas Directory 2004*, Arab Petroleum Research Center, Paris.

APRC (2005), *Arab Oil and Gas Directory*, APRC, Paris.

British Petroleum (2005), *Statistical Review of World Energy 2005*, BP, London.

Cedigaz (2005), *Natural Gas in the World*, Institut Français du Pétrole, Rueil-Malmaison.

Global Insight World Market Research Centre, www.worldmarketsanalysis.com

IHS (2004), IHS Energy databases, http://www.ihs.com/energy/index.html.

National Commercial Bank (2003), "The Saudi Automobile Sector: Growth Dynamics", *The NCB Economist.*

Oil and Gas Journal (2004), PennWell Corporation, Oklahoma City, 20 December.

Samba (2005), "The Saudi Economy at Mid-Year 2005", Samba Financial Group, Riyadh.

Saudi Arabia Monetary Agency (SAMA) (2004), *Annual Report*, SAMA, Riyadh, http://www.sama.gov.sa

Saudi Aramco, *Facts and Figures*, http://www.saudiaramco.com.

Saudi Electricity Company (SEC) (2004), *Turn on to New Investment Opportunities in Saudi Electricity Company and Water & Electricity Joint Ventures*, SEC, Saudi Arabia.

Saudi Ministry of Economy and Planning (2004), Central Department of Statistics, Riyahd.

Simmons, Matthew R. (2002), "The World's Giant Oil Fields", in *Hubert Centre Newsletter*, Colorado School of Mines, Colorado, January.

C

Simmons, Matthew R. (2005), *Twilight in the Desert*, John Wiley & Sons, Inc., Hoboken, New Jersey.

United Nations Department of Economics and Social Affairs (UNDESA), Population Division (2004), *World Population Prospects: The 2004 Revision*, United Nations, New York.

United States Geological Survey (USGS) (2000), *World Petroleum Assessment 2000*, USGS, Washington, DC.

World Bank (2005), "A Water Sector Assessment Report on the Countries of the Cooperation Council of the Arab States of the Gulf", Report No. 32539-MNA, World Bank, Washington, DC, 31 March.

CHAPTER 17: United Arab Emirates

Abu Dhabi National Oil Company (ADNOC) (2005), *ADNOC's Five Year Achievements Report*, ADNOC, Abu Dhabi.

Al Yabhouni, Ali Obaid (2005), *UAE Oil and Gas Potential and Capacity Expansion*, presentation at the 3rd Joint OPEC-IEA Workshop, Kuwait City, 15 May.

Arab Petroleum Research Center (APRC) (2005), *Arab Oil and Gas Directory 2005*, Arab Petroleum Research Center, Paris.

Cedigaz (2005), *Natural Gas in the World*, Institut Français du Pétrole, Rueil-Malmaison.

Central Bank of the UAE (2004), *Annual Report 2003*, www.uaecb.gov.ae/Annual/ann2003E.pdf.

Ghanem, Shihab M. (2001), "Industrialization in the UAE", in Ibrahim Al Abed and Peter Hellyer (eds.), *The United Arab Emirates: A New Perspective*, Trident Press, London.

Environmental Research and Wildlife Development Agency (ERWDA) (2005), official website, http://www.erwda.gov.ae/eng.

International Monetary Fund (2005), *United Arab Emirates: Staff Report for the Article IV Consultation*, Country Report No 05/269, IMF, Washington, DC.

Ministry of Economy and Planning (2005), official website, www.uae.gov.ae/mop/E_home.htm.

Ministry of Information and Culture (2005), *The United Arab Emirates Yearbook 2005*, Trident Press, London.

Oil and Gas Journal (2004), PennWell Corporation, Oklahoma City, 20 December.

United Nations Conference on Trade and Development (UNCTAD) (2004), *World Investment Report 2004*, United Nations, Geneva.

United States Geological Survey (USGS) (2000), *World Petroleum Assessment 2000*, USGS, Washington, DC.

World Bank (2005), "A Water Sector Assessment Report on the Countries of the Cooperation Council of the Arab States of the Gulf", Report No. 32539-MNA, World Bank, Washington, DC, 31 March.

C

The Online Bookshop

International Energy Agency

All IEA publications can be bought online on the IEA Web site:

www.iea.org/books

You can also obtain PDFs of all IEA books at 20% discount.

Books published before January 2004
- with the exception of the statistics publications -
can be downloaded in PDF, free of charge,
on the IEA Web site.

IEA BOOKS

Tel: +33 (0)1 40 57 66 90
Fax: +33 (0)1 40 57 67 75
E-mail: books@iea.org

International Energy Agency
9, rue de la Fédération
75739 Paris Cedex 15, France

CUSTOMERS IN NORTH AMERICA

Turpin Distribution
The Bleachery
143 West Street, New Milford
Connecticut 06776, USA
Toll free: +1 (800) 456 6323
Fax: +1 (860) 350 0039
oecdna@turpin-distribution.com
www.turpin-distribution.com

You can also send your order to your nearest OECD sales point or through the OECD online services:

www.oecdbookshop.org

CUSTOMERS IN THE REST OF THE WORLD

Turpin Distribution Services Ltd
Stratton Business Park,
Pegasus Drive, Biggleswade,
Bedfordshire SG18 8QB, UK
Tel.: +44 (0) 1767 604960
Fax: +44 (0) 1767 604640
oecdrow@turpin-distribution.com
www.turpin-distribution.com

IEA PUBLICATIONS, 9, rue de la Fédération, 75739 PARIS CEDEX 15
PRINTED IN FRANCE BY STEDI
(61 2005 26 1P1) ISBN 92-64-1094-98 - 2005